Microprocessors

PRINCIPLES AND APPLICATIONS

Microprocessors

PRINCIPLES AND APPLICATIONS

SECOND EDITION

Charles M. Gilmore

**LumenX Company
A Division of
Alltrista Corporation**

GLENCOE

McGraw-Hill

New York, New York   Columbus, Ohio   Woodland Hills, California   Peoria, Illinois

Cover photograph: COMSTOCK Inc./Michael Stuckey. Safety section photographs: *Top left:* Charles Thatcher/Tony Stone Worldwide; *Top right:* Lou Jones/The Image Bank; *Lower right:* © Cindy Lewis.

ACKNOWLEDGMENTS

The *Basic Skills in Electricity and Electronics* series was conceived and developed through the talents and energies of many individuals and organizations.

The original, on-site classroom testing of the texts and manuals in this series was conducted at the Burr D. Coe Vocational Technical High School, East Brunswick, New Jersey; Chantilly Secondary School, Chantilly, Virginia; Nashoba Valley Technical High School, Westford, Massachusetts; Platt Regional Vocational Technical High School, Milford, Connecticut; and the Edgar Thomson, Irvin Works of the United States Steel Corporation, Dravosburg, Pennsylvania. Postpublication testing took place at the Alhambra High School, Phoenix, Arizona; St. Helena High School, St. Helena, California; and Addison Trail High School, Addison, Illinois.

Early in the publication life of this series, the appellation "Rainbow Books" was used. The name stuck and has become a point of identification ever since.

In the years since the publication of this series, extensive follow-up studies and research have been conducted. Thousands of instructors, students, school administrators, and industrial trainers have shared their experiences and suggestions with the authors and publishers. To each of these people we extend our thanks and appreciation.

Library of Congress Cataloging-in-Publication Data

Gilmore, Charles Minot.—
   Microprocessors : principles and applications / Charles M.
Gilmore. —2nd ed.
     p.   cm.—(Basic skills in electricity and electronics)
   Includes index.
   ISBN 0-02-801837-0
   1. Microprocessors.   I. Title.   II. Series.
QA76.5.G5143 1995
004.16—dc20
                                   95-32670
                                      CIP

Microprocessors: Principles and Applications
Second Edition

Send all inquiries to:
Glencoe/McGraw-Hill
8787 Orion Place
Columbus, OH 43240

ISBN 0-02-801837-0

Printed in the United States of America.

4 5 6 7 8 9 10 11 12   027/043   04 03 02 01 00

TRADEMARKS

Contents

■

Preface

Microprocessors: Principles and Applications, Second Edition is an introductory text on microprocessors, microcomputers, their associated subsystems and software, and related careers in this area of electronics. Although the subject matter is less than two decades old, the material required to introduce the subject properly grows significantly each year.

The first edition of Microprocessors: Principles and Applications was nearly two-thirds larger than its predecessor, Introduction to Microprocessors, reflecting the significant advances in microprocessors and the use of microprocessors during the early and mid-1980s. Although the pace of radical introductions (16- and 32-bit microprocessors versus 8-bit microprocessors, for example) has slowed somewhat, there have been significant increases in product complexity in the time between the publication of Microprocessors: Principles and Applications and Microprocessors: Principles and Applications, Second Edition. Many of these changes are reflected in discussions of product speed, capacity, and complexity. Other changes are reflected in the addition of new material. Fortunately, the basic principles are beginning to stabilize, and a student who has a good command of, for example, the basic Intel X86 architecture can easily extend this information to the post-Pentium processors, which are in the works and will soon be on the market. Emphasis is given to the application of microprocessors in the construction of personal computers, a device used only by a few hobbyists when the first text was being prepared. However, this should not be considered a text on the PC. The PC is just one common example of how advanced microprocessors are used. Nearly all other sections have been expanded to cover currently used technologies.

A fundamental objective of this text is to give the student both a solid theoretical and a practical introduction to the microprocessors and support devices used to create the extensive array of microprocessor-based devices found in consumer and industrial electronics today. Since the personal computer is a common microprocessor-based product which involves virtually every aspect of the technology, it is frequently referenced in the text.

This text is written for students who have completed fundamental electricity, electronics, and digital circuits courses. However, this text does not presume an in-depth knowledge in these subjects. The mathematics required in the text assumes the student has completed a first-year algebra course.

After completing a course in microprocessor technology using this text, students are prepared for a number of career paths. First, they will have sufficient background to extend electronic service training to the repair and maintenance of microprocessor-based products. Second, they will have a solid background with which to enter advanced studies, leading to careers as technologists or design-engineering technicians. Third, students completing such a course are well equipped to pursue advanced studies, leading to a career in programming microprocessors.

This text employs a multiple-segment approach to learning the subject; the topics of microprocessors and their related devices are interwoven. The first segment, Chapters 1 to 7, provides the history of microprocessors, an introduction to the specialized mathematics used by microprocessors, an overview of microprocessor architecture, and an introduction to the concepts of I/O devices.

The second segment, Chapters 8 to 10, introduces three major categories of real-life microprocessors: two 8-bit general-purpose microprocessors, the Z80 and the 6802; and 8-bit single-chip microprocessor, the 8051; and two advanced (16- and 32-bit) microprocessor families, the Intel X86 and the Motorola 68XXX. It is important to note that the coverage of the advanced microprocessors is not at the depth given to the 8-bit devices. The objective is to give the student a good overview of these extremely complex devices. This overview should suffice for most needs aside from detailed design or software development using an advanced microprocessor. This segment serves two purposes. First, it introduces real-life devices students will encounter in practical use. Second, it serves to cement the more theoretical concepts introduced in the first segment.

The third segment, Chapters 11 to 13, significantly expands the student's knowledge of microprocessor memory, mass storage, and I/O devices. Emphasis is placed on devices and standards commonly found in commercial devices, especially the personal computer. This segment is intended to be a very practical introduction to these topics.

The fourth segment, Chapters 14 and 15, introduces the concepts of programming, operating systems, and system support software. These two chapters are a departure from the hardware orientation found in the balance of the text. They are included for two reasons. First, it is essential for any person involved with microprocessor hardware to understand the fundamentals of programming, because programming is an equally important and sophisticated discipline. Second, many students when exposed to programming find their interest piqued and may wish to further explore career opportunities in this fascinating field. If so, they have a great deal of the necessary hardware background required for this field of study.

The fifth segment, Chapters 16 and 17, introduces career-oriented topics related to servicing microprocessor-based products and participating in the design and development of microprocessor-based products. In many ways, servicing of microprocessor-based products is different than servicing conventional electronic products. Chapter 16 attempts to show the student that this can lead to an interesting career choice. Chapter 17 is intended to pique a career interest in design engineering support efforts. The format walks the student through a complete product design project, introducing the participants and their individual tasks on the way.

The sixth, and last, segment, Chapter 18, is a brief look at some of the latest technological introductions in the field of microprocessors and related technology. It is intended that this chapter can be updated and supplemented from time to time to keep this text as abreast with relevant technological changes as possible.

Wherever possible, theoretical concepts are supported by real-life examples using the 8-bit and advanced microprocessors introduced in this text along with their associated peripheral devices. Where possible, the examples are illustrated with simplified schematic diagrams and short programming examples. Although some of the examples may be slightly simplified for clarity, most are taken from real situations. The hardware introduced in this text is expected to have a long life in practical use.

As noted earlier, *Microprocessors: Principles and Applications*, Second Edition, evolved from its predecessors, *Microprocessors: Principles and Applications* and *Introduction to Microprocessors*. This evolution has been supported by many events and people. Many comments have been received from educators who have used the first text, and many additional comments were received from reviewers during the preparation of this edition. Many thanks are owed to these people for their efforts. Additionally, many thanks go to my son John for his contributions, questions, and review of detail. I hope this work will help his career in electronics.

As usual, I look forward to comments and suggestions from students and teachers involved in this continuously evolving field.

Charles M. Gilmore

Safety

Electric and electronic circuits can be dangerous. Safe practices are necessary to prevent electrical shock, fires, explosions, mechanical damage, and injuries resulting from the improper use of tools.

Perhaps the greatest hazard is electrical shock. A current through the human body in excess of 10 milliamperes can paralyze the victim and make it impossible to let go of a "live" conductor or component. Ten milliamperes is a rather small amount of electrical flow: It is only *ten one-thousandths* of an ampere. An ordinary flashlight uses more than 100 times that amount of current!

Flashlight cells and batteries are safe to handle because the resistance of human skin is normally high enough to keep the current flow very small. For example, touching an ordinary 1.5-V cell produces a current flow in the microampere range (a microampere is one-millionth of an ampere). This amount of current is too small to be noticed.

High voltage, on the other hand, can force enough current through the skin to produce a shock. If the current approaches 100 milliamperes or more, the shock can be fatal. Thus, the danger of shock increases with voltage. Those who work with high voltage must be properly trained and equipped.

When human skin is moist or cut, its resistance to the flow of electricity can drop drastically. When this happens, even moderate voltages may cause a serious shock. Experienced technicians know this, and they also know that so-called low-voltage equipment may have a high-voltage section or two. In other words, they do not practice two methods of working with circuits: one for high voltage and one for low voltage. They follow safe procedures at all times. They do not assume protective devices are working. They do not assume a circuit is off even though the switch is in the OFF position. They know the switch could be defective.

As your knowledge and experience grow, you will learn many specific safe procedures for dealing with electricity and electronics. In the meantime:

1. Always follow procedures.
2. Use service manuals as often as possible. They often contain specific safety information.
3. Investigate before you act.
4. When in doubt, *do not act*. Ask your instructor or supervisor.

General Safety Rules for Electricity and Electronics

Safe practices will protect you and your fellow workers. Study the following rules. Discuss them with others, and ask your instructor about any you do not understand.

1. Do not work when you are tired or taking medicine that makes you drowsy.
2. Do not work in poor light.
3. Do not work in damp areas or with wet shoes or clothing.
4. Use approved tools, equipment, and protective devices.
5. Avoid wearing rings, bracelets, and similar metal items when working around exposed electric circuits.
6. Never assume that a circuit is off. Double-check it with an instrument that you are sure is operational.
7. Some situations require a "buddy system" to guarantee that power will not be turned on while a technician is still working on a circuit.
8. Never tamper with or try to override safety devices such as an interlock (a type of switch that automatically removes power when a door is opened or a panel removed).
9. Keep tools and test equipment clean and in good working condition. Replace insulated probes and leads at the first sign of deterioration.
10. Some devices, such as capacitors, can store a *lethal* charge. They may store this charge for long periods of time. You must be certain these devices are discharged before working around them.

11. Do not remove grounds and do not use adaptors that defeat the equipment ground.

12. Use only an approved fire extinguisher for electrical and electronic equipment. Water can conduct electricity and may severely damage equipment. Carbon dioxide (CO_2) or halogenated-type extinguishers are usually preferred. Foam-type extinguishers may also be desired in some cases. Commercial fire extinguishers are rated for the type of fires for which they are effective. Use only those rated for the proper working conditions.

13. Follow directions when using solvents and other chemicals. They may be toxic, flammable, or may damage certain materials such as plastics.

14. A few materials used in electronic equipment are toxic. Examples include tantalum capacitors and beryllium oxide transistor cases. These devices should not be crushed or abraded, and you should wash your hands thoroughly after handling them. Other materials (such as heat shrink tubing) may produce irritating fumes if overheated.

15. Certain circuit components affect the safe performance of equipment and systems. Use only exact or approved replacement parts.

16. Use protective clothing and safety glasses when handling high-vacuum devices such as picture tubes and cathode-ray tubes.

17. Don't work on equipment before you know proper procedures and are aware of any potential safety hazards.

18. Many accidents have been caused by people rushing and cutting corners. Take the time required to protect yourself and others. Running, horseplay, and practical jokes are strictly forbidden in shops and laboratories.

Circuits and equipment must be treated with respect. Learn how they work and the proper way of working on them. Always practice safety; your health and life depend on it.

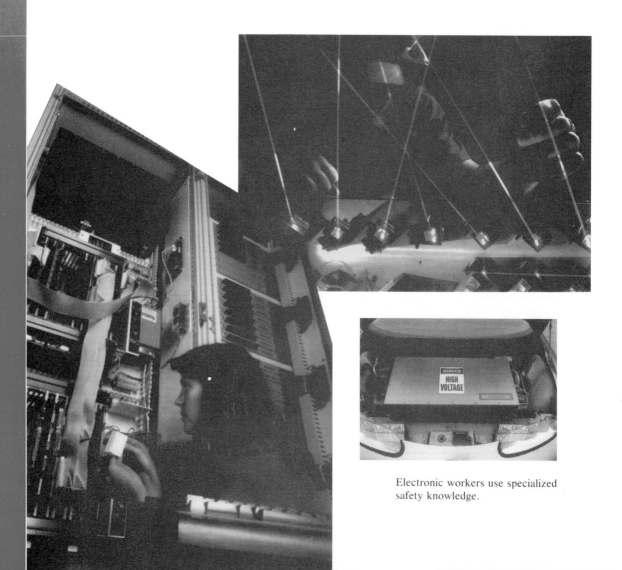

Electronic workers use specialized safety knowledge.

This book is dedicated to my father James Keith Gilmore, 1915–1992. As a father, geologist and engineer, his inspiration led me to pursue a career in the sciences and the "passing fad" of electronics.

CHAPTER 1

What Is the Microprocessor?

∎

CHAPTER OBJECTIVES

This chapter will help you to:

1. *Define* the terms architecture, bit, integrated circuit, read-only memory (ROM), and random-access memory (RAM).
2. *Explain* the differences between a microprocessor and a microcomputer.
3. *Identify* the two chief functions of the microprocessor.
4. *Compare* the computing power of two different microprocessors.
5. *Calculate* the memory size of a microprocessor.

———

The microprocessor was not just born. It is the combination of solid-state technological development and the advancing computer technologies which came together in the early 1970s; it could not have been developed until both these technologies were ready. With the low cost of a solid-state device and the flexibility of a computer, the microprocessor is a product which performs both control and processing functions.

∎

1-1 A BRIEF HISTORY

To understand how the microprocessor came about, we must follow the growth of two major technologies: digital computers and solid-state circuits. These two technologies came together in the early 1970s, allowing engineers to produce the microprocessor.

The *digital computer* is a set of digital circuits controlled by a program that makes it do the job you want done. The program tells the digital circuits how to move and process data. It does this by using the digital computer's calculating logic, memory circuits, and input/output devices. The way the digital computer's logic circuits are put together to build the calculating logic, memory circuits,

and input/output devices is called its *architecture*.

The microprocessor has architecture similar to the digital computer's. In other words, the microprocessor is like the digital computer because both do computations under program control. Consequently, the history of the digital computer will help us understand the microprocessor. The history of solid-state circuits will also help, because the microprocessor is a solid-state circuit—a large-scale or very large-scale integrated (VLSI) microcircuit.

The chart in Fig. 1-1 shows the major events in the two technologies as they developed over the last five decades. Use the chart as you read the following account of how the technologies developed from the days of World War II.

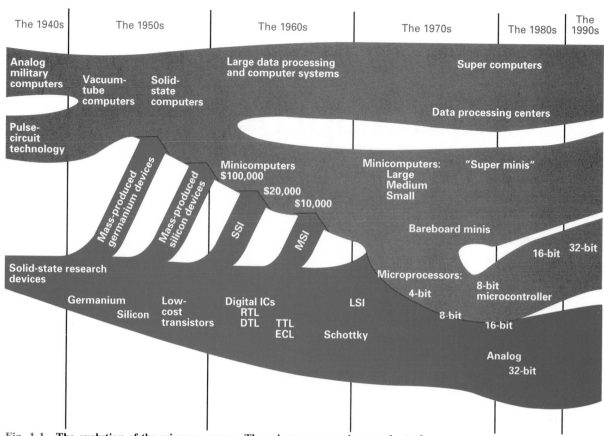

The 1940s | The 1950s | The 1960s | The 1970s | The 1980s | The 1990s

Fig. 1-1 The evolution of the microprocessor. The microprocessor is a product of computer and semiconductor technology. Linked from the mid-1950s, these technologies merged in the early 1970s in a product called the microprocessor.

During World War II, scientists developed computers for military use. In the latter half of the 1940s, digital computers were developed to do scientific and business work. Electronic circuit technology also advanced during World War II. Radar work increased the understanding of fast digital circuits called *pulse circuits*. After the war, scientists made great progress in solid-state physics. Scientists at Bell Laboratories invented the transistor, a solid-state device, in 1948.

In the early 1950s, the first general-purpose digital computers appeared. They used vacuum tubes for the active electronic components. Vacuum-tube modules were used to build basic logic circuits such as gates and flip-flops. By assembling gate and flip-flop modules, scientists built the computer's calculating logic and memory circuits. Vacuum tubes also formed part of the machines built to communicate with the computer—the *I/O (input/output) devices*.

From studying digital circuits, you know that building even a simple adder circuit takes quite a few gates. Most digital computers, of course, have a number of adders. Building a digital computer requires many circuits. Since all their circuits were made of bulky vacuum tubes, early digital computers were huge. Because the vacuum tubes were hot, these early computers required air conditioning. The vacuum tubes were also unreliable by today's standards. Vacuum tubes made the early computers expensive to run and maintain. The drawbacks of the vacuum tube were holding back the development of the digital computer.

Nevertheless, the early computers did introduce the important idea of *program storage*. The computers of the late 1940s and early 1950s had used patch-cord programming. Using a patch panel, the programmer actually wired in the steps telling the computer what to do with data. Nothing was stored in memory except data.

Later designs provided computers with program storage. This means that the steps telling the computer what to do with data are stored as digital words in the computer's memory. The only way of knowing that some words are program steps instead of data is to check their location in memory. The idea of program storage was an important, basic addition to the computer's architecture.

Solid-state circuit technology also made great strides during the 1950s. The knowledge of semiconductors increased. The use of silicon lowered costs, because silicon is much more plentiful than germanium, which had been the chief material for making the early semiconductors. Mass production methods made transistors common and inexpensive.

Naturally, the designers of digital computers jumped at the chance to replace vacuum tubes with transistors, and in the late 1950s, they began doing so. But *logic circuits*, although now made of transistors, were still discrete. That is, each logic circuit was built from a number of components, such as individual transistors. But these first solid-state computers were already much smaller, much cooler, and much more reliable than vacuum-tube computers.

In the early 1960s, the art of building solid-state computers advanced in two directions. Moving in one direction were the giants like IBM, Burroughs, and Honeywell. They were building huge solid-state computers that still required large, air-conditioned rooms. These computers were very complicated. They could process large amounts of data. These large data processing systems were used for commercial and scientific applications.

These big computers were still very expensive. In order to pay for themselves, they had to be run 24 hours a day, 7 days a week. Two different methods were developed for getting the maximum use out of these expensive machines: the *batch mode* and the *timesharing mode*. In the batch mode, only one large job is run at a time, and one job is run immediately after another. In the timesharing mode, the large computer is used to do many jobs "at once" by working on a part of each job in turn.

Moving in a new direction were some younger, smaller companies. They began building small computers, about the size of a desk. These *minicomputers* were not as powerful as their larger relatives, but they were not as expensive either. And they still performed many useful functions.

Minicomputers quickly proved useful in the laboratory. Scientists found that *dedicated computers*—computers used for only a single kind of job—had real value. Instead of running many different kinds of jobs on one of the giant computers, scientists ran each kind of job on a separate, dedicated minicomputer.

The idea that a single device with the computer's architecture could be tied up on a single job was a major change. No one could afford to dedicate a computer until the low-cost minicomputer appeared. Often a dedicated minicomputer did a short job and was turned off until that job had to be done again.

Solid-state circuitry continued to develop along with the digital computer. But now the two technologies were moving closer together. Computers use many of the same few logic circuit designs again and again. The need for large numbers of these circuits began to drive the semiconductor industry to develop new products.

By the early 1960s, the semiconductor industry found a way to put a number of transistors on one silicon wafer. The transistors are connected together with small metal traces. When the transistors are connected together, they become a circuit which performs a function, such as a gate, flip-flop, register, or adder. This new technology created basic semiconductor building blocks. The building blocks or circuit modules made this way are called an *integrated circuit (IC)*.

By the mid-1960s, small- and medium-scale integration (SSI and MSI) produced major families of digital logic. These gave digital circuit designers a wide range of building blocks to work with. The technology of ICs pushed in two directions. There was a push to develop low-cost manufacturing techniques. At the same time, there was a push to develop circuits that were more complex.

The use of ICs let minicomputers become more and more powerful for their size. The desk-sized minicomputer of the 1960s became as powerful as a room-sized computer of the late 1950s. New $10,000, drawer-sized minicomputers were as powerful as the older $100,000, desk-sized minicomputers. Some computers even began to appear in "bare-board" form. That is, printed circuit boards with all the logic needed to make a computer processor were for sale. The buyer of such a board had to obtain a power supply and other necessary equipment.

As we all know, IC technology has progressed further since the mid-1960s. The late 1960s and early 1970s saw *large-scale integration (LSI)* become common. Large-scale integration was making it possible to produce more and more digital circuits in a single IC. By the 1980s *very large-scale integration (VLSI)* gave us ICs with over 100,000 transistors.

Most of the early large-scale ICs performed

Logic circuits

Batch mode

Timesharing mode

Minicomputers

Dedicated computers

Integrated circuit (IC)

Large-scale integration (LSI)

Very large-scale integration (VLSI)

special functions. But a few LSI circuits were produced to perform universal functions. Memory devices are a good example.

The development of the electronic calculator shows the dramatic improvements in large-scale integration. The first electronic calculators required 75 to 100 individual IC packages. Special LSI replaced most of these ICs with five to six LSI circuits. By the mid-1970s, LSI had reduced the calculator to a single circuit.

After the calculator was reduced, the next natural step was to reduce the architecture of the computer to a single IC. Designers soon achieved this step, and the resulting circuit was called the *microprocessor*.

The microprocessor made possible the manufacture of powerful calculators and many other products. Like the earlier dedicated minicomputer, the computerlike architecture of the microprocessor could be programmed to carry out a single task. Products like microwave ovens, telephone dialers, and automatic temperature-control systems became commonplace. With a cheap computer available in the form of an IC, such products became practical.

Since the early 1970s, the main effort has been to improve the microprocessor's architecture. Every improvement in architecture increases the microprocessor's speed and computing power.

The early microprocessors processed digital data 4 bits (4 *bi*nary dig*its*) at a time. That is, they used a *4-bit word*. These microprocessors were slow and did not compare to minicomputers. But new generations of microprocessors came fast. The 4-bit microprocessors grew into 8-bit microprocessors, then into 16-bit microprocessors, and then into 32-bit microprocessors. Microprocessors' *instruction sets*—the instructions that microprocessors can carry out—increased in size and sophistication. Some microprocessors soon equaled or surpassed the capabilities of modest minicomputers. During the early 1980s, complete 8-bit microprocessor systems (microprocessors with memory and communications ability) were developed. These *microcontrollers*, or single-chip microprocessors, have become popular as the basis of controllers for keyboards, VCRs, TVs, microwave ovens, smart telephones, and a host of other industrial and consumer electronic devices. Today it is not uncommon to find two, three, or more microcontrollers in a single consumer product.

Self-Test

Answer the following questions.

1. The basic design, that is, the way the data moves about inside the microprocessor and how the calculations are done, is called the microprocessor's
   a. Packaging
   c. Module
   b. Architecture
   d. Power supply
2. A microprocessor's program is stored in memory along with its data. The only way you can tell the difference between program steps and data is by
   a. The word length
   b. The word type
   c. The place you put the program steps
   d. Seeing if the microprocessor will execute the steps
3. The minicomputer introduced the idea of a separate processor for each task. The microprocessor used to implement a telephone dialer
   a. Is an example of multiple tasks in one processor
   b. Is the same idea as the minicomputer
   c. Proves stored programming is not possible
   d. Will be made possible at some future date
4. The IC is used a great deal in computer applications because
   a. Computers use the same logic over and over
   b. All computers must run very fast
   c. Vacuum tubes generate too much heat even though they are faster
   d. All of the above

1-2 WHAT IS A MICROPROCESSOR?

The word "microprocessor" tells us something about the device it names. The microprocessor uses the same type of logic that is used in a digital computer's *central processing unit* (CPU). Because of its resemblance to the CPU, and because it is constructed with microcircuit (integrated circuit) technology, we say it is a microprocessing unit in microcircuit form. Like the CPU, the microprocessor has digital circuits for data handling and computation under program control. In other words, the microprocessor is a *data processing unit*.

Unlike an ordinary, full-scale CPU, the microprocessor has digital logic made up of one large-scale (or very large-scale) IC. Since LSI and VLSI circuits are also called microcircuits, it is easy to see why the microprocessor has its name.

Data processing is the microprocessor's main function. Data processing includes both computation and data handling. Computation is performed by logic circuits that make up what is usually called the *arithmetic logic unit* (ALU). These logic circuits enable us to use functions that cause data changes. Among these functions are Add, Subtract, AND, OR, Compare, Increment, and Decrement.

The ALU cannot perform any of these functions without data to operate on. If the ALU is to add two numbers, for example, then each of the numbers must be put in the right place beforehand. The ALU cannot itself move data from place to place. Instead, the ALU merely performs an operation on whatever data it finds in certain places, and it leaves the result in the same place.

You may find it helpful to think of the ALU as a blindfolded juggler. The juggler can do amazing tricks, but only after being handed objects by someone else. Being blindfolded, the juggler cannot find the objects unaided. Like a blindfolded juggler waiting to be handed objects, the ALU must wait for data to be placed in certain places.

How then does the ALU get the data that it operates on? The microprocessor has other logic circuits, outside the ALU, that handle data. This data-handling logic moves data into place so that the ALU can process the data. After the operation, the data-handling logic moves the data elsewhere.

While a juggler cannot change the objects juggled, the ALU can perform operations that actually change data, and the microprocessor's data-handling logic moves the data from place to place as necessary. But what tells the ALU how to process the data? What tells the ALU which of the possible operations to perform?

In order to process data, the microprocessor must have control logic which tells the microprocessor how to decode and execute the *program*—a set of instructions for processing the data. The control logic steps the microprocessor through the stored program steps (instructions) in memory. It calls (fetches) them one at a time. After the instruction is fetched, the microprocessor's control logic decodes the instruction. Then the control logic carries out (executes) the decoded instruction.

Because the instructions are stored in memory, you can change them when you want to.

When you change the microprocessor's instructions, you change what it does to the data. The instructions that you store in memory determine what the microprocessor will do. This is a very important point for you to understand about microprocessors.

To review: The microprocessor's purpose is to process data. To do this, it must have logic to process and handle data, and control logic. The processing logic moves data from place to place and performs operations on the data. The control logic determines which circuits process the data. The microprocessor operates in the following steps. First, the microprocessor fetches (gets) an instruction. Then the control logic decodes what the instruction says to do. After decoding, the microprocessor executes (carries out) the instruction. These steps are called the *fetch-and-execute cycle,* or the *fetch/ execute cycle*. For each instruction in memory, the microprocessor goes through one fetch-and-execute cycle.

Besides fetching and executing instructions, the control logic also performs other major tasks. The control logic controls how the microprocessor works with all of the outside circuits (memory, input, and output) connected to the microprocessor. Powerful though the microprocessor is, you should keep in mind that it can do nothing by itself. The microprocessor must have the aid of other circuits. Some memory circuits are required to store the program instructions. Circuits are also needed to move data into and out of the microprocessor; these circuits are called *input/output* (I/O) circuits. Storage of data requires additional memory. The microprocessor also needs a power supply.

For example, look at even the simplest of hand-held microprocessor-based games. Such games need an input such as a joystick or keyboard to get data into the microprocessor and a display to get answers out. A battery is necessary to supply power. And the whole system must be put into a package.

This point is important for telling the difference between the microprocessor and the microcomputer: The microprocessor is the heart of many products, but the microprocessor is never a complete, working product by itself. Some single-chip microcomputer ICs do have circuits for input/output, data storage, or program storage, but these circuits still require input/output devices and power. The microprocessor's control logic can, however, control the other necessary parts when they are added.

Self-Test

Answer the following questions.

5. A microprocessor does not include _____ circuits.
   a. Logic
   c. Memory
   b. Computational
   d. All of the above

6. The microprocessor's fetch/execute cycle is used to get and carry out
   a. Logic work
   c. Arithmetic work
   b. Instructions
   d. ALUs

7. The ALU is used to do
   a. Addition
   c. Fetching
   b. Data moves
   d. All of the above

8. You can change what a microprocessor will do by
   a. Changing the instructions in memory
   b. Adding more inputs
   c. Adding more outputs
   d. Increasing memory size

9. The microprocessor generates signals to control the _____ circuits.
   a. Memory
   c. Output
   b. Input
   d. All of the above

10. A microprocessor is not a stand-alone device. That is, to operate, it requires at least
    a. Memory
    c. An output
    b. An input
    d. All of the above

1-3 WHAT IS A MICROCOMPUTER?

Frequently the words "microprocessor" and "microcomputer" are used to mean the same thing, but in fact these similar words have different meanings. As you learned in the previous section, the microprocessor is an IC. It includes circuits for two purposes: data processing and control.

The *microcomputer* is a complete computing system built around a microprocessor. A complete computing system has a microprocessor-based CPU, and it has memory and input/output functions. Figure 1-2 shows a complete microcomputer system. This microcomputer is a *personal computer (PC)*; however, all microcomputers are not PCs. That is, some microcomputer systems are not personal computers but, rather, dedicated computational systems based on a microprocessor. Note that the system has a card cage containing a CPU card, a random-access memory card, a disk controller card, and a video (I/O) card. The microcomputer also has a video display and a keyboard. The video display gives an alphanumeric display (output) on a *cathode-ray tube*

(CRT). That is, it displays both letters and numbers on a screen that is like a television's picture tube. The keyboard provides alphanumeric input. The floppy disks are connected to the disk controller card. They provide a way to store programs and data and to interchange programs and data between different microcomputers; they are a form of memory called *mass storage*. This system also has a power supply and packaging. We say this system is *self-contained*.

Figure 1-3 shows the system in a block diagram. Note the extra circuits needed to make a microprocessor into a microcomputer.

All the cards in the microcomputer are connected by a *bus*. The bus consists of many signal lines that let the different cards "talk" to one another. The word "bus" is short for *omnibus*, a Latin word meaning "to all." All the cards communicate through the bus signals because each uses the same set of signals. Usually any microcomputer card will work if plugged into any slot on the bus. A physical diagram of a bus is shown in Fig. 1-4 on page 8.

Returning to Fig. 1-3, you can see that the microprocessor is on the CPU card. The CPU card also has a clock generator, which generates timing signals for the microprocessor. A *read-only memory* (ROM) stores a few program instructions which are used to input other programs from the floppy disk. Like all the other cards, the CPU card has a number of ICs at the bottom. These ICs electrically *interface* (connect) the CPU card circuits to the bus, which usually requires higher current levels than those produced by most logic chips.

Besides the ICs that interface with the bus, the *random-access memory* (RAM) card has RAM ICs. This card has all the microcomputer's working data- and program-storage space.

The memory card also has the input/output function, which is implemented by a *universal asynchronous receiver-transmitter (UART)* and a clock IC. The UART converts parallel data to serial so the microcomputer can talk to a printer, modem, or other remote devices.

A fourth card used in this microcomputer is for video display. Besides the ICs that interface with the bus, the front-panel card has circuits to drive a video display. The video display actually displays the contents of a special section of memory (RAM); therefore this card includes the video RAM.

The power supply of this microcomputer system is another separate component. It is connected to all the boards by the bus.

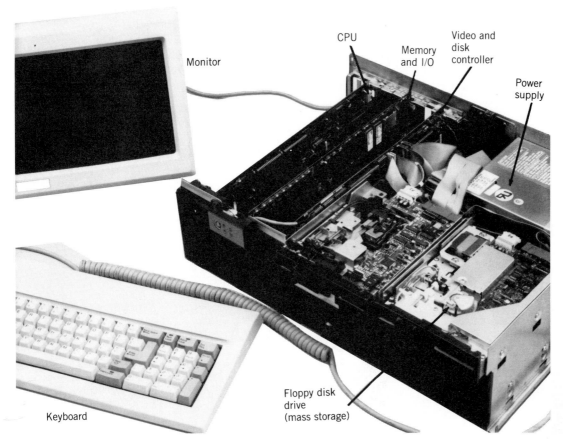

Fig. 1-2 A microprocessor-based microcomputer. This product, which includes a self-contained video display, keyboard, and mass storage subsystem, is built around the Intel 80286 microprocessor. It also makes extensive use of 256-kbit dynamic RAMs and high-density EPROMs.

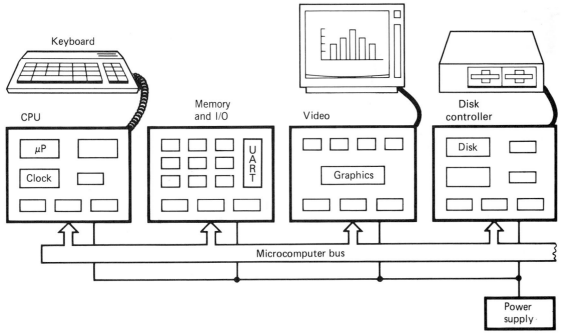

Fig. 1-3 A microcomputer system block diagram. This block diagram shows the electrical makeup of the microcomputer pictured in Fig. 1-2. The microprocessor (μP) is shown on the left-hand card.

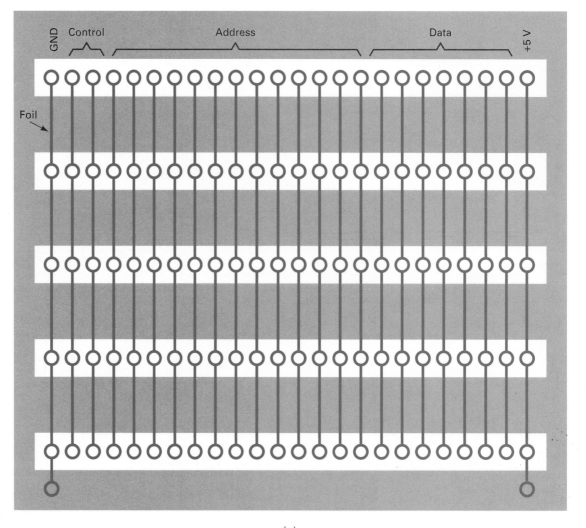

(a)

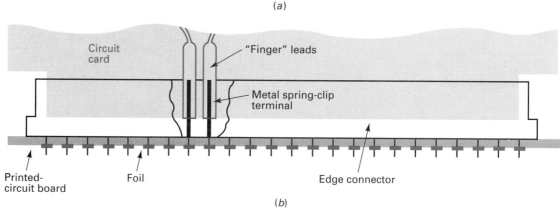

(b)

Fig. 1-4 **A microcomputer bus. (*a*) An electrical schematic of a typical microcomputer bus. The edge connector of each card is connected to that of the next card by a series of foils on the bus printed circuit board. (*b*) Edge view showing the bus printed circuit board's edge connector, the printed circuit board, and the foil, which actually carries the bus signals.**

As you can see from this description of the system shown in Fig. 1-3, the microprocessor is a small, but vital, part of a microcomputer. The microcomputer, on the other hand, is a complete system. To repeat: A complete system must have at least a CPU, memory, and some form of input/output.

Self-Test

Answer the following questions.

11. A microcomputer system has at least _____ circuits.
    a. Memory (ROM or RAM)
    b. Input/output
    c. CPU
    d. All of the above
12. The microcomputer's instructions are stored in the _____ circuits.
    a. Memory (ROM or RAM)
    b. Input/output
    c. CPU
    d. All of the above
13. The microcomputer's bus is a well-defined set of signal lines that let the different parts of the microcomputer system
    a. Communicate
    b. Get rid of heat
    c. Talk to a serial device
    d. Use power
14. The microprocessor must have timing signals to make the circuits work. These signals originate in a microprocessor support circuit called the
    a. ROM   c. Clock
    b. RAM   d. CPU
15. The microcomputer is a complete computing system. The microprocessor is the IC that
    a. Generates video signals
    b. Controls the system power supply
    c. Contains the microcomputer's CPU
    d. Lets the system talk to serial devices

1-4 WHAT IS THE POWER OF A MICROPROCESSOR?

Almost all microprocessors are made on silicon die (often called *ICs*). These ICs are about 1/4 inch (in.), or 0.64 centimeters (cm), on a side. Since the IC with the microprocessor circuits usually comes inside a 16- to 64-pin package, we cannot see the IC. But seeing the IC is unimportant. The appearance of the IC tells us little about the power of the microprocessor "printed" on the IC.

What we mean by the *power* of a microprocessor is its capacity to process data. There are three main measures of the power of a microprocessor: the length of the microprocessor's data word; the number of memory words that the microprocessor can address; and the speed

with which the microprocessor can execute an instruction.

Microprocessors are most often compared in terms of the lengths of their *data words*. Each microprocessor works on a data word of fixed length. Having to handle only one word length makes the design of the processor simpler.

Word lengths of 4 bits, 8 bits, 16 bits, and 32 bits are the most common today. Soon, 64-bit microprocessors will become common too. Data words of four different lengths are shown in Fig. 1-5 on page 10.

The 8-bit data word is so common that it has been given the special name *byte*. Because the byte is so commonly used, 16-bit microprocessors often have instructions that let them process their 16-bit data word in two 8-bit bytes. In Fig. 1-6 on page 11, you can see that a 16-bit data word is really made up of two 8-bit words. These words are called the *Upper byte*, or *Hi byte* (bits 8 through 15), and the *Lower byte*, or *Lo byte* (bits 0 through 7). Likewise, a 32-bit word is made up of 4 bytes.

Often the byte is used to speak of the size of part of a microcomputer. For example, you might speak of a microprocessor program that has 4000 bytes.

The byte is also used as a sort of common denominator for measuring microprocessor size. The byte is used as a measure rather than the actual data word length because the size of a word varies from one processor to another. However, a byte is always 8 bits. For the same number of bytes, an 8-bit processor has half as many words as a 4-bit microprocessor and twice as many words as a 16-bit microprocessor. For example: 4000 bytes on an 8-bit microprocessor equals 4000 words; on a 4-bit microprocessor, 4000 bytes equals 8000 words; on a 16-bit microprocessor, 4000 bytes equals 2000 words.

The 4-bit microprocessor was the first one developed. Microprocessors of this word length are still popular in some types of work. Four bits is the length needed for a *binary-coded decimal* (BCD) number. In some applications, including calculators, simple consumer products, and toys, the microprocessor deals only with BCD numbers. The 4-bit microprocessor is ideal for those applications. Another reason for the continuing use of the 4-bit microprocessor is its extremely low cost.

The 8-bit microprocessor is also both common and inexpensive. The 8-bit word length was the next developed after the 4-bit, because (1) the 8-bit word length is twice 4 bits, (2) the

ICs

Power

Data words

Byte

Upper byte (Hi byte)

Lower byte (Lo byte)

Binary-coded decimal (BCD)

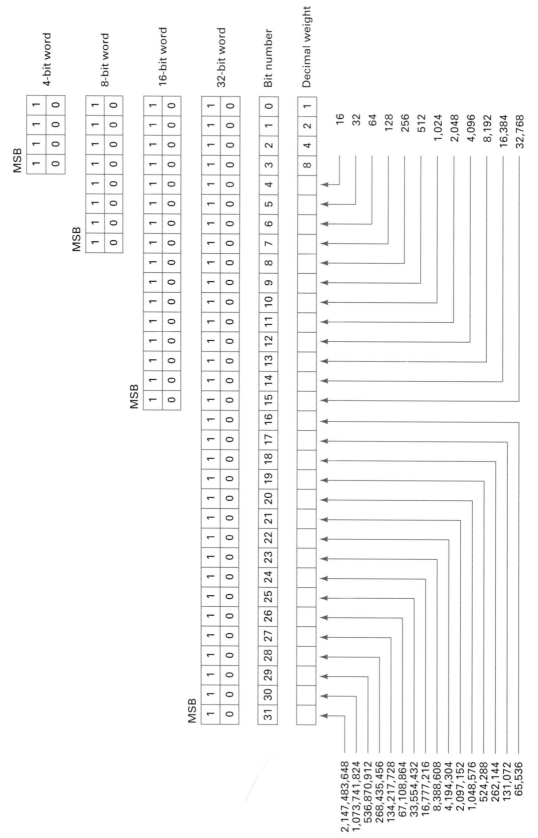

Fig. 1-5 Digital words of 4, 8, 16, and 32 bits. Note that the bit positions are numbered from right to left starting with bit 0. Bit 0 is the least significant bit, and the bit weight increases to the left. Each bit can be a binary 1 or a 0.

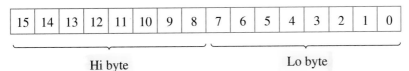

| 15 | 14 | 13 | 12 | 11 | 10 | 9 | 8 | 7 | 6 | 5 | 4 | 3 | 2 | 1 | 0 |

Hi byte   Lo byte

Fig. 1-6   A 16-bit digital word showing the High- and Low-byte breakdown.

American Standard
Code for
Information
Interchange (ASCII)

Address

Address range

8-bit word length allows two BCD numbers for each CPU data word, and (3) the 8-bit length can hold all the data needed for one character in the *American Standard Code for Information Interchange* (ASCII, pronounced "ask-key"). ASCII characters are used widely in data processing to represent numbers, letters, and many special symbols.

Most of the early 16-bit microprocessors were implementations of standard 16-bit minicomputers in LSI form. Examples are the Digital Equipment Corporation LSI-11, a copy of the PDP-11 minicomputer; the Data General MicroNova, a copy of the Nova minicomputer; and the Texas Instruments 9900, a copy of the 990 minicomputer. More recent 16-bit microprocessors have their own architecture, not taken from a minicomputer. These new 16-bit microprocessors form the basis for the popular personal computers (PCs) first introduced by IBM in 1981. Although originally the PC used 16-bit microprocessors, most PCs today use microprocessors with variations on a 32-bit architecture.

Each time the microprocessor's word length doubles, the processor becomes more powerful. Greater word lengths have required improved LSI technology. For example, the LSI used to develop one of the early 16-bit microprocessors had over 60,000 transistors on a single chip. The LSI used to develop some of the new 32-bit microprocessors uses a similar-sized chip, but it contains over 250,000 transistors. Superior technology has resulted in other improvements in the 16-bit microprocessors besides their greater word length.

Another common measure of microprocessor power is the number of memory bytes that the microprocessor can address. Here, too, the length of the data word plays an important role. The length of the data word in memory is the same as the length of the data word used by the microprocessor. A 4-bit microprocessor, for example, stores 4-bit words in memory.

Each word in memory is assigned a location number or *address*. When a word is needed from memory, the computer gets the word by referring to a memory address. Memory addresses start at 0 and end at some large binary number, the value of which varies from one processor to another; the larger the number of memory addresses, the greater the microprocessor's power.

Figure 1-7 shows the memory-addressing power of a single 4-bit word. As you can see, 4 bits can address 16 words in memory. Put differently, the 4-bit word has an *address range* of 16 words. We number these 16 words from 0 to 15.

A single 8-bit word has an address range of 256 memory words. A 16-bit word has an address range of 65,536 memory words.

Most microprocessors can use more than a single word to address memory. Therefore the memory-address range is not limited by the length of the microprocessor's data word. Memory addresses can be 32 or more bits long. These microprocessors can address billions of memory words.

Figure 1-8 shows the memory-address ranges of some commonly available micropro-

Binary address	Memory contents (4 bits long)
1 1 1 1	Data word 15
1 1 1 0	Data word 14
1 1 0 1	Data word 13
1 1 0 0	Data word 12
1 0 1 1	Data word 11
1 0 1 0	Data word 10
1 0 0 1	Data word 9
1 0 0 0	Data word 8
0 1 1 1	Data word 7
0 1 1 0	Data word 6
0 1 0 1	Data word 5
0 1 0 0	Data word 4
0 0 1 1	Data word 3
0 0 1 0	Data word 2
0 0 0 1	Data word 1
0 0 0 0	Data word 0

Fig. 1-7   A 16-word memory addressed by a 4-bit word.

cessors with 4-, 8-, 16-, and 32-bit data words. Common address ranges for 4-bit microprocessors are 4096 and 8192 memory words. Eight-bit microprocessors often have an address range of 65,536 memory words. The address ranges of 16-bit microprocessors extend from 32,768 to 4,194,304 memory words, and the range of 32-bit microprocessors extends from 4 billion to 32 billion memory words, with even greater ranges possible with memory management techniques.

Referring to these large numbers is often necessary when we work with microprocessors. Since the numbers are not round figures, they are difficult to remember and awkward to say. A shorthand has developed to simplify describing these memory sizes. The shorthand figures are shown in parentheses in Fig. 1-8. The 4096-word addressing range, for instance, is called a 4K range. The "K," from the prefix "kilo," meaning "thousand," is a scientific abbreviation for 1000. The expression 4K represents 4096 rounded off to the nearest thousand. Expressions for the other memory sizes have been shortened in the same way. When the memory sizes get extremely large, the letter "M," from the prefix "mega," meaning "million," is used as shorthand for a million, or the letter "G" from the prefix "giga," meaning "billion," is used as a shorthand for a billion.

Memory sizes of 64 kilobytes are very common for 8-bit microprocessors. The newer 16-bit microprocessors have address ranges extending into megabytes.

In talking about a microcomputer's memory size, you must always take care to explain whether you are talking of bytes or words. Throughout this text we will speak of memory size in bytes. Since a byte is always 8 bits, a 65,536-word memory on an 8-bit microcomputer and a 32,768-word memory on a 16-bit microcomputer are really the same size. That is, they both have 65,536 bytes.

If this comparison is unclear, then compare the memory sizes in bits. Multiply the number of words in each case by the number of bits per word. That is, 65,536 words $\times$ 8 bits per word = 524,288 bits, and 32,768 words $\times$ 16 bits per word = 524,288 bits.

A third common measure of microprocessor power is the speed with which the microprocessor executes an instruction. Speed is determined by the time it takes the microprocessor to complete the *fetch/execute cycle* for one program step.

Some microprocessors are 20 to 100 times faster than others. Each one has an oscillator circuit which sets its pace. This oscillator circuit is called the microprocessor's *clock*. Slow microprocessors may use a clock that runs at a few hundred kilohertz (kHz). It takes such a microprocessor 10 to 20 microseconds (μs; the Greek letter "mu" stands for one-millionth) to execute one instruction. On the other hand, some microprocessors use clocks that run at 10, 25, 33, 50, or even 100 megahertz (MHz). These microprocessors may execute an instruction in only a few tenths of a microsecond.

Data word length	4 bit	8 bit	16 bit	32 bit
Memory address range	4096 (4K)			
	8192 (8K)			
		65,536 (64K)		
			32,768 (32K)	
			65,536 (64K)	
			1,048,576 (1M)	
			2,097,152 (2M)	
			4,194,304 (4M)	
				4,294,967,296 (4G)
				34,359,738,367 (32G)

Fig. 1-8 **Word size and memory addressing range of some commonly available microprocessors.**

The microprocessor's speed is related to its maximum clock frequency. Sometimes people compare microprocessors simply in terms of clock frequency. Comparisons are more meaningful, however, when they find out how long a given operation will take on the different processors. Short programs, called *benchmark programs*, are written to make such comparisons easier. Different microprocessors are timed while executing the same *benchmark program*.

Self-Test

Answer the following questions.

16. One of the most common measurements of a microprocessor's power is its
    a. Chip size   c. Number of pins
    b. Word length   d. All of the above

17. An 8-bit word is a very common length. It is called a(n)
    a. Byte   c. ALU
    b. CPU   d. Address

18. A 16-bit word length is used by the 80286 microprocessor. If an 80286 addresses 32 kilowords of memory, its memory will have _____ bits of data.
    a. 32,768   c. 384,000
    b. 262,144   d. 524,288

19. Many 8-bit microprocessors address 65,536 memory locations. Usually this is shown as 64K. Why do you think the term 64K rather than 65K is used as shorthand for 65,536? *26*

20. Two identical microcomputer systems have different clock rates. System A has a 5-MHz clock, and system B has a 1-μs clock. Which system executes an instruction faster? Why? *System A Its faster 5MHz vs 1/s*

SUMMARY

1. The microprocessor is a large-scale or very large-scale IC that uses the architecture of the general-purpose digital computer.

2. The architecture of the digital computer and the capability of LSI were brought together in the early 1970s. The computer had developed from the post-World War II period, as had solid-state technology.

3. Two important concepts were developed as the computer evolved. First came the concept of program storage. Second came the concept of many independent low-cost processors to do many independent tasks.

4. The microprocessor is a microprocessing unit that has control and computational functions in microcircuit form.

5. The microprocessor's logic circuits do both data-handling and data processing functions. The data processing functions are carried out by the ALU (arithmetic logic unit). The logic also fetches instructions and executes them (decodes them and carries them out).

6. A microcomputer is a complete computing system based on a microprocessor. It has a microprocessor, memory, and input/output functions. It also has a power supply and packaging.

7. The microcomputer's bus allows each part of the microcomputer system to communicate with the other parts by using a com-

mon set of signals. Often the signals are a high-powered extension of the microprocessor's internal bus.

8. The microprocessor's data word is a common measurement of its size. Word lengths of 4, 8, 16, and 32 bits are common. The 8-bit word is so common it is called by the special name *byte*.

9. The number of memory locations addressed by the microprocessor is also used to describe the microprocessor. Most 8-bit microprocessors address 65,536 memory locations. Sixteeen-bit microprocessors have addressing ranges in the millions of words, and 32-bit microprocessors have addressing ranges in the billions of words.

10. A shorthand notation is used in specifying the number of bytes. The symbol "K" is used to say "times 1000," and numbers in the kilobyte range that are powers of 2 are rounded to the nearest thousand. The symbol "M" means "times 1 million," and numbers in the megabyte range that are powers of 2 are rounded to the nearest million. The symbol "G" means "times 1 billion," and numbers in the gigabyte range that are powers of 2 are rounded to the nearest billion.

11. The speed of a microprocessor is measured by how long it takes to fetch and execute an instruction. Both the instruction time

and the clock frequency are used to describe a microprocessor's speed. We can also compare the speeds of different microprocessors by measuring the time they take to execute a standard or "benchmark" program.

CHAPTER REVIEW QUESTIONS

Answer the following questions.

1-1. Microprocessor architecture is taken from the architecture of a
   a. Calculator LSI chip
   b. Patch-paneled programmed computer
   c. General-purpose digital computer
   d. Vacuum-tube CRT

1-2. The semiconductor technology that finally made the microprocessor possible was a development of
   a. CPUs   *c.* LEDs
   b. LSI   *d.* Germanium transistors

1-3. The microprocessor uses the stored-program concept. This means the instructions are stored in the _____ along with the data.
   a. Serial I/O   *c.* Power supply
   b. CPU   *d.* Memory

1-4. Using the computer's architecture to solve small problems has been made possible only because the microprocessor is so _____ compared to the general-purpose digital computer.
   a. Fast   *c.* Powerful
   b. Inexpensive   *d.* All of the above

1-5. What does the abbreviation CPU stand for? *Central processing unit.*

1-6. The microprocessor's data-handling functions let it
   a. Perform a logic AND function
   b. Move data between microcomputer parts
   c. Make the computation using the data
   d. Operate more quickly than it could with only computational functions

1-7. To process one of your instructions, the microprocessor must go through a fetch/execute cycle. The execute parts are carried out in the
   a. ALU   *c.* I/O
   b. Memory   *d.* ROM

1-8. A microprocessor is at the heart of a microcomputer system. However, the microcomputer must also have _____ circuits.
   a. Input   *c.* Output
   b. Memory   *d.* All of the above

1-9. You can safely say that the true microprocessor will
   a. Always run at almost 1 MHz
   b. Have a minimum of 16 kilobytes of memory
   c. Always be a single chip
   d. Never work by itself

1-10. The main purpose of a microcomputer's bus is to
   a. Allow the industry to build standard products
   b. Allow different parts of the microcomputer system to communicate using a well-defined signal path
   c. Be a standard mechanical connection for the microcomputer system
   d. Be sure that signals of 3.58 MHz and higher are properly transmitted

1-11. Often the microcomputer's bus is a set of parallel conductors interconnecting a number of circuit-card connectors. Why do you think it is built this way? *to allow easier communication.*

1-12. The 8-bit data word is a very popular length. If two 8-bit data words are

used to address a memory location, this means the memory has _____
8-bit locations.
  a. 16K   c. 64K
  b. 32K   d. 128K
1-13. A data word _____ bits long is used on many microprocessors.
  a. 4   c. 16
  b. 8   d. All of the above
1-14. An 8-bit data word is often called
  a. Very popular   c. A logic circuit
  b. A byte   d. A CPU
1-15. A 16-bit microprocessor must have _____ memory locations.
  a. 65,536   c. No fixed number of
  b. 64K   d. Millions of
1-16. The microprocessor's fetch/execute cycle time depends on its
  a. Address range   c. Bus size
  b. Clock frequency   d. All of the above

CRITICAL THINKING QUESTIONS

1-1. Although some microprocessors become the heart of a personal computer, most microprocessors sold today are used to make the control functions of a dedicated electronic device work (microwave oven control panels, VCR controls, electronic games, etc.). Why is the microprocessor, instead of the digital integrated circuit, popular for this use?

1-2. Figure 1-1 shows the band for minicomputers growing narrower, and the band for microcomputers growing larger, as we move into the 1990s. What is happening here?

1-3. In this chapter we learned that a microprocessor must have both input and output circuits (and memory) before it can become a microcomputer. Why do you need both input and output circuits? Can you think of a product which does not have both input and output?

1-4. You are balancing your checkbook. You have a check register containing the record of checking activity, the bank statement, a pencil, scratch paper, and a calculator. Compare each of these items to the microprocessor's memory, data-handling circuits, and ALU.

1-5. There are two PCs in your lab. One is a 486 DX33, and the other is a 486 SX25. Not knowing what DX and SX mean but assuming that the 33 and 25 refer to clock speed, can you tell which is the most "powerful"? Why?

1-6. Some 386 microprocessors have a 16-bit data word; other versions of the 386 have a 32-bit data word. What difference does this make in terms of how much memory you buy for your PC if the manufacturer of the program you want to run suggests that the computer have 4 Mbytes of memory?

Answers to Self-Tests

1. b
2. c
3. b
4. a
5. c
6. b
7. a
8. a
9. d
10. d

11. d
12. a
13. a
14. c
15. c
16. b
17. a
18. d
19. Because 64 is 2^6, and therefore 64,000 is

thought of as the round number closest to 65,536.
20. System A, because the clock time is 200 nanoseconds (ns) for a 5-MHz clock, and 200 ns is five times as fast as 1 μs, which is 1000 ns.

CHAPTER 2

The Decimal and Binary Number Systems

■

CHAPTER OBJECTIVES

This chapter will help you to:

1. *Define* base, weight, radix point, and binary point.
2. *Express* decimal numbers in scientific notation.
3. *Identify* the least and most significant bit or digit in a number.
4. *Convert* decimal numbers to binary, and binary to decimal.
5. *Explain* the use of the hexadecimal number system.
6. *Convert* binary and decimal numbers to hexadecimal numbers.

───────

The microprocessor is built with digital logic and therefore manipulates binary data. We often ask the microprocessor to perform arithmetic functions. Often, we want the input and output information in the familiar decimal numbers. Therefore, to use the microprocessor, we must be familiar with converting decimal numbers into binary and with converting binary numbers into decimal.

■

2-1 THE DECIMAL NUMBER SYSTEM

The decimal number system is the most commonly used number system in the world. It uses ten different characters to show the values of numbers. Because the system uses ten different characters, it is called the *base 10 system*. The *base* of a number system tells you how many different characters are used. The mathematical term for the base of a number system is *radix*.

The ten characters used in the decimal number system are

0, 1, 2, 3, 4, 5, 6, 7, 8, 9

Looking at the characters that make up the decimal number system, you see that the character "0" is one of the ten. This is very important to remember. All counting in computer, and therefore microprocessor, systems starts at 0.

What do we do when there are more than ten objects to be counted? As we all know, we add more numbers to the left of the original column. We can say that the next number to the left tells us how many times we have completely used the column to the right.

What does the number 23 mean? We can look at this as (a) we found enough objects to use all the characters (0 through 9) in the right-hand column once (with the left-hand column equal to 0); (b) we found enough objects to use all the characters (0 through 9) a second time (with the left-hand column equal to 1); and (c) with the left-hand column equal to 2 we had enough objects that we used the first, second, third, and fourth characters in the system. The fourth character is 3.

This way of counting is shown in Fig. 2-1. Note that this shows the decimal number, how we got it, and the number of objects shown by each number. Frequently we drop insignificant

From page 16:

Base 10 system

Base

Radix

On this page:

Weight

Scientific notation

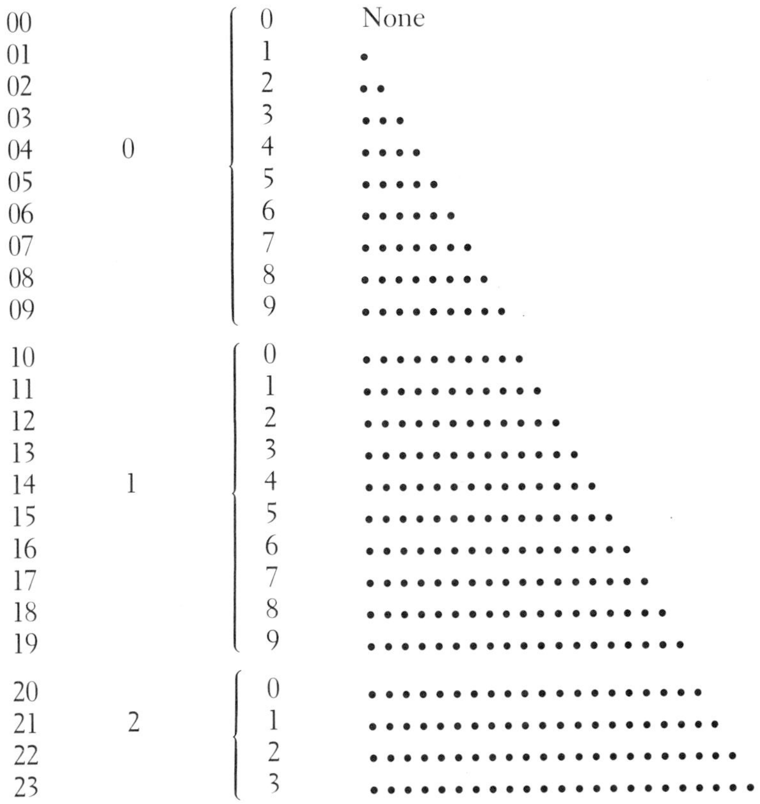

00		0	None
01		1	•
02		2	••
03		3	•••
04	0	4	••••
05		5	•••••
06		6	••••••
07		7	•••••••
08		8	••••••••
09		9	•••••••••
10		0	••••••••••
11		1	•••••••••••
12		2	••••••••••••
13		3	•••••••••••••
14	1	4	••••••••••••••
15		5	•••••••••••••••
16		6	••••••••••••••••
17		7	•••••••••••••••••
18		8	••••••••••••••••••
19		9	•••••••••••••••••••
20		0	••••••••••••••••••••
21	2	1	•••••••••••••••••••••
22		2	••••••••••••••••••••••
23		3	•••••••••••••••••••••••

Fig. 2-1 Counting 23 objects. For each of the first nine objects we have a unique character to show the number of objects. We use the characters 0 through 9. After the ninth object we begin to repeat the count characters.

0s. These are 0s to the left of the most significant digit or to the right of the right-most nonzero fraction digits. In this figure, an insignificant 0 is shown in the left-hand column beside the first set of numbers 0 through 9.

Normally we say that each column is given a *weight*. The weights are powers of 10. The first column is the ones column, the second is the tens column, the third column is the hundreds column, the fourth column is the thousands column, and so forth. This is shown in Fig. 2-2. We do not usually show the insignificant 0s. They are shown in Fig. 2-2 so that you will know where they are.

Using *scientific notation*, we can show the first column weight as 10^0, the second as 10^1, the third as 10^2, the fourth as 10^3, and so forth. You can see that the weight of each column is the base of the number system raised to a power. The power is the column's position.

As an example, let's look at the number 6321, which is used in Fig. 2-2. Assigning a weight to each column, we represent this as

$$10^3 \quad 10^2 \quad 10^1 \quad 10^0$$
$$6 \quad\quad 3 \quad\quad 2 \quad\quad 1$$

We can evaluate this number as

$$(6 \times 10^3) + (3 \times 10^2) + (2 \times 10^1)$$
$$+ (1 \times 10^0) = 6321$$

As an additional example, let's look at the number 1,260,523:

$$10^6 \quad 10^5 \quad 10^4 \quad 10^3 \quad 10^2 \quad 10^1 \quad 10^0$$
$$1 \quad\quad 2 \quad\quad 6 \quad\quad 0 \quad\quad 5 \quad\quad 2 \quad\quad 3$$

We can evaluate this number as

$$(1 \times 10^6) + (2 \times 10^5) + (6 \times 10^4)$$
$$+ (0 \times 10^3) + (5 \times 10^2) + (2 \times 10^1)$$
$$+ (3 \times 10^0) = 1,260,523$$

100,000s	10,000s	1000s	100s	10s	1s	$\frac{1}{10}$s	$\frac{1}{100}$s
0	0	6	3	2	1	0	0

Fig. 2-2 Assigning a decimal weight to each column of a decimal number. Whether you use them or not, the insignificant 0s are always there.

$$10^3 \quad 10^2 \quad 10^1 \quad 10^0 \qquad 10^{-1} \quad 10^{-2} \quad 10^{-3}$$
$$6 \quad\; 3 \quad\; 2 \quad\; 1 \quad . \quad\; 5 \quad\;\; 6 \quad\;\; 4$$
$$6 \times 10^3 + 3 \times 10^2 + 2 \times 10^1 + 1 \times 10^0 + 5 \times 10^{-1} + 6 \times 10^{-2} + 4 \times 10^{-3} = 6321.564$$

Fig. 2-3 Evaluating a decimal mixed number by using scientific notation. Note that the zero power of any number is equal to 1.

In the decimal number system, we show fractions by putting characters to the right of the decimal point. The *decimal point* separates the integer and fraction parts of a number. The more general mathematical term for the decimal point is the *radix point*.

Each column to the right of the decimal point is also given a weight. The column immediately to the right of the decimal point has a weight of $\frac{1}{10}$. The next column to the right has a weight of $\frac{1}{100}$, the next column a weight of $\frac{1}{1000}$, and so forth. As you know from studying scientific notation, these may also be expressed as negative powers of 10. That is, $\frac{1}{10}$ may be expressed as 10^{-1}. (See Fig. 2-3.)

Self-Test

Answer the following questions.

1. You will find positive powers of 10 on the _____ side of a decimal number's radix point.
   a. Right-hand
   (b.) Left-hand
   c. Opposite
   d. Wrong
2. To evaluate the value of a column, you _____ the column's weight.
   a. Add its number to
   b. Divide its number by
   (c.) Multiply its number by
   d. Subtract its number from
3. The expression

   $$(2 \times 10^4) + (7 \times 10^3) + (3 \times 10^2)$$
   $$+ (2 \times 10^1) + (9 \times 10^0)$$
   evaluates as __27329__ .

4. Negative powers of 10 are used to show _____ column weights.
   a. Half
   b. Integer
   c. Whole
   (d.) Fraction

2-2 THE BINARY NUMBER SYSTEM

In many ways the binary number system is simpler than the decimal number system. The binary number system has only two characters. The binary number system is used in digital electronics because digital circuits have only *two states* (two signal levels). Most of the time, 0 and 1 are the two characters used. Sometimes other characters are used, but any two figures may be used. Some other common figures which are used in special situations are shown in Fig. 2-4.

Comparing a decimal number with a binary number, you see that the binary number uses more columns, because there are only two characters in the binary number system. The decimal number system can show 10 different counts in each column (0–9). The binary system can show only two in each column (0 or 1).

Like the decimal system, the binary number system assigns a weight to each column. In the decimal number system, each column is assigned a weight that is a power of 10. This is done because the decimal number system is the base 10 system. The binary number system is the *base 2 system*. The assigned weights are powers of 2. The first 13 column weights are shown in Fig. 2-5. As you might expect, the size of the equivalent number is quite small even though there are many columns.

In the following example, we evaluate a binary number to find its decimal value.

$$101101_2$$

$$(1 \times 2^5) + (0 \times 2^4) + (1 \times 2^3)$$
$$+ (1 \times 2^2) + (0 \times 2^1) + (1 \times 2^0) = 45_{10}$$

The binary number 101101 is the same as the decimal number 45.

Note that when we are using both binary and decimal numbers, the decimal numbers are followed by the subscript 10 and binary numbers are followed by the subscript 2. For example, there is quite a difference between 101_2 and

One state	The opposite state
0	1
Off	On
Space	Mark
Open	Closed
Low	Hi

Fig. 2-4 A number of different ways used to show which state a binary or logic circuit is in.

Binary point

Digit

Bit

Least significant
bit (LSB)

Most significant
bit (MSB)

Most significant
digit (MSD)

Least significant
digit (LSD)

$$\begin{array}{cccccccccccccc} 4096 & 2048 & 1024 & 512 & 256 & 128 & 64 & 32 & 16 & 8 & 4 & 2 & 1 \\ 2^{12} & 2^{11} & 2^{10} & 2^9 & 2^8 & 2^7 & 2^6 & 2^5 & 2^4 & 2^3 & 2^2 & 2^1 & 2^0 \end{array}$$

Fig. 2-5 The powers of 2 and their decimal equivalents.

101_{10}. In the first example we have the binary representation of the decimal number 5. In the second example we have the decimal representation of the decimal number one hundred one.

Like the decimal system, the binary number system uses a radix point. It is called the *binary point*. The binary point separates the integer part of the number from the fraction part of the number. It serves the same purpose in binary mixed numbers as the decimal point does in decimal mixed numbers. Each column to the right of the binary point is assigned a fraction weight.

As in the decimal system, the column weight is shown as a negative power of the number system's base. In the binary number system the weights are the fractions ½, ¼, ⅛, ¹⁄₁₆, ¹⁄₃₂, and so forth. These can be expressed as 2^{-1}, 2^{-2}, 2^{-3}, 2^{-4}, 2^{-5}, and so forth. The first six negative powers of 2 and their decimal equivalent are shown in Fig. 2-6.

$$\begin{array}{cccccc} 0.5 & 0.25 & 0.125 & 0.0625 & 0.03125 & 0.015625 \\ 2^{-1} & 2^{-2} & 2^{-3} & 2^{-4} & 2^{-5} & 2^{-6} \end{array}$$

Fig. 2-6 The negative powers of 2 and their decimal equivalents.

In the decimal number system each column or place is referred to as a *digit* or, more precisely, a *decimal digit*. In the binary number system each place is referred to as a *bit*. The term "bit" stands for binary digit. "Bit" is a very commonly used (and misused) term in digital electronics.

Often when referring to binary numbers you will hear the terms "LSB" (*least significant bit*) and "MSB" (*most significant bit*). These are very much like the terms we use when speaking of decimal numbers. In decimal numbers we refer to the *most significant digit* (MSD) and the *least significant digit* (LSD). The LSB is the bit with the least weight. The MSB is the bit with the greatest weight. Normally, binary numbers are shown with the MSB as the leftmost bit.

Self-Test

Answer the following questions.

5. In the decimal number 2364 the character 2 is the
   a. LSD   c. LSB
   b. MSD   d. MSB
6. Is the number 1210110 a binary number? Why? No, 2 is not binary
7. What does "LSB" stand for? Least significant bit
8. "Bit" is the abbreviation for binary digit .
9. The base of the binary number system is
   a. 10_{10} c. 2_{10}
   b. 1_2 d. 0_{10}
10. Which is the larger number, 1111_2 or 11_{10}? Why? 11_{10} decimal is more then binary
11. The value 2^{-3} is the same as
    a. ¹⁄₁₆   c. ¼
    b. ⅛   d. ½

2-3 BINARY-TO-DECIMAL CONVERSION

Often when you are working with microcomputers, you have to change binary numbers to their decimal equivalents. There are many different reasons for doing this. Perhaps the most common reason is the need to display the number for people who cannot read binary numbers. You must use the binary-to-decimal conversion process to make the results meaningful for these people.

The binary-to-decimal conversion process is simple: We add the decimal weights of all bits that are a "1." The following two examples demonstrate this process.

Convert 11001100_2 to decimal.

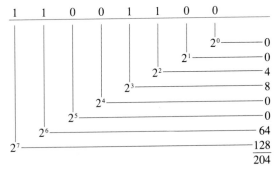

$$11001100_2 = 204_{10}$$

To convert the binary number 101.011 to decimal, we do the same:

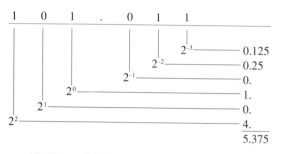

$$101.011_2 = 5.375_{10}$$

Self-Test

Answer the following question.

12. Evaluate the following binary numbers to find their decimal values:

   a. 10110110   182
   b. 01010000   80
   c. 00011110   30
   d. 10000000   128
   e. 11111111   255
   f. 10000001   129
   g. 01111110   126
   h. 00111111   63
   i. 0.0101   .3125
   j. 100.011   4. .375

2-4 DECIMAL-TO-BINARY CONVERSION

When working with a microcomputer, you will often want to make the computer use decimal numbers. To do this, you must input decimal numbers, convert them to binary numbers, process the binary numbers with the microprocessor, and then convert the numbers back to decimal. This means we need a way to convert decimal numbers to binary numbers.

The process for converting integer decimal numbers into binary numbers is a specific case of the general process for converting a number in one base to a number in another base. Suppose you wanted to convert the decimal number 10 to a binary number. Converting a decimal integer to a binary integer is done by using the following procedure. Follow the example as you read the procedure.

1. The number to be converted is divided by the same base of the number system it is to be converted into. In this case, the decimal number 10 is divided by 2, the base of the binary number system. In a division by 2, the remainder must be either 1 or 0.

This remainder becomes the least significant bit of the new number.

$$\begin{array}{r} 5 \\ 2\overline{)10} \\ \underline{10} \\ 0 \end{array} \quad \text{LSB} = 0$$

2. The result from the division in step 1 is also divided by 2. The remainder is either 1 or 0. This becomes the next more significant bit of the resulting number.

$$\begin{array}{r} 2 \\ 2\overline{)5} \\ \underline{4} \\ 1 \end{array} \quad \text{Next bit} = 1$$

3. Once again, the result of the previous division is divided by 2. The remainder is the next more significant bit.

$$\begin{array}{r} 1 \\ 2\overline{)2} \\ \underline{2} \\ 0 \end{array} \quad \text{Next bit} = 0$$

4. This process continues until the result of the division is 0. The remainder from the last division, a 1, is the most significant bit of the binary number.

$$\begin{array}{r} 0 \\ 2\overline{)1} \\ \underline{0} \\ 1 \end{array} \quad \text{MSB} = 1$$

Assembling these four remainders in order gives the binary integer 1010.

The following two examples illustrate the procedure:

$$57_{10} = ?_2$$

Division	Remainder	
$2\overline{)57}$ with 28	1	LSB
$2\overline{)28}$ with 14	0	
$2\overline{)14}$ with 7	0	

$$\frac{3}{2)7} \quad 1$$

$$\frac{1}{2)3} \quad 1$$

$$\frac{0}{2)1} \quad 1 \qquad \text{MSB}$$

Therefore $\quad 57_{10} = 111001_2$.

$$134_{10} = ?_2$$

Division	Remainder	
$\frac{67}{2)134}$	0	LSB
$\frac{33}{2)67}$	1	
$\frac{16}{2)33}$	1	
$\frac{8}{2)16}$	0	
$\frac{4}{2)8}$	0	
$\frac{2}{2)4}$	0	
$\frac{1}{2)2}$	0	
$\frac{0}{2)1}$	1	MSB

Therefore $\quad 134_{10} = 10000110_2$

The procedure we have learned to convert decimal to binary numbers is for integers. Fractions must be handled separately. But the procedure for fractions is very similar to that for integers. Once the fraction and the integer are converted, the results are combined as the right-hand and left-hand numerals around the binary point.

Converting a decimal fraction to a binary fraction is done by using the following procedure. The example shows how to convert the decimal number 0.375 to a binary number. Follow the example as you read the procedure.

1. The fraction to be converted is multiplied by the base of the number system it is to be converted into. In this case, the decimal fraction 0.375 is multiplied by 2, the base of the binary number system.

$$2 \times 0.375 = 0.75$$

2. If the result of the multiplication is less than 1, the most significant bit of the new binary number is a 0. If the result is greater than 1, the most significant bit of the new binary number is a 1.

$$0.75 < 1$$

Therefore the MSB is 0.

3. The fraction from the previous multiplication is again multiplied by 2. Note: This is only the fraction portion. It does not include the integer portion if the result was greater than 1.

$$2 \times 0.75 = 1.5$$

4. If the result of the multiplication is less than 1, the next most significant bit of the new binary number is a 0. If the result of the multiplication is equal to or greater than 1, the next most significant bit of the new binary number is 1.

$$1.5 > 1$$

Therefore, the next most significant bit of the binary number is 1.

5. This process continues either until the result of the multiplication is exactly 1 or until you have sufficient accuracy.

$$2 \times 0.5 = 1.0$$

The next bit, which will be the LSB, is 1. The binary fraction is 0.011.

You will not always be able to reach a result of exactly 1 when you multiply repeatedly by 2. Therefore, you stop when you get the accuracy you want—that is, when you have enough bits in your binary fraction for your needs. The integer result of this last step becomes the least significant bit of the binary number.

The following examples illustrate this procedure.

$$0.34375_{10} = ?_2$$

Multiplication		Integer Result	
$2 \times 0.34375 = 0.6875$		0	MSB
$2 \times 0.6875 = 1.375$		1	
$2 \times 0.375 = 0.75$		0	
$2 \times 0.75 = 1.5$		1	
$2 \times 0.5 = 1.0$		1	LSB

Therefore $0.34375_{10} = 0.01011_2$

$$0.3_{10} = ?_2$$

Multiplication	Integer Result
$2 \times 0.3 = 0.6$	0
$2 \times 0.6 = 1.2$	1
$2 \times 0.2 = 0.4$	0
$2 \times 0.4 = 0.8$	0
$2 \times 0.8 = 1.6$	1
$2 \times 0.6 = 1.2$	1
$2 \times 0.2 = 0.4$	0
$2 \times 0.4 = 0.8$	0
$2 \times 0.8 = 1.6$	1
$2 \times 0.6 = 1.2$	1
$2 \times 0.2 = 0.4$	0

This conversion will repeat forever. We shall cut it off after 8 bits of resolution. Therefore, $0.3_{10} = 0.01001100$, to 8 bits of resolution for the fraction.

Self-Test

Answer the following questions.

13. Converting the integer portion of a number from one radix to another radix is done by using a process which uses
    a. Addition   *c.* Multiplication
    b. Subtraction   *d.* Division
14. Converting the fraction part of a number from one radix to another radix is done by using a process which uses
    a. Addition   *c.* Multiplication
    b. Subtraction   *d.* Division
15. Convert the following decimal numbers to binary numbers.
    a. 23 *10111*   **g.** 63 *111111*
    b. 105 *1101001*   **h.** 29 *11101*
    c. 32 *100000*   **i.** 12.125 *1100.001*
    d. 15 *1111*   **j.** 16.375 *10000.011*
    e. 206 *11001110* **k.** 5.015625 *101.*
    f. 128 *10000000* **l.** 2.5 *10.1*

2-5 THE HEXADECIMAL NUMBER SYSTEM

The *hexadecimal numbering system* refers to the base 16 number system. In the base 16 number system there are 16 different characters. In the hexadecimal numbering system we use the characters 0, 1, 2, 3, 4, 5, 6, 7, 8, 9, A, B, C, D, E, and F.

Why is the hexadecimal numbering system used? We know that no electronic system uses 16 different levels in the way that binary electronics uses two different levels. The hexadecimal numbering system is not required by machines. It is a convenience for people. The hexadecimal numbering system is used in the world of computers, microcomputers, and microprocessors as a shorthand technique.

Stop and think of some of the common uses for binary numbers. You can see that there is a very real problem because of their length. For example, many microprocessors use an 8-bit word. That is, when they work with a binary number, it has 8 bits. An 8-bit number can represent only the decimal values 0 through 255. This 8-bit number has just as many characters as the decimal numbers for any value from 0 to 99,999,999.

Clearly, the binary number that represents a large decimal number like 99,999,999 is very long. In fact, the binary version of 99,999,999 is 27 bits long! Such a long number is very difficult to read. The hexadecimal numbering system is used as a shorthand method to reduce the length of binary numbers.

Sixteen, as we know, is the fourth power of 2. That is, $16 = 2^4$. Each of the 16 hexadecimal characters (0 through F) can be represented by a 4-bit binary number. The 4-bit binary numbers are 0000 to 1111. This means that one hexadecimal character can serve as a shorthand notation for a 4-bit binary number. The relationships between the hexadecimal, binary, and decimal numbers are shown in Fig. 2-7.

Because the word *hexadecimal* is such a mouthful, the term *hex* is often used in speaking of hexadecimal numbers.

The conversion from a binary number to a hexadecimal number is simple. The binary number is collected into groups of 4 bits. This grouping starts at the least significant bit (the number to the left of the binary point). Each of the groups of 4 bits is then converted into its equivalent hex number. *Note:* Grouping binary numbers into 4-bit units makes the

Hexadecimal number	Binary number	Decimal number
0	0	0
1	1	1
2	10	2
3	11	3
4	100	4
5	101	5
6	110	6
7	111	7
8	1000	8
9	1001	9
A	1010	10
B	1011	11
C	1100	12
D	1101	13
E	1110	14
F	1111	15
10	1 0000	16
11	1 0001	17
12	1 0010	18
13	1 0011	19
14	1 0100	20
15	1 0101	21
16	1 0110	22
17	1 0111	23
18	1 1000	24
19	1 1001	25
1A	1 1010	26
1B	1 1011	27
1C	1 1100	28
1D	1 1101	29
1E	1 1110	30
1F	1 1111	31
20	10 0000	32
32	11 0010	50
40	100 0000	64
6E	110 1110	110
80	1000 0000	128

Fig. 2-7 **Hexadecimal (hex) numbers and their binary and decimal equivalents.**

binary numbers easier to read, just as commas help us read multidigit numbers.

The following two examples demonstrate the conversion of binary numbers into hexadecimal numbers. To convert the binary number 10101011111101 to a hexadecimal number, we first group by fours

$$0010 \quad 1010 \quad 1111 \quad 1101$$

Note that we can add insignificant 0s as necessary to make groupings that are 4 bits long. Now we replace each group with the equivalent hexadecimal number.

$$2 \quad A \quad F \quad D$$

Therefore, 10101011111101_2 is $2AFD_{16}$.

Converting 11000111_2 to hex, we have

$$1100 \qquad 0111$$
$$C \qquad\quad 7$$

That is, $11000111_2 = C7_{16}$.

Again, you can see that hexadecimal numbers are much easier to read or say than binary numbers. We now know that

$$10101011111101_2 = 2AFD_{16}$$

and that

$$11000111_2 = C7_{16}$$

The hexadecimal number is much easier to say and remember than the binary number. You will become quite familiar with hex numbers. Always remember that hex numbers are just ways of showing binary numbers. The microprocessor works on bits, not hexadecimal characters!

Converting from binary fractions to hexadecimal fractions follows similar rules. The fraction bits are grouped in fours starting at the fraction's MSB (this is the number to the right of the binary point). Each grouping of four is converted into its equivalent hex number. Insignificant (trailing) 0s are added to the binary bits if necessary.

For example, to convert the binary fraction 0.0101101 to hex, we group by fours:

$$0. \quad 0101 \quad 1010$$

Now we replace each group with the equal hexadecimal number:

$$0. \quad 5 \quad A$$

That is, 0.0101101_2 is $0.5A_{16}$.

Converting 1101.0111_2 to hexadecimal, we have

$$1101 \quad . \quad 0111$$
$$D \quad . \quad 7$$

Therefore, 1101.0111_2 is $D.7_{16}$. Be sure to note that

$$13.24_{16} \neq 13.24_{10}$$

Don't confuse base 16 and base 10 (decimal) numbers. One way to avoid mixing these numbers is to be careful how you say them. For example, when you are reading 13.24_{16} say "one three point two four," not "thirteen point twenty-four." Thirteen is a decimal number! Do this whether reading aloud or reading in your head.

As you can see from the examples, the binary number is considerably shortened by conversion to hex. For the 8-bit number, which is frequently used in microprocessors, the range of numbers is from $0000\ 0000_2$ to $1111\ 1111_2$. This reduces to the far simpler 00_{16} to FF_{16}.

Self-Test

Answer the following questions.

16. The hexadecimal system uses the characters _____ to represent binary numbers.
   a. 0 through 10   *c.* 0 through 7
   b. 0 through F   *d.* 0 and 1
17. Does the number 01C34 belong to the binary, decimal, or hexadecimal number system? Why? Hexadecimal The Limit of usage etc.
18. Hex shorthand reduces the length of a binary number by a factor of
   a. 4   *c.* 2
   b. 3   *d.* 1
19. Convert the following binary numbers to hexadecimal notation:
   a. 101   5   *f.* 01010101 55
   b. 11111111 FF   *g.* 00000010 2
   c. 11101101 Ed   *h.* 111.111 7.E75
   d. 00000000 0   *i.* 0110.0110 6.375
   e. 10000000 80   *j.* 1000.0001 8.0625

2-6 DECIMAL AND HEXADECIMAL CONVERSIONS

When you work with microprocessors, you will have to convert decimal numbers to binary numbers and binary numbers to decimal numbers. Often you will want to show these binary numbers in the hexadecimal shorthand form. This means you will want to convert hexadecimal numbers back to their binary form. You may also want to convert a hexadecimal number into its decimal equivalent.

When you are converting hexadecimal numbers to decimal form, you use the same procedure that you used for binary numbers. That

is, you assign a weight to each column of the number. Then you multiply these weights by the number in the column and add the results of the multiplications. The following examples show conversions from hexadecimal to decimal.

$$27A.54_{16} = ?_{10}$$

Assign weights:

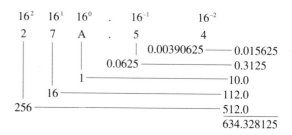

Therefore $27A.54_{16} = 634.328125_{10}$.

The procedure is really quite simple. As we saw with the binary fractions, the numbers can easily get quite long. Often you will wish to round numbers off.

Let's look at converting a number from decimal to hexadecimal. You may use a procedure similar to the one you used to convert decimal numbers into binary numbers. We learned that that procedure is a general-purpose one. For converting from decimal to hex, instead of dividing (multiplying in the case of a fraction) the number by 2, we divide (or multiply) the number by 16. The remainders (or integer products) are used to make up the final number.

In the following examples, you see hexadecimal numbers created from decimal numbers. The process is different for the integer and fraction parts of a decimal number. In both cases, the original decimal number must be split at the radix point into its integer and its fraction. Then separate conversions must be done on both parts. The parts are then put back together in the hexadecimal form.

$$634.328125_{10} = ?_{16}$$

The Integer Part

Division	Remainder
39	
16)634	
48	
154	
144	
10	$10_{10} = A_{16}$ LSD

Division	Remainder

$$\begin{array}{r} 2 \\ 16\overline{)39} \\ 32 \\ \hline 7 \end{array} \qquad 7_{10} = 7_{16}$$

$$\begin{array}{r} 0 \\ 16\overline{)2} \\ 0 \\ \hline 2 \end{array} \qquad 2_{10} = 2_{16} \qquad \text{MSD}$$

Hence, $634_{10} = 27A_{16}$. Note that the remainders must be converted into hexadecimal form.

The Fraction Part

Multiplication	Integer Result	
$16 \times 0.328125 = 5.25$	5	MSD
$16 \times 0.25 \quad = 4.0$	4	LSD

Therefore, in this example, $0.328125_{10} = 0.54_{16}$ and $634.328125_{10} = 27A.54_{16}$.

This example shows us how to get hexadecimal numbers from decimal numbers. Remember, the reason we do this is to express the number in a shorthand form.

Converting $27A.54_{16}$ to its binary form, we see that

$$1001111010.010101_2$$

is

```
        2    7     A  .  5     4₁₆
       /    /      \     \     \
     0010  0111  1010 . 0101  0100
```

which is

$$634.328125_{10}$$

Self-Test

Answer the following questions.

20. Converting decimal numbers into hexadecimal numbers uses the same procedure as converting
    a. Binary numbers into octal
    b. Hexadecimal numbers into binary
    c. Decimal numbers into binary
    d. Binary numbers into hexadecimal
21. For the following decimal numbers, give the binary, decimal, and hexadecimal forms:
    a. 126
    b. 4
    c. 16
    d. 63
    e. 101
    f. 12
    g. 1
    h. 127
    i. 9.25
    j. 64.015625
22. A real number may have both integer and fraction parts. How would you expect to convert a real decimal number into a real base 4 number? What characters would you use?

SUMMARY

1. The decimal number system uses the characters 0 through 9. We use it every day. Each column of numbers is assigned a weight that is a power of 10. The decimal point separates the integer and fraction parts of a mixed number.
2. The binary number system uses two characters, usually 1 and 0. Each character position is called a *bit*, for binary digit. The column weights are powers of 2. The binary point separates the integer and fraction portions of mixed binary numbers. *LSB* and *MSB* refer to the least significant bit and the most significant bit of a binary number.
3. Binary numbers are converted to decimal numbers by adding the column weights for those columns that have a 1.
4. Decimal numbers are converted to binary numbers by a division and multiplication process. Division is used for the integer part of the conversion, and multiplication is used for the fraction part of the process.
5. The hexadecimal number system uses the characters 0 through F. It is called "hex" for short. Hex is a shorthand way to represent long binary numbers. For conversion, the binary numbers are grouped by fours and then replaced by the equivalent hex characters.
6. Converting from a decimal number to a hexadecimal number is done in the same way that a decimal number is converted into a binary number.

Answer the following questions.

2-1. The base of a number system is
    a. The highest number which can be used
    b. The number of characters used in the system
    c. 8 in the decimal system
    d. F in the hexadecimal system

2-2. In the number 426, the character 4 is the
    a. MSD   *c.* MSB
    b. LSB   *d.* LSD

2-3. In scientific notation the weight of the thousands column is
    a. 10^0 *c.* 10^2
    b. 10^1 *d.* 10^3

2-4. The binary number system uses the characters
    a. 0 through 9   *c.* 0 through F
    b. 0 and 1   *d.* 0 through 7

2-5. Complete the following:
    a. $2^0 = 1$ **h.** $2^7 = 128$
    b. $2^1 = 2$ **i.** $2^8 = 256$
    c. $2^2 = 4$ **j.** $2^9 = 512$
    d. $2^3 = 8$ **k.** $2^{10} = 1024$
    e. $2^4 = 16$ **l.** $2^{11} = 2048$
    f. $2^5 = 32$ **m.** $2^{12} = 4096$
    g. $2^6 = 64$

2-6. An 8-bit binary number can represent _____ different decimal numbers.
    a. 8   *c.* 256
    b. 99,999,999   *d.* 1024

2-7. A bit is a
    a. Binary increment   *c.* Least significant number
    b. Binary digit   *d.* Most significant number

2-8. Convert the following binary numbers to decimal numbers:
    a. 1111   15   **h.** 0000 0001   1
    b. 1111 1111   255   **i.** 0111   7
    c. 0100   4   **j.** 0011 1111   63
    d. 0000 0000   0   **k.** 0.0001   .0625
    e. 1101   13   **l.** 101.11111
    f. 1101 1101   221   **m.** 1110.1110
    g. 0001   1   **n.** 00.01011

2-9. Convert the following decimal numbers to binary numbers:
    a. 128   10000000   **h.** 29   11101
    b. 12   1100   **i.** 255   11111111
    c. 75   1001011   **j.** 1024
    d. 256   100000000   **k.** 10.5   1010.1
    e. 31   11111   **l.** 16.015625
    f. 30   11110   **m.** 12.03
    g. 56   111000   **n.** 17.7

2-10. Convert the binary numbers in question 2-8 to a table of equivalent hexadecimal numbers.

2-11. Convert the decimal numbers in question 2-9 into a table of equivalent hexadecimal numbers.

2-1. You could suppose that the decimal number system came into being because human beings first started to count with their 10 fingers and thus originated the base 10 number system. Humans also have two fists. Why do you suppose the decimal number system, rather than the binary number system, developed?

2-2. Why do scientists and engineers use scientific notation?

2-3. If you make a mistake in writing a number, which causes the least error, a mistake near the LSD or a mistake near the MSD?

2-4. Can you think of any situations when the average person uses binary numbers rather than decimal numbers? What are these situations?

2-5. Why do computer programmers prefer to use hex numbers to represent binary values rather than converting the binary number to decimal and then using the decimal number?

2-6. Can you think of any practical reason why the hexadecimal number system uses the characters 0 to F rather than some other set of 16 characters?

Answers to Self-Tests

1. *b*
2. *c*
3. 27, 329
4. *d*
5. *b*
6. No, because the character 2 is not a binary character.
7. Least significant bit
8. Binary digit
9. *c*
10. 1111_2, because it is equal to 15_{10} which is larger than 11_{10}.
11. *b*
12. a. 182   f. 129
    b. 80   g. 126
    c. 30   h. 63
    d. 128   i. 0.3125
    e. 255   j. 4.375
13. *d*
14. *c*
15. a. 10111   g. 111111
    b. 1101001   h. 11101
    c. 100000   i. 1100.001
    d. 1111   j. 10000.011
    e. 11001110   k. 101.000001
    f. 10000000   l. 10.1
16. *b*
17. Hexadecimal, because the character C is not used in the other systems.
18. *a*
19. a. 5   f. 55
    b. FF   g. 02
    c. ED   h. 7.E
    d. 00   i. 6.6
    e. 80   j. 8.1

20. *c*
21.

	Binary	Decimal	Hexadecimal
a.	01111110	126	7E
b.	100	4	4
c.	10000	16	10
d.	111111	63	3F
e.	1100101	101	65
f.	1100	12	C
g.	1	1	1
h.	01111111	127	7F
i.	1001.01	9.25	9.4
j.	1000000.000001	64.015625	40.04

22. Divide the integer part of the decimal number by 4. The remainder becomes the LSD. Keep dividing the results by 4 until the result is 0. The last remainder is the MSD. The fraction part is multiplied by 4. The integer part of the result becomes the MSD of the fraction. The fraction parts are again multiplied by 4 until the desired precision is achieved. The base 4 number system uses the characters 0, 1, 2, and 3.

CHAPTER 3

Processor Arithmetic

CHAPTER OBJECTIVES

This chapter will help you to:

1. *Solve* binary arithmetic problems.
2. *Express* signed numbers using 2's complement numbers.
3. *Explain* multiple-precision arithmetic and floating-point arithmetic.
4. *Define* the term shift and add.

One of the most common functions we ask of a microprocessor is to perform simple arithmetic—often many times and very quickly. However, the microprocessor cannot perform decimal arithmetic; it must use binary numbers. Because the microprocessor uses binary arithmetic, it is important to understand the processes it uses for addition, subtraction, multiplication, and division and to understand how it handles large numbers and negative numbers.

3-1 BINARY ADDITION

Binary addition and decimal addition are done in the same general way. In either number system, when we add two numbers we start with the right-hand column. This column represents the least significant number. Any result that is greater than a single place causes a carry. We add the carry to the numbers in the next more significant column. An operation in any column may generate a carry.

In the decimal number system, a result which is greater than 9 generates a carry. In the binary system, a result which is greater than 1 generates a carry.

Figure 3-1 will help you see both the similarities and the differences between binary and decimal addition. The same addition problem is shown first in the decimal number system and then in the binary number system.

In the decimal version, we add 95 to 99. The addend is 99, and the augend is 95. We add 9 to 5 in the right-hand column, the result is 4 with 1 to carry. The carry is shown in Fig. 3-1 as an extra 1 at the top of the tens column. In that column, the carried 1 is added to 9 plus 9. The result is 9 with 1 to carry. The carried 1 is added to the 0s in the hundreds column. The

result in the hundreds column is 1. The sum of 99 and 95 is 194.

Figure 3-2(*a*) is a table summarizing the rules for decimal addition. We could have used this table to find the sum of 99 and 95. We can find the result of adding any two digits by finding where the row (representing the augend) and the column (representing the addend) cross. If the row and the column cross in the shaded area, then the two digits generate a carry of 1 into the next more significant column. In the decimal number system, this procedure is so familiar that we may forget the rules we are using. We just do decimal addition automatically.

Since we are unaccustomed to doing binary arithmetic, it may seem awkward at first. But

Decimal addition	Terms	Binary addition
11	Carry	1111 1110
099	Addend	0110 0011
095	Augend	0101 1111
194	Sum	1100 0010

Fig. 3-1 Decimal and binary addition. When the sum of the addend and the augend cannot be expressed as a single digit, a carry is generated. Carries are shown as a 1 in the carry row of the next column.

	+	Addend									
		0	1	2	3	4	5	6	7	8	9
Augend	0	0	1	2	3	4	5	6	7	8	9
	1	1	2	3	4	5	6	7	8	9	0
	2	2	3	4	5	6	7	8	9	0	1
	3	3	4	5	6	7	8	9	0	1	2
	4	4	5	6	7	8	9	0	1	2	3
	5	5	6	7	8	9	0	1	2	3	4
	6	6	7	8	9	0	1	2	3	4	5
	7	7	8	9	0	1	2	3	4	5	6
	8	8	9	0	1	2	3	4	5	6	7
	9	9	0	1	2	3	4	5	6	7	8

(a)

+		Addend	
		0	1
Augend	0	0	1
	1	1	0

(b)

Fig. 3-2 **The addition tables. Results in the shaded areas mean there is also a carry. (*a*) Decimal addition. (*b*) Binary addition.**

$$\begin{matrix} 1 \\ 1 \\ +1 \\ \hline 11 \end{matrix} \quad = \quad \begin{matrix} 1 \\ +1 \\ \hline 10 \\ +1 \\ \hline 11 \end{matrix}$$

Fig. 3-3 **Adding three binary 1s. This is done as two separate addition problems. All addition problems can be reduced to adding groups of two numbers (both binary and decimal).**

it is really much simpler than decimal arithmetic. Figure 3-2(*b*) summarizes the rules for binary addition. We can find the result of adding any 2 bits by finding where the row (representing the augend) and the column (representing the addend) cross. If the row and the column cross in the shaded area, then the two bits generate a carry of 1 into the next more significant column. As you can see, the shaded area in Fig. 3-2(*b*) is only a single square. Figure 3-2(*b*) is much simpler than Fig. 3-2(*a*) because there are only two characters, 0 and 1, in the binary system.

Now let's return to Fig. 3-1 and add the binary equivalents of 99 and 95, 0110 0011 and 0101 1111. Starting with the least significant bit, we see that both the addend and the augend are 1s. From our binary addition table, we can see that the result is 0 with 1 to carry.

The carry is shown in Fig. 3-1 as an extra 1 at the top of the 2s column. In that column, the carried 1 is added to 1 plus 1. We are adding three 1s.

The procedure for adding three 1s is shown in Fig. 3-3. There we see that the addition of three 1s can be broken down into two separate addition problems. Like other binary addition problems, these can be solved by using the rules summarized in our binary addition table. Adding 1 to 1, we get 0 with 1 to carry. That is, we get 0 in the ones place and 1 in the 2s place, or 10. Now we add 1 to 10. This gives us 1 in the 1s place and no carry to add to the 1 in the 2s place. Thus, the binary sum of 1 and 1 and 1 is 11.

In our larger addition problem in Fig. 3-1, the sum in the 2s place is 11, or 1 with 1 to carry. The carry is shown as an extra 1 in the 4s place. The carried 1 is added to the 0 and the 1 that are already in the 4s place. This gives us 0 with 1 to carry to the 8s place. The carry is shown as a extra 1 in the 8s place. The carried 1 is added to the 0 and the 1 that are already in the 8s place.

This gives us 0 with 1 to carry to the 16s place. The carry is shown as an extra 1 in the 16s place. The carried 1 is added to the 0 and the 1 that are already in the 16s place. This gives us 0 with 1 to carry to the 32s place. The carry is shown as an extra 1 in the 32s place. The carried 1 is added to the 1 and the 0 that are already in the 32s place. This gives us 0 with 1 to carry to the 64s place. The carried 1 is added to the two 1s that are already in the 64s place. The result of adding three 1s, as we have seen, is 1 with 1 to carry. The carried 1 is added to the two 0s that are in the 128s place. This gives us 1 in the 128s place, with nothing to carry. Thus, the binary sum of 0110 0011 and 0101 1111 is 1100 0010.

Once you have added a few binary numbers, you will see that binary addition is easy. The only difficulty is that binary numbers need lots of bits to express large numbers. Consequently there are many columns and many carries in a binary addition.

We deliberately chose an 8-bit representation of the binary numbers in our examples. We did this because many microprocessors represent binary numbers as 8-bit words or multiples of 8-bit words (16 bits, 32 bits, 64 bits, etc.). Although neither the addend nor the augend in this example required 8 bits, we still showed all 8 bits. We did this because we cannot shorten a microprocessor word length just because there are insignificant leading 0s.

We do get rid of insignificant 0s for most decimal addition problems. That is, when we are writing a problem on paper, we only show the number of digits necessary to express significant numbers. However, in this problem we did have a carry into the hundreds column. That is why this figure shows the two leading 0s. We must always remember that the leading 0s are really there.

Self-Test

Answer the following questions.

1. Add the following binary numbers:

    a. 0 0 0 0   0 1 0 1
       0 0 0 1   0 0 0 1

    b. 1 0 0 0   1 0 0 1
       0 0 0 0   1 1 1 1

    c. 0 1 1 1   1 1 1 1
       0 1 1 1   1 1 1 0

    d. 0 1 0 1   0 1 0 1
       1 0 1 0   1 0 1 0

    e. 1 0 1
       0 1 1

    f. 1 0 0 1
        0 1 1

    g. 0 1 0 0   0 0 1 0   0 1 1 0   1 1 0 0
       0 1 0 1   1 1 1 0   1 0 0 1   0 1 1 0

    h. 0 1 1 1   1 1 1 1   1 1 1 1   1 1 1 1
       0 0 0 1   0 1 1 1   1 0 1 1   1 0 0 1

2. Express the following decimal number addition problems in 8- or 16-bit binary numbers and do the additions.

    a. 101
        16

    b. 225
       168

    c. 398
       132

    d. 56
       10

    e. 86
       25

    f. 289
       493

3. What is the maximum number of bits needed to hold the result of adding two 8-bit numbers?

4. What is the difference between the three numbers 101, 0101, and 0000 0101? How would this affect any binary sum?

3-2 BINARY SUBTRACTION

We do *binary subtraction* in the same way we do decimal subtraction. As with decimal addition and binary addition, we simply have different sets of rules for combining the numbers.

Figures 3-4(a) and (b) show the subtraction table for decimal and binary arithmetic. In Fig. 3-4(a) we can see the rules for decimal subtraction. In this figure any subtraction in which the subtrahend is larger than the minuend is shown in the shaded area. This means that the subtraction causes a borrow. That is, we borrow a 1 from the next more significant column in the minuend.

Looking at Fig. 3-4(b), we can see that the same is true for binary subtraction. That is, when the subtrahend is larger than the minuend, we generate a borrow. In binary subtrac-

		Subtrahend									
		0	1	2	3	4	5	6	7	8	9
Minuend	0	0	9	8	7	6	5	4	3	2	1
	1	1	0	9	8	7	6	5	4	3	2
	2	2	1	0	9	8	7	6	5	4	3
	3	3	2	1	0	9	8	7	6	5	4
	4	4	3	2	1	0	9	8	7	6	5
	5	5	4	3	2	1	0	9	8	7	6
	6	6	5	4	3	2	1	0	9	8	7
	7	7	6	5	4	3	2	1	0	9	8
	8	8	7	6	5	4	3	2	1	0	9
	9	9	8	7	6	5	4	3	2	1	0

(a)

		Subtrahend	
		0	1
Minuend	0	0	1
	1	1	0

(b)

Fig. 3-4 The subtraction tables. Results in the shaded areas mean there is a borrow.

tion, the only such case is when we subtract 1 from 0. The result is 1, and this operation causes us to borrow a 1.

Figure 3-5 shows a decimal and a binary subtraction problem. We can see the borrow, the minuend, the subtrahend, and the difference.

In the decimal problem, we start with the far right-hand column. Nine subtracted from 9 gives 0; no borrow is required. In the next column, 4 subtracted from 0 gives 6, but we must borrow 1 from the hundreds column. In the hundreds column we only had 1 to start with. Then we borrow 1 from a minuend that is 1; the result is a minuend that is 0. From the minuend of 0 is subtracted a subtrahend of 0 in the hundreds column, giving 0 as a result ($0 - 0 = 0$).

In the binary subtraction in Fig. 3-5, we start with the far right-hand column. Here we see that a 1 subtracted from a 1 is 0. No borrow is generated. In the 2s column a 0 is subtracted from a 0, and no borrow is generated. In the 4s column, 0 from 1 is 1, with no borrow. The same is true in the 8s column. In the 16s column, 1 is subtracted from 0. The result is 1, but we must borrow 1 from the 32s column. When we borrow a 1 from the minuend in the 32s column, the result is a minuend that is 0. The subtraction therefore is 1 from 0, which leaves us with 1 and again results in a borrow. In the 64s column we must borrow 1 from the minuend, and therefore the minuend becomes 0. The subtraction is therefore 0 from 0, and the result is 0 with no borrow. Because we are dealing with 8-bit numbers, we must complete the subtraction in the 128s column even though it is all insignificant 0s. Of course, the result is 0.

Once again you will find binary subtraction relatively easy to do. The rules are simple, and it just takes practice to become familiar with how it is done.

Decimal subtraction	Terms	Binary subtraction
1	Borrow	110 0000
109	Minuend	0110 1101
49	Subtrahend	0011 0001
060	Difference	0011 1100

Fig. 3-5   A subtraction problem in decimal and binary arithmetic. The first binary borrow in bit 5 resulted when a 1 was subtracted from 0. The borrow is shown as a 1 in the bit 6 column and borrow row.

Answer the following questions.

5. Subtract the following binary numbers:

   a. 0 1 0 1
      0 0 0 1

   b. 1 0 0 1
       1 1 1

   c. 1 1 1 1
      1 0 0 1

   d. 1 0 0 1
      0 0 0 1

   e. 1 0 0 1   0 1 1 0
      0 1 1 0   1 0 0 1

   f. 1 1 1 1   0 0 0 1
      0 0 0 0   0 0 1 1

   g. 1 0 0 1   0 0 1 1   1 0 0 1   1 0 0 1
      0 1 0 1   1 1 0 1   1 1 1 1   0 0 0 0

6. Express the following subtraction problems in either 4-, 8-, or 16-bit binary numbers and do the subtractions.

   a.   15
       − 12

   b.   8
       − 2

   c.   255
       − 24

   d.   52
       − 36

   e.   29
       − 12

   f.   136
       − 108

7. What causes a borrow?
8. What is the difference between a borrow and a carry?

3-3 TWO'S COMPLEMENT NUMBERS

When you first look at binary numbers, it is very easy to think that the only numbers you can express with them are positive integers. We have already found, however, that a binary point is commonly used. So, we know that we can represent binary fractional numbers. How can we express binary negative numbers?

Many different methods have been used to represent negative numbers. Most of them do not work well with the binary electronics of the arithmetic and logic unit. These unsuccessful methods did, however, lead to the development of a system that works well.

One of the first systems was called the *sign and magnitude system*. In this system, the most significant bit of the binary number is used as a sign bit. So, a "0" in the most significant bit means that the number is positive. A "1" in the most significant bit means that the number is negative. Examples of this method are shown in Fig. 3-6. In both cases, the 7-bit binary number 0011100 has a magnitude of 28. The sign bit (the eighth bit) is either "0" (for positive) or "1" (for negative).

Another method that uses "0" in the most significant bit to mean "positive" and "1" to mean "negative" is the 1's complement representation. In the 1's complement method, all bits of a negative number are complemented. That is, all 1s are changed to 0s, and all 0s are changed to 1s. This process changes the sign bit as well, thus giving a 1 in the MSB for negative numbers. Although generating the 1's complement is simple, the 1's complement method has a number of disadvantages. It is difficult to work with, and 0 can be represented by either all 0s or all 1s. That is, there are two ways of saying "zero."

Figure 3-7 shows an example of the 1's complement expression of a binary number.

The 2's complement method of representing negative numbers is now very commonly used in microprocessor systems. It solves the problem of the double representation of 0. The 2's complement is made by taking the 1's complement and adding 1. You can see from Fig. 3-8 that this allows 8-bit binary numbers to represent the decimal numbers from −128 to +127, including 0. In the figure, we can see the two common methods used to represent binary numbers in microprocessors. These are the unsigned binary and the 2's complement systems.

In the left-hand column, we show the 8-bit binary numbers from 0000 0000 to 1111 1111.

Binary	Decimal
0 0 0 1 1 1 0 0	+ 28
1 1 1 0 0 0 1 1	− 28

Fig. 3-7 The 1's complement method of representing positive and negative binary numbers. The 1's complement complements each bit. Both 00000000 and 11111111 are 1's complement 8-bit numbers representing decimal zero.

In the right-hand column we show the decimal equivalents using the unsigned binary system. These go from 0 to +255.

The center column shows the decimal equivalent using the 2's complement system. Here we see the binary numbers from 0000 0000 to 0111 1111 represent the decimal numbers from 0 to +127. Negative numbers are indicated by a "1" in the eighth bit. So, 1000 0000 represents −128. Negative numbers continue until 1111 1111 is reached. It represents a −1.

Using the 2's complement method, we express 256 different values. There are 127 positive numbers, 0, and 128 negative numbers. As we noted before, taking the 2's complement of a number is quite simple. Just take the 1's complement and add 1. This is shown in Fig. 3-9. You can see that we have generated the 2's complement of 4. The result checks with the table in Fig. 3-8.

There is also a second shorthand method for generating the 2's complement of a binary number. Start at the least significant bit. As long as the least significant bits are 0s, copy them directly. When you reach the first 1, copy it directly. Each bit after that is simply inverted. That is, you take the 1's complement of all bits following the first 1 starting from the least significant bit.

Using 2's complement notation, we can express both positive and negative numbers. Because the logic to do 2's complement arithmetic and the logic for unsigned binary arithmetic are identical, the microprocessor's logic becomes simple. However, we must remember which system we are using. *That is, the bit patterns will always be the same, but interpreting the answers depends on our knowing which system we are using.* If we add two unsigned numbers, we get a result that is another unsigned number. It is expressed as a binary bit pattern. This bit pattern can be read either as a 2's complement negative number or as an unsigned number.

If an addition or subtraction problem generates a negative number and the input to this

Sign bit →

Binary	Decimal
0 0 0 1 1 1 0 0	+ 28
1 0 0 1 1 1 0 0	− 28

Fig. 3-6 The sign and magnitude method of representing positive and negative binary numbers. In both cases the 7-bit binary number 0011100 has a magnitude of 28. The sign (in the eighth bit) is either 0 (+) or 1 (−).

8-bit binary number	Decimal number from 2's complement	Decimal number from unsigned binary
0000 0000	+ 0	0
0000 0001	+ 1	1
0000 0010	+ 2	2
0000 0011	+ 3	3
.	.	.
.	.	.
.	.	.
.	.	.
0111 1100	+124	124
0111 1101	+125	125
0111 1110	+126	126
0111 1111	+127	127
1000 0000	−128	128
1000 0001	−127	129
1000 0010	−126	130
1000 0011	−125	131
.	.	.
.	.	.
.	.	.
.	.	.
1111 1100	− 4	252
1111 1101	− 3	253
1111 1110	− 2	254
1111 1111	− 1	255

Fig. 3-8 Comparison of 8-bit binary numbers and their decimal representations. The binary number does not change. What changes is how you think of the number.

computation is intended as 2's complement notation, then the output must be in 2's complement notation. If the result of this computation leaves a logic "1" in the most significant bit, then the result is being expressed as a negative 2's complement number.

If you wish to find the magnitude of a negative 2's complement result, you must store the fact that it is negative and take the 2's complement. That is, you must take the 1's complement of the result and add 1.

Binary	Explanation
0000 0100	4_{10}
1111 1011	1's complement of 4_{10}
1	Add 1
1111 1100	2's complement of 4_{10}

Fig. 3-9 Taking the 2's complement. First find the complement of each bit by changing 1s to 0s and 0s to 1s, then add 1.

The simplicity of the 2's complement system is shown by a subtraction problem. Subtracting is simply taking the 2's complement of the subtrahend and adding it to the minuend. The difference is expressed in 2's complement form. That is, if the difference is positive, then the most significant bit is 0, the magnitude is expressed as a binary number, and the final carry must be discarded. If the difference is negative, the most significant bit is 1 and the number is expressed in 2's complement form. As indicated above, a negative result requires an additional 2's complement operation to show the magnitude in simple binary form.

Two problems that illustrate 2's complement subtraction are shown in Fig. 3-10 on page 34. In Fig. 3-10(a), we subtract the number 23 from 58. To do the subtraction, 23 is put into 2's complement form. This is done by taking the 1's complement and adding 1. We then add the 2's complement of 23 to the binary 58. We discard the final carry, and the result

```
Binary      Explanation

0001 0111   23₁₀

1110 1000   1's complement of 23₁₀
0000 0001   Add 1
1110 1001   2's complement of 23₁₀
```

Decimal arithmetic	Binary arithmetic	Explanation for binary arithmetic
58	0011 1010	58_{10}
-23	1110 1001	Add 2's complement of 23_{10}
35	1 0010 0011	Difference (35_{10})

└─ Discard carry for positive results

(a)

```
Binary      Explanation

0010 0010   34₁₀

1101 1101   1's complement of 34₁₀
0000 0001   Add 1
1101 1110   2's complement of 34₁₀
```

Decimal arithmetic	Binary arithmetic	Explanation for binary arithmetic
26	0001 1010	26_{10}
-34	1101 1110	Add 2's complement of 34_{10}
-08	1111 1000	Difference (in 2's complement form because MSB is 1)

(b)

To find the absolute value:

```
Binary      Explanation

1111 1000   Difference in 2's complement

0000 0111   Take 1's complement
0000 0001   Add 1
0000 1000   Result is 8₁₀
```

(c)

Fig. 3-10 Two binary subtraction problems using 2's complement arithmetic. (a) First, find the 2's complement of 23. Then add it to 58, thus performing the subtraction. (b) The same process with the subtrahend larger then the minuend produces a negative unit. (c) We find the absolute value of the result from part (b) using the 2's complement process.

is a binary number that has a 0 in the most significant bit. Therefore we can treat the magnitude as that of a unsigned binary number.

In Fig. 3-10(b) we subtract 34 from 26. To perform this subtraction, we take the 2's complement of 34. This is done by inverting each bit and adding 1. The 2's complement of 34 is then added to 26. The result is a negative number. This is shown by the 1 in the most significant bit. If we want to express the magnitude as an unsigned binary number, we must take the 2's complement of the difference. This is done by taking the 1's complement of the difference and adding 1. The result is an unsigned binary 8; it is shown in Fig. 3-10(*c*).

Self-Test

Answer the following questions.

9. Today, the most common way to show a negative binary number is to show its
   a. 1's complement
   b. 2's complement
   c. Sign and magnitude
   d. Absolute value
10. Complement the following binary numbers:
    a. 1011 0110   **f.** 0000 0001
    b. 0110 1011   **g.** 1111 1111
    c. 101   **h.** 0101 0101
    d. 0000 1100   **i.** 1111 0000 1111 0000
    e. 1   **j.** 1100 1100 1100 0011

11. Treat each of the binary numbers in question 10 as if it were either an 8- or 16-bit 2's complement number. Indicate which numbers are positive and which are negative.
12. Express the following as 8-bit 2's complement numbers:
    a. 64   **g.** 32
    b. −56   **h.** −32
    c. 12   **i.** 256
    d. 0   **j.** 16
    e. −128   **k.** −100
    f. 127   **l.** −4

13. Why can the same logic be used to add or subtract both 2's complement and unsigned binary numbers? Give an example to show your point.

3-4 BINARY MULTIPLICATION

As with addition and subtraction, decimal multiplication and *binary multiplication* are very much the same. Each is a quick way to add one number to itself many times. For example, multiplying 7 by 5 is simply a quick way of adding 7 to itself five times.

In multiplication, we call one number the multiplicand and the other the multiplier. We use a digit-by-digit method in multiplication. We often generate a carry in decimal multiplication, but we normally solve it in our heads.

```
     17   Multiplicand
  ×  12   Multiplier
     34   1st partial product
     17   2nd partial product
    100   Carry
    204   Total product
```

Fig. 3-11   The parts of a decimal multiplication problem. There will be a partial product for each digit in the multiplier. Normally, we do not show the carry. This is done in our heads; however, it is still there.

After the entire multiplicand is multiplied by the least significant digit of the multiplier, the result is called the first partial product. The second partial product is generated after the multiplicand is multiplied by the second least significant digit in the multiplier. This process continues until we have generated all the needed partial products.

Because each partial product has been created by a multiplier that is 10 times greater than the previous one, each partial product is shifted to the left by one decimal place. Then all the partial products are added to generate a final product. Of course, the process of adding all the partial products may also generate carries, all of which must be included in the additions.

An example of decimal multiplication is shown in Fig. 3-11. In this problem we multiply 17 by 12. Because there are two digits in the multiplier, we generate two partial products. Adding these two partial products together generates a carry in the hundreds column. The product is 204.

In Fig. 3-12 we show the binary multiplication table. Of course, we all learned our decimal multiplication table many years ago. It is a large table with 10 places on each side. It gives 100 different products.

The binary multiplication table shown in Fig. 3-12 is quite simple because there are only two binary characters. You can see that the multiplication of two binary digits will never generate a carry.

Fig. 3-12   The binary multiplication table. Multiplication of two binary digits does not generate a carry.

Let's use this binary multiplication table to multiply 17 by 12, but in binary form. This is shown in Fig. 3-13. The multiplicand and the multiplier are simply 8-bit binary expressions of the numbers 17 and 12.

The first thing you will note in Fig. 3-13 is that there are eight partial products. This is to be expected, because there are 8 bits in the multiplier.

Because the rules of binary multiplication are so simple, the multiplication is easy to follow. The first partial product is all 0s. This is because the least significant bit in the multiplier is 0. If we look at the binary multiplication table in Fig. 3-12 we see that 0 times 0 is 0. The second partial product is just the same: all 0s.

The third partial product is an exact copy of the multiplicand. The only difference is that it is shifted over three binary places, because it is the product of the third-place multiplier and the multiplicand.

The fourth partial product is again a copy of the multiplicand, this time shifted over four places.

The fifth, sixth, seventh, and eighth partial products are all 0s, because the fifth, sixth, seventh, and eighth multiplier bits are 0s.

In adding all the partial products, we have generated a final (total) product. These additions resulted in no carries. Since they could have caused a carry, we left room for one.

The result has 16 places, but the 8 most significant bits are all 0s. Therefore, we may use the 8 least significant bits of the final product for the result. We know that binary numbers of 8 bits or less represent decimal numbers smaller than 255.

Because binary multiplication is so simple, a very simple method has been developed. This method is called *shift and add*. The shift-and-add method works as follows:

1. The first partial product is taken. If the least significant bit of the multiplier is 0, the result is 0. If the LSB of the multiplier is 1, the result is an exact copy of the multiplicand.
2. Each time another bit of the multiplier is worked on, the multiplicand is shifted one bit to the left.
3. Each time there is a 1 in the multiplier bit being worked on, we add the multiplicand in its shifted position to the result we already have.
4. The sum at the end of all the shifts and adds is the product.

Binary
arithmetic | Terms

00010001	Multiplicand 17_{10}
00001100	Multiplier 12_{10}
00000000	1st partial product
00000000	2nd partial product
00010001	3rd partial product
00010001	4th partial product
00000000	5th partial product
00000000	6th partial product
00000000	7th partial product
00000000	8th partial product
000000000000000	Carry
0000000011001100	Product 204_{10}

(a)

00010001	
00001100	
10001	Multiplicand shifted left 2 times
+10001	Multiplicand shifted left 3 times
11001100	Sum of shifted multiplicands

(b)

Fig. 3-13 **Multiplying 17 by 12 with binary multiplication. Both numbers are shown as 8-bit numbers as they are in an 8-bit microprocessor. The result is a 16-bit binary number.**

Looking at Fig. 3-13(a), we can see that this is how this problem was solved. The multiplicand was not shifted for the first partial product. It was not shifted because we were working on the least significant bit of the multiplier. Since the least significant bit of the multiplier was 0, the partial product was 0. The multiplicand was shifted one bit to the left for the second multiplier bit. But again no addition took place, because the second bit of the multiplier was also 0. The multiplicand was shifted one more bit to the left. This time the multiplicand was added to the result, because the third bit in the multiplier was a 1. For the fourth multiplier bit, the multiplicand was shifted one bit to the left for a third time. The multiplicand was again added to the result, because the fourth bit of the multiplier was a 1. No further additions took place, because the fifth, sixth, seventh, and eighth bits of the multiplier were 0. The shift-and-add method is shown in simpler form in Fig. 3-13(b).

You can easily see how the shift and add method works in binary multiplication. This simple method is made possible because multiplying a binary number by 0 yields 0 and multiplying a binary number by 1 yields the number itself.

Self-Test

Answer the following questions.

14. Multiplication is the process that lets you do repetitive _____ quickly.
15. The most commonly used procedure for multiplying creates a _____ for each bit or digit in the multiplier.
16. Multiply the following:

    a. 1 0 1
    　　0 1 1

    b. 0 1 1 0
    　　0 1 1 1

    c. 0 1 1
    　　0 1 1

    d. 0 1 0 1　1 1 0 1
    　　0 0 1 0　1 1 0 1

    e. 0 0 0 1　1 0 1 1
    　　1 1 1 1　1 1 0 0

    f. 0 1 1 1
    　　1 0 0 0

    g. 1 0 0 1
    　　1 0 1 0

17. In a multiplication problem, the magnitude of each partial product is greater than the previous one by a factor that is the same as the _____ .
18. In binary multiplication, the partial product is _____ if the multiplier bit is 1.
19. In binary multiplication, the partial product is _____ if the multiplier bit is 0.
20. The binary multiplication procedure is called _____ .

3-5 BINARY DIVISION

Division is the reverse of multiplication. That is, we subtract one number from another until we cannot subtract it anymore. The number of times that we are able to subtract the first number tells us how many times it can be divided into the second number. We can see that this is the reverse of adding one number repeatedly to get a result.

Multiplication by repeated addition and division by repeated subtraction are shown in Fig. 3-14. Here we see that five successive additions of 7 result in the answer 35. In the division column, we can see that 0 will result if we subtract 7 from 35 five times.

The process of division is a little more difficult than that of multiplication. If we look at a decimal long division problem, we see that division requires more understanding of what we are doing.

For example, let's look at the problem shown in Fig. 3-15(a) on page 38. This example is just the reverse of the one used in Fig. 3-13 showing binary multiplication. That is, the

	Multiplication	Division
	$7 \times 5 = ?$	$35 \div 7 = ?$
$+$	0	35
	7　1	$-$ 7　1
	7	28
$+$	7　2	$-$ 7　2
	14	21
$+$	7　3	$-$ 7　3
	21	14
$+$	7　4	$-$ 7　4
	28	7
$+$	7　5	$-$ 7　5
	35	0

Fig. 3-14 **Multiplication by repeated addition and division by repeated subtraction.**

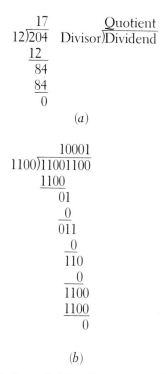

Fig. 3-15 Long division. (*a*) A numerical example using decimal arithmetic. (*b*) A numerical example using binary arithmetic to do the same division as in part (*a*).

product (204) is the dividend, the multiplier (12) becomes the divisor, and the multiplicand from Fig. 3-13 (17) becomes the quotient.

Let's look at how we do this division. We begin by using the process of inspection. We inspect the problem and guess that 12 will go into 20 once. We then try it. We subtract 12 from 20, and indeed the remainder (8) is less than the divisor. If it was greater than 12, we would try our guess again. We combine the remainder from the first difference with the next digit of the dividend, and again we guess the result.

Now let us turn to Fig. 3-15(*b*). Here, we face the same problem expressed in binary numbers. The first inspection is easy: 1100 will go into 1100 exactly once. We place the trial 1 in the quotient, perform a multiplication, and then subtract to complete the test. The difference (in this case 0) is less than the divisor, so we can continue. We bring down the next bit in the dividend. This division, too, is obvious. We know that 1100 goes into 1 zero times. This process continues until the division is complete.

You can see that this way of doing division cannot be made into a machine operation as easily as multiplication was. There are too many guesses and tests. However, an inge-

nious way of doing *binary division* in a microprocessor has been developed. In the following paragraphs, we will see how to do the same job as the binary long division in Fig. 3-15(*b*).

The procedure for binary division is actually quite simple, because each bit of the quotient can be only a 1 or a 0. Once again we will use a shifting process, this time taken from binary long division.

Before we can begin dividing 204 by 12, there is one step we must do. We must take the 2's complement of 12. This is done so that we can use a binary adder for both subtraction or addition. The 2's complement of 12 is

0	1 1 0 0	Binary 12
1	0 0 1 1	1's complement of 12
0	0 0 0 1	Add 1
1	0 1 0 0	2's complement of 12

sign bits   magnitude bits

Now that we have have the 2's complement of 12, we can start doing the division.

Just as in long division, we will test to see whether the divisor will go into the same number of the dividend's most significant bits. Of course, the microprocessor can't take a guess. Therefore, we start by actually subtracting the divisor from the dividend's most significant bits. If the divisor won't go into that part of the dividend, we can always add the subtracted bits back. We will know if the divisor won't go, because the subtraction will produce a negative number. That is, the sign bit will be a 1.

Trying the first subtraction, we have

0	1 1 0 0 1 1 0 0	Dividend
1	0 1 0 0 0 0 0 0	Subtract 12*
0	0 0 0 0 1 1 0 0	First result

↑
0 here means 1st bit of quotient is 1   Quotient = 1XXXX

If the divisor will go into the appropriate part of the dividend, then the sign bit will be 0. This means that the result of the division is a positive number. As you can see, our first test has worked. We know that the first bit in the quotient is a 1.

---

*Note: In this and the following examples, subtractions are done by adding the 2's complement of 12 as found earlier in the process.

Now let's try the next step. We will again try to subtract the divisor. But we must shift the first result. We shift the first result so that the next operation will generate the second bit in the quotient. Shifting the first result, we have

0 | 0 0 0 0 1 1 0 0   First result
0 | 0 0 0 1 1 0 0   Shifted first result

We are now ready to do another subtraction. This subtraction gives us

0 | 0 0 0 1 1 0 0   Shifted first result
1 | 0 1 0 0 0 0 0 0   Subtract 12
1 | 0 1 0 1 1 0 0   Second result
↑

1 here means 2d bit
of quotient is 0   Quotient = 10XXX

Looking at this subtraction, we can see a 1 in the sign bit, indicating that a negative number has been generated. A negative number means we have tested for 1 in the quotient, and the test has failed. This means we must do two things. First, we must make this bit in the quotient a 0. In this case, the second bit of the quotient becomes a 0. Second, we must correct for this wrong test. That is, we must add back the divisor, because the negative sign of the result shows that the divisor should not have been subtracted in the first place. Adding back the divisor, we have

0 | 0 1 0 1 1 0 0   Second result
0 | 1 1 0 0 0 0 0   Add back 12
0 | 0 0 0 1 1 0 0   Shifted first result
                       (again)

We are now ready to shift again. Shifting, we have

0 | 0 0 0 1 1 0 0   Shifted first result
                       (again)
0 | 0 0 1 1 0 0   First result shifted
                       twice

From this you can see we are back to the first result but have shifted it twice. That is, each time we test, we shift.

Once again, we will try to see if we can make the divisor go into the result. Again, we will

subtract a binary 12 from the shifted result. Subtracting, we have

0 | 0 0 1 1 0 0   First result shifted
                       twice
1 | 0 1 0 0 0 0   Subtract 12
1 | 0 1 1 1 0 0   Third result
↑

1 here means 3d bit
of quotient is 0   Quotient = 100XX

Looking at the third result, we again see a negative number. Once again, this tells us the shifted result was not equal to or bigger than the divisor. This means that the third bit in the quotient is also a 0. It also means that we must add 12 back to the shifted result. Adding 12 back, we have

1 | 0 1 1 1 0 0   Third result
0 | 1 1 0 0 0 0   Add back 12
0 | 0 0 1 1 0 0   First result shifted
                       twice

Now that we have corrected for our mistake, we must shift the result again. This gives us

0 | 0 0 1 1 0 0   First result shifted
                       twice
0 | 0 1 1 0 0   First result shifted
                       three times

Once again, we will try to subtract the divisor. This gives

0 | 0 1 1 0 0   First result shifted
                       three times
1 | 0 1 0 0 0   Subtract 12
1 | 1 0 1 0 0   Fourth result
↑

1 here means 4th bit
of quotient is 0   Quotient = 1000X

The result of this subtraction is also a negative number. This means the fourth bit of the quotient is also a 0. Again, we must correct by adding 12 back to the result.

1 | 1 0 1 0 0   Fourth result
0 | 1 1 0 0 0   Add back 12
0 | 0 1 1 0 0   First result shifted
                       three times

We now have the first result shifted three times.

Again, we shift this result. This gives us the first result shifted 4 bits. It is

0	0 1 1 0 0	First result shifted three times
0	1 1 0 0	First result shifted four times

Again we subtract the divisor from this shifted result. As you can see, we are now going to subtract 12 from 12. That is,

0	1 1 0 0	First result shifted four times
1	0 1 0 0	Subtract 12
0	0 0 0 0	Fifth result

↑

0 here means 5th bit quotient is 1   Quotient = 10001

The fifth result is a positive number. Therefore, the fifth bit of the quotient is a 1.

We can stop here because we have our answer. So, 1100 1100 divided by 1100 is 10001.

You have seen that the process for division is a little more difficult than the process for multiplication. But this kind of division is a process that can be done by simply following the rules a step at a time. As you will see, this is all-important when working with microprocessors. *If you can find a set of rules to solve a problem, the microprocessor will execute them faithfully.* However, you must be sure that the rules cover all possible situations.

Let's take a look at one more example to be sure we understand the process. The example is shown in Fig. 3-16. In Fig. 3-16(a) we see a long division problem, 35 divided by 5. In Fig. 3-16(b) we see the long division done with binary numbers. In Fig. 3-16(c) we generate the 2's complement of the divisor (5). In Fig. 3-16(d) the division is done by using the subtract-and-shift-left method we have just learned.

In the first step the test fails. Therefore, the quotient is 0. We must add back the divisor. This gives us the dividend again. The addition is done in the second step. In the third step we shift the result (the dividend). Then we subtract the divisor from the shifted dividend. This time, the result is a positive number. This means the next bit in the quotient is a 1 and the subtraction process was correct. That is,

$$\frac{7}{5\overline{)35}}$$

(a)

```
        0111
101)100011
    000
    1000
     101
     00111
      101
      0101
       101
       000
```

(b)

0 1 0 1	5_{10}
1 0 1 0	1's complement
0 0 0 1	Add 1
1 1 0 1	2's complement

(c)

Binary arithmetic	Quotient	Process
0 1 0 0 0 1 1		Dividend 35_{10}
1 0 1 1 0 0 0		2's complement of 5_{10}
1 1 1 1 0 1 1	0	1st result
1 1 1 1 0 1 1		1st result
0 1 0 1 0 0 0		Add back 5_{10}
0 1 0 0 0 1 1		Dividend
1 0 0 0 1 1		Dividend shifted 1 time
1 0 1 1 0 0		2's complement of 5_{10}
0 0 1 1 1 1	1	2nd result
0 1 1 1 1		2nd result shifted 1 time
1 0 1 1 0		2's complement of 5_{10}
0 0 1 0 1	1	3rd result
0 1 0 1		3rd result shifted 1 time
1 0 1 1		2's complement of 5_{10}
0 0 0 0	1	4th result

(d)

Fig. 3-16 Dividing 35 by 5. (a) Long division with decimal numbers. (b) Binary long division. (c) Taking the 2's complement of 5. (d) The subtract-and-shift-left method.

we do not have to add back the divisor. All we need to do is shift the result.

This is done in the fourth step. We also subtract 5 from the shifted result. Once again, this gives a positive number. This means that the third bit of the quotient is a 1 and we should shift the third result.

In the fifth step we shift the third result and subtract 5. This also leaves us with a positive number, and therefore the quotient's fourth bit is a 1. At this point, we stop the division process because a zero (or less than zero) result shows the process is complete.

If we carry on the division process, all future quotients bits are 0. Usually we know how many quotient bits we are looking for and where the binary point is. For example, we might divide a 16-bit dividend by an 8-bit divisor until we have an 8-bit quotient. If we know where the dividend and divisor binary points are, then we automatically know where the quotient's binary point is.

Self-Test

Answer the following questions.

21. Division is just repeated
    a. Addition   *c.* Multiplication
    b. Subtraction   *d.* (All of the above)
22. Explain why division needs a somewhat more complicated set of rules for a processor to follow than multiplication does.
23. Perform the following division problems. Use the procedure outlined in this section. Check your work with either binary or decimal long division. All numbers are positive.

    a. $101\overline{)11110}$ **d.** $11\overline{)1001}$

    b. $111\overline{)10101}$ **e.** $1001\overline{)101101}$

    c. $100\overline{)101000}$ **f.** $1100\overline{)10000100}$

3-6 MULTIPLE-PRECISION ARITHMETIC

When working with a microprocessor, we will often find that the microprocessor's word length does not allow enough precision. This means that we need a way to express larger numbers so that the microprocessor can work with them.

For example, the very popular 8-bit microprocessor lets us use 2's complement numbers from +127 through 0 to −128.

Obviously, for most work these numbers are nowhere near large enough. Using two 8-bit words to represent a 2's complement number gives us a range of +32,767 through 0 to −32,768. We are limited to a precision of about 1 part in 60,000. That is, our precision is ± 0.0015 percent.

For many applications this *double precision* is good enough. However, there are applications that require even greater precision. Some applications require *triple precision*. Triple precision gives us 1 sign bit and 23 magnitude bits.

When we are using triple-precision representation in an 8-bit system, the range of numbers is from +8,388,607 through 0 to −8,388,608. This gives a much better precision than six decimal digits. That is, the number has a precision of greater than 1 part per million.

As usual, we do not get something for nothing. Using *multiple-precision notation* requires the microprocessor to store more data and to do more work each time a calculation is done.

For example, suppose you are going to use triple precision on an 8-bit system. You can no longer simply call two numbers from memory, add them with the ALU, and store them at a location reserved for results. First, you must call the least significant byte of each number. The 2 bytes are added, and the result is stored. Any carries generated must be saved. The middle bytes of the two numbers are then called. These are added, together with the carry from the previous addition, and the result is stored in a space reserved for the middle byte of the result. Finally, the most significant bytes of the two words are called and added, and the carry from the previous addition is added to the sum. The result of this addition is stored in the place reserved for the most significant byte.

As you can see, this addition process takes three times as long as a simple 8-bit addition, and of course it takes three times as much storage space.

Multiple-precision arithmetic can be done for all four arithmetic operations. That is, we have multiple-precision addition, subtraction, multiplication, and division.

Self-Test

Answer the following questions.

24. Explain why multiple precision is used.
25. What is the range of a double-precision 2's complement number in a 16-bit processor?

26. Why does multiple-precision arithmetic mean extra work?
27. You are using an 8-bit microprocessor. What precision is needed to handle each of these numbers?
    a. 1568
    b. −10,264,329
    c. 22,438
    d. −129
    e. 12,348
    f. −1,000,274

Memory location	Contents	
M + 3 (fourth)	+/—	7-bit exponent
M + 2 (third)	+/—	7-bit hi mantissa
M + 1 (second)	8-bit mid mantissa	
M (first)	8-bit low mantissa	

Fig. 3-17 **Storing a floating-point number. The number here is in the form** $\pm$ XXXXXXXXXXXXXXXXXXXXX $\times$ $2^{\pm \text{XXXXXXX}}$ **where X is an unknown binary digit (bit).**

3-7 FLOATING-POINT ARITHMETIC

The use of multiple-precision numbers does not solve all our problems. For example, the numbers we have discussed so far are all integers. We have not seen any way to take care of fractions. Nor have we seen any way to represent very large or very small numbers.

These problems are solved by using *floating-point arithmetic*. In floating-point arithmetic, the microprocessor keeps track of where the decimal point is. The microprocessor does this by using scientific notation.

We have all used scientific notation in the past. Using scientific notation requires that all numbers be shown as signed decimal fractions lying between 0.1 and 1. Note: The number 1.0 is not used. The number 1 is shown as 0.1 X 10^1. These signed decimal fractions are multiplied by a signed power of 10. For example, if we wish to present the number 50 in scientific notation, we would show it as

$$0.5 \times 10^2$$

we would show the number −750 as

$$-0.75 \times 10^3$$

To show a very small number, such as 0.00105, we would write

$$0.105 \times 10^{-2}$$

When using floating-point notation, the microprocessor stores numbers in the same way. That is, the microprocessor stores both a signed mantissa and a signed exponent. The *mantissa* is the number that lies between 0.1 and 1, and the *exponent* is the power of 10.

Let's see how an 8-bit microprocessor stores numbers in floating-point notation. Figure 3-17 shows that four memory locations are used. The first memory location has the low-order byte of the mantissa. The second memory location has the middle byte of the mantissa. The third memory location has the high seven bits of the mantissa and the sign bit. We can see that the first 3 bytes of this floating-point package are a triple-precision number. A fourth memory location has a signed 7-bit exponent.

Using this floating-point arithmetic we can represent numbers from $+ (2^{23} - 1) \times N^{127}$ to $- 2^{23} \times N^{127}$, including fractional values as small as $\pm 1 \times N^{-128}$. The value of N is usually 2. However, we can use 10 for some special purposes. Obviously, this is a very wide range of numbers. *Note:* Even though the range of numbers is very large, we still only have $\pm$ 23 bits of precision.

Once a microprocessor system has been set up to operate in floating-point notation, all arithmetic operations are usually passed through the floating-point package. The floating-point package usually contains a series of special programs. These programs let us place a number in four memory locations using a floating-point format like the one shown in Fig. 3-17. Once we have placed the number in this *floating-point accumulator*, we can call programs that will perform mathematical operations on the number in the floating-point accumulator and numbers in another set of memory locations.

Most floating-point packages include more than programs to do addition, subtraction, multiplication, and division. Many include programs that do such operations as squares, square roots, sines, cosines, tangents, and logarithms.

Obviously, once a floating-point package has been written, it is used over and over again. Often the floating-point package is treated almost as if it were an extension of the hardware itself. Don't be confused when someone speaks of the "floating-point accumulator." It may not be a piece of hardware. As we have learned, the floating-point accumulator can be just four memory locations that are worked on by the floating-point programs.

There are floating-point hardware accessories which can be added to the microprocessor. These "math chips" are usually special VLSI circuits which go with a particular microprocessor. They are faster than a software floating-point package, but they increase the amount of hardware in the microprocessor-based system.

The floating-point package works much more slowly than the microprocessor's natural instructions. In fact, a floating-point addition may take 10 to 20 times longer than an addition instruction for a microprocessor's hardware. The reason is that the floating-point ADD subroutine executes 20 or 30 instructions for each operation. Not only does this program always have to work with a triple-precision number, but it also must keep track of the exponent on each calculation.

The floating-point package is almost always used whenever the microprocessor is controlling a system into which people can enter a wide range of numbers. One of the most common applications for the floating-point package is when the microprocessor is executing a higher-level language such as BASIC. All higher-level languages, such as BASIC and FORTRAN, include their own floating-point packages. Usually, these floating-point packages are very sophisticated when compared with the ones you might find in a microprocessor control system designed to perform a few limited functions.

Self-Test

Answer the following questions.

28. You would use floating-point arithmetic to represent very _____ numbers.
29. Express the following numbers in scientific notation:
    a. 12   f. −1000
    b. 222.3   g. −0.000101
    c. −0.334   h. 100
    d. 1,256,000   i. 22,000,000
    e. 0.0000125   j. 0.000000021
30. The two numbers used in the floating-point system are called the _____ and the _____ .

SUMMARY

1. Binary and decimal addition are alike; they just use different number systems.
2. When you add two digits and the resultant sum cannot be expressed as a single digit, the result includes a carry.
3. There are only four possible results from adding two binary digits. The binary addition rules are

    $0 + 0 = 0$

    $0 + 1 = 1$

    $1 + 0 = 1$

    $1 + 1 = 1$ plus 1 carry

    A carry is added to the next more significant column. A carry is added according to the rules of addition.
4. Computer arithmetic uses fixed-length binary numbers. Insignificant (leading) 0s fill up the required number of bits.
5. Binary and decimal subtracting processes are alike. Only the number systems are different. If the subtrahend is larger than the minuend, subtracting causes a borrow.
6. The rules for binary subtraction are

    $0 - 0 = 0$

    $1 - 0 = 1$

    $0 - 1 = 1$ and 1 borrow

    $1 - 1 = 0$

    A borrow must be subtracted from the next column to the left. The borrow is subtracted according to the rules of subtraction.
7. To complement a binary number, change all the 1s to 0s and all the 0s to 1s. This is called a 1's complement.
8. The most common way to express a negative binary number is to show it as a 2's complement number. In 2's complement notation, a 0 in the number's most significant bit means it is a positive number. A 1 in the number's most significant bit means it is a negative number.
9. The 2's complement is made by taking the 1's complement and adding 1. An 8-bit 2's complement binary number can represent decimal numbers from + 127 to −128.

10. A short method of generating 2's complement numbers is
    a. Start at the LSB.
    b. Copy the 0s directly until you reach the first 1.
    c. Copy the first 1 directly.
    d. Complement each bit after that.

11. The logic to do binary arithmetic is the same for both unsigned and 2's complement numbers. To know the value of the answer, we must know which notation we are using. If the result of a 2's complement arithmetic operation is negative, take the 2's complement of the result to get to its absolute value.

12. Both binary and decimal multiplications are just quick ways to do a lot of additions. The process of multiplication is repeated for each digit of the multiplier. Each multiplication results in a partial product. The partial products are added together to form the final product. Each partial product is shifted to the left one digit before adding.

13. The rules for binary multiplication are
    $0 \times 0 = 0$
    $0 \times 1 = 0$
    $1 \times 0 = 0$
    $1 \times 1 = 1$

    Binary multiplication generates a great many partial products. The partial product is a copy of the multiplicand if the multiplier bit is a 1. The partial product is all 0s if the multiplier bit is a 0.

14. The binary multiplication procedure is shift and add. The multiplicand is shifted left and added to the previous result each time there is a 1 in the multiplier. Each 0 in the multiplier causes a shift but no add.

15. Division is really just repeated subtraction. Division is more complicated than multiplication because we guess the partial results and then test our guess. In binary division, we first guess the result is a 1. If the division will go into the most significant remaining bits of the dividend, then we keep the 1 in the quotient. If not, then we change the quotient's bit to a 0. We use the 2's complement arithmetic for division because we must both add and subtract during the process.

16. Often the microprocessor's word length does not give enough precision. We use multiple words to increase precision. A triple-precision number in an 8-bit microprocessor has 23 magnitude bits and one sign bit. Using multiple precision requires extra data handling and keeping track of any carries or borrows generated by the operations on less significant bytes. Multiple-precision arithmetic is used for all kinds of mathematical operations.

17. Floating-point arithmetic is used to represent very large or very small numbers. It keeps track of the radix point. Floating-point arithmetic uses scientific notation. In floating-point arithmetic, the mantissa lies between 0.1 and 1.

18. The mantissa is multiplied by a power of 10 or 2. The power is called the number's exponent. Most microprocessor software packages have a program called the floating-point package which often has many floating-point instructions for advanced arithmetic operations.

CHAPTER REVIEW QUESTIONS

Answer the following questions

3-1. If you add two digits and the result cannot be expressed as one digit, you have caused a
    a. Borrow   *c.* Carry
    b. Sum   *d.* Multiplicand

3-2. Fill in the table, showing the results of binary addition:
    a. $0 + 0 =$ _____ + a carry of _____
    b. $0 + 1 =$ _____ + a carry of _____
    c. $1 + 0 =$ _____ + a carry of _____
    d. $1 + 1 =$ _____ + a carry of _____

3-3. Computer arithmetic is done with fixed-length numbers. Unused leading digits are expressed as
    a. Carries   *c.* Borrows
    b. 1s   *d.* 0s

3-4. If an addition generates a carry, it must be added to _____ .

3-5. Express the following as 8- or 16-bit binary numbers and find the sums:

    a. 12,525   **d.** 1296
        621   151

    b. 2048   **e.** 56,274
         64   32,768

    c.   99   **f.** 128
      107   256

3-6. Fill in the table, showing the results of binary subtraction:
    a. $0 - 0 =$ _____ with a borrow of _____
    b. $0 - 1 =$ _____ with a borrow of _____
    c. $1 - 0 =$ _____ with a borrow of _____
    d. $1 - 1 =$ _____ with a borrow of _____

3-7. A borrow must be subtracted from the _____ .

3-8. Complete the following binary subtractions. What are the decimal results?

    a. 0 1 0 1 1 1   **d.** 0 0 0 1 0 1 1
        0 0 1 0 1 1   0 0 0 0 1 1 1

    b. 0 1 1 0 1 1 1   **e.** 0 1 1 1 1 1 1 1
        0 0 0 1 0 1 1   0 1 1 1 1 1 1 0

    c. 0 1 0 1 0 1   **f.** 0 0 0 0 1 1 1
        0 0 0 0 0 1   0 0 0 0 0 1 0

3-9. If you need to express an integer binary number in both positive and negative forms, you would probably use _____ notation.
    a. Scientific   *c.* 2's complement
    b. 1's complement   *d.* Sign and magnitude

3-10. Two's complement, 1's complement, and sign and magnitude all express a negative number with a(n) _____ .

3-11. What is the range of a 16-bit 2's complement number?

3-12. What is the range of a 32-bit 2's complement number?

3-13. Show the following numbers in 8-bit 2's complement form:
    a. $+ 12$ **g.** $- 70$
    b. $+ 16$ **h.** $- 127$
    c. $- 15$ **i.** 0
    d. $+ 125$ **j.** $- 1$
    e. $- 100$ **k.** $- 123$
    f. $+ 64$ **l.** $+ 127$

3-14. Complete the following subtraction problems, using 2's complement arithmetic. If the answer is negative, express the number both in 2's complement form and in absolute value binary form.

    a. 0 1 0 1   **d.** 0 1 0 1 1 1 0 0
        0 1 1 1   0 0 1 1 1 0 1 1

    b. 0 1 1 1 0 1 1 1   **e.** 0 1 1 0 1 1 0 1 1 0 1 1 1 1 1 1
        0 0 1 1 0 1 0 1   0 1 1 1 1 1 0 1 1 1 0 0 0 1 0 1

    c. 0 1 1 0 1 0 1 1   **f.** 0 1 1 1 1 1 1 1 1 0 0 0 0 0 0 0
        1 1 1 1 1 1 1 1   0 0 0 0 1 1 0 1 0 1 1 0 1 1 0 0

3-15. Explain why the shift-and-add process works so well for binary multiplication.

3-16. Fill in the following binary multiplication table:
    a. $0 \times 0 =$ _____   **c.** $0 \times 1 =$ _____
    b. $1 \times 0 =$ _____   **d.** $1 \times 1 =$ _____

3-17. Perform the following binary multiplications using the shift and add process:

a. 0 0 0 0 1 0 0 1
     0 0 0 0 0 1 0 1

b. 0 1 0 0 0 1 1 0
     0 0 0 1 0 1 0 1

c. 0 1 0 1 0 1 0 1
     0 1 0 0 0 1 0 1

d. 0 1 0 1 1 1 1 0
     1 0 0 0 0 1 0 1

e. 0 1 1 1 0 1 1 1 0 1 1 0 1 1 1 0
     0 0 0 0 1 0 1 0 0 1 1 1 0 1 0 1

f. 1 0 0 0 0 0 0 0 0 0 0 1 0 1 1 1
     0 0 1 0 0 0 0 0 0 0 0 0 0 1 1 1

3-18. The result of multiplying the multiplicand by 1 of the multiplier bits is called a(n) _____ .

3-19. Divide the following, using the method a microprocessor would use.

a. $01010\overline{)1100100}$

b. $0100\overline{)10000}$

c. $D\overline{)9C}$

d. $0100\overline{)11110}$

3-20. Multiple-precision arithmetic is used to improve
  a. Addition
  b. Carrying
  c. Resolution
  d. Precision

3-21. A triple-precision number in an 8-bit microprocessor uses _____ magnitude bit(s) and _____ sign bit(s).

3-22. Floating-point representation uses multiple-precision arithmetic and _____ notation.

3-23. You would not expect a floating-point package to be able to call the _____ function.
  a. Addition
  b. Sine
  c. Log
  d. Reset

CRITICAL THINKING QUESTIONS

3-1. Binary arithmetic and decimal arithmetic are a lot alike except that, with binary arithmetic, we seem to use insignificant 0s and pay a lot more attention to the rules. Why is this so?

3-2. What would be the impact of a decision to use 1's complement arithmetic in microprocessors?

3-3. The example of multiple-precision numbers showed 3 bytes for the mantissa and 1 byte for the exponent. Is there a place for 2-byte mantissas and 4-byte mantissas? What happens if you are using a 16-bit microprocessor?

3-4. You have seen that the rules of binary arithmetic let us perform the four basic mathematical functions by using addition, and later you will find out that addition is really all that a microprocessor's hardware can do. Subtraction, for example, is just the addition of 2's complement numbers. If a microprocessor can only do addition, how can it possibly compute the sine of an angle?

Answers to Self-Tests

1. **a.** 0001 0110
   b. 1001 1000
   c. 1111 1101
   d. 1111 1111
   e. 1000
   f. 1100
   g. 1010 0001 0000 0010
   h. 1001 0111 1011 1000

2. **a.** 0110 0101
      0001 0000
      0111 0101

   b. 1110 0001
      1010 1000
      11000 1001

   c. 0000 0001 1000 1110
      0000 0000 1000 0100
      0000 0010 0001 0010

d. 0 0 1 1 1 0 0 0
  0 0 0 0 1 0 1 0
  ‾‾‾‾‾‾‾‾‾‾‾‾‾‾‾
  0 1 0 0 0 0 1 0

e. 0 1 0 1 0 1 1 0
  0 0 0 1 1 0 0 1
  ‾‾‾‾‾‾‾‾‾‾‾‾‾‾‾
  0 1 1 0 1 1 1 1

f. 0 0 0 0 0 0 0 1 0 0 1 0 0 0 0 1
  0 0 0 0 0 0 0 1 1 1 1 0 1 1 0 1
  ‾‾‾‾‾‾‾‾‾‾‾‾‾‾‾‾‾‾‾‾‾‾‾‾‾‾‾‾‾
  0 0 0 0 0 0 1 1 0 0 0 0 1 1 1 0

3. 9 bits

4. The only difference is that 0101 and 0000 0101 have insignificant leading 0s. It would not affect any sum.

5. **a.** 0100
   b. 0010
   c. 0110
   d. 1000
   e. 0010 1101
   f. 1110 1110
   g. 0011 0101 1010 1001

6. **a.** 1 1 1 1
   1 1 0 0
   ‾‾‾‾‾‾‾
   0 0 1 1

   b. 1 0 0 0
   0 0 1 0
   ‾‾‾‾‾‾‾
   0 1 1 0

   c. 1 1 1 1 1 1 1 1
   0 0 0 1 1 0 0 0
   ‾‾‾‾‾‾‾‾‾‾‾‾‾‾‾
   1 1 1 0 0 1 1 1

   d. 0 0 1 1 0 1 0 0
   0 0 1 0 0 1 0 0
   ‾‾‾‾‾‾‾‾‾‾‾‾‾‾‾
   0 0 0 1 0 0 0 0

   e. 0 0 0 1 1 1 0 1
   0 0 0 0 1 1 0 0
   ‾‾‾‾‾‾‾‾‾‾‾‾‾‾‾
   0 0 0 1 0 0 0 1

   f. 1 0 0 0 1 0 0 0
   0 1 1 0 1 1 0 0
   ‾‾‾‾‾‾‾‾‾‾‾‾‾‾‾
   0 0 0 1 1 1 0 0

7. A borrow is caused by subtracting a larger digit (bit) from a smaller digit (bit).

8. A borrow comes from subtraction, and a carry comes from addition. Both are arithmetic overflows.

9. *b*

10. **a.** 0100 1001
    b. 1001 0100
    c. 010
    d. 1111 0011
    e. 0
    f. 1111 1110
    g. 0000 0000
    h. 1010 1010
    i. 0000 1111 0000 1111
    j. 0011 0011 0011 1100

11. **a.** Negative
    b. Positive
    c. Positive
    d. Positive
    e. Positive
    f. Positive
    g. Negative
    h. Positive
    i. Negative
    j. Negative

12. **a.** 0100 0000
    b. 1100 1000
    c. 0000 1100
    d. 0000 0000
    e. 1000 0000
    f. 0111 1111
    g. 0010 0000
    h. 1110 0000
    i. 256 cannot be expressed as an 8-bit number. It can be expressed as the 16-bit number 0000 0001 0000 0000.
    j. 0001 0000
    k. 1001 1100
    l. 1111 1100

13. The same logic can be used because the binary relationships don't change. All that changes is how you look at the numbers. For example, 1100 can represent either 12 or the 2's complement of 4 in the following two illustrations of binary arithmetic:

           0 1 1 1   7
        + 1 1 0 0   + 12
        ‾‾‾‾‾‾‾‾‾   ‾‾‾‾
        1 0 0 1 1   19

    or   0 1 1 1   7
           1 1 0 0   − 4
        ‾‾‾‾‾‾‾‾‾   ‾‾‾‾
        1 0 0 1 1   3

14. Additions

15. Partial product

16. **a.**   1 0 1
             0 1 1
            ‾‾‾‾‾‾
             1 0 1
           1 0 1
         0 0 0
         ‾‾‾‾‾‾‾‾‾
         0 1 1 1 1

    b.   0 1 1 0
               0 1 1 1
              ‾‾‾‾‾‾‾
               0 1 1 0
             0 1 1 0
           0 1 1 0
         0 0 0 0
         ‾‾‾‾‾‾‾‾‾‾‾
         0 1 0 1 0 1 0

c.
```
      0 1 1
      0 1 1
      0 1 1
    0 1 1
  0 0 0
  0 1 0 0 1
```

d.
```
                    0 1 0 1 1 1 0 1
                    0 0 1 0 1 1 0 1
                    0 1 0 1 1 1 0 1
                  0 0 0 0 0 0 0 0
                0 1 0 1 1 1 0 1
              0 1 0 1 1 1 0 1
            0 0 0 0 0 0 0 0
          0 1 0 1 1 1 0 1
        0 0 0 0 0 0 0 0
        0 0 0 0 0 0 0 0
  0 0 0 1 0 0 0 0 1 0 1 1 0 0 1
```

e.
```
                  0 0 0 1 1 0 1 1
                  1 1 1 1 1 1 0 0
                  0 0 0 0 0 0 0 0
                0 0 0 0 0 0 0 0
              0 0 0 1 1 0 1 1
            0 0 0 1 1 0 1 1
          0 0 0 1 1 0 1 1
        0 0 0 1 1 0 1 1
      0 0 0 1 1 0 1 1
    0 0 0 1 1 0 1 1
  0 0 0 1 1 0 1 0 1 0 0 1 0 1 0 0
```

f.
```
      0 1 1 1
      1 0 0 0
      0 0 0 0
    0 0 0 0
  0 0 0 0
  0 1 1 1
  0 0 1 1 1 0 0 0
```

g.
```
      1 0 0 1
      1 0 1 0
      0 0 0 0
    1 0 0 1
  0 0 0 0
  1 0 0 1
  0 1 0 1 1 0 1 0
```

17. Number's base
18. The same as the multiplicand.

19. 0
20. Shift and add.
21. *b*
22. Because in division you must guess the correct quotient and then test to see if your guess was right. The processor does not guess.
23. **a.** 110
    b. 11
    c. 1010
    d. 11
    e. 101
    f. 1011
24. Multiple precision is used when you cannot express a number with digits using one word.
25. $-2,147,483,648$ to $+2,147,483,647$
26. Because you must work on it one word at a time. This means you must store many intermediate results and keep track of carries and borrows.
27. **a.** Double
    b. Quadruple
    c. Double
    d. Double
    e. Double
    f. Triple
28. Large or small.
29. **a.** 0.12×10^2
    b. 0.2223×10^3
    c. -0.334×10^0
    d. 0.1256×10^7
    e. 0.125×10^{-4}
    f. -0.1×10^4
    g. -0.101×10^{-3}
    h. 0.1×10^3
    i. 0.22×10^8
    j. 0.21×10^{-7}
30. Exponent; mantissa.

CHAPTER 4

Basic Microprocessor Architectural Concepts

■

CHAPTER OBJECTIVES

This chapter will help you to:

1. *Identify* several important differences in microprocessor architectures.
2. *Predict* the relationship of word length to performance and cost.
3. *Calculate* the RAM and ROM required to perform a given function.
4. *Explain* the MIPS rating of a microprocessor.
5. *Explain* the terms parallel processing, coprocessing, cache memory, and pipelining.
6. *Define* the terms register, addressing mode, and assembler.

Over the years quite a number of different microprocessors have been developed. Each new one focuses on a particular application which it performs very well. The differences between these microprocessors, and therefore what makes one microprocessor better for a particular application than another one, arise from the differences in their architectures.

■

4-1 WHAT IS THE MICROPROCESSOR'S ARCHITECTURE?

Earlier in this book, we described the microprocessor's architecture. We said it is "the general way in which the computations are done." That is a very broad definition of a microprocessor's architecture. In this chapter, we will look for a specific answer to the question. We will look at a number of examples to help understand microprocessor architectural concepts. Although we know there is no precise definition of architecture, these examples should help our understanding.

We all know that the person who designs buildings is an architect. If we follow this line of thinking a bit further, we realize that while all buildings perform similar functions—that is, they keep the weather out—they do it in different ways. The same is true of microprocessors. There are many different ones. Most try to do the same job, but the internal structure which they use to do the job is different. That is, their *architecture* is different.

The basic job of the microprocessor is to process data. The data processing is done using a general-purpose integrated circuit. The design of the integrated circuit lets the user program the function as one would program a general-purpose digital computer. In this, all microprocessors are alike. How are they different?

For some microprocessors, the architectural differences are small. For others, the differences are large. For example, the Intel 8051 8-bit single-chip microprocessor and the Intel 486 32-bit microprocessor will both add two numbers. The way they add the two numbers and the parts they use to add the two numbers are quite similar. However, if you ask a programmer to make each of these microprocessors compute the cosine of an angle, the way

From page 49:

Architecture

On this page:

Microprocessor design

Registers

Word length

they perform this task and the parts they use will be very different. Each of them can do the job, one with great ease and one with extreme difficulty. They have very different architectures.

In this chapter, we will look at some of the major sections of a *microprocessor design* which contribute to architectural differences. We will also try to understand where and when these architectural features are important and when they are not. Like a building, the architectural features of a microprocessor cost money. Greek columns cost a great deal more than wooden poles with the bark left on. However, poles with the bark left on do not look right holding up the roof of a courthouse. Likewise, Greek columns are really not necessary to hold up a shed covering hay. There are places for 4-bit microprocessors, and there are places for 32-bit microprocessors. To use the microprocessor intelligently, we must understand the differences.

What are some of the major architectural differences between microprocessors? We briefly looked at three major differences in the first chapter. They are

- The length of the microprocessor's data word
- The size of the memory which the microprocessor can directly address
- The speed which the microprocessor can execute instructions

Although these are some of the major architectural differences between microprocessors, there are other important ones which you must know if you are to use the microprocessor with understanding. Some of these additional architectural differences are

- The number of *registers* available to the programmer
- The different types of registers available to the programmer
- The different types of instructions available to the programmer
- The different types of memory addressing modes available to the programmer
- The different types of support circuits available to the system designer
- Compatibility with readily available development, system and applications software
- Compatibility with hardware development systems

As you can see from this list, there are many differences between microprocessors. These differences make some microprocessors better for some applications, and some other microprocessors better for other applications. Some microprocessors are better because they appeal to the user more than another which might work equally well in the same situation.

In the following sections, we will look at these differences in more detail. You will find, however, that as the microprocessor's *word length* grows its other features tend to grow. That is, microprocessors with a long word also tend to have extensive instruction sets, many registers, a wide variety of addressing modes, and lots of applications software. This means that the choices may not be as great as one might first think when looking at the above lists of microprocessor architectural differences.

Self-Test

Answer the following questions.

1. Select the one microprocessor characteristic which is not an architectural feature.
   a. Data word length d. Instruction set
   b. Processing speed e. Manufacturer
   c. Memory size f. Number of registers
2. As a general rule, you would not expect the cost of a microprocessor to increase with
   a. Increased data word length
   b. A slower-speed processor
   c. Increased memory addressing
   d. More registers
3. The microprocessor got its basic architecture from
   a. Greek buildings
   b. The digital computer
   c. Silicon integrated circuits
   d. The electronic calculator
4. The architecture of the microprocessor can briefly be described as the _____ have been put together so that the microprocessor function can be done.
   a. Way the logic circuits
   b. External logic parts which
   c. Type of silicon transistors which
   d. Register types which

4-2 WORD LENGTHS

The microprocessor started as a 4-bit device. It has progressed to an 8-bit, a 16-bit, a 32-bit, and now a 64-bit device. If 16- and 32-bit microprocessors can outperform the 4- and

8-bit microprocessors, why do the smaller devices remain in existence? What is the reason that the shorter-word-length architecture is still used?

The answer to this question lies in the definition of performance. Our initial idea of performance is usually speed—how quickly the microprocessor will perform its intended function. This thinking tells us that a microprocessor with a longer word length will solve more problems faster. Therefore, a longer word length should give a better solution to all problems.

However, one must first ask, do I need to solve a variety of problems? Suppose you are trying to build a simple microprocessor-controlled toy. Solving problems where you need to calculate the cosine of an angle are the farthest thing from your mind. On the other hand, the cost of the product is very close to your heart. You can pay as little as 50 cents for a 4-bit microprocessor which will do the job. This means that the dollar you would add to the product for an extra 4 or 12 bits of data latch to support a bigger microprocessor becomes a lot of money.

In this example, the real performance issue is cost, not horsepower. In almost every case,

the longer the data word, the more the microprocessor and its support components cost. It may only be pennies, but the pennies add up. In other words, the choice of microprocessor word length should be made to get the job done without overkill. Figure 4-1 shows three typical board-level microprocessor-based systems. From this you can see that the number of support circuits typically used with a microprocessor grows with the number of bits in the microprocessor's data word.

What are some of the application areas that the various data word lengths fall into? As we have seen, the 4-bit microprocessors are very good for low-cost applications. The typical 4-bit microprocessor costs well under $1 in volume and requires very few support parts. As noted before, they tend to be simple microprocessors designed for simple applications. You will frequently find the 4-bit microprocessor in

- Toys: robots, remote-controlled cars, hand-held games
- Calculators: financial, scientific, data base
- Power tool controllers: speed controls, sequencers, measurement devices

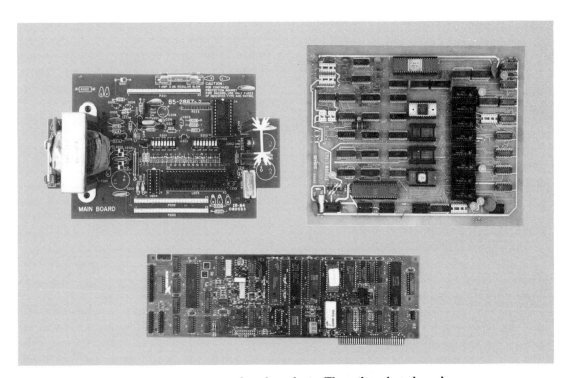

Fig. 4-1 Three board-level microprocessor-based products. These three boards each have a microprocessor and the support components which let the microprocessor-based system talk to external data input and output devices. The 4-bit microprocessor uses far fewer support components than the 8-bit microprocessor on the right. The 8-bit microprocessor uses fewer support components than the 16-bit microprocessor underneath.

- Simple, intelligent consumer product controllers: microwave ovens, telephone dialers, smart thermostats, shortwave scanners, TV remote controls
- Computer peripherals: keyboard scanners, simple printers, clocks

All of these examples have two common features. First, all of the end products are low cost. Second, none of the applications is particularly demanding. For example, you may say that the scientific calculator must perform some difficult calculations such as sines, cosines, or logarithms, but these calculations can be done very slowly in a hand-held calculator. The 4-bit microprocessor is quite capable of either high-speed simple operations or of complex operations at a slower speed. It just cannot do both at once.

What are some applications which need the power of the 8-bit microprocessor? Where do we need additional speed, a greater word length, a more extensive instruction set? We must also ask, what applications can afford a little more money for the microprocessor?

The typical 8-bit microprocessor costs $2 to $4. It must be combined with another $5 to $10 of support circuits before a working system is built. If an 8-bit single-chip microprocessor is used, only a limited number of support circuits are needed. But, the cost of the 8-bit single chip microprocessor is about $4 to $5. So, for an application which really needs the 8-bit microprocessor, you must pay three to five times as much as for the 4-bit microprocessor.

Some applications which use 8-bit microprocessors include:

- Toys: video games, programmable robots
- Complex, intelligent consumer product controllers: VCR control and programming, security systems, lighting system controllers
- Computer peripherals: video displays, higher-speed printers, modems, plotters, disk controllers, communications controllers
- Industrial controllers: robotics, substation data gathering, process control, sequence control, machine tool control
- Instruments: logic analyzers, communications analyzers, disk drive testers, digital oscilloscopes, smart voltmeters

The common feature of these products is a much higher product cost to the user. The end customer knows that the product is worth more and expects to pay more than for the products in the 4-bit microprocessor applica-

tions list. That is why these products can afford the extra cost of an 8-bit microprocessor.

We also find that the tasks that the products are asked to do are more complex. This means that the microprocessor's job is going to be harder. The microprocessor must process more data and do it faster. In many cases, the processing must be done fast enough for a machine rather than for a human being.

For example, the microprocessor used in a hand-held game only needs to respond as quickly as you can press the keys on the game's keyboard. On the other hand, the microprocessor which controls a dot matrix printer attached to your microcomputer may receive the data to be printed at a rate of 1000 characters a second. Because the printer cannot hold all the data the microcomputer might send it, the printer must send the microcomputer control signals. These signals tell the microcomputer when to send the data, and when to hold the data because the printer's memory is full and waiting until more printing is done. At a data rate of 1000 characters per second, this signal to the microcomputer must happen very quickly. This is a machine-to-machine response time, not a machine-to-human response time.

To date, 16-bit, 32-bit, and 64-bit microprocessors have found one high-volume application and a number of lower-volume applications. They are the heart of the *personal computer* (PC). First introduced by IBM in the summer of 1981, the personal computer was based on the *Intel 8088* microprocessor. Internally, the 8088 microprocessor is a 16-bit processor, but externally it uses an 8-bit bus. This feature allowed the first PCs to be developed using lower-cost external parts. However, the 8088's internal 16-bit architecture makes the PCs upwardly compatible with newer versions of the PC which are based on the 16-bit 80286 microprocessor or the 32-bit 386, 486, and Pentium microprocessors.

The introduction of a microcomputer based on these microprocessors created a large base of development systems and application software. It has also caused many manufacturers to produce PC-compatible machines and PC-compatible accessories. The great rush to become involved with personal computing through the use of a PC or PC compatible has produced a wide variety of applications for PCs.

Most of the applications use a complete PC rather than just one of the microprocessors in dedicated hardware. There are, however, applications, especially within the industrial, scientific, and medical applications areas, where special products are being built that use these microprocessors. One area is robotics. The control of a five- or six-axis-of-freedom robotic arm requires a sophisticated controller. The 16-bit and 32-bit microprocessors are ideal for this application.

In addition to the Intel family of microprocessors, the *Motorola 68000* family is used as the basis for a number of advanced personal computers. The Apple Macintosh is based on this microprocessor family. The 68000 is also the basis of a number of industrially oriented microcomputer systems. These systems use an advanced systems software package known as Unix. The microcomputers created for these industrial computers are frequently used as *computer-aided design* (CAD) or *computer-aided manufacturing* (CAM) workstations. Thirty-two-bit microprocessors are needed in these applications because of the great use of graphic displays. A great deal of computing power is required, for example, to display a two- or three-dimensional drawing in color.

In the above examples, you can see that the 16-bit and 32-bit microprocessors are mainly used as the microprocessors to build microcomputers. Again, there is some use of these microprocessors in nonmicrocomputer systems, but the use is limited. There is a reasonable amount of use of these microprocessors at the card level. This means that the printed circuit cards such as the CPU card or memory (RAM) card from a PC may be used. Most of these uses are in industrial applications where a full microcomputer is not required to do the job. Basically, these applications use only the parts of the microcomputers (CPU card, I/O card, memory card, etc.) needed to do the job and no more.

The *32-bit microprocessors* are outgrowths of the 16-bit microprocessors. Both the Intel and the Motorola families have full 32-bit parts: the 80386 and 68020 series microprocessors. Although these two products are quite capable of being used in special-purpose designs, their main applications have been to create 32-bit microcomputers. Both products are used to create extensions of personal computer designs.

Why does the world need 32-bit microcomputers? If you only look at the need for a 32-bit data word, the biggest reasons become speed for certain calculations and speed of data handling. Why is the 32-bit microprocessor faster in these situations? A 32-bit word gives the user a resolution of 1 part in 4 billion. The 8-bit word only has a resolution of 1 part in 256, and the 16-bit word has a resolution of 1 part in 65,536. The great resolution of the 32-bit word is in many cases sufficient for complex commercial, industrial, and scientific calculations. Figure 4-2 compares 8-bit, 16-bit, and 32-bit words. As you can see, the 32-bit word has enormous data-handling capability when compared with either the 8-bit or 16-bit data words.

The use of the 32-bit word means that the calculations can be performed in a single- rather than multiple-precision operation. Multiple-precision calculations are much slower than single-precision calculations. Therefore, if the end application can use the resolution provided by a 32-bit word, the 32-bit microprocessor is much faster than either an 8-bit or 16-bit microprocessor doing the same calculation.

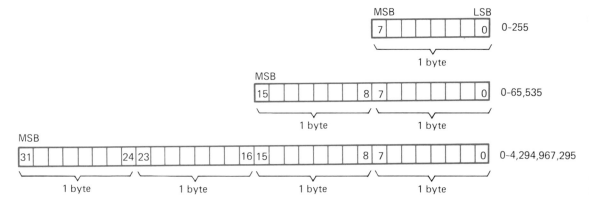

Fig. 4-2 Comparing the 32-bit data word to the 8- and 16-bit data words. With resolutions of one part in 256 and 65,536, respectively, the 8- and 16-bit data words are small compared with the 32-bit word's resolution of one part in 4,294,967,296.

The 32-bit microprocessor is also faster at data handling. Each time that the 32-bit microprocessor takes data from its memory, it gets four times as much data as the 8-bit microprocessor does doing the same operation. That is, the 8-bit microprocessor gets data from memory 1 byte at a time. The 32-bit microprocessor gets data 4 bytes at a time. This means that the same speed 32-bit microprocessor handles data four times faster than the 8-bit microprocessor. As we will see later, increasing the microprocessor's speed is difficult. Therefore the speed increase achieved by doubling the word length is a very real improvement.

Does this mean that the *64-bit microprocessor* is just another extension of this design? Yes, it does. When microcomputer applications are slowed down by the time that the 32-bit microprocessor takes to get data from external devices, a 64-bit microprocessor becomes a real solution. As we examine the internal architecture of some 32-bit microprocessors, we will see that the 64-bit architecture is already in use.

Self-Test

Answer the following questions.

5. Each new generation of microprocessors has a data word which has been _____ the length of the previous generation.
   a. One-half   d. More than
   b. 2^3 e. Double
   c. Three times   f. Less than

6. If the application you are working with requires an 8-bit microprocessor rather than a 4-bit microprocessor, you can expect that the cost of the parts which go with the microprocessor will be
   a. More expensive
   b. About the same
   c. Less expensive
   d. Dependent entirely on the application

7. You have been called out to service a microprocessor-based industrial product. The type of microprocessor is not specified, but you know the application performs very high speed 14-bit data transfers and that the customer had many of the systems for a number of years. Based on this knowledge you presume you must service a _____ -bit microprocessor system.
   a. 4   d. 16
   b. 8   e. 32
   c. 12   f. 64

8. The different microprocessor architectural features which tell how it performs in a particular application depend on its
   a. 16-bit characteristics
   b. Application
   c. 8-bit characteristics
   d. Word length

4-3 ADDRESSABLE MEMORY

As we discussed earlier, the amount of memory which a particular microprocessor can address is also part of its architecture. The early microprocessors had very limited memory addressing capability compared with current designs. The earliest 4-bit microprocessor, the Intel 4004, could only address 16,384 memory locations. It only had 14 memory address lines ($2^{14} = 16,384$). Most 8-bit microprocessors use 16 address lines and can therefore address 65,536 memory locations ($2^{16} = 65,536$).

Many of the single-chip microcomputers can only address the number of memory locations which are actually on the chip. Often, this is only 1 to 8K locations of ROM (*read-only memory*) for program storage and 64 to 512 RAM (*random-access memory*) locations for storing variable data. Other single-chip microcomputers are capable of addressing all of their internal memory locations and can also address enough memory locations outside the chip to give them a full 64-kbyte (65,536) memory address space.

How much memory-addressing capability is needed? Once again, we will find that the more addressing capability the microprocessor has, the greater the associated costs may be. The rule is still to avoid overkill. Therefore, we must ask the question, "How much memory do the various applications require?"

In Sec. 4-2, we looked at a number of applications which use 4-bit microprocessors. How much memory do applications like these need?

Before we look at an application, we must review the need for ROM and RAM. We know the microprocessor stores two different kinds of information in its memory. First, it stores the program steps it is to perform. Second, it stores variable data that it receives or generates from working on the data.

For example, let's look at a microprocessor which is performing the function of a digital alarm clock. Figure 4-3(*a*) shows a block diagram of such a system. The program steps are the instructions which tell the microprocessor how to tell time. They also tell it how to compare the current time to the alarm set time and how to ring the alarm if a match is found. These program steps must be stored in memory before the microprocessor knows how to start acting as a digital alarm clock.

Before the microprocessor can begin to work as a digital alarm clock, you must tell it at what time it is to start and at what time the alarm is to ring. You must enter data called the start time and data called the alarm time. This data will be in hours, minutes, and seconds, and it will have an AM or PM flag. The microprocessor will then follow the program and calculate new data—the current time. The program will also perform a check every second to see if the current time equals the alarm time. If the current time does equal the alarm set time and the alarm is turned on, the microprocessor rings the alarm.

How will this information be stored in the microprocessor's memory? We do not need to change the programming information. That is, once the program to calculate time is designed and proven to work properly, it can be permanently stored in the microprocessor's memory. However, the set time, the current time, the alarm time, the alarm On/Off flags, and clock mode flags are all data that can change. The set time changes each time the user starts the clock. The current time changes once each second. The alarm time changes each time the user wants to ring the alarm at a new time and the alarm On/Off flag changes each time the user wants to activate the alarm function. The

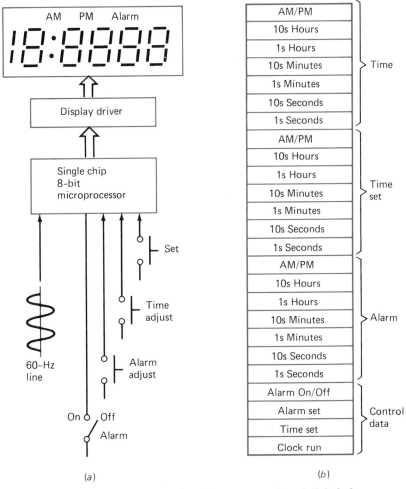

(a)　　　　　　　　　　(b)

Fig. 4-3　(a) The block diagram of a microprocessor-based digital alarm clock. The major parts are the display, the display drivers, the microprocessor, and the control switches. An input from the 60-Hz line is counted to give the microprocessor a time signal. (b) A ''memory map'' of the variable data used by the digital alarm clock. Each element of data uses 1 word in memory. The memory map shows us how the microprocessor's RAM locations are used.

mode flags change each time the user is setting the clock.

Because the program never changes, we can store this information in ROM. Because the time and alarm set data, the current time data and the flags change, this information must be stored in read-write memory. In today's terminology, this is usually called RAM. We can now ask the question, "How much ROM and how much RAM does my application require?"

In this example, the application program is the set of microprocessor instructions which tell the microprocessor how to perform the digital alarm clock function. A typical digital alarm clock application can be written in less than 1K instructions. Therefore, the microprocessor in this application will require a ROM which can store about 1000 instructions.

How much RAM will be required? To answer this question, we must look at the data which changes. The changeable data for our function is shown in Fig. 4-3(b). If we allow one memory location for each unit of information we store (4-bit BCD data), each time group requires 7 bytes (2 for hours, 2 for minutes, 2 for seconds, and 1 for the AM-PM flag). Additionally, we need four locations to store the control (Alarm On/Off, Alarm Set On/Off, Time SET On/Off and Run On/Off) data. This means we need a total of 25 RAM locations. In a real-life application, a few more RAM locations are needed for some programming requirements, but a 4-bit microprocessor with 64 RAM locations will do the job.

We find that our microprocessor-based digital alarm clock needs a 4-bit microprocessor with 64 RAM locations and 1000 ROM locations. In this application, we have no need for a microprocessor which has a memory address architecture that gives us greater capability. In other words, an 8-bit microprocessor with 64-kbyte addressing capability is truly overkill.

Let's now look at an application where we depend on a microprocessor architecture which has extensive memory-addressing capability.

Some of the microprocessors in use are very powerful. In fact, for many applications, they are so powerful that a number of users can use the microprocessor at one time. To do this, we need a microprocessor with extensive memory-addressing capability.

Even though it appears that many users are working with the microcomputer at the same time, the microprocessor is really only able to do one thing at a time. How does the microprocessor support multiple users? The answer

to this question is that the microprocessor has a *control program* which sees that each user is given a small slice of processing time. Once one user's time slice is over, the control program stores the current condition of the user's application and switches in the next user's application. In order to make the switch from one user to the next user very quickly, each user's application permanently resides in the microprocessor's memory. The control program only selects which area of memory it is currently executing instructions from.

Many of the application programs in use today, such as word processors, spreadsheets, and data bases, require a minimum of 128 kbytes of memory. If we are to build a multiprogramming system such as the one described above, and the system is to support four users, we require a minimum memory-addressing capability of 128 kbytes times four users, or 512 kbytes. Fortunately, applications of this sort are very common for both the Z80 8-bit microprocessor and the 8088 16-bit microprocessor. The Z80 architecture allows the processor to address 64 kbytes of memory. Unfortunately, this is not enough memory for even a single user! However, the 8088 has a 1-Mbyte (1,048,576 bytes) addressing capability. Therefore, this 16-bit architecture provides us with twice the memory-addressing capability we need. Figure 4-4 shows a memory map for such a system.

As we learned in Sec. 4-2, the 8-bit microprocessor is used in many control applications. The size of control application programs varies widely; however, it is common to find such control programs with sizes between 4 and 32 kbytes. The RAM requirements also vary widely; however, the typical control application does not usually have extensive requirements for read-write storage. Often, you will find such 8-bit control microprocessor applications with either 2 kbytes or 8 kbytes of RAM.

When the 8-bit microprocessor is used as a general-purpose computer, it is best to have as much RAM as possible. Of course, the microprocessor needs some initial program to get started. This initial program is called a *boot program*. The boot program must be in a small ROM, which may be replaced by RAM once the microprocessor has been booted. Once booted, the microprocessor program loads additional programs from mass storage into the RAM. From that point forward, the microprocessor can continue to execute programs from RAM until power is lost.

1024K

Other system-
control
memory

640K

User 4
memory space
128 kbytes

512K

User 3
memory space
128 kbytes

384K

User 2
memory space
128 kbytes

256K

User 1
memory space
128 kbytes

128K

Multiuser
system-control
program

0K

Fig. 4-4   A memory map for an 8088 multiuser system. This system has the multiuser control program in the lowest 128 kbytes and has four 128-kbyte partitions set aside for four different users.

From these two examples, we see that there are two common ROM-RAM combinations used with the 8-bit microprocessors. In the first example, we learned that control systems typically use a large amount of ROM to store the permanent control program and a small amount of RAM to store variable data. In the second example, we learned that general-purpose systems use the maximum possible amount of RAM. This is done because the general-purpose microcomputer does not execute any fixed program. The general-purpose microcomputer is designed to load executable programs from a mass storage device into RAM whenever the user decides to use a particular program. However, when the microcomputer is first powered up, there is no program stored in the RAM. Therefore, the first executable program, called the boot program, must be stored in ROM.

In the first example, we see that the amount of memory-addressing space required by 8-bit control applications can vary from a few kilobytes to the full 64-kbyte addressing limit of the microprocessor. In the second example, we see that the amount of memory-addressing space required by 8-bit general-purpose applications is usually the full 64-kbyte addressing limit of the microprocessor.

In a similar manner, we find that the memory-addressing requirements for micropro-

cessors used in 16-, 32-, and 64-bit applications can vary as well. When these microprocessors are used in a general-purpose application such as the basis for a PC, the maximum amount of memory possible is desirable. Many of the newer applications programs written for the PC require many hundreds of kilobytes of memory just to hold the application program alone. These applications programs may well require additional hundreds of kilobytes of memory to accommodate the data generated by or used by the applications program.

The 8088 architecture, as you have learned, allows this microprocessor to address 1 Mbyte of memory. When the 8088 is used in the PC, some of the memory address space is devoted to RAM. Other sections are devoted to the ROM, which contains the initialization (boot) programs and other permanent programs (called the BIOS) which the microcomputer always needs to be able to access. Figure 4-5 shows how the 8088 memory space is used in a PC. Such a diagram is called a *memory map*. This shows the user the microprocessor's memory architecture.

When 16-bit microprocessors are used in control applications, the memory requirements are not usually as great as they are in the general-purpose application. If the application and system programs are fixed, then a great

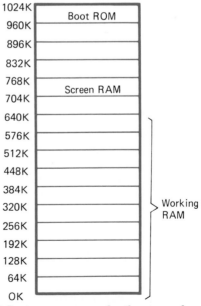

Fig. 4-5   A memory map for the personal computer (PC). The maximum working RAM is 640 kbytes. Note that the 8088 microprocessor can address 16 times as much memory as the 8-bit microprocessors.

Clock speed

Millions of
instructions per
second (MIPS)

Large-scale
integrated circuit
(LSI)

deal of the memory space may be occupied by ROM. If the application programs change frequently, then the memory space will be occupied by RAM so that the user can change the program which is to be executed. This, for example, is common in process control computers which may be used to monitor and control a number of different processes. In this example the 16-bit microprocessor architecture may be chosen to provide rapid calculations and access to a large program-addressing space.

Two major 32-bit microprocessors, the 68020 and 80386, have architectures which allow them to address 4 Gbytes of memory (1 Gbyte equals 1 billion bytes). These architectures are intended to support large multiprogramming applications or scientific applications which generate very large amounts of data.

Self-Test

Answer the following questions.

9. The typical 8-bit microprocessor can address _____ different memory locations.
   a. 14K   d. 512K
   b. 32K   e. 1M
   c. 64K   f. 4M

10. If you add a calendar function to the digital alarm clock, you increase the number of RAM locations that are needed. Explain why more locations are required. Approximately how many additional locations do you think might be required for the calendar function?

11. Does adding the calendar function in question 4-10 change the type of microprocessor you need to do the job? Explain your answer in terms of the microprocessor's architectural characteristics you have learned about so far.

12. Explain why the memory-addressing requirements are different for a microprocessor used in a control application and one used in a general-purpose computing application.

4-4 THE MICROPROCESSOR'S SPEED

The speed with which a microprocessor executes its instructions is often used as a way to compare the performance of two microprocessors. Speed may be measured in two different ways. The first is the microprocessor's *clock speed*. This is the frequency at which the microprocessor's clock oscillator operates. Typically this is expressed in megahertz (1 megahertz (MHz) equals 1 million cycles per second). A second method of rating a microprocessor's speed is the number of instructions it can execute per second. Since the job we are asking the microprocessor to accomplish is the execution of instructions, this becomes a good measure of its processing speed. Processing speed is usually expressed in *millions of instructions per second* (MIPS).

As is frequently the case, no measure of a microprocessor's power is exact. If you use MIPS, you must ask what instructions the processor is being asked to execute. Some instructions take longer than others to execute. Therefore, the MIPS rating for a microprocessor can only tell you the average processing rate for a fairly large program.

The data in Fig. 4-6 shows the clock frequency and the MIPS rating for some common microprocessors. As you can see, the MIPS rating depends on both the clock frequency and the word size. As the clock frequency is increased, the microprocessor executes more instructions per second. However, you should note that the number of instructions that a Z80, for example, executes in 1 s only improves by 50 percent as we increase the clock speed from 4.0 MHz (Z80A) to 6.0 MHz (Z80B). However, if we can double the word size and keep the clock frequency the same, the MIPS rate is nearly doubled. This is so because the average instruction length is greater than 1 byte. This means that, on the average, the 8-bit microprocessor must access memory at least twice to fetch a complete instruction. On the other hand, the 16-bit microprocessor only needs one memory access to fetch the average instruction. If both microprocessors perform a memory access in the same amount of time, then the 16-bit microprocessor will be twice as fast as the 8-bit microprocessor.

The frequency at which the microprocessor's clock operates depends largely on the semiconductor manufacturing technology used to build the microprocessor. The semiconductor manufacturing processes which are used to build a *large-scale integrated* (LSI) circuit are very much like the processes used to develop a photograph. The frequency at which the microprocessor will operate and the complexity of the microprocessor depend on how many transistors can be put on the IC. Microprocessors and other LSI circuits which must oper-

Microprocessor	Clock frequency	Internal word length	Average processing MIPS
8051	12 MHz	8-bit	0.58
Z80A	4 MHz	8-bit	0.3
Z80B	6 MHz	8-bit	0.45
286	16 MHz	16-bit	3.0
486DX2-66	66 MHz	32-bit	32
Pentium	66 MHz	32-bit	64

Fig. 4-6   Comparing the clock frequencies and MIPS (millions of instructions per second) rating for different microprocessors. *Note: The MIPS rating depends on the average number of processor cycles the microprocessor uses for an instruction and the machine cycle time (the time needed for the simplest fetch/execute cycle).*

N-channel metal-oxide semiconductor (NMOS)

Complementary metal-oxide semiconductor (CMOS)

Very large scale integration (VLSI)

ate at a high frequency must be made with a semiconductor manufacturing process that makes very small parts. The processes used to build the very fast and very dense microprocessors today are able to make parts as close as a few tenths of one millionth of an inch.

Most microprocessors are constructed using one of two semiconductor technologies. These two technologies are NMOS and CMOS. *NMOS* stands for *N-channel metal-oxide semiconductor. CMOS* stands for *complementary metal-oxide semiconductor.*

NMOS processors typically allow clock frequencies from a few megahertz to nearly 20 MHz. The NMOS process produces the less expensive devices, but the NMOS devices typically draw more power (current) and therefore run hotter. Devices produced using the CMOS technology are much lower in power, will operate over a wider temperature and voltage range, and, as noted, cost more. There are some *VLSI (very large scale integration)* circuits which have so many circuits on them that they cannot be produced by NMOS technology because they would run so hot they would burn up. These devices must be produced by CMOS technology. The Intel 386 is an example of such a device.

CMOS processors typically will operate over a greater frequency range than the NMOS devices. In both cases, an increase in operating frequency requires a like decrease in the size of the VLSI's internal parts. As the requirement for processor speed increases, the need for very difficult to make submicrometer semiconductors increases. This means that increased processor frequencies are becoming harder and harder to come by. Clock frequencies of 20 to 100 MHz are common, and a few microprocessors are reaching clock frequencies near 300 MHz.

Self-Test

Answer the following questions.

13. The Intel 8051 single-chip microcomputer uses a 12-MHz clock. One 8051 machine cycle (the time to complete the simplest fetch/execute cycle) requires 12 clock periods. Therefore, a program written to use only simple fetch/execute cycles would run at a speed of _____ MIPS.
    a. 12   c. 2
    b. 6   d. 1
14. Explain your answer to question 13.
15. You are working with a microprocessor system which is intended for use in an automotive environment. List three characteristics of CMOS microprocessors which would lead you to use them over an NMOS device for this application.
16. If MIPS stands for millions of instructions per second, what do you think KIPS stands for? Do you think KIPS is coming into more common usage or going into less common usage?

4-5 OTHER MICROPROCESSOR ARCHITECTURAL CHARACTERISTICS

Some other microprocessor architectural techniques being used to increase speed include

- Parallel processing
- Coprocessing
- Cache memory techniques
- Pipelining techniques
- Wider buses

Although some of these techniques are only just beginning to be used, others are quite advanced. This section will briefly introduce you to these techniques.

Parallel processing means just what it says. It is a description of any microprocessor architecture which permits two parts of the computing process to occur at the same time. One of the most obvious implementations of parallel processing involves the use of two microprocessors in the same product. To be considered parallel processing, both the microprocessors must be working on a job which could be done by a single microprocessor but which can be done faster with the two microprocessors. Parallel processing is also used within the microprocessor itself. In the course of designing the control logic to execute an instruction, the microprocessor architect looks for a way to divide the execution process into two sections. The architect looks for processes which if done by a single element of the control logic would have to be done one after the other, but which can be done at the same time if two pieces of control logic are provided. Figure 4-7 shows a parallel processing configuration using two microprocessors. An advanced microprocessor architecture may include multiple execution units within the microprocessor design. This form of parallel processing is called a *superscalar architecture*.

Coprocessing is very much like parallel processing. The coprocessor is a separate processor which works with the main processor to perform special functions. Because the coprocessor is typically limited to special function processing, it can be built to perform those particular functions much faster than the general-purpose processor might.

Many conventional microprocessors are supported with a separate IC which performs a coprocessing function. For example, the 386 is supported with the 80387 numeric data coprocessor. The 80387 is wired in parallel with the 386 microprocessor. When the 80387 is in place, the combined system has extra high-level mathematical instructions which let it perform multiple-precision addition, subtraction, multiplication, and division much faster than can be done with the 386 alone. Additionally, the 80387 coprocessor gives the 386 other mathematical functions such as logarithms, trigonometric functions, and square roots. A 386 system using an 80387 numeric data coprocessor is shown in Fig. 4-8.

Other coprocessors used within a microcomputer system may be no more than dedicated microprocessors assigned a specific task such as managing the serial I/O, disk drives, and video display.

Part of the microprocessor speed limitation which faces every microprocessor architect is the speed with which the memory can supply the data the microprocessor requested. One technique to overcome the problems of slow memory is to use a *cache memory*. A cache memory is a very fast memory. This fast memory may be located very close to the microprocessor, or it may be included within the microprocessor on new microprocessor architectures. To use the cache memory, an instruction fetch, for example, not only fetches the instruction at the addressed memory location but also fetches enough of the next location's contents to fill the cache memory. After the initial fetch, the microprocessor will always look in the cache memory first because it is a much faster-responding memory. If the instruction is not found in the cache memory, the microprocessor then looks in the slower main memory for the instruction.

To offer a speed improvement, the cache memory must have more hits than misses. That is, the cache memory only offers a speedup if it can offer enough cached instructions to overcome the time lost to instruction fetches which must be looked for twice—first in the cache memory and then in the main memory, because they were not in the cache memory. Figure 4-9 shows a bus-oriented microprocessor system which uses a 16-kbyte cache memory to provide a speed improvement over operations with the main 4-Mbyte memory.

Pipelining is another architectural technique

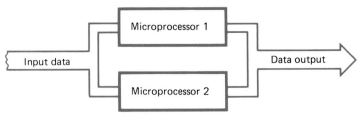

Fig. 4-7 A system which uses two microprocessors in parallel to process incoming data. Special circuits may be needed to split the data between the two microprocessors and to combine the data once they are finished with the data processing.

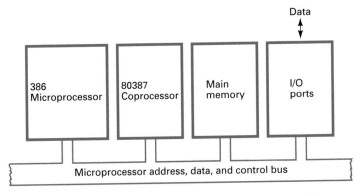

Fig. 4-8 A 386 with an 80387 numeric data coprocessor. The 80387 is assigned the tasks of special arithmetic processes, which it does in parallel with the work that the 386 is doing.

where the name describes the function it performs. Like the assembly line, the pipelined architecture processes a number of instructions at one time. Each instruction is processed a little at each station just like a product is built a little at each station on an assembly line. The use of a pipelined architecture allows the designer to build faster, more dedicated logic at each station and therefore the microprocessor completes the desired process quicker.

As you have learned earlier, when the microprocessor has the ability to process wider data, it is naturally faster. Wider data buses are used inside the microprocessor as well as in the microcomputer design. We have seen examples of this architectural technique when microprocessors use double-wide accumulators. The double-wide accumulator allows the user to process certain functions faster because it eliminates the need to perform an intermediate storage in memory.

In summary, we find there are many ways to speed up the microprocessor. If possible, we just make it go faster by increasing the microprocessor's clock frequency. This is not done without expense, because parts which work at higher frequencies cost more. If we cannot

speed up the microprocessor's clock frequency, there are a number of architectural techniques which add hardware to the microprocessor but which result in faster operation.

Again, remember that performance costs money. If you don't need the speed, don't select a high-speed microprocessor. You will be paying for performance you don't need.

Self-Test

Answer the following questions.

17. To improve both execution speed and reliability, some microcomputer systems are built to run with multiple CPU cards. The more cards that are plugged into the system, the faster it runs. On the other hand, should a CPU card fail, the system only slows down, it does not quit entirely as a single CPU system does. This system uses
    a. Coprocessing
    b. Pipelining
    c. Parallel processing
    d. Cache memory
    e. A wide bus
    f. Z80 microprocessors

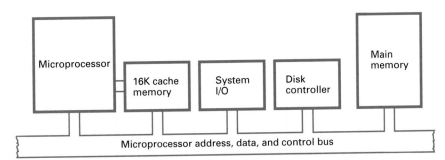

Fig. 4-9 A microprocessor system which uses a 16-kbyte cache memory to speed up memory operations. Without the cache memory, all memory operations for this system must work with the 4-Mbyte memory connected to the system's bus.

18. One procedure frequently used in scientific calculations is array processing. This algebraic technique is useful in solving complex problems with many variables. Often, scientific mini- and microcomputers have a special array processor added to the main CPU to assist with solving the array problems. This is an example of
    a. Coprocessing
    b. Pipelining
    c. Parallel processing
    d. Cache memory
    e. A wide bus
    f. Z80 microprocessors

19. As we will learn in Chap. 12, the disk is a common mass storage device. Its speed of access is limited by the rotational speed of the disk. Therefore, it is common to read large sections of the disk into a special area of memory. Before the disk is accessed, this memory is checked to see if the information may already be in the fast memory. If it is, one disk access is avoided. This is an example of
    a. Coprocessing
    b. Pipelining
    c. Parallel processing
    d. Cache memory
    e. A wide bus
    f. Z80 microprocessors

20. When an 8-MHz Z80 is compared with an 8-MHz 80286, the 80286 is twice as fast for simple data transfers. This is an example of
    a. Coprocessing
    b. Pipelining
    c. Parallel processing
    d. Cache memory
    e. A wide bus
    f. Z80 microprocessors

21. A special Z80-based system uses three microprocessors to perform a signal analysis function. Each microprocessor performs a part of the signal analysis function so that when the signal data has been through all three microprocessors the signal is fully analyzed. It is diagrammed in Fig. 4-10. This is an example of
    a. Coprocessing
    b. Pipelining
    c. Parallel processing
    d. Cache memory
    e. A wide bus
    f. Z80 microprocessors

4-6 THE MICROPROCESSOR'S REGISTERS

The microprocessor's registers are an important and distinctive part of its architecture. The registers are temporary data storage devices found inside the microprocessor. Some have special, single-purpose functions and others are used for general purpose data storage. In the next chapter, we will learn that there are certain basic registers in every microprocessor. We will learn that a microprocessor may have many registers beyond these basic registers. We look at registers that are used in a number of different microprocessors. At this time, it is only important to understand that the number and type of registers in any particular microprocessor will be different from those in another microprocessor.

The number of registers in the microprocessor is important because the more registers that the programmer has available, the more registers there are to assign to specific tasks. If the microprocessor architecture does not have enough registers to allow the programmer to assign registers to a specific task, the programmer must transfer the data in the register into a temporary memory location for storage when the task is complete. The data is then retrieved from the temporary location when the task is going to use the register again. Each of the data transfers from a register to memory or from memory to a register are wasted instructions.

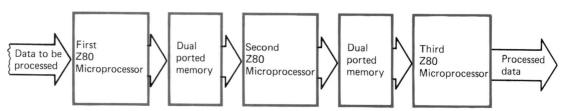

Fig. 4-10 **The system described in question 4-21. This system uses three Z80 microprocessors. Each performs a part of the data processing work and passes a partial result to the next microprocessor through the common dual-ported memory. The final fully processed data comes from the third microprocessor.**

The types of registers available to the programmer can also represent a performance issue. There are many ways to determine what memory location the microprocessor should be addressing. Certain types of registers can be provided which make the job of selecting the memory address for program data easier than could be done without these registers. For example, the *index register* allows the programmer to add a word to the current index register value to compute a memory address value. In a microprocessor without an index register, the memory address value must be computed using conventional registers, which use more instructions to get the same job done.

There are, as you will learn, many different types of registers in a microprocessor. Not all microprocessors have all the registers, and there are some types of registers which do not exist in a microprocessor that has other types of registers. That is, you cannot have both types in one microprocessor. As with any other performance improvement, additional registers cost money. Registers take space and cause other increases in the microprocessor's complexity. Both of these factors may call for additional parts to be used outside the microprocessor as support.

Self-Test

Answer the following questions.

22. Increasing the number of registers available to the programmer increases the programmer's
    a. Clock speed
    b. Productivity
    c. Number of languages
    d. External chip count
23. A microprocessor with a large number of registers probably _____ than a microprocessor with fewer registers.
    a. Uses less power
    b. Is more efficient
    c. Costs more
    d. Uses more memory

4-7 THE MICROPROCESSOR'S INSTRUCTIONS

Like the registers, the number and kind of the microprocessor's instructions are a measure of the microprocessor's power. Additional microprocessor instructions can only be in the microprocessor if the microprocessor has the necessary logic circuits to execute the instruction. This means that more instructions mean a bigger microprocessor, and bigger microprocessors are more complex.

For example, comparing the instruction sets of an 8-bit single-chip microprocessor—the Intel 8051- and the popular 8-bit Zilog Z80—we find that the 8051 has 111 different instructions and the Z80 has 178. What additional instructions do we get when we use a microprocessor instead of a single-chip microcomputer?

When we examine the two instruction sets, we find that the Z80 offers some extended capabilities. First, it is able to perform a number of 16-bit data transfers between register pairs and between register pairs and memory. To do this, the microprocessor must have special control logic to recognize a register pair instruction and to execute the instruction in two parts (the High byte and the Low byte). Second, the Z80 instruction set includes a series of block transfer and block search instructions. These instructions let the programmer move data in blocks instead of 1 byte at a time. The programmer specifies the beginning and ending addresses for the block and then the start address where the block of data is to move. The Z80 can also search a selected block of memory for specified data.

Both the 16-bit data transfer instructions and the block move instructions are made possible by the addition of special logic to the Z80's control logic and memory addressing logic. Needless to say, the 8051 does not have this logic: thus it does not have the instructions. For the programmer with a need to perform 16-bit data transfers or block moves, the Z80 has a significant advantage over the 8051.

On the other hand, because of the special circuits, the 8051 does have some instructions that the Z80 does not have. The 8051 has a built-in serial port and therefore has the instructions to receive or transmit data via this serial port. The 8051 costs less than half as much to put into a system as the Z80 does, and it is more powerful in that respect than the Z80.

Self-Test

Answer the following questions.

24. When we compare the Z80 with the 8051, we find that the 8051 does have a few instructions not found in the Z80. These instructions are in the 8051 to
    a. Use more control logic
    b. Implement block data moves
    c. Fill memory

d. Support its special internal hardware features

25. You would expect the Intel 80286 16-bit microprocessor to have _____ the Z80 8-bit microprocessor.
    a. Twice as many instructions as
    b. The same number of instructions as
    c. One-half as many instructions as
    d. More instructions than

4-8 MEMORY ADDRESSING ARCHITECTURE

Part of the programming flexibility for each microprocessor is the number and different kinds of ways the programmer can refer to data stored in memory. At first, one would think that all memory reference instructions in an 8-bit microprocessor would simply be a byte of instruction followed by 2 bytes of memory address. Although most 8-bit microprocessors do have such a direct addressing mode, it is not the most efficient. In the 8-bit microprocessor, for example, such a direct addressing instruction takes 3 bytes. If one uses the indexed addressing discussed in Sec. 4-4 the complete instruction becomes 1 byte of instruction and 1 byte which is either added to or subtracted from the index register. Of course, indexed addressing is only available to the programmer if an index register is available.

In general, it is fair to assume that the more complicated the microprocessor's architecture, the more memory addressing modes there will be. For example, Fig. 4-11 shows the different kinds of addressing modes available to three

of the five microprocessors studied in this book. As you can see, as the complexity of the microprocessor grows, the number of different addressing modes also grows.

Once again, the need for a wide variety of addressing modes depends on the job you are doing with the microprocessor. For simple control applications which have, for example, 2 kbytes of ROM and 4 kbytes of RAM, you may well not require great flexibility in the addressing. On the other hand, when you are building the control system for a sophisticated data gathering and signal analysis instrument, you may need extensive memory-addressing capability to build and then analyze large arrays of signal data in memory.

Self-Test

Answer the following questions.

26. Why does the 8-bit microprocessor take longer to fetch and execute a directly addressed instruction than it does to fetch and execute an indexed instruction?

27. Which application might make greater use of highly flexible addressing modes, the 4-bit microprocessor-based scientific calculator or the same 4-bit microprocessor used to support a hand-held video game? Why?

4-9 THE MICROPROCESSOR'S SUPPORT CIRCUITS

The microprocessor's architecture does not just include its internal structure; it also includes other circuits and devices intended to

Addressing mode	Microprocessor		
	6800	Z80	8088
Implied	X	X	X
8-bit	X	X	X
16-bit direct	X	X	X
8-bit immediate	X	X	X
16-bit immediate	X	X	X
8-bit relative	X	X	X
8-bit indexed	X	X	X
16-bit indexed			X
Bit		X	X
8-bit indirect			X
16-bit indirect		X	X
16-bit computed			X
8-bit I/O		X	X
16-bit I/O			X

Fig. 4-11 Memory and I/O addressing modes found on three microprocessors of increasing complexity. The 6800 and Z80 are 8-bit microprocessors and the 8088 is an advanced 8-bit (external) microprocessor with an internal 16-bit structure.

work with the microprocessor. For example, to be useful, the microprocessor must communicate with the outside world. As we will learn in detail in Chap. 13, this is often done with a special I/O subsystem called *serial I/O*.

What does the microprocessor use to perform the serial I/O function? Although the serial I/O function can be built from gates and flip-flops, it almost never is. Instead, the architects of microprocessor systems use a special LSI circuit known as a *UART*.

There are two different kinds of UARTs available to the microprocessor system architect. During the design process one can use a general-purpose UART or, if the microprocessor's architecture includes a special UART developed for that microprocessor, that UART may be used. The special UART often includes many bus- and system-related features which make it much easier than the universal UART to connect to the microprocessor.

Special support circuits included in the microprocessor's architecture are not limited to UARTs. The table in Fig. 4-12 shows the different types of support circuits available to support the Z80 and the 80286. As you can see,

these support parts are a major portion of these microprocessors' architectures. Often, the support circuits are so important that they will cause the microprocessor system designer to choose one microprocessor over another.

Self-Test

Answer the following questions.

28. Why would you expect the number of integrated circuits used to implement a particular microprocessor-based application to be fewer or greater if general-purpose LSI circuits are used instead of microprocessor-specific circuits?
29. Which implementation in the above question would you expect to cost more? Why?

4-10 MICROPROCESSOR DEVELOPMENT AND MAINTENANCE SYSTEMS

As we saw in Sec. 4-9, not all of the microprocessor's architectural features are limited to those inside the chip itself. The systems used

Z80 Support circuits		
Device number	Device description	Package
8410	Direct memory access controller	40-Pin DIP
8420	Parallel input/output controller	40-Pin DIP
8430	Counter timer circuit	28-Pin DIP
8440	Serial input/output controller	40-Pin DIP
8470	Dual-channel asynchronous receiver transmitter	40-Pin DIP
8530	Serial communications controller	40-Pin DIP

8088/80286 Support circuits		
Device number	Device description	Package
8087/80287	Arithmetic coprocessor	40-Pin DIP
8116	Dual baud rate clock generator (programmable)	18-Pin DIP
8202	Dynamic RAM controller	40-Pin DIP
8224	Clock generator/driver	16-Pin DIP
8250	Asynchronous communications element	40-Pin DIP
8253	Programmable interval timer	24-Pin DIP
8272	Floppy disk controller	40-Pin DIP

Fig. 4-12 Some of the different support circuits available for the popular Z80 8-bit microprocessor and the popular 8088/80286 16-bit microprocessors. These circuits are implemented using LSI techniques, so quite complex circuits can be found in one IC.

Micro[...]
develo[...]
system[...]

Assem[...]

Cross [...]

In-circ[...]
(ICE)

24. Cache memory is a special memory which is used to store a section of data so that it can be accessed quickly by the processor. The effectiveness of cache memory depends on how many times the processor is able to make a hit in the cache memory.

25. Pipelining describes an architectural feature which allows the process to be worked on in assembly line fashion. This means that the process is broken up into substeps which are then worked on by multiple processors.

26. Processors which have a wider bus than other processors have increased data-handling capacity because they move more data per microprocessor cycle.

27. The number of registers that a microprocessor has is an important feature. Registers give the programmer flexibility to hold and process data without having to address memory. Registers can also give the programmer memory-addressing flexibility.

28. A microprocessor will have additional instructions to support additional features. These instructions allow the programmer to use and control these features. The more complex the microprocessor, the more complex its instruction set is likely to be.

29. One of the other features offered by microprocessors is advanced memory addressing. The instructions and registers allow the programmer additional programming flexibility.

30. A list of a microprocessor's architectural features should include the support circuits which go with it. These circuits allow the designer to create a complex system by using fewer parts.

31. The available microprocessor development tools are an important part of the microprocessor's architecture. The development system tools include the programming software and the special instruments which allow the developer/maintainer to see what is happening inside the microprocessor.

CHAPTER REVIEW QUESTIONS

Answer the following questions.

4-1. Briefly describe what we mean by the microprocessor's architecture.

4-2. Can the architecture of one microprocessor be different from that of another microprocessor if they both can perform an add-with-carry function? Explain your answer.

4-3. List the three major architectural features which separate one microprocessor from another.

4-4. List four other architectural features which differ from one microprocessor to another.

4-5. List three popular word lengths used in today's microprocessors.

4-6. If low cost is your objective, would you be more likely to select a 4-bit or an 8-bit microprocessor? Why?

4-7. List three applications commonly used for 4- and 8-bit microprocessors by data word length.

4-8. What is the major application that you will find the 16-, 32-, or 64-bit microprocessor used in?

4-9. Both CAM and CAD systems use sophisticated graphic displays. Typically, you will find that these applications use either _____ or _____ -bit microprocessors.

4-10. If you are working with a 32-bit microprocessor, you would expect to find that it has _____ or more bytes of main memory-address space.
   a. 16K   *c.* 1M
   b. 64K   *d.* 4M

4-11. If you were asked to service a moderate-cost industrial controller which had been installed in 1980, you would expect to find it based on an 8-bit microprocessor. Why?

4-12. If you were asked to service an expensive CAE workstation built in 1994, which performed complex mathematical functions on large data arrays, you would expect to find that it is based on a 64-bit microprocessor. Why?

4-13. As the microprocessor word length increases, it takes _____ (more/ fewer) microprocessor cycles to perform a data transfer operation.

4-14. If the data you are to store in memory can change while the microprocessor is performing its function, you must use _____ (ROM/RAM) for this data.

4-15. Usually, a microprocessor-based product which does not have any mass storage device to load a program from uses _____ (ROM/RAM) to store the program steps.

4-16. If the microprocessor-based system loads the program steps into RAM from a mass storage device, a ROM is used to hold the microprocessor's _____ program.

4-17. We can compare the speed of two microprocessors by comparing their _____ frequency or by comparing their instruction execution rate, which is given in _____ .

4-18. The two semiconductor technologies typically used to build microprocessors are _____ and _____ .

4-19. Typically, processors built with _____ have lower power than those built with _____ , which is known for its simpler parts and, therefore, lower cost.

4-20. Give a brief description of what is meant by parallel processing, coprocessing, cache memory, pipelining, and bus width.

4-21. Briefly explain how each of the architectural features listed in the question above can improve the operation of a microprocessor or a microprocessor-based system.

4-22. How can additional registers help the programmer write more efficient programs?

4-23. You expect that a microprocessor with additional features will have additional _____ to allow the programmer to use and control those features.

4-24. Generally speaking, you would expect the more complex microprocessors to have _____ (more/fewer) addressing modes than the less complex microprocessors.

4-25. Explain why the external support circuits available for a particular microprocessor are also a part of its architectural features.

4-26. Explain why the richness of the microprocessor's development tools are of interest to the person looking for a full-featured microprocessor.

CRITICAL THINKING QUESTIONS

4-1. The phrase *a microprocessor's architecture* does not refer to the physical package in which the IC is installed. Why?

4-2. Can you describe some of the microprocessor applications that make the best use of a microprocessor with a 32- or 64-bit word length? Why do they use these long words?

4-3. Describe some microprocessor applications that do not need to use a microprocessor with an ability to address large amounts of memory.

4-4. Suppose that you are purchasing a replacement microprocessor-based device to sample data once per second, compute, and display a result. Why is it most likely that you will not spend extra money to get a replacement unit with a 33-MHz clock instead of the 20-MHz clock which was on the original device?

4-5. Until recently the world of supercomputers consisted of a few multi-million-dollar computers which used very fast super-cooled digital circuits to create the ALU. Today, many supercomputer jobs are being done by systems which use massively parallel processing techniques. What do you think the term *massively parallel processing* means?

4-6. As the length of a microprocessor's data word increases, you would expect the size of the microprocessor's instruction set (the number of instructions available to the programmer) to increase. Why?

Answers to Self-Tests

1. *e*	4. *a*	7. *d*
2. *b*	5. *e*	8. *b*
3. *b*	6. *a*	9. *c*

10. More RAM locations are required because you must store the day, month, and year data. This is data which changes so it must be stored in RAM. If you allow 1 byte for each numeral which can be stored, you will need at least 6 bytes to store DD, MM, and YY.

11. No it probably will not. The microprocessor-based alarm clock only used a limited amount of ROM and RAM. The addition of the calendar function will not cause the programming to exceed the addressing capability of the microprocessor. Additionally, it will not cause the microprocessor to make too many calculations and exceed its speed capabilities.

12. In a control application, you usually know how big the program is and how big the variable data will be. In a general-purpose microcomputer, the size of the memory used by the application programs depends entirely on the program being written. Therefore, in a control application, you can usually use only enough addressing capability to do the specific job it is intended to do, and that is usually less than the addressing capability of most modern microprocessors. However, in a general-purpose application, you can never have enough memory and therefore can use all the microprocessor's addressing capability.

13. *d*

14. The microprocessor's clock runs at 12 million times per second or 12 MHz. The microprocessor's internal logic requires 12 of these cycles to complete a simple fetch/execute cycle, or, in other words, it can complete 1,000,000 fetch/execute cycles in a second, giving an execution speed of 1 MIPS.

15. The CMOS microprocessor, because it uses little power, would cause minimal drain on a car battery if the device is left on when the car is not running to charge the battery. Additionally, the CMOS microprocessor operates well over a wide range of temperatures, and a car can get very cold in winter and very hot in summer. The supply voltage in an automobile can vary over a wide range, especially when the car is being started, and can contain a lot of noise as motors start and stop, relays pull in, and other devices operate. A CMOS microprocessor works much better than an NMOS microprocessor in this environment.

16. KIPS is kilo (thousands) of instructions per second. The term is only used with older processors and is therefore becoming less common.

17. *c*	20. *e*	23. *c*
18. *a*	21. *b*	24. *d*
19. *d*	22. *b*	25. *d*

26. The microprocessor uses a 3-byte instruction to directly address a memory location, and this takes three processor cycles to fetch the instruction. On the other hand, the microprocessor only needs 2 bytes for an indexed instruction.

27. The calculator application can have some modes where a great deal of data is being worked with at one time, and therefore the microprocessor will be using all of its available addressing modes.

28. The general-purpose LSI circuits will usually have to be supported with some SSI and MSI logic circuits to make them work with the microprocessor. On the other hand, the special-purpose LSI circuits designed to work with the particular microprocessor should not need any extra logic to connect them to the microprocessor.

29. If the special-purpose circuits are for a high-volume microprocessor, they will cost about the same as the general-purpose circuits. Therefore the implementation with the general-purpose circuits will cost more.

30. The in-circuit emulator allows the person developing or maintaining the microprocessor-based system to see what is going on in the microprocessor as the programs are executing.

31. The programming languages are used to develop the programs which give the microprocessor-based system the intelligence to solve the problem it was designed to solve.

CHAPTER 5

Inside the Microprocessor

■

CHAPTER OBJECTIVES

This chapter will help you to:

1. *Identify* the major parts of a microprocessor block diagram.
2. *Explain* the difference between a microprocessor block diagram and its programming model.
3. *Explain* the basic functions of the microprocessor's arithmetic logic unit, registers, control logic, and internal data bus.
4. *Describe* the primary registers found in most microprocessors and *explain* their functions.
5. *Identify* the seven status register bits common to most microprocessors.
6. *Define* the terms subroutine, stack, last-in-first-out operation, decrements, and clock.

Most microprocessors use a fairly common internal construction consisting of the ALU, the registers, and the control logic. Once you understand how these parts go together, you can begin to understand how to use programming to make those parts work together to solve a problem. In later chapters, we will learn that there are differences in just how many of these parts are in certain microprocessors and how they are interconnected.

■

5-1 THE MICROPROCESSOR BLOCK DIAGRAM AND PROGRAMMING MODEL

The microprocessor *block diagram* and the microprocessor *programming model* show the user how a specific microprocessor is constructed. The block diagram shows a microprocessor's logic functions for data processing and data handling. The block diagram also shows how each of these logic functions is connected to all the others. Referring to a block diagram often makes it easier to see how to use a microprocessor in hardware and system design situations. The programming model assists you in the programming process. The difference is that the programming model shows only those parts of the microprocessor which the programmer can change.

In this chapter, we will use the block diagram shown in Fig. 5-1 on page 72 and the programming model shown in Fig. 5-2 on page 73. The 8-bit microprocessor shown is only an example. It is a subset of a commercially available microprocessor, which we will study later. We will work with a number of commercially available microprocessors in later chapters. This subset microprocessor will only be used to introduce some fundamental microprocessor concepts.

When you work with microprocessors, you use the block diagrams and programming models provided by the microprocessor's manufacturer. The block diagram makes it easier to understand the architecture of a microprocessor. The programming model makes it easier to understand the microprocessor in a programming environment. Normally, you will use the

From page 71:

Block diagram

Programming model

On this page:

Programming model

Arithmetic logic unit (ALU)

block diagram to become familiar with the architecture of a new microprocessor. You will use the programming model regularly as you write software for the microprocessor.

Returning to Fig. 5-1, we find three major logic devices in the microprocessor's block diagram: (1) the ALU, (2) several registers, and (3) the control logic. Also shown is the microprocessor's system for transmitting data from one of these logic devices to another: the microprocessor's internal data bus. In Fig. 5-2 the programming model only shows some of the registers. Again, these registers are the only ones a person programming the microprocessor can control.

In the following sections, we will look at the major parts of the microprocessor. As each part is discussed, a block diagram with color shows the part. If the part also appears in the programming model, a color *programming model* is also shown.

Self-Test

Answer the following questions.

1. Which diagram is more detailed, the programming model or the block diagram? Why?
2. If the microprocessor you are using has dual accumulators, you would expect both of the accumulators to appear in the microprocessor's
   a. Block diagram
   b. Descriptive literature
   c. Programming model
   d. All of the above

5-2 THE ALU

One of the microprocessor's major logic functions is the *arithmetic logic unit (ALU)*. It contains the microprocessor's data processing logic. Figure 5-3 on page 74 shows the ALU

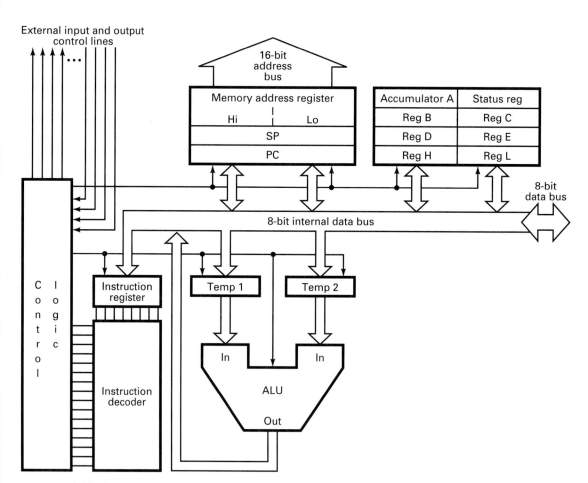

Fig. 5-1 **A block diagram for the microprocessor used as a model in this chapter. This 8-bit microprocessor is used in this chapter to illustrate basic microprocessor features. These features will be found in many different forms as the more complex actual microprocessors are examined in later chapters.**

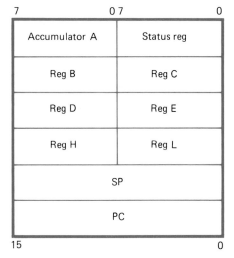

7	0	7	0
Accumulator A		Status reg	
Reg B		Reg C	
Reg D		Reg E	
Reg H		Reg L	
SP			
PC			

15   0

Fig. 5-2 A programming model for the 8-bit microprocessor used in this chapter. The programming model only shows those registers which the programmer can use to store data or which the programmer must access to have the right data for a program operation.

in color to make it stand out from the other parts of the microprocessor.

Notice that the ALU has two *input ports*. On the diagram they are labeled IN. The ALU also has one output port, labeled OUT. An input port is made up of the logic circuits used to get a data word into a logic device. An *output port* is made up of the logic circuits used to get a data word out of a logic device. Most logic devices have one or more input ports and a single output port. The simplest analogy, for example, is the AND gate. It has two inputs and an output. These are also called the AND gate's input ports and output port.

Both of the input ports are buffered by a temporary register (shown as Temp 1 and Temp 2 on the diagram); that is, each port has a register that temporarily stores one data word, holding the word for the ALU.

The microprocessor's internal bus is connected to the ALU's two input ports through the two temporary registers. This allows it to take data from any device on the microprocessor's internal data bus. Often, the ALU gets its data from a special register called the accumulator. The ALU's single output port allows it to send a data word over the bus to any device connected to the bus. Frequently, data is sent to the accumulator. When the ALU adds two data words, for example, one of the two words is placed in the accumulator. After the addition is performed, the resulting data word is sent to the accumulator and stored there.

The ALU works on either one or two data words, depending on the kind of operation performed. The ALU uses input ports as necessary. Since addition requires two data words, for example, an addition operation uses both ALU input ports.

Complementing a data word, conversely, uses only one input port. To complement a data word all of the word's bits that are logic 1 are set to logic 0, and all of the word's bits that are logic 0 are set to logic 1. As you can see, the ALU needs to work on only one word to perform a complement operation. That is why the complement operation uses only one input port.

The ALU is used whenever it is necessary to change or test a data word. The list of ALU functions—the exact ways in which the ALU can change or test data—varies from one microprocessor to another. Some ALUs have many functions. Other ALUs have only a few. ALU functions are part of the microprocessor's architecture.

The ALUs of most microprocessors can perform all the functions listed below:

Add	Complement
Subtract	Shift right
AND	Shift left
OR	Increment
Exclusive OR	Decrement

We will look at specific ALU functions in detail when we study particular microprocessors. The most important point to remember now is that any instruction that changes data must use the ALU. Because the ALU processes data, but does not store any data, it may not be part of some programming models.

Self-Test

Answer the following questions.

3. A microprocessor's block diagram is used to
   a. Describe the detailed gate and flip-flop logic used to construct the microprocessor
   b. Describe how the microprocessor's logic is externally connected to the memory and I/O devices
   c. Show the logic devices which you can use on your data to solve a problem
   d. All of the above
4. Which of the following is not an ALU function?
   a. Add   c. Power-up sequence
   b. Shift left   d. Complement

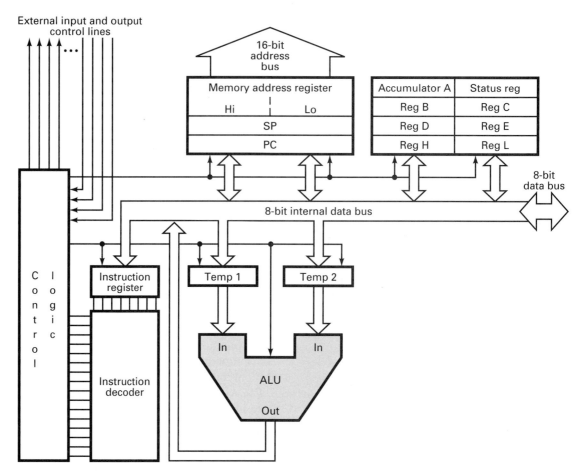

Fig. 5-3 The microprocessor's ALU. The microprocessor's arithmetic and logic unit can input data from one or two sources and outputs a single data word.

5. The ALU has two inputs. In the microprocessor used in this chapter they are connected to the
   a. Program counter c. Control logic
   b. Internal data bus d. Memory address register
6. The ALU's main job is to
   a. Perform addition
   b. Act as an output for the accumulator
   c. Logically or arithmetically modify data words
   d. All of the above

5-3 THE MICROPROCESSOR'S REGISTERS

Registers are a prominent part of the block diagram and the programming model of any microprocessor. They are a major logic function. This is true whether a microprocessor has only a few registers or many. Figure 5-4 shows our microprocessor with the registers in the block diagram in color. The other parts, will be discussed later in this chapter.

Note that not all of the registers are shown on the programming model. This is because the programmer cannot control the data in the omitted registers even though they are used in the programming process.

The microprocessor's registers can each temporarily store a word of data. You may wish to think of a register as a data latch with enough bits to store one word.

Some microprocessor registers serve a special purpose. That is, they hold data used in a specific job. Other registers serve general purposes. The general-purpose registers are shown as Reg B, Reg C, Reg D, and Reg E in Fig. 5-4.

General-purpose registers are available for any use in the programmer's imagination.

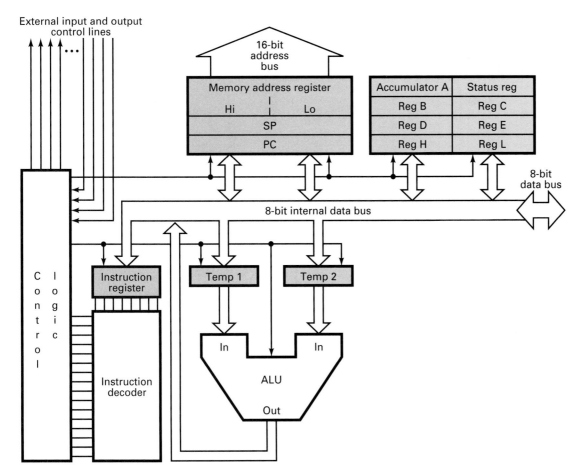

Fig. 5-4 The microprocessor's registers. The microprocessor's registers are shown outlined in black. The general-purpose registers (Reg B, Reg C, Reg D, Reg E, Reg H, and Reg L) may be found on some microprocessors and not on others.

Imaginative use of general-purpose registers often allows one programmer to do a job another programmer using the same microprocessor cannot do. The number of general-purpose registers varies from none to many, depending on the particular microprocessor.

Both the number and the use of all registers in a microprocessor depend on the architecture. But there are some basic types of registers that almost all microprocessors have. All additional registers are included to make life easier for the programmer.

The basic registers found in most microprocessors are the accumulator, the program counter, the stack pointer, the status register, the general-purpose registers, the memory address register and logic, the instruction register, and the temporary data registers.

In the next few sections, we will look at each of these basic registers in turn. As you read,

pay careful attention to how each of these registers affects the microprocessor's data flow.

Later you may be surprised to find that some of these registers do not appear in microprocessor programming models. The reason is that the programmer cannot do anything to change what happens with them. You must remember, however, that all the registers are present in the microprocessor and are serving a purpose. Otherwise you may not be able to understand how each register affects the microprocessor's data flow. In working with microprocessors, understanding data flow is essential.

The following sections are divided into two parts. The first seven sections cover the working registers shown in the programming model (Fig. 5-2). The second section covers those registers which the programmer cannot change, but which one must know about to effectively use the microprocessor.

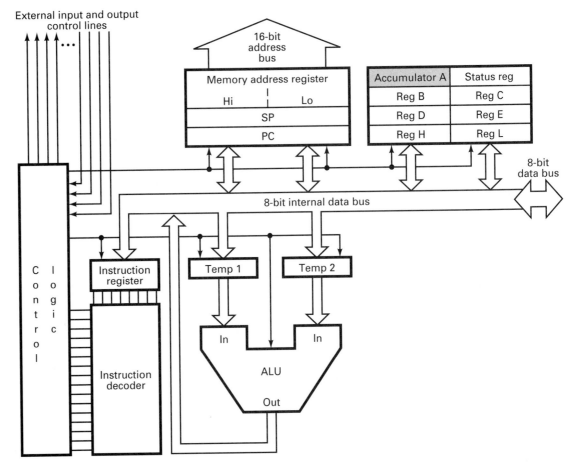

Fig. 5-5(a)   The accumulator, the microprocessor's most important working register. The accumulator can take data from the ALU or the bus and can place data on the bus or in the ALU.

5-4 THE ACCUMULATOR

When we looked at the ALU in Sec. 5-2, we had a glimpse of the accumulator's importance. In fact, the *accumulator* is the microprocessor's major register to hold data for manipulation. Most arithmetic and logic operations on data use both the ALU and the accumulator. Whenever the operation combines two words, whether arithmetically or logically, the accumulator contains one of the words. (The other word may be contained in another register or in a memory location.)

The accumulator is the microprocessor's major working register. Therefore, the accumulator is always a major part of the microprocessor's programming model as illustrated in Fig. 5-5(b).

For example, consider how the microprocessor adds two words. Let's call them word A and word B. First, word A is placed in the accumulator. The addition is then executed by adding word B (the contents of a memory location) to word A (in the accumulator). The resulting sum (C) is placed in the accumulator, replacing word A.

The result of an ALU operation is usually placed in the accumulator. It is important to understand that the accumulator's original contents are lost because they are overwritten.

Another kind of operation using the accumulator is the programmed data transfer. A programmed data transfer moves data from one place in the microcomputer to another. Data movements between an I/O port and a memory location, or between one memory location and another, are examples of programmed data transfers. When a programmed data transfer is executed, the transfer takes place in two stages. First, the data is moved from its *source* to the accumulator. Then the data is moved from the accumulator to its *destination*.

We have seen how the microprocessor uses the ALU to combine accumulator data with other data. The microprocessor also works directly on data in the accumulator.

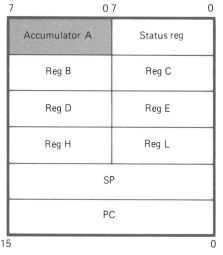

	7	0 7	0
	Accumulator A	Status reg	
	Reg B	Reg C	
	Reg D	Reg E	
	Reg H	Reg L	
	SP		
	PC		
15			0

Fig. 5-5(*b*) **The accumulator shown in the programming model. This important register is always part of a programming model.**

An example is clearing the accumulator. This operation sets all the accumulator's bits to logic 0. Other accumulator instructions set all the accumulator's bits to logic 1, shift the data to the right or to the left, complement the data in the accumulator, or perform other operations. We will learn about other accumulator instructions when we look at individual microprocessor instruction sets.

No other register is as versatile as the accumulator. We can perform only more limited operations with the other registers. To do real work on a piece of data, we must first move it to the accumulator.

Figure 5-5(*a*) shows that the accumulator receives data from the microprocessor's internal data bus. The accumulator also sends its contents to other devices using the internal data bus. Note that the path from the accumulator's output to the ALU is buffered by a temporary register. The need for this temporary register is discussed in Sec. 5-11.

The accumulator works on the same word length as the microprocessor's data word. That is, an 8-bit microprocessor has an 8-bit accumulator.

Some microprocessors, however, have *double-length accumulators*. Here, the double-length accumulator may be treated either as one device or as two separate accumulators. Treating the double-length accumulator as one device makes it possible to do a mathematical operation that introduces a carry into the second word. Let us consider the example of multiplying two 8-bit words. The result is a 16-

bit word, which the double-length accumulator can hold.

Some microprocessors have *multiple accumulators*. One, for example, might have one accumulator called accumulator A and another called accumulator B. A microprocessor that has two accumulators needs two different instructions to place the ALU's output in a specific accumulator. One instruction places the data in accumulator A. Another instruction places the data in accumulator B. Similarly, there are two Clear instructions—one to clear accumulator A and one to clear accumulator B.

What is the advantage of a microprocessor's having more than one accumulator? It is in the ability to perform accumulator-to-accumulator operations. Data can be stored temporarily in one accumulator while the other is used to do a different job. When the data in the first accumulator is needed again, there is no need to move the data because it is already in an accumulator.

By contrast, a microprocessor with one accumulator requires working on the data in the accumulator and then storing the result in memory or in another register. Often, the programmer wants to do this anyway. In other cases, however, having two accumulators can eliminate the need for many operations.

Self-Test

Answer the following questions.

7. Most arithmetic and logic operations in a microprocessor perform the operation between the contents of a memory location or a register and the
   a. Accumulator
   b. Program counter
   c. Memory address register
   d. Instruction register

8. The accumulator is connected to the other logic devices of the microprocessor by the microprocessor's internal data bus. The accumulator can
   a. Only input data
   b. Only output data
   c. Both input and output data
   d. Only talk to the ALU

9. When we say that we have cleared or reset the accumulator, we really mean that we have
   a. Set the accumulator's contents to all logic 0s
   b. Shorted out the accumulator's outputs
   c. Set the accumulator contents to all logic 1s
   d. Stopped using the accumulator

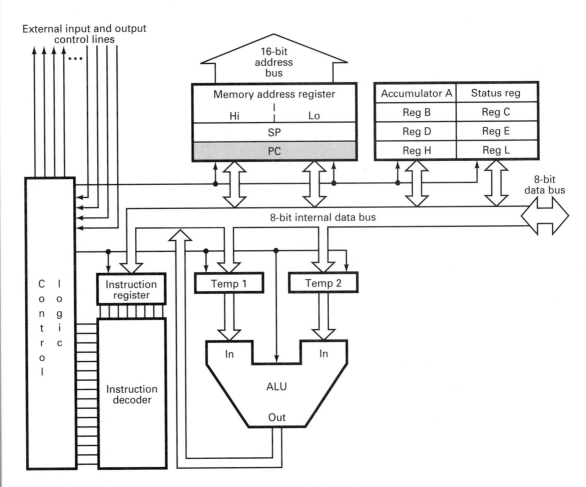

External input and output control lines

Fig. 5-6 **The program counter. Both the High byte and the Low byte of the program counter are bidirectionally connected to the microprocessor's internal data bus. However, all operations on the program counter work with it as a full 16-bit register.**

10. You are using a 16-bit microprocessor. It has a double-length accumulator. You expect this accumulator to be able to store the results of ALU operations. These results can be up to _____ bytes long.
    a. 1   c. 3
    b. 2   d. 4

11. A programmed data transfer between input port 40_{16} and memory location $0E9C_{16}$ uses the _____ to temporarily hold the data.
    a. Program counter
    b. ALU
    c. Memory location $0E9C_{16}$
    d. Accumulator

5-5 THE PROGRAM COUNTER

The *program counter* is one of the most important registers in the microprocessor. As you know, a program is a series of instructions stored in the microprocessor's memory. These instructions tell the microprocessor exactly how to solve a problem. Each instruction is simple and is exact. But all the instructions must occur in the right order if the program is to work correctly. The program counter's job is to keep track of what instruction is being used and what the next instruction will be.

Often the program counter is much longer than the microprocessor's data word. For example, in most 8-bit microprocessors that address 64K of memory, the program counter is 16 bits long. There is good reason for having such a long program counter. In such a general-purpose microcomputer, any one of the 65,536 different memory locations can contain program steps. That is, the program can start anywhere and stop anywhere in the memory address range of location 0 through location 65,535. The program counter must be 16 bits long in order to be able to point to any one of these different memory locations.

Remember that wherever the program instructions are located, they must be in order.

In Figs. 5-6 and 5-7, the program counter is shown in color. Notice that the block diagram (Fig. 5-6) shows the program counter connected to the microprocessor's internal data bus. In theory, the program counter could receive program address data from any other logic function connected to the internal data bus. In practice, however, the program counter generally gets its data from the microcomputer's memory.

The program counter cannot perform the variety of operations the accumulator can. The program counter has fewer special instructions than the accumulator has.

Before the microprocessor can start executing a program, the program counter has to be loaded with a number. *This number is the address of the memory location containing the first program instruction.* Usually this is a simple number such as all zeros or all ones. This memory location *must* always contain a program step because the first instruction the microprocessor executes will come from this memory location. All other locations can contain either instructions or data as the programmer chooses.

Note the memory address register and the 16-bit address bus, shown in Fig. 5-6. The address of the memory location containing the first program instruction is sent from the program counter to the memory address register. The contents of the program counter and the memory address register are now the same.

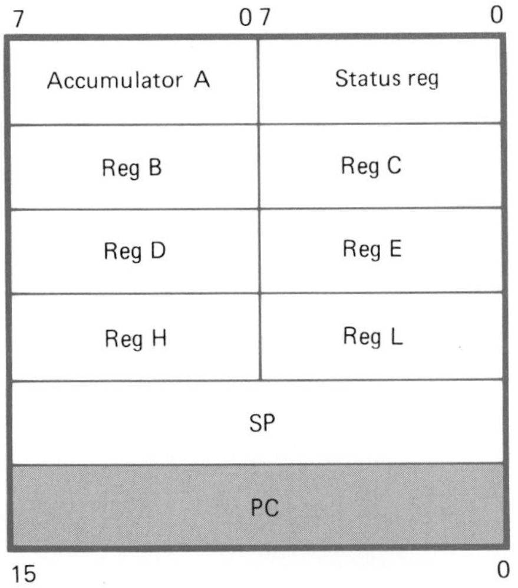

Fig. 5-7 **The program counter in the programming model. Note that the program counter for this 8-bit microprocessor is shown as a 16-bit register and is grouped with the other 2-byte registers.**

Section 5-9 gives more information on the memory address register. For now, all you need to know is that the memory address register, holds the same length binary number as the program counter.

The memory address of the first instruction is sent to the memory circuits over the 16-bit *address bus.* The memory then puts the contents of the addressed location on the data bus. These contents are an instruction. The memory sends the instruction back to a special register in the microprocessor—the *instruction register,* described more fully in Sec. 5-10.

For now, the important point to note is this: Once the microprocessor fetches an instruction from memory, the microprocessor automatically *increments* the program counter. The program counter is incremented just as the microprocessor is starting to execute the instruction fetched immediately before.

The program counter now points to the *next* instruction. *The program counter always points to the next instruction throughout the time the current instruction is being executed.* This is an important concept to remember, because there are times when you may need to use the current value of the program counter. To know what the current value is, you must know that the program counter is pointing to the next instruction, not to the current one.

This does not mean that every instruction must follow the last instruction in memory. One of the instructions in the program can be to load the program counter with a new value. When this "new program counter value" instruction is executed, the next instruction is taken from the new address rather than the next address in sequence. You may wish to execute part of a program that is not in sequence with the main program. For example, there may be a part of the program that must be repeated many times during the execution of the entire program. Rather than writing the repeated part of the program again and again, the programmer need write that part only once, but return to it—going out of sequence—many times. The part of the program to be done out of sequence is called a *subroutine.* When we change the program counter so that the next instruction to be executed is not the next sequential instruction in memory, we say that this is a *branch* to a new location in the program. Once the program counter is set to the starting address of the subroutine, the program counter then increments through the subroutine in order until it finds an instruction that tells it to return to the main program. Subroutines are explained in greater detail in Chap. 14.

Self-Test

Answer the following questions.

12. A 16-bit microprocessor has a memory address range of 2^{20} (1,048,576). You would expect the program counter in this microprocessor to be _____ bits long.
    a. 4   *d.* 20
    b. 8   *e.* 22
    c. 16   *f.* 32
13. The program counter is one of a microprocessor's _____ registers.
    a. Special-purpose
    b. General-purpose
    c. Memory
    d. All of the above

14. Except when it is fetching an instruction, the program counter is pointing to the _____ program instruction.
    a. Last   *c.* Current
    b. Next   *d.* Subroutine
15. Once the program counter is loaded with a starting address, it will
    a. Skip around, pointing to different memory locations as it thinks you need them
    b. Go to a lower memory location after it executes each instruction
    c. Increment to the next program instruction once the current instruction is fetched
    d. Go directly to location 65,536

5-6 THE STATUS REGISTER

The register that can make the difference between a simple calculator and an authentic computer is the status register. The *status*

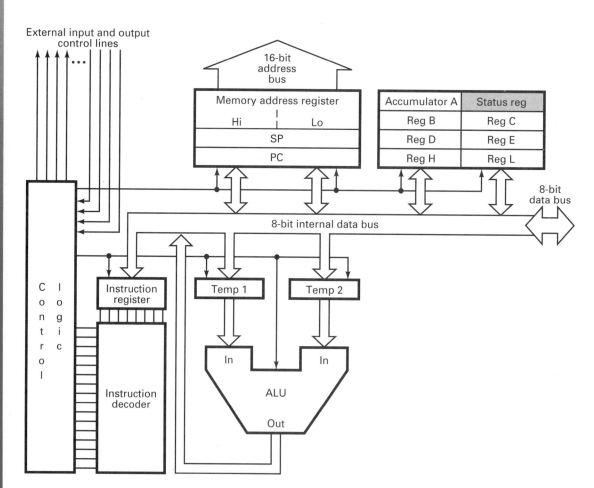

Fig. 5-8   The status or flag register. Its bits are set according to operations which happen in the ALU. Data operations with many different devices in the microprocessor set the status register's bits; however, you should remember that it is operations that happen through the ALU which really set the status register's bits.

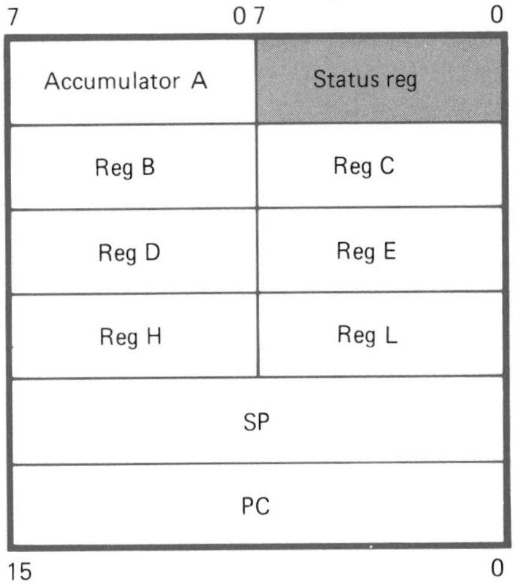

Flag register

Conditional branching

Conditional instructions

Carry bit

Negative bit

Zero bit

Decrement

Fig. 5-9 The status register in the programming model. Note that the status register is shown as an 8-bit register even though only 7 bits are used in this particular example. For microprocessors which use more that 8 status bits, the status register may be shown as a part of the 16-bit register grouping.

register is used to store the results of certain tests performed during the execution of a program. Sometimes the status register is called a *flag register*. ALU operations and certain register operations may set or clear one or more bits in the status register.

Storing the results of tests lets you write programs that branch. When the program branches, it starts at a new location. That is, the program counter is loaded with a new starting value. In *conditional branching*, the branch happens *only* if the result of a certain test comes out as required. The status register holds the results of such tests. Figures 5-8 and 5-9 show the status register.

The ability to branch sets the computer architecture apart from a simple calculator. Since it lets you test the condition of bits in the status register, you have the ability to write programs which carry out different actions depending on which bits are set and which bits are cleared. This lets us say the microprocessor can make decisions based on these conditions. No simple calculator can make decisions.

Status bits lead to a new set of microprocessor instructions. These instructions permit the execution of a program to change course on the basis of the condition of a status bit. The usual method of using these special instructions is to write the program so that the program counter is loaded with a new value

when a certain status bit is set. The *conditional instructions* are discussed in detail in Chap. 7.

As already noted in this section, ALU operations set the status register's bits. The status register stores the results of tests for conditions that are generated by ALU operations. Mathematical operations create the most common conditions indicated by the status register. For example, mathematical operations can produce a carry bit, or they can result in a logic 0 (cleared register), or both.

If two 8-bit numbers are added, for example, and their sum is greater than 1111 1111, then a carry is generated. When a carry is generated, the microprocessor sets the status register's *carry bit*.

If we add 1110 1110 and 0111 0000, we get

$$
\begin{array}{r}
1110 \quad 1110 \\
0111 \quad 0000 \\
\hline
1 \quad 0101 \quad 1110
\end{array}
$$

carry   8-bit positive result

This operation sets the status register's carry bit to logic 1. If we add 0001 1111 and 0100 0001, we get

$$
\begin{array}{r}
0001 \quad 1111 \\
0100 \quad 0001 \\
\hline
0 \quad 0110 \quad 0000
\end{array}
$$

carry   8-bit positive result

and the carry bit remains a logic 0.

If we add 1110 1110 and 1111 0000, we get

$$
\begin{array}{r}
1110 \quad 1110 \\
+ \quad 1111 \quad 0000 \\
\hline
1 \quad 1101 \quad 1110
\end{array}
$$

carry   8-bit negative result

This sets the carry bit *and* the *negative bit* to logic 1.

If the result of an operation sets all the bits of the accumulator to logic 0, the status register's *zero bit* is set.

In many microprocessors, general-purpose register operations can also set the status register's zero bit. For example, we often set the value of a register—let's call it register D—at some specific number and then *decrement* the register (reduce the value of the number in the register) each time we pass through a designated point in the program.

The result, each time the designated point is passed, is that register D is decremented and the status register's zero bit is checked. If, after register D is decremented, register D's value is logic 0, the status register's zero bit is set. If decrementing register D does not cause a logic 0 there, then the status register's zero bit is not set. The program that tests for the logic 0 in register D continues until the status register's zero bit is set.

Figure 5-10 shows a short program using the status register to test a decrementing register. First we set register B to 1100. Then we decrement the register B. After decrementing the register, we test the status register for a logic 1 in the zero bit. If there is no logic 1 in the zero bit, we decrement and test again. When a logic 1 is found in the zero bit, the program stops.

1. Load register B with 1100.
2. Decrement register B 1 count.
3. Is status register's zero bit set?
4. No. Go back to step 2.
5. Yes. Stop.

Fig. 5-10 A simple five-step program showing how the status register is used. The instruction in the third step tests the status register's zero bit, which was set by the instruction in the second line.

Some of the common status register bits are:

1. *Carry/borrow:* This bit indicates that the last operation caused either a carry or a borrow. The carry bit is set when two binary numbers are added and generate a carry from the eighth bit. A borrow is generated when a larger number is subtracted from a smaller number.
2. *Zero:* The zero bit is set when the operation causes all of a register's bits to be logic 0. This happens not only when you decrement the register, but also when any operation causes the register's bits all to become logic 0.
3. *Negative:* The status register's negative bit is set when a register's most significant bit is logic 1. In 2's complement arithmetic a logic 1 in the register's most significant bit means that the number in the register is negative.
4. *Intermediate carry:* The status register's intermediate carry bit is set when an addition in the first 4 bits caused a carry into the fifth bit. This is often referred to as a half carry. It is frequently used to assist in the conversion of BCD numbers into binary numbers.
5. *Interrupt flag:* The status register's interrupt flag is set when the programmer desires to enable (turn on) the interrupt function (see Chap. 6). It is cleared when the programmer wants to disable (turn off) the interrupt function. If the microprocessor has multiple interrupts, there may be multiple interrupt flags.
6. *Overflow:* The microprocessor status register's overflow bit is set when both an arithmetic carry and a signed carry occur in 2's complement arithmetic operations.
7. *Parity:* The microprocessor status register's parity bit is set when the results of an operation leave the indicated register with an odd number of 1s.

Most microprocessors use these 7 status bits. Many microprocessors, however, also use additional status register bits. The use of other status bits is not standardized. Some nonstandard status bits in use require a thorough understanding of binary arithmetic, because these status bits are set only when specific arithmetic operations occur.

Other nonstandard status bits are set for nonregister or non-ALU operations. These status bits are used to indicate that certain microprocessor functions are turned on or off. Such status bits can indicate the hardware status of certain microprocessor options and are examined before the use of those options.

In some microprocessors, you can clear or set all the status bits by using a special microprocessor instruction. In others, however, you can only read the status register's value. To understand how nonstandard status bits are used in a particular microprocessor, you must study the manufacturer's data sheets.

The status register's length depends on the number of status bits used by a particular microprocessor. In general, those bits not used are permanently set to logic 1.

Figure 5-11 shows the status word as it is used in a typical 8-bit microprocessor. The least significant bit is permanently set to logic 1. The status word is treated just like the other registers. As a result, the data in the status register can be placed on the microprocessor's internal data bus, but the status register cannot receive data from the microprocessor's internal data bus. In this example, the status register is a read-only register.

Stack

Push

Pop

Last-in-first-out
(LIFO) operation

Stack pointer

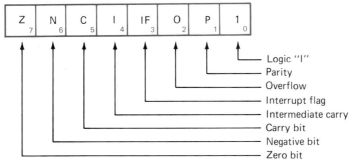

Fig. 5-11 **The status word of a typical microprocessor. For this microprocessor, there are only 7 status bits. The remaining bit is permanently set to a logic 1.**

Self-Test

Answer the following questions.

16. Add the following 8-bit binary numbers. After adding these numbers, indicate how the result sets the zero (Z), negative (N), and carry (C) bits.

    Example

```
      1 0 1 0 1 0 1 0
      1 1 1 1 1 1 1 1   Z N C
    ┌─────────────────────────┐
    │1  1 0 1 0 1 0 0 1  0 1 1 │
    └─────────────────────────┘
```

    a. 0 0 0 0 1 1 1 1
        1 1 1 1 0 0 0 0

    b. 0 0 1 1 1 0 1 1
        1 1 0 0 0 1 0 1

    c. 1 1 1 1 1 1 1 1
        1 1 1 1 1 1 1 1

    d. 0 0 0 0 0 0 0 1
        1 1 1 1 1 1 1 0

    e. 0 1 0 1 0 1 0 0
        1 1 0 0 1 1 0 0

    f. 0 0 0 0 0 0 0 1
        0 1 1 1 1 1 1 1

    g. 0 0 0 0 1 1 1 1
        0 0 0 1 0 0 0 0

    h. 1 1 0 0 0 0 0 0
        1 0 0 0 0 0 0 1

17. You want to increment a register three times and cause the status register's zero bit to be set on the third increment. What is the 8-bit starting number? What are the three steps?

18. The status register's carry bit can also indicate

    a. A zero   *c.* A borrow
    b. A negative   *d.* All of the above

5-7 THE STACK POINTER

The stack pointer is another of the important registers which the programmer uses on a regular basis. In the two earlier sections, you learned that the program counter may be temporarily reloaded with a new value. Frequently this new value is only loaded after the program tests the status register. This process, which is discussed more extensively in Chaps. 7 and 14, is a very powerful programming tool.

How does the programmer keep track of the return address? The programmer keeps this information in a special area of memory called the *stack*. The programmer *pushes* (puts) information onto the stack in order and *pops* (takes) information from the stack in reverse order. This is to say that the last information placed on the stack is the first information off the stack. This kind of stack operation is called *LIFO (last-in-first-out) operation.*

You can think of the stack just as you would a stack of 3 in. × 5 in. cards with notes. You place the first card on the table, the second card on the first, the third card on the second, etc. When you pick the cards up, you pick the top card up first, then the next, then the next, and so on. This is called a LIFO stack. In the microprocessor, the 3 in. × 5 in. cards are replaced with memory locations.

How does the programmer keep track of the memory address which is the top of the current stack? The programmer uses a special register called the *stack pointer*. The stack pointer, like the program counter, must have enough bits to address any location in memory. Therefore, the stack pointer is usually very similar to the program counter. The stack pointer is shown in color in the block diagram of Fig. 5-12 and in the programming model of Fig. 5-13.

Like the program counter, *the stack pointer automatically points to the next available lo-*

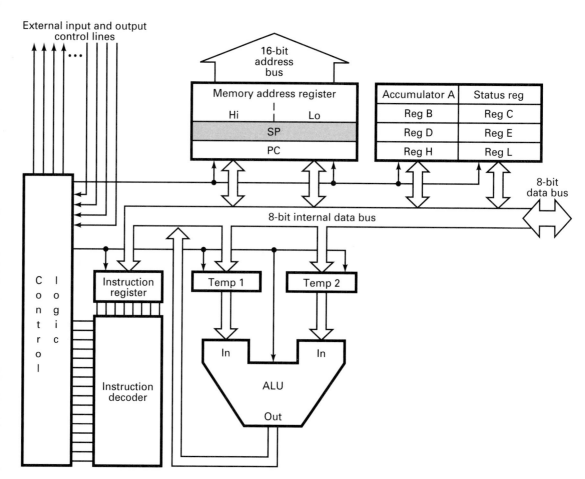

External input and output
control lines

Fig. 5-12 The stack pointer. This 16-bit register points to the address of the *next* available memory location on the stack. This memory location is called the top of the stack.

cation. In most microprocessors, the stack pointer *decrements* (points to the next lower memory address) after it is used. This allows the programmer to build the stack down in memory. Frequently, stack operations are 2-byte operations. This means that the stack pointer decrements by two memory address locations each time it is used.

The stack pointer starts at a memory location initially set by the programmer. This process is called *stack pointer initialization.* If the stack pointer is not initialized, the stack starts at a random memory location. An uninitialized stack pointer can allow the stack data to write over other important data, write over the program itself or to write to an address where there is no RAM. If any of these events occur, the stack is destroyed. This will eventually cause the program to fail.

Many good programmers use the stack a great deal. Additionally, many high-level languages rely on having a large stack to store intermediate operations. Earlier, we learned that

the stack operates as a LIFO device. That is, the last data placed on the stack is the first data off the stack, or the first data placed on the stack is the last data off the stack. To ensure the correct stack data is taken from the stack, the stack pointer must keep an exact record of the memory address used for the stack. Any error in the stack pointer causes a loss of the stack data. Therefore, it is important that the programmer know exactly what is happening to the stack pointer at all times. This is why this job is assigned to a special register and not left to one of the many general-purpose registers.

Self-Test

Answer the following questions.

19. You are using a 16-bit microprocessor. It can directly address 4,194,304 memory locations. You would expect the stack

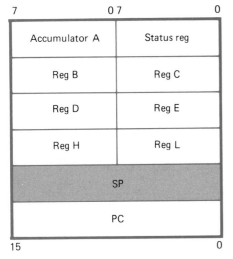

Fig. 5-13 The stack pointer in the programming model. Grouped with the other 16-bit specific-purpose registers this autodecrementing register provides the programmer with a continuously available pointer to a special storage area called the stack. The stack pointer is automatically decremented one or two memory locations each time one or two words are put on (pushed onto) the stack. The stack pointer is automatically incremented one or two memory locations each time one or two words are taken from (popped off) the stack.

pointer in this microprocessor to be _____ bits long.

   a. 4   *d.* 20
   b. 8   *e.* 22
   c. 16   *f.* 32

20. You set the starting value of your stack pointer in your 8-bit microprocessor at memory address 20FFH. You have pushed two program counter values onto the stack and one 8-bit offset number. The stack pointer, which decrements each time the stack is used, is now pointing to the next available memory address for stack operations. This address is at memory location _____ .

21. Draw a diagram showing the memory locations starting at address 20FFH and ending at the address you chose as the answer to question 5-20. Show how the memory locations are used by the stack operations in question 5-20. Presume that the program counter values are stored Low byte first.

22. Except when the stack pointer is actually being used, it is pointing to the _____ memory address.

   a. Next program
   b. Last used stack
   c. Next available stack
   d. Current

23. The stack pointer is one of the microprocessor's _____ registers.

5-8 THE MICROPROCESSOR'S GENERAL-PURPOSE REGISTERS

In addition to the six basic registers, most microprocessors have other registers for general programming use. These other registers are called *general-purpose registers*. On some microprocessors, general-purpose registers serve only as simple storage areas. On other microprocessors, however, the general-purpose registers are as powerful as an accumulator. General-purpose registers achieve such power if the ALU can put its data into them.

The microprocessor used in this chapter has six general-purpose registers (Fig. 5-14 on the next page) called the B, C, D, E, H, and L registers. Since the ALU of our microprocessor does not put its data into these registers, they lack the power of an accumulator. Nevertheless, many instructions do use these general-purpose registers.

For many operations, these registers are six identical 8-bit registers. The choice of which one to use for a certain job depends simply on which register is available and most convenient. Usually, the H and L register pair is only used for special memory-pointing operations. It can have limited instructions.

Operations with these registers affect the status register. Therefore, any of them can be used as a decrementing or incrementing counter. What happens if we use the D register, for example, as a decrementing counter? When the D register decrements to 0, the status register's zero bit is set to logic 1.

Together, the BC, DE, and HL registers have a unique function: they can operate as special-purpose 16-bit registers. We then call them *register pairs*. When used as register pairs, the two registers act as a single, 16-bit register. This is shown in the programming model (Fig. 5-15 on page 86), as the registers are shown side by side, indicating that they can be used as a 16-bit register if needed.

Our microprocessor has an addressing mode that places the contents of the HL register pair in the memory address register. This mode enables us to do simple register arithmetic on a 16-bit address. For example, we can increment the HL register pair and then use its contents to point to a memory location.

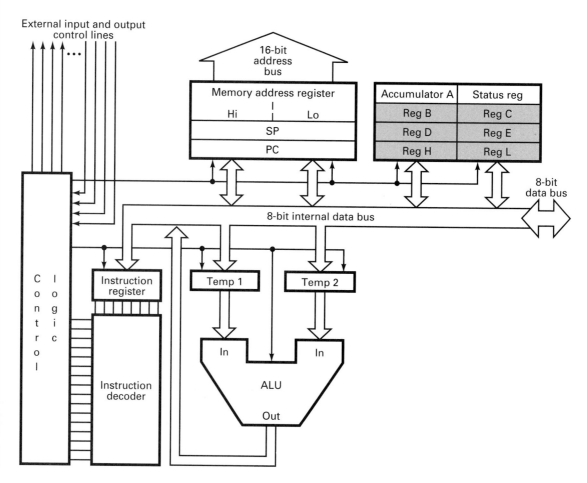

Fig. 5-14 The general-purpose registers. These registers can be either single 8-bit registers or can be used as 16-bit register pairs (e.g., register pair BC, register pair DE, etc.).

Fig. 5-15 (a) The B, C, D, E, H, and L general-purpose registers as a part of the programming model. (b) When used as a 16-bit register, for example, this programming model shows that the B register holds the low-order bits (bits 0 to 7) and the C register holds the high-order bits (bits 8 to 15).

Self-Test

Answer the following questions. Note: All questions in this section refer to the microprocessor discussed in this chapter.

24. The B, C, D, and E registers can be used as
    a. A program counter
    b. A memory address register
    c. General-purpose registers
    d. A DC register pair
25. The HL register pair is used as a memory pointer because
    a. It is near the program counter
    b. It is near the memory address register
    c. It can be used as two independent 8-bit registers
    d. 16 bits will address all of memory
26. If you need to use a register as a 16-bit decrementing counter, you will probably use the
    a. D register   c. B register
    b. C register   d. BC register pair

5-9 THE MEMORY ADDRESS REGISTER AND LOGIC

Every time the microprocessor addresses the microprocessor's memory, the *memory address register* points to the memory location the processor wants to use. That is, the memory address register holds a binary number. That number is the address of a memory location. The memory location pointed to (addressed) is the one identified by the binary number this register holds.

The output of the memory address register drives the 16-bit address bus. This output is used to select a memory location or, in some cases, to select an input/output port. Often, the memory address register is combined with other logic used to help with the memory addressing task. This logic varies with different microprocessors and their different ways of addressing the memory attached to the microprocessor.

During the fetch cycle, as described in Sec. 5-5, an instruction is taken from memory. The contents of the memory address register and the contents of the program counter are the same at this time. That is, the memory address register points to the instruction word being fetched from memory. Once the instruction is decoded, the program counter increments.

The memory address register does not increment. During the execute cycle, the contents of the memory address register depend on the instruction being executed. If this instruction requires that the microprocessor address memory, then the memory address register is used for a second time during this one instruction.

The execution of some instructions does not require addressing memory. An example is the Clear accumulator instruction. If the instruction being executed does not require addressing memory, then the memory address register is used only during the fetch cycle.

In most microprocessors, the memory address register is the same length as the program counter. Like the program counter, the memory address register must have enough bits to address any location in the microprocessor's memory. For most 8-bit microprocessors, the memory address register is 16 bits long. The length of the memory address register in 16-, 32-, and 64-bit microprocessors varies from 20 to 32 bits depending on the particular microprocessor.

A 16-bit memory address register may be divided into two separate registers, each with an independent connection to the microprocessor's data bus. If so, these two registers are called the High-byte and the Low-byte memory address registers. Figure 5-16 on page 88 shows the memory address register in color.

As Fig. 5-16 shows, the memory address register is connected to the microprocessor's internal bus. Consequently, the memory address register can be loaded from several different sources. Most microprocessors have instructions to load the memory address register from the program counter, from a general-purpose register, or from memory.

Additionally, the memory address register also receives data from the stack pointer, the index register, and other devices. Usually these data transfers happen as a part of an instruction and are not directly under the programmer's control.

Some instructions permit setting the memory address register to a new number by computation: the new number is computed from the program counter's value plus or minus a number that is part of the instruction itself. In this case the computation is done by the microprocessor's ALU. The term for this kind of memory addressing is *offset addressing*.

Self-Test

Answer the following questions.

27. The memory address register points to
    a. The memory's contents
    b. The memory location
    c. A memory register
    d. A CPU location

28. A certain 16-bit microprocessor can address 4,194,304 locations. You would expect this microprocessor to have a memory address register which is _____ bits long.
    a. 8   d. 22
    b. 16   e. 24
    c. 20   f. 32

29. The memory address register is connected to the microprocessor's internal data bus so that it can be loaded from
    a. The program counter
    b. The general-purpose register
    c. Memory
    d. All of the above

30. The memory address register's outputs drive the microprocessor's
    a. Accumulator
    b. Internal data bus
    c. Memory address bus
    d. Instruction decoder input

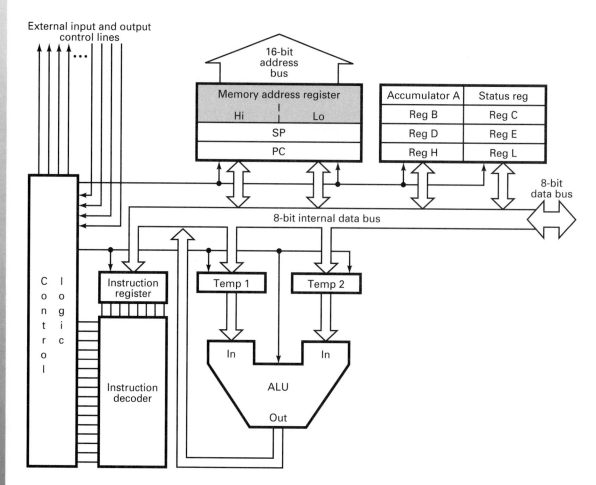

Fig. 5-16 The memory address register. This 16-bit register is broken into the High and Low bytes. Both bytes have independent connections to the microprocessor's internal data bus.

5-10 THE INSTRUCTION REGISTER

The *instruction register* holds the instruction the microprocessor is currently executing. This is the instruction register's only job, and the instruction register performs it automatically. The instruction register is loaded by starting the microprocessor on a *fetch/execute* cycle, also called an *instruction cycle*.

As we have seen earlier, the instruction cycle consists of a fetch cycle and an execute cycle. Except for loading the instruction into the instruction register during the fetch cycle, the programmer can make no other use of the instruction register. As shown in Fig. 5-17, the instruction register is connected to the microprocessor's internal data bus, but it can only receive data. The instruction register cannot place data on the internal bus.

Nevertheless, the instruction register plays a very important role in the microprocessor.

The instruction register is important because its output always drives the part of the control logic known as the instruction decoder.

Remember the sequence of the fetch/execute cycle. First an instruction is fetched. Then the program counter points to the next instruction in memory. Whenever an instruction is fetched, a copy of it is taken from the instruction's memory location. The copy is placed on the microprocessor's internal data bus and carried to the instruction register.

Then the instruction is executed. During execution, the instruction decoder reads the contents of the instruction register. The decoder decodes the instruction in order to tell the microprocessor exactly what to do to carry out the instruction. The instruction decoder is discussed in Sec. 5-12, which is about the microprocessor's control logic.

The length of the instruction register varies from one microprocessor to another. In some

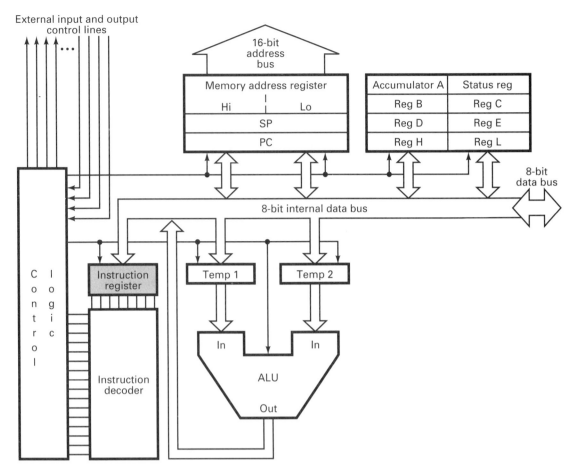

External input and output
control lines

16-bit
address
bus

Memory address register

Hi | Lo

SP

PC

Accumulator A | Status reg

Reg B | Reg C

Reg D | Reg E

Reg H | Reg L

8-bit
data bus

8-bit internal data bus

Control logic

Instruction
register

Instruction
decoder

Temp 1

Temp 2

In | In

ALU

Out

**Fig. 5-17   The instruction register. This register holds the program instruction telling
the microprocessor how to operate during its execute cycle.**

microprocessors, the instruction word is as
long as the data word. In others, the instruction word may be as short as 3 or 4 bits.

Self-Test

Answer the following questions.

31. During the execution of an instruction, the
instruction register holds the _____
instruction.
   a. Previous   *c.* Next
   b. Current   *d.* All of the above
32. The instruction register's length depends on
   a. The microprocessor's architecture
   b. Whether the microprocessor is an 8-bit
        or 16-bit design
   c. The size of the memory being ad-
        dressed
   d. The microprocessor's speed
33. The instruction register is loaded with the
contents of the memory location pointed
to by the

   a. Accumulator
   b. CPU
   c. Previous instruction
   d. Program counter

5-11 THE TEMPORARY DATA REGISTERS

Figure 5-18 on page 90 shows the *temporary
data registers* in color. Each of the temporary
data registers has enough bits to store one
data word. *Note:* This microprocessor's block
diagram shows two temporary registers, however, a microprocessor may use more.

   The need for the temporary registers arises
because the ALU has no data storage of its
own. The ALU is constructed entirely of combinational logic. Because the ALU has no storage, any data applied to its input immediately
appears at its output. The data that appears at
the ALU's output has been modified by the
ALU operation determined by the program.

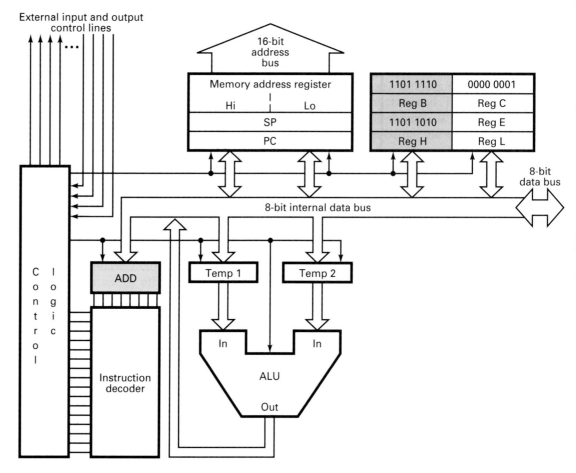

Fig. 5-22 **The accumulator and register D are loaded with data and all bits in the status register are cleared. (The LFB is a permanent logic 1.) Note that the instruction register contains the ADD instruction. At this time, neither register D nor the accumulator are connected to any other device via the internal data bus.**

Remember that the data bus moves data words, not bits. In a 16-bit microprocessor, for example, all the movements on the internal data bus transfer 2 bytes of data (16 bits), and in 32- and 64-bit microprocessors, the internal bus moves data 4 or 8 bytes at a time respectively.

Self-Test

Answer the following questions.

40. The control logic guides the microprocessor through the steps to carry out a program. It also does housekeeping functions such as
    a. Temporary storage
    b. The power-up sequence
    c. Adding two numbers
    d. Transferring data

41. The microprocessor's bus is bidirectional. This means that
    a. All data flows in two directions
    b. Data can flow in the direction needed to complete the transfer
    c. Each device has two input ports
    d. Each device must have both an input and an output port

42. All the microprocessor's logic functions are connected to its internal data bus. The microprocessor's _____ tells them to talk to the bus, to listen to the bus, or to do nothing.
    a. Accumulator   c. CPU
    b. Control logic   d. Instruction register

43. The microprocessor control logic is driven by a high-frequency square wave called the microprocessor's clock. The purpose of the microprocessor's clock is to
    a. Provide time-of-day information
    b. Operate the ALU's data storage mechanism
    c. Provide a timing signal to sequence the control logic
    d. Be sure the data in registers is kept current

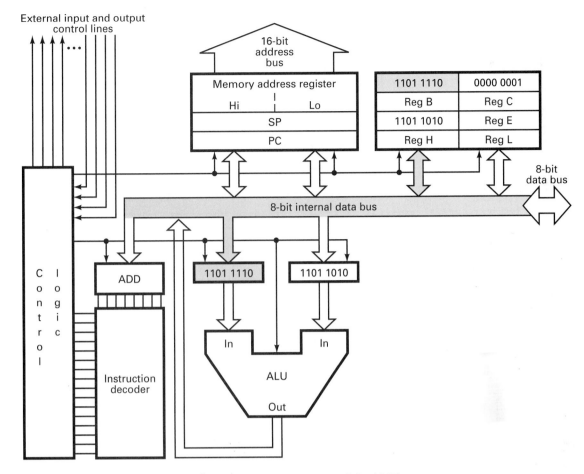

External input and output control lines

16-bit address bus

Memory address register

Hi | Lo

SP

PC

1101 1110	0000 0001
Reg B	Reg C
1101 1010	Reg E
Reg H	Reg L

8-bit data bus

8-bit internal data bus

Control logic

ADD

1101 1110

1101 1010

In | In

ALU

Out

Instruction decoder

Fig. 5-23 **The accumulator's contents are transferred to one of the ALU's temporary registers.**

SUMMARY

1. A microprocessor has three major logic functions: the ALU, the registers, and the control logic.
2. A logic device has ports through which input or output data pass.
3. The ALU has two input ports and one output port. One input port comes from the microprocessor's internal data bus. The other input port comes from the accumulator. An ALU works on either one or two data words. The ALU is used to arithmetically or logically change or test data.
4. A register lets you temporarily store a word of data. Some microprocessor registers are general-purpose, but a few are very special purpose.
5. All microprocessors have the following basic registers
   a. Accumulator
   b. Program counter

   c. Status register
   d. Stack pointer
   e. General-purpose registers
   f. Memory address register and logic
   g. Instruction register
   h. Temporary register

   These registers are needed to make a microprocessor work. However, a programmer may not have access to all of them.

6. The accumulator works with the ALU. It is the microprocessor's major register for data manipulation. The accumulator's contents are usually one half of an arithmetic or logic operation. The accumulator is the microprocessor's most versatile register. The accumulator's length is the same as that of the microprocessor's data word. Some sophisticated microprocessors have an accumulator which is twice as long as their data word. Some microprocessors

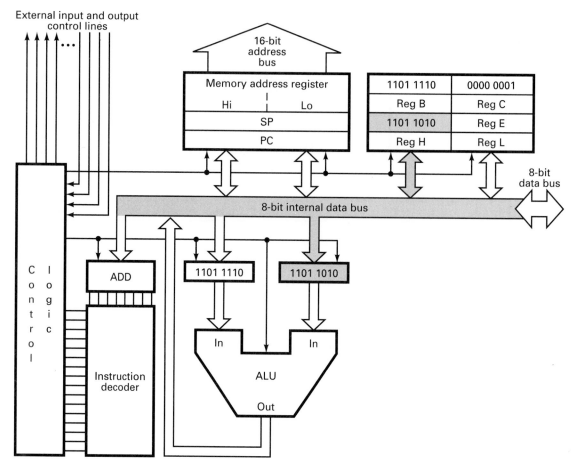

Fig. 5-24 **The contents of register D are transferred to the other of the ALU's temporary registers.**

have two accumulators. This feature lets you do accumulator-to-accumulator arithmetic and logic operations.

7. The program counter keeps track of exactly what instruction is to be executed next. The program counter's length must be long enough to address any memory location.

8. A program can start at any memory location and end at any memory location. However, the program instructions must be in their proper order, no matter where they are located.

9. a. When a microprocessor first starts up it always gets its first instruction from the same memory location.

   b. Each program instruction step is then executed in sequence unless a special instruction changes the sequence.

   c. The program counter points to a memory location. The control logic fetches an instruction from this memory location.

   d. Once the instruction is fetched, the mi-

croprocessor increments the program counter and starts to execute the instruction.

   e. The incremented program counter now points to the *next* program instruction.

   f. The memory address register points to each memory location that the microprocessor wants to use when it wants to use it.

   g. The memory address register drives the microprocessor's address bus.

   h. The memory address register is long enough to address every memory location in the microprocessor's main memory.

10. The status register records the results of certain register operations.

   a. Often the data in the status register is used to control how the rest of the program will execute. They give the program decision-making information.

   b. At a minimum, the status register has a zero, a negative, a carry, a half-carry, a parity, an overflow, and an interrupt bit.

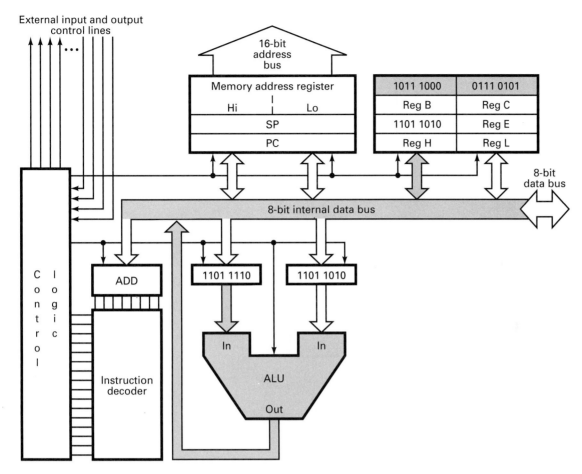

External input and output control lines

16-bit address bus

Memory address register

Hi | Lo

SP

PC

1011 1000	0111 0101
Reg B	Reg C
1101 1010	Reg E
Reg H	Reg L

8-bit data bus

8-bit internal data bus

Control logic

ADD

Instruction decoder

1101 1110

1101 1010

In   In

ALU

Out

Fig. 5-25   The ALU is told to ADD. The ALU's output is connected to the accumulator by the internal data bus, so the contents of the accumulator are overwritten with the new result. The accumulator's original contents are lost. The status register's negative and carry bits are set.

c. Often a status register has additional bits or flags. These are peculiar to the particular microprocessor.

d. Some status bits may be used to indicate the status of certain programmable options. The bit tells you whether the option is on or off.

e. Because the microprocessor has a status register, it has conditional instructions that change the course of the program only if the proper bit is set.

11. The stack pointer points to a memory location where temporary data is stored. Each time the stack is used to store data the stack pointer is decremented so that it points to the next available location.

12. Many double-width registers are broken into a High and a Low word. The High word and Low word are each the same length as a data word.

13. a. The instruction register holds the binary word that tells the microprocessor what to do to carry out the current instruction.

b. When the instruction is fetched, a copy is taken from its memory location in the program and placed in the instruction register.

c. During execution, the instruction decoder and control logic read the word in the instruction register.

14. The temporary registers hold data at the ALU's input while the ALU's combinational logic operates on the data.

15. The microprocessor's control logic decodes the instruction and tells the other logic devices what to do and in what order to carry out the instruction. The microprocessor's control logic timing comes from the clock. Usually the clock is a two-phase, nonoverlapping signal.

16. The data path between all the microprocessor's logic devices is called its data bus.

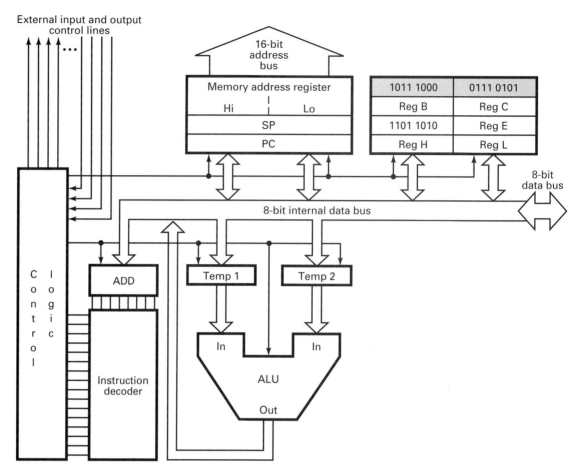

Fig. 5-26 The operation is complete. The new data is in the accumulator and the status register, and the microprocessor waits for the next instruction.

The data bus is bidirectional. All the logic devices are always connected to the bus. However, they must wait for a signal from the control logic before they talk to or listen to the bus. Data on the bus moves from a source to a destination.

CHAPTER REVIEW QUESTIONS

Answer the following questions.

5-1. List the three major logic functions found in a microprocessor's block diagram.

5-2. What is the purpose of the programming model? How does it compare to the block diagram?

5-3. The logic functions are interconnected by the microprocessor's _____ .

5-4. Briefly explain what the arithmetic-logic unit does. That is, what is its effect on the microprocessor's data?

5-5. List four typical specific functions performed by the ALU.

5-6. The ALU has two input ports. Does it always use both inputs? Why?

5-7. What does a microprocessor's register do?

5-8. The microprocessor's registers are
   a. Connected to the bus

b. Single-word storage elements

c. General- or special-purpose

d. All of the above

5-9. Which of the following is a user-programmable register?

  a. Memory address register

  b. Accumulator

  c. Program counter

  d. All of the above

5-10. The accumulator is used for

  a. Pointing to memory locations

  b. Pointing to the next instruction

  c. The storage part of arithmetic or logic operations

  d. Holding the instruction during execution

5-11. The result of an ALU operation is usually placed in the

  a. Program counter   c. Instruction register

  b. Accumulator   d. Temporary register

5-12. You would expect the accumulator in a 16-bit microprocessor to be at least _____ bits long.

  a. 8   d. 24

  b. 12   e. 32

  c. 16

5-13. A double-length accumulator for the microprocessor in question 5-12 would be _____ bits long.

  a. 8   d. 24

  b. 12   e. 32

  c. 16   f. 64

5-14. Why would you want a microprocessor to have a double-length accumulator?

5-15. Some microprocessors have two accumulators for added flexibility. Give one example of this added flexibility.

5-16. The program counter keeps track of which program instruction is to be executed. It always points to the _____ instruction.

5-17. The program counter in a typical 8-bit microprocessor is _____ bits long. This means it can address 65,536 different memory locations.

  a. 4   c. 16   e. 32

  b. 8   d. 24   f. 64

5-18. Why does the program counter need to be able to address any memory location even though a program cannot use all of memory?

5-19. When you power up a microprocessor, the reset command always starts the program counter at

  a. The same location

  b. Memory location 0000

  c. Memory location FFFF

  d. A random memory location

5-20. List the three common status register bits. Briefly explain what setting each of these three bits means.

5-21. The status register is tested by certain program instructions. The results of these tests are used to _____ .

5-22. The stack pointer is a _____ purpose register which points to the memory address for the next stack entry.

5-23. When data is pushed onto the stack, the stack pointer value is _____ (decremented/incremented).

5-24. Typically, you could expect the general-purpose registers in a 16-bit microprocessor to be _____ bits long.

  a. Either 8 or 16   c. 12

  b. Either 16 or 32   d. 64

5-25. The memory address register's output drives the _____ bus.

5-26. The data on the 16-bit address bus is used to select a desired _____ .

5-27. You would expect a 16-bit microprocessor that can address 1,048,576 memory locations to have a _____ -bit memory address register.

5-28. The instruction register holds the program's instruction while it is being executed. The instruction register's input is connected to the _____ , and its output is connected to the _____ .

5-29. The microprocessor has a two-phase cycle as each instruction is called and carried out. This is called the microprocessor's _____ cycle.

5-30. During the _____ cycle an instruction is moved from program memory into the instruction register.

5-31. During the _____ cycle the instruction is decoded and carried out.

5-32. The instruction register is _____ bits long.
  a. 4   c. 16
  b. 8   d. No fixed number of

5-33. Explain why the microprocessor uses a temporary register. Would you expect the temporary register to be in the same place for different microprocessors? Why?

5-34. The microprocessor's clock provides the control logic with its basic
  a. Register   c. Timing
  b. Reset   d. Instruction

5-35. The job of the microprocessor's control logic is to _____ .

5-36. The microprocessor's internal data bus is
  a. Bidirectional
  b. As wide as a data word
  c. Used to interconnect the logic devices
  d. All of the above

5-37. Name one logic device which has bidirectional communications with the internal data bus. Name one that does not.

CRITICAL THINKING QUESTIONS

5-1. You are going to design and build a microprocessor-based device. Explain why you would want to have both the block diagram and the programming model for the microprocessor.

5-2. If you were going to use the very simplest (and probably therefore the least expensive) microprocessor, how would you expect the block diagram to be different from the one used in this chapter?

5-3. Why do you think that most microprocessors use a number of general- and special-purpose registers rather than simply storing data in memory?

5-4. Program counters normally increment as each instruction is executed. Is there any reason you can think of why they couldn't decrement? If they did decrement, what would have to change to make the microprocessor work?

5-5. Some microprocessors do not use external memory for the stack but, rather, have a few internal memory locations (registers) for the stack. What are the limitations of using this architecture?

5-6. Some microprocessors have a wider internal data bus than external bus (for example, a 32-bit internal data bus and a 16-bit external data bus). What has to happen for this architecture to work? What are the advantages and disadvantages of this architecture?

1. The block diagram. Because it includes many parts of the microprocessor which the programmer cannot change. The programming model only includes those parts that the programmer uses directly.
2. *d*
3. *c*
4. *c*
5. *b*
6. *c*
7. *a*
8. *c*
9. *a*
10. *d*
11. *d*
12. *d*
13. *a*
14. *b*
15. *c*
16.

	Result	Z	N	C
a. 1	1 1 1 1 1 1 1 1	0	1	0
b. 1	0 0 0 0 0 0 0 0	1	0	1
c. 1	1 1 1 1 1 1 1 0	0	1	1
d.	1 1 1 1 1 1 1 1	0	1	0
e. 1	0 0 1 0 0 0 0 0	0	0	1
f.	1 0 0 0 0 0 0 0	0	1	0
g.	0 0 0 1 1 1 1 1	0	0	0
h. 1	0 1 0 0 0 0 0 1	0	0	1

17. The three steps will be

    1 1 1 1 1 1 0 1   start
    1 1 1 1 1 1 1 0   1st increment
    1 1 1 1 1 1 1 1   2d increment
    0 0 0 0 0 0 0 0   3d increment

    The zero bit will be set in the third increment. You could use the following:

    0 1 1 1 1 1 0 1   start
    0 1 1 1 1 1 1 0   1st increment
    0 1 1 1 1 1 1 1   2d increment
    1 0 0 0 0 0 0 0   3d increment

    This time you would check the status register's negative bit.

18. *c*
19. *e*
20. 20FA
21. Address contents

Address	contents
20FF	1st PC Low
20FE	1st PC High
20FD	2d PC Low
20FC	2d PC High
20FB	8-bit no.
20FA	Next

22. *c.*
23. Special purpose
24. *c*
25. *d*
26. *d*
27. *b*
28. *d*
29. *d*
30. *c*
31. *b*
32. *a*
33. *d*
34. *c*
35. *b*
36. One possible series of steps is
    a. Load offset word into accumulator
    b. Load accumulator into one of the temporary registers
    c. Load program counter's lower byte into the other ALU temporary register
    d. Store result (sum) in accumulator
    e. Load program counter's upper byte into memory address register's upper byte
    f. Load accumulator into memory address register's lower byte
37. *c*
38. *a*
39. *c*
40. *b*
41. *b*
42. *b*
43. *c*

CHAPTER 6

An Introduction to Microprocessor Instructions

■

CHAPTER OBJECTIVES

This chapter will help you to:

1. *Define* the terms instruction set, op code, direct addressing, indirect addressing, and mnemonic.
2. *Explain* the two things an instruction must do.
3. *Name* and *explain* the nine common types of microprocessor instructions.
4. *Recognize* the basic form of a mnemonic-address expression and equation.
5. *Differentiate between* general-purpose and arithmetic and logical expressions.
6. *Explain* the operation of a subroutine.
7. *Name* and *explain* the basic microprocessor addressing modes.

────────

The microprocessor's instructions tell it what to do and where to get the data it is to act on as it executes the instruction. The complete collection of these instructions is called the *microprocessor's instruction set*. Although the instruction set for each microprocessor is different from that of other microprocessors, there is a lot that is the same.

■

6-1 WHAT IS AN INSTRUCTION SET?

A microprocessor instruction is a binary word. When it is read as an instruction, this binary word tells the microprocessor to do one simple task. No other binary word tells the microprocessor to do this task. Most microprocessor instructions let you move or process data. The data may be in memory, or it may be in one of the microprocessor's registers. A few other microprocessor instructions let you do "housekeeping" functions. These instructions let you control certain microprocessor instructions.

When we speak of the microprocessor *instruction set,* we are talking about all the instructions the microprocessor knows how to

use. It is the complete set of instructions that allows the microprocessor to manipulate data by moving it or processing it, and to perform housekeeping functions. Each microprocessor has its own, unique instruction set. If a microprocessor's instruction set is identical to the instruction set of another microprocessor, then they are the same.

The microprocessor's instruction word is the same length as its data word. That is, the instruction word in an 8-bit microprocessor is 8 bits long. The instruction word in a 16-bit microprocessor is 16 bits long. However, a complete instruction may take one, two or three words. Therefore, an 8-bit microprocessor's instruction may be 8, 16, or 24 bits long. Usually 16- and 32-bit microprocessors use only one- or two-word instructions. This means that the instruction is from 16 to 64 bits long. Later in this chapter, we will see how all these bits are used.

During the instruction fetch cycle, the instruction is sent to the microprocessor's instruction register, decoder, and control logic. These logic functions determine which instruction it is, and send control signals to the other logic in the microprocessor. This logic carries out the instruction.

The instruction is placed in the instruction register during the microprocessor's fetch cycle. It is during the execute cycle that the microprocessor's decoding control logic makes the processor do what the instruction tells it to. The diagram shown in Fig. 6-1 illustrates the microprocessor's fetch/execute cycle.

If we look more closely at a microprocessor instruction, we find that the instruction must do two things. First, the instruction must tell the microprocessor what to do. That is, it must give the microprocessor a command. For example, the instruction must tell the microprocessor to add, clear, move, shift, or do any other action that it can do.

Second, the instruction must give the microprocessor some address information. That is, it must tell the microprocessor the location of the data to work on. For example, the instruction may say add from memory to the accumulator, clear the accumulator, move data from register A to register B, shift the accumulator, and so on. You can see that each example not only tells the microprocessor what to do but also tells the microprocessor which logic function it is to work on.

An instruction can be broken into two parts, called the *op code* (*operation code*) and the *address*. This is shown in Fig. 6-2. The op

code tells the microprocessor what to do, and the address tells the microprocessor where to take the action. Often you will find that the first word of a multiword instruction is the op code. Either the second word or the second and third words are address words.

But, this does not mean that single-word instructions do not have addresses. All instructions have some form of address which is either stated or implied. We will look more closely at addressing modes later in this chapter.

This chapter introduces the nine major kinds of instructions. We will find that most microprocessors have more than nine different instructions. Typically, today's microprocessors have a minimum of 75 to 100 instructions. Some of the more powerful microprocessors have hundreds of them.

There are many more op codes than there are basic instruction types, and there are many more instructions than there are op codes. This occurs because each of the basic instruction types has many different op codes. Each op code is a variation on the basic instruction type. Also, an op code can be combined with a number of different addressing modes. All of these combinations result in the microprocessor having a great number of possible instructions available to the programmer. This expansion of the basic instruction type is shown in Fig. 6-3 on the next page.

The microprocessor's addressing modes are the way the microprocessor identifies the location of the data. As we will see, the normal microprocessor has many different ways it can identify data in a particular location. Each different way may be used at a different time, depending on other circumstances.

For example, most microprocessors have a *Clear op code*. What is cleared depends on its instruction set. For some microprocessors, there may only be a single Clear instruction: Clear accumulator. For other microprocessors, there may be a number of Clear instructions, such as Clear accumulator A, Clear accumulator B, Clear register A, Clear register B, Clear register C, Clear register D, or

From page 102:

Housekeeping functions

Instruction set

On this page:

Operation code (op code)

Address

Clear op code

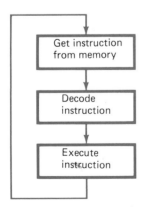

Fig. 6-1 The microprocessor's fetch/execute cycle. This is the basic cycle that the microprocessor goes through to get the next instruction in the program and to carry out that instruction.

Fig. 6-2 The op code and address parts of an instruction word. The op code (operation code) tells the microprocessor what to do. The address tells the microprocessor where the data it is to act on is located.

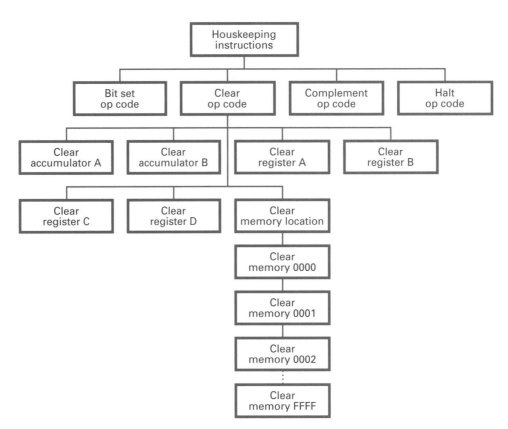

Fig. 6-3 **The hierarchy of instructions. This chart shows that the seven Clear instructions are different addressing modes in the Clear group, which is a part of the housekeeping instruction group.**

even Clear memory location n. In this example, you can see that seven different addresses turn one op code, Clear, into seven different instructions. The seventh instruction (Clear memory location n) actually becomes $FFFF_{16}$ instructions because each memory location has a different address. This expansion of the op code is illustrated in Fig. 6-3.

In a similar example, the microprocessor may have an instruction which tells the microprocessor to load register B with the data from a particular memory location. The op code is load register B. The address is a particular memory location. You may, however, use a number of different instructions to do this job.

For example, the memory location may be contained in the instruction. This is called *direct addressing*. There may also be an instruction which tells the microprocessor to perform the load register B using the data in the memory location which is pointed to by the HL register pair. This is called *indirect addressing*. On the other hand, another version of this instruction may tell the microprocessor to load register B with the data pointed to by the index register plus or minus a certain 8-bit offset.

This is called an *indexed instruction*. In this example, the one op code, load register B, is expanded into three instructions by the various addressing modes.

Self-Test

Answer the following questions.

1. The microprocessor's op code tells the microprocessor
   a. What to do
   b. Where to do it
   c. What to do and where to do it
   d. None of the above
2. A microprocessor instruction may be _____ as long as its data word.
   a. Just   *c.* Three times
   b. Twice   *d.* Any of the above
3. The address portion of the microprocessor's instruction tells the microprocessor
   a. What action to take using the op code
   b. Where the data is located that the instruction is to act on when it is executed
   c. Where to find the instruction
   d. How long the instruction will be

4. The number of times a microprocessor's instruction set uses a particular op code depends on
   a. The op code length
   b. How many different address modes the op code uses
   c. The amount of microcomputer memory
   d. All of the above

6-2 MNEMONICS

A microprocessor instruction is a binary number. Even single-byte binary numbers are hard to remember. It would be very difficult to remember instructions consisting of 2-, 3-, or 4-byte binary numbers. You have learned how the hexadecimal numbering system is used as a shorthand method to express large binary numbers. We could use the hexadecimal shorthand when writing instruction codes, but there would still be a problem. The numbers do not give us any feeling for the instructions they stand for. Instructions in hexadecimal are difficult to learn and even more difficult to use.

To solve this problem, we have mnemonics (pronounced nee-*mon*-iks). A *mnemonic* is an abbreviation that reminds us of what it stands for. In most instruction code mnemonics, the op code is abbreviated to three letters. For example, the mnemonic for Clear is usually CLA. If the microprocessor has two accumulators, the two Clear accumulator instructions might be CLA A and CLA B. The op code is CLA in both cases. The address is either accumulator A or accumulator B. Likewise, the mnemonic for Load is LD, and the instruction Load register B might have the mnemonic and address of LD B. You should understand that there are no right or wrong mnemonics. There are, however, some mnemonics which have become very popular (such as CLA for Clear) through common usage.

Obviously, when numeric data or a memory location goes to the op code, a number is still used. For example, the JMP (meaning Jump) needs an address to jump to. A jump instruction might read

JMP 10CA

Here, the address is expressed as a four-digit hexadecimal number, meaning memory location 0001000011001010_2. The mnemonic JMP is much more meaningful than an actual op code such as $A3_{16}$ (11000011_2). This mix of mnemonics and numbers makes instructions easy to remember. Later, in the chapter on pro-

gramming, we will learn even the absolute hexadecimal values for addresses may be replaced with meaningful words or symbolic addresses, so the instruction might read

JMP START

where START is the symbolic address given to the hexadecimal address 10CA.

We will use mnemonics when writing programs. We will find that mnemonics are not only handy when we are thinking about microprocessor instructions, but they are also built into the microprocessor's assembly language. An *assembler* is a program that converts mnemonics and symbolic addresses into binary instructions and binary addresses.

This text uses mnemonics for the various instructions that we study. Of course, somewhere there must be a table that matches mnemonics with the actual binary numbers that are the op codes. Remember, *the microprocessor does not process mnemonics, op codes, or hexadecimal numbers; it works on binary words*. Therefore, all of our shorthand notation must be converted into these binary numbers sometime before the instructions can actually be loaded and executed. The set of binary numbers which make up a program is called a *machine language* program. Very few people work in machine language because it is so difficult to understand.

Usually the op codes and the addresses are shown in hexadecimal. The manufacturer's literature for most microprocessors includes a card with an instruction set summary. Such cards include both a mnemonic and the op code. The exact op code is not needed until you are ready to program a particular microprocessor. Many programmers are not even aware of the actual op code, because that process is included in the programming language.

Self-Test

Answer the following questions.

5. As we have learned, both A3H and JMP mean "jump" for a certain microprocessor. Which of these two is the op code, and which is the mnemonic? Why?
6. Many mnemonics are almost self-explanatory. What would you *guess* the following mnemonics to mean?

CLA	MOV
INC	AND
NOP	HLT
ADD	DEC

7. Some mnemonics are four letters long instead of three. Should this change the length of the instruction the mnemonics represent? Why?

8. One of the assembler's jobs is to convert assembly language source code into binary object code or machine code for the target microprocessor. To do this, you would expect the assembler to have a table where it can look up the microprocessor's _____ for each mnemonic.
   a. Op code   c. Hexadecimal code
   b. Address   d. All of the above

9. A microprocessor has a unique binary word for each of its instructions. Would you expect a microprocessor to have a unique mnemonic for each of its instructions? Why?

6-3 THE MICROPROCESSOR'S BASIC INSTRUCTION TYPES

As we learned in Sec. 6-2, the op code tells the microprocessor what it is to do. The op code is followed by an address which tells the microprocessor where the action is to take place. As we learned earlier, there are many different types of op codes. Different manufacturers try to describe the op codes or instruction types in different groupings.

For most major microprocessors, we can group microprocessor instructions into nine different categories. Although some manufacturers will group them somewhat differently, it is usually relatively easy to see these different types. The nine different types of instructions are:

Data transfers
Exchanges, block transfers, and searches
Arithmetic and logic
Rotates and shifts
General-purpose and CPU control
Bit set, bit reset, and bit test
Jumps
Calls, returns, and restarts
Input and output

In the following paragraphs, we learn what basic functions a microprocessor performs when it executes each of these instruction types. Later, as we study several commercially available microprocessors, we will see the exact implementation that particular manufacturers give these general instruction types.

Often the manufacturers use a mnemonic-address expression and an algebraic equation to describe what the instructions will do. We will use similar mnemonic-address expressions and equations here to describe the instructions. They will be similar to the ones used by various microprocessor manufacturers, but there are some slight differences between the notation used by one manufacturer and another.

The terms *load, move, data transfer,* and *store* are used by various manufacturers for the same type of instructions. In this text, we call them "data transfer" instructions. Data transfer instructions allow the programmer to *copy* data from register to register, and between registers and memory locations. The data transfer instructions come in four forms: register-to-register, register-to-memory and memory-to-register, immediate, and register-to-stack.

The simplest instructions are the register-to-register instructions. For example, the Z80 instruction LD B,E is a single-byte instruction. The instruction tells the Z80 to copy 8 bits of data currently in register E into register B. This operation can be described with the mnemonic-address expression

$$LD\ r,r'$$

and the equation

$$r \leftarrow r'$$

where r and r′ are two different registers. It is important to understand these operations are copies, *not* moves. That is, the data in the source is copied into the destination, and the original data stays in the source.

The register-to-memory and memory-to-register instructions must use some form of addressing. That is, the address of the memory location used in the transfer must be part of the instruction. As you will learn in the section on memory addressing, there are a number of different addressing modes which can be used. For example, the mnemonic-address expression

$$LD\ r,(HL)$$

and the equation

$$r \leftarrow (HL)$$

tells us the register r is loaded from the memory location *pointed* to by register pair HL.

This is called an *indirect instruction* because the instruction does not contain the address of the memory location. The memory location address must have been put into the register pair called out in the instruction. Indirect addressing is shown by the parentheses around HL in the instruction's address.

The data transfer instructions include the load immediate instructions. These allow the programmer to load any of the microprocessor's registers with a single instruction. The data to be loaded follows as the word after the op code word. For example, the mnemonic-address expression

$$LD\ r,\ n$$

and the equation

$$r \leftarrow n$$

tells us that register r is loaded with the value n, which is the second word on the instruction.

The *exchange instructions* are really a special version of the data transfer instructions (sometimes called *swap*) used to exchange the data in one set of registers with the data in another set of registers. The registers used depend on the particular microprocessor. The exchange instructions do not leave the original data in the source registers. The source data in both sets of registers is overwritten with data from the other register set. For example, the Z80 has an exchange instruction which can be described by the mnemonic-address expression

$$EX\ DE,HL$$

and the equation

$$DE \leftrightarrow HL$$

Executing this exchange instruction swaps the data in the DE register pair with the data in the HL register pair. Normally, the exchange instructions are grouped with the block transfer instructions because they use some of the same capabilities of the microprocessor.

The block transfer and search instructions are very helpful when you need to work with a lot of data and to use as few instructions as possible. The *block transfer instructions* allow the programmer to move a block of data from one group of memory locations to another. For example, the block transfer instruction might be used to move a 256-byte block of data starting at hex address 0100 to the memory locations starting at hex address 0400.

The *search instruction* allows the programmer to search a block of data for a word or even for a series of words. The search instruction is another instruction which allows the programmer to replace a commonly used routine

with a single instruction. As you can see, the block transfer and search instructions use the microprocessor's capability to address a block of data.

The arithmetic and logical instructions allow the programmer to modify the data in the registers or sometimes the data in memory. This is done by arithmetically or logically combining the data in one of the registers with the data in the accumulator.

Most microprocessors have a variety of *arithmetic instructions*. They offer a number of different forms of add and subtract. Arithmetic instructions usually have versions which use the carry bit and versions which do not use the carry bit. For example, the mnemonic-address expression

$$ADC\ (HL)$$

and the equation

$$A \leftarrow A + (HL) + CY$$

tell us that executing this instruction replaces the accumulator's current contents with a value which is the sum of the accumulator's current contents, the contents of the memory location *being pointed to* by the HL register pair, and the contents of the status register's carry bit (CY).

The arithmetic instructions include the ability to increment and decrement any register. The mnemonic-address expression

$$DEC\ m$$

and the equation

$$m \leftarrow m - 1$$

tell us that the current value of memory location m is replaced by the value minus 1.

The *logical instructions* typically include AND, OR and Exclusive OR (XOR). These are very important instructions for creating masks or otherwise isolating some specific data from a complete data word. For example, the mnemonic-address expression

$$AND\ n$$

and the equation

$$A \leftarrow A \cap n$$

tell that the current value of the accumulator is replaced with the current value ANDed *bit for bit* with the value n. Some microprocessors also have an instruction which performs the compare function. The *compare function* per-

Exchange instructions

Swap

Block transfer

Search instruction

Arithmetic instructions

Logical instructions

Compare function

Memory address	Memory contents	Comments
0000	IN	Input status register
0001	00	
0002	ANI A	Mask all but Data Available bit. Is Data Available bit set?
0003	02	
0004	JZ	No, try status word again.
0005	00	
0006	00	
0007	IN	Yes, input data word
0008	01	
0009	RET	Leave routine

(a)

Memory address	Memory contents	Comments
0000	IN	Input status register
0001	00	
0002	ANI A	Mask all but Data Available bit. Is Data Available bit set?
0003	02	
0004	JZ R	No, jump back 6 bytes and try status word again
0005	FA	
0006	IN	Yes, input data word
0007	01	
0008	RET	Leave routine

(b)

Fig. 6-17 A program listing for the program flowcharted in Fig. 6-16. The first listing (a) used the jump if zero direct addressing instruction. (b) Using the jump if zero relative addressing instruction.

```
0 0 0 0   0 1 1 0                Binary 6
1 1 1 1   1 0 0 1                Complement of 6
0 0 0 0   0 0 0 1                Add binary 1
1 1 1 1   1 0 1 0                2's complement of 6
   F        A                    Hexadecimal offset
```

In Fig. 6-17(a) and Fig. 6-17(b), you can see that both programs occupy almost the same number of memory locations. The program that uses relative addressing, however, is 1 byte shorter, because the jump direct instruction at memory location 0004H in Fig. 6-17(a) takes 3 bytes.

Figure 6-18 shows the program from Fig. 6-17(b). Here, the program starts at memory location 01AEH. None of the instruction mnemonics in the listing nor any of the addresses are different. Jumping back six locations relative to the program counter, even though the

Memory address	Memory contents	Comments
01AE	IN	Input status register
01AF	00	
01B0	ANI A	Mask all but Data Available bit. Is Data Available bit set?
01B1	02	
01B2	JZ R	No, jump back 6 bytes and try status word again
01B3	FA	
01B4	IN	Yes, input data word
01B5	01	
01B6	RET	Leave routine

Fig. 6-18 A program listing for the program flowcharted in Fig. 6-16 but relocated to start at memory location 01AE. Note that the memory contents are identical to the memory contents shown in Fig. 6-17(b).

program counter now points to memory location 01B4H, still takes you back to the input status register instruction (now at memory location 01AEH).

On the other hand, moving the program from 6-17(a) would require a code change. The JZ0000 instruction would have to have its address changed so that the instruction reads JZ01AE.

As you have seen, relative addressing frees the programmer from worrying about where the program's absolute address is. But the programmer must be good at hexadecimal arithmetic to compute the relative addresses. Using an assembler relieves the programmer of the need to calculate relative addresses. The assembler does all the calculations itself.

Self-Test

Answer the following questions.

33. The memory addressing mode that takes the least time is
    a. Direct addressing
    b. Extended direct addressing
    c. Immediate addressing
    d. Inherent addressing
34. When you use immediate addressing in an 8-bit microprocessor, you expect the second byte to be
    a. A memory address between 0 and 255
    b. 8 bits of data
    c. Easily referred to by many instructions
    d. All of the above
35. A 16-bit microprocessor uses a 22-bit address field. That is, it can address

4,194,304 memory locations. A direct address instruction can address any memory location in a single instruction. The first instruction word is the op code. How many bytes are used to make up its longest instruction? Why?

36. Indirect register addressing is so named because the register in the instruction
    a. Is where the data would be stored
    b. Tests the data directly
    c. Points to the memory location
    d. Is only used for memory locations 0 to 255

37. Rank the three fastest memory addressing modes by speed in order of increasing execution times.

38. Explain why you think direct addressing or register indirect addressing is the simplest form of memory addressing.

39. Given the following index register values, and index mode instructions, explain what will happen when each instruction is executed.

Index register	Instruction
a. 01F0	ADD A,IX + 00
b. 0005	LD A,IX + 05
c. 0011	LD A,IX + 00
d. 00FF	INX
e. A1AE	SUB A,IX + AE
f. A105	AND A,IX + A0
g. FFFF	DEX
h. 0AEE	CLA

40. An instruction using relative addressing is always two bytes long. Relative to the instruction's memory location, what is its addressing range?

41. For the following relative address instructions, tell what will happen as each instruction is executed.

Instruction Address	Instruction	
a. 010F	LD	A
0110	JZ	R
0111	F1	
0112	RET	
b. 00DE	JMP	R
00DF	01	
00E0	CLA	A
00E1	CLA	B
c. 0021	INX	
0022	JNZ	R
0023	0D	
0024	DEX	
d. 0FAB	ADD	B
0FAC	JNC	R
0FAD	8D	
0FAE	RET	

42. Often the microprocessor's single-byte addressing range (256 memory locations) is called a *page*. Why do you think the relative addressing mode is said to address a "floating" page?

SUMMARY

1. A microprocessor instruction is a unique binary word which, when interpreted by the microprocessor as an instruction, tells it to perform a certain specific operation.

2. The microprocessor's instruction set is the complete collection of all of the instructions the microprocessor knows how to perform. Its instruction set is unique to that particular model of microprocessor.

3. A microprocessor's instruction word is the same length as its data word. However, a microprocessor may use a number of instruction words to make up a complete instruction.

4. The microprocessor's instructions are decoded and carried out when an instruction is loaded into the microprocessor's instruction register by the fetch cycle. During the execute cycle, the instruction decoder and the control logic are used to make the microprocessor perform the function indicated by the instruction.

5. A microprocessor instruction contains two pieces of information. First, it tells the microprocessor what to do. Second, it tells the microprocessor the address of the data or target of the instruction.

6. The two parts of the instruction are called the op code (operation code) and the address.

7. Even single-word instructions have both an op code and an address; however, the address may be implied rather than stated.

8. There are nine basic types of instructions. However, there are many more op codes because many of the op codes are variations on the basic instructions.

9. The microprocessor's instruction set has many more instructions than it has different op codes because each op code may be used a number of times. The variations using a single op code reflect the microprocessor's different addressing modes.

10. A microprocessor instruction is just a binary number. We call the actual binary instruction the *machine code*. Often, we use a wordlike abbreviation to remind us of the instruction's function. These are called instruction *mnemonics*.

11. Mnemonics must be interpreted into machine code by a human or by a computer program. The computer program which converts mnemonics into machine executable instructions is called an assembler.

12. We may class microprocessor instructions by type. One grouping of microprocessor instructions is data transfer; exchanges, block transfer, and search; arithmetic and logic; rotates and shifts; general-purpose; bit operation; jumps; calls and returns; and input/output.

13. The data transfer instructions allow the programmer to move data from register to register and between registers and memory locations. The source data is copied into the destination.

14. The exchange instructions are used to swap data between sets of registers. The block transfer instructions are used to move a block of data from one set of locations to another set of locations. The search instructions allow the programmer to search a block of memory locations for a particular set of data.

15. The arithmetic and logical instructions allow the programmer to modify the data in a register (and sometimes in a memory location) using the arithmetic or logical operators: add, subtract, add with carry, subtract with carry, AND, OR, and Exclusive OR (XOR).

16. The arithmetic and logical instruction group also includes instructions which let

you increment or decrement the current value contained in a register or memory location. They also include instructions which allow the programmer to compare the value of a register or memory location to the value in another register or memory location. The compare instruction only affects the microprocessor's status register.

17. The rotate and shift instructions let you move the data in a register or in a memory location to the right or to the left. A rotate instruction is a closed loop instruction. That is, the data that is moved out the end is put back in at the other end. The shift instruction loses the data that is moved out of the last bit location.

18. The microprocessor's general-purpose instructions usually address the microprocessor itself. They include the functions of halting the microprocessor, enabling the interrupts, disabling the interrupts, complementing a register, etc.

19. The bit test, bit set, and bit reset instructions give the programmer the ability to work at the bit level with the data which is in a register, memory location, or other location. This is an important instruction group when you are working with data from external hardware.

20. The jump instructions are used to permanently change the program counter's value. When you change the program counter's value, you start executing a different part of the program. The new starting address is given as a part of the jump instruction.

21. There are two different kinds of jump instruction—unconditional and conditional. Executing an unconditional jump instruction simply changes the program counter's value. Executing a conditional jump instruction only changes the program counter's value if the conditional test is met.

22. The call instructions are used to start execution of a subroutine. The call instructions are different from the jump instructions because executing the call instruction saves the current program counter value on the stack. Then program execution begins at the program step specified by the call instruction.

23. When the current value of the program counter is saved on the stack, executing a return instruction loads the program counter with the value saved on the stack.

This allows the program to continue execution from where it left off.

24. Both call instructions and return instructions may be either unconditional or conditional.

25. A subroutine which has been started by a call instruction can have a call instruction. Executing this second call instruction starts a subroutine within a subroutine called nesting. Most microprocessors can support many nesting levels.

26. The input and output instructions are used to move data into and out of the microprocessor's I/O ports. I/O instructions are only used on microprocessors which have separate I/O capability.

27. The microprocessor's addressing modes are the different ways the microprocessor uses to access memory locations. All instructions have some kind of an addressing mode.

28. The addressing mode in which the address is contained in the instruction itself is called inherent addressing. Instructions using inherent addressing only require one word.

29. The immediate addressing mode uses two-word instructions. The first word is the instruction (op code) and the second word is the immediate data to be acted on.

30. An instruction which uses direct addressing has the actual address in the second or second and third instruction words.

31. Register indirect addressing is another single-word instruction. When you use register indirect addressing, the instruction word names a register which points to the desired memory location.

32. Indexed addressing uses a special register called the index register. The instruction's second word is a data value which is added to the index register's current value. The sum is used as the memory address for the instruction.

33. The relative addressing mode is like indexed addressing; however, the data value is added to the program counter's current value. The sum is used as the instruction's address. Relative addressing allows a programmer to create position-independent code.

CHAPTER REVIEW QUESTIONS

Answer the following questions.

6-1. Why is a microprocessor instruction a unique binary word?

6-2. How does the microprocessor tell the difference between an instruction word and a data word of the same value?

6-3. What is the microprocessor's instruction set?

6-4. How can an instruction in a 16-bit microprocessor be 32 bits long? Why can't the instruction be 24 bits long?

6-5. Briefly explain what happens after an instruction is loaded into the microprocessor's instruction register.

6-6. A microprocessor instruction can be broken into two parts. What are they, and what function does each part perform?

6-7. What kinds of instructions do not address anything?

6-8. Why are there many more op codes than there are basic instruction types?

6-9. Why are there more instructions than there are op codes?

6-10. Why do we use a mnemonic instead of the actual binary op code?

6-11. What do we call the computer language used to convert instruction mnemonics into machine readable code?

6-12. Briefly explain what functions the data transfer instructions perform.

6-13. Briefly explain what functions exchange instructions perform.

6-14. Briefly explain what functions block move instructions perform.

6-15. Briefly explain what functions the arithmetic instructions perform.

6-16. Briefly explain what functions the logical instructions perform.

6-17. Briefly explain what functions the shift and rotate instructions perform.

6-18. Give an example of two different general-purpose instructions.

6-19. Briefly explain what functions the bit-oriented instructions perform.

6-20. Briefly explain how the jump instruction works and give an example of both a conditional jump and an unconditional jump.

6-21. Explain the steps the microprocessor goes through when a call instruction is executed.

6-22. Explain the purpose of each step listed as an answer to question 6-21.

6-23. What is the difference between a conditional call instruction and an unconditional call instruction?

6-24. Briefly, using an example, explain subroutine nesting.

6-25. Where are the I/O instructions used? What are they used for?

6-26. Which addressing modes put the instruction and the address in a single word?

6-27. Where is the data located in an instruction using immediate addressing?

6-28. What is the difference between immediate addressing and direct addressing?

6-29. Explain how register indirect addressing works. Why can it be a single-word instruction?

6-30. What is the difference between indexed addressing and relative addressing?

CRITICAL THINKING QUESTIONS

6-1. Suppose that you examine the contents of memory location 1AE4H and find that it contains $1010\ 0011_2$. Without knowing anything else, can you tell whether this is the jump instruction, the binary equivalent of 163_{10}, or the 2's complement representation of -35_{10}? Why?

6-2. Suppose that you examine the contents of memory location 0000H and find that it too contains $1010\ 0011_2$. Without knowing anything else, can you tell whether this is the jump instruction, the binary equivalent of 163_{10}, or the 2's complement representation of -35_{10}? Why?

6-3. Long ago the computer industry turned away from machine language programming and started using assemblers and other high-level languages. Why do you think this happened?

6-4. In some situations the data from an external device has some bits which are set in a random fashion. Often an AND immediate instruction is used to mask out the random bits. Explain what is happening here.

6-5. Why do 32- and 64-bit microprocessors not need three- and four-word instructions when they are using direct addressing?

6-6. There are times when it is necessary to write a program very carefully in order to use the smallest amount of memory possible. What conditions would cause this to be necessary?

Answers to Self-Tests

1. *a*
2. *d*
3. *b*
4. *b*
5. A3H is the op code, and JMP is the mnemonic. A mnemonic is a wordlike abbreviation, such as

JMP. An op code is the actual binary instruction which A3 represents.

6. CLA = Clear
   MOV = Move
   INC = IncrementAND = Logic AND
   NOP = No Operation

HLT = Halt
ADD = Add
DEC = Decrement

7. No. The mnemonic is only an abbreviation which must be matched to its proper op code.

8. *a*

9. Yes. The mnemonic must also be unique if it is to be assigned to a unique op code.

10. d

11. a. Load the accumulator (A) with the number n.
    b. Load the memory location pointed to by the register pair HL with the contents of register r.
    c. Load the register IY with the two-word data nn.
    d. Put the current contents of the stack location into register L and the current contents of the next stack location into register H.

12. The exchange instruction swaps the data contained in one register or memory location with the data in another register or memory location. When the exchange instruction is executed, the original data in each register is written over by the data from the other register. When a load instruction is executed, the data is copied from the source to the destination.

13. The instruction register. It is placed in the instruction register so that the instruction may be interpreted by the instruction decoder.

14. The block transfer instruction moves data located in one series of memory locations into another series of memory locations.

15. c

16. a. Add the number n to the current value in the accumulator and leave the results in the accumulator.
    b. Perform an add with carry so that the data contained in the memory location pointed to by the IX register plus the offset (n) are added to the data in the accumulator and the results

are left in the accumulator.
    c. Subtract the value of the data in register r from the current value of the accumulator and leave the results in the accumulator.
    d. Subtract the value of the data in memory location m from the current value of the accumulator; then subtract the current value of the status register's carry bit, and leave the results in the accumulator.
    e. Increment the register pair HL. That is, add 1 to the present value contained in the HL register pair.
    f. Decrement the BC register pair. That is, subtract 1 from the present value contained in the BC register pair.

17. A register is loaded with a preset value, and at a certain point in the program the register is decremented. After the register is decremented, the status register is tested to see if decrementing the register caused its value to become 0. Each time the register is decremented to 0, we say that one "tick" of the clock has passed. The accuracy of the "clock" depends on the stability of the microprocessor's clock and the uniformity of the decrementing operation.

18. a. AND the contents of the memory location pointed to by the HL register pair with the current contents of the accumulator and leave the results in the accumulator.
    b. OR the contents of register r with the contents of the memory location pointed to by the register pair HL and leave the results in that memory location.

    c. AND the contents of register r with the current contents of the accumulator and leave the result in the accumulator.
    d. Perform an exclusive OR between the hexadecimal value 0F and the current contents of the accumulator. Leave the result in the accumulator.

19. a. 0010 1000
    b. 1111 1010
    c. 1110 0110
    d. 1111 0000
    e. 1010 1010
    f. 0000 0000
    g. 1111 1111
    h. 1010 1010
    i. 0101 0101
    j. 1010 1010

20. An OR B is used to pack the two words and make the accumulator contain 1001 1000.

21.	C	ACCUMULATOR
	1	10101000
22.	0	01011110
23.	0	11111110
24.	1	11111100

25. a. 1100 0111
    b. 0000 1111
    c. 1110 1111
    d. 1000 1111
    e. 0000 0000
    f. 1111 1111
    g. 0101 0101
    h. 1010 1010
    i. 1111 0000
    j. 1110 0111

26. The bit set, bit reset, and bit test instructions are used to operate on the individual bits of a data word. This is very useful when external hardware modifies a single or a few bits of a data word, therefore giving each bit a separate and independent meaning.

27. 006A, negative. The instruction is to jump to

memory location 006A if the last operation was not negative, that is, did not set the negative bit.

28. 012D, negative. If the status register's negative bit is set, no jump happens; therefore, the next instruction following the jump instruction is executed. If the jump instruction is at memory location 012A, then the two direct address bytes are at memory locations 012B and 012C. This means that the next instruction must start at memory location 012D.

29. Each time there is a subroutine called, 3 bytes are used for the call instruction. One extra is used for the return instruction. This means that there will be 10 extra bytes used if the subroutine is called three times. Therefore if the routine is 4 bytes or longer, it consumes 12 bytes if it is repeated three times. At 4 bytes, more memory is used.

30. It means that while one subroutine is executing, another is called. If nesting is five levels deep, then there must be five different versions of the program counter stored on the stack. Because the program counter has 16 bits, it takes 2 bytes to store each value of the program counter. The five levels will take 10 bytes.

31. A jump instruction permanently changes the program counter. A call instruction leaves information on the stack so that the original

program execution sequence can be resumed.

32. The in instruction is used to move data from an I/O port into the accumulator. An out instruction is used to move data from the accumulator to an I/O port. The in and out instructions are only used on microprocessors which use a separate address space for interfacing.

33. d

34. d

35. 6. The first 2 bytes make up the op code. Using two 16-bit words (an additional 4 bytes) addresses 4 gbytes of memory. However, a single 16-bit word only addresses 64K memory locations which is not enough, so the next step is two words.

36. c

37. Inherent, indirect, immediate

38. Register indirect addressing is simple because one only has to keep a value in the pointing register. On the other hand, direct addressing is simple because the desired address is contained in the instruction.

39. a. The contents of memory location 01F0 are added to the accumulator.
    b. The accumulator's contents are stored at memory location 000A.
    c. The contents of memory location 0011 are loaded into the accumulator.
    d. The index register now contains 0100.
    e. The contents of memory location A25C are

subtracted from the accumulator.
    f. The contents of memory location A1A5 are ANDed to the accumulator.
    g. The index register now contains FFFE.
    h. The accumulator is cleared. This is not an indexed instruction.

40. Using complement notation a single byte can represent -128 to $+127$. Because the program counter is pointing two memory locations higher than the relative instruction, the relative addressing range is -126 to $+129$.

41. a. The program will jump back to memory location 0103 if a 0 is loaded by the LD A instruction. Otherwise the return instruction will be executed.
    b. Register B is cleared. The CLA A is not executed.
    c. If incrementing the index register does not cause a 0, then the program jumps to memory location 0031. If incrementing the index register causes a 0, the index register is decremented.
    d. If the ADD does not cause a carry, the program jumps to 0F3B. If there is a carry, then return.

42. It is called a floating page because the range of -126 to $+129$ bytes is 256 bytes long (a page), and because it is relative to the program counter, the page moves as the program counter increments.

CHAPTER 7

Communicating with the Microprocessor

∎

CHAPTER OBJECTIVES

This chapter will help you to:

1. *Define* the terms port, programmed data transfer, interface, interrupt, and servicing.
2. *Identify* the basic purpose of an I/O port.
3. *Differentiate between* microprocessor-to-machine and microprocessor-to-human interfaces.
4. *Compare* the two ways that an I/O port can be connected to the microprocessor.
5. *Explain* the two methods the microprocessor uses to service multiple I/O devices.
6. *Map* the interrupt sequence on a flowchart.

The microprocessor is nothing if it cannot communicate with an outside device. To do this, the microprocessor uses its I/O ports. It gets data to process through the input port, and it delivers the results of its processing through its output port. Some microprocessors use a separate addressing method to reach an I/O port; to some the I/O port is just a memory location. There are certain activities which the microprocessor does not need to do until an external device needs attention. To deal with this, there is a special input to the microprocessor called the *interrupt*. External devices use the interrupt to get the microprocessor's attention.

∎

7-1 THE NEED FOR MICROPROCESSOR I/O

The microprocessor must be able to communicate if it is to be of value. The communications may be with another machine or with a human. If the microprocessor cannot communicate with the outside world, it is of no value. It must receive data from somewhere, and when it is finished processing the data, it must send the data somewhere. Microprocessor-based products and systems do not just swallow data, nor do they randomly generate data. All systems have data that is input and data that is output.

Although a microprocessor application may seem to just generate data and not receive it,

in reality it does receive data to work on. For example, think of the test box shown in Fig. 7-1 on page 128. It is designed to check a printer without having a computer connected. This *microprocessor-based test box* is battery operated. It plugs into the printer's input connector and generates electrical signals which cause the printer to print

THE QUICK BROWN FOX JUMPS OVER THE LAZY DOG'S BACK 1234567890

This test box has *output data,* but where is the input data? The answer is, you generate a single bit of input data when you turn on the box. This data tells the box to start generating the text (the output data). Although this is a very simple input, it is input data. With further thought, you realize that data was placed

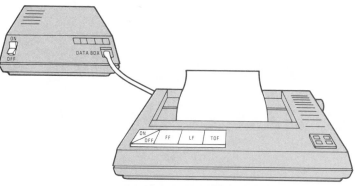

Fig. 7-1 A simple microprocessor-based test box. This device is used to test a printer without having a computer attached. It demonstrates the need for such a system to have input and output.

in the box (probably in an EPROM). This data contains the test words to be sent when the power on command (data) is received.

In fact, this test box might well have several sources of *input data*. First, it needs an input to tell it how fast to send the data to the printer. Second, it needs an input to tell it when to wait while the printer prints the data which has been sent. Third, the box may need to send an alternative message, and therefore it will need an input to tell it which message to send. As shown in Fig. 7-2, this simple test box, which is a very common type of microprocessor-based tester, has many data inputs.

This example shows that even a simple microprocessor-based system has input data and output data. Sometimes you need to look with some imagination to find all the data inputs and, perhaps, all of the data outputs, but you will find them.

If the microprocessor must be able to receive data from the outside world and send data to the outside world, how does it do this? The microprocessor uses an *I/O port* to send and receive data. A *port* is a hole which allows things to get into or out of something. Some micro-

processor ports are only input ports, and some are only output ports. However, it is common to simply call them all I/O ports.

In most ways, you can think of the microprocessor's I/O ports as working like memory locations. The difference is that instead of RAM or ROM, the addressed I/O location has a data register. The data register is used to hold the information you are sending or receiving. Usually, two registers are used; one for output data and one for input data.

Figure 7-3 diagrams how memory and I/O are different. As you can see, the data registers are dual-ported with two ports for each register. One of the ports is connected to the microprocessor's data bus, and the other port is the connection to the outside world. On the other hand, a memory location normally only has one port, the one which connects the memory location to the microprocessor's data bus. *Note:* There are some special *dual-ported memory systems* which are used for special functions.

Often, *special-purpose logic* is attached to the input or output register, as shown in Fig. 7-4. The logic connects the I/O register's

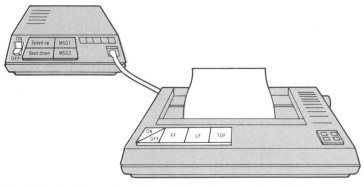

Fig. 7-2 An advanced version of the printer test box. This box uses many different inputs to "set up" the test for the particular printer being tested.

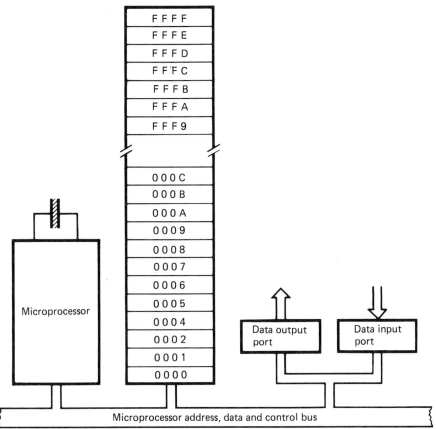

Fig. 7-3 A microprocessor system with both memory and I/O. Note that the I/O uses data registers for both input and output. Also note that the registers are dual-ported. That is, you can read data into the register from one side and read data out of the register from the other side. You read data from and write data to a memory location from the same side or port.

outside world port to the outside world. This logic converts the data word being exchanged with the microprocessor into a different electrical form and helps control the transfer. There are many different types of circuits used for this job. We will study some of the different types used with I/O ports in Chap. 13.

Usually, the data is transferred from the I/O port into one of the microprocessor's registers. Often, the accumulator is used for this exchange of data. For example, a data word at the I/O port may be first transferred into the accumulator and then transferred from the accumulator to a memory location. This opera-

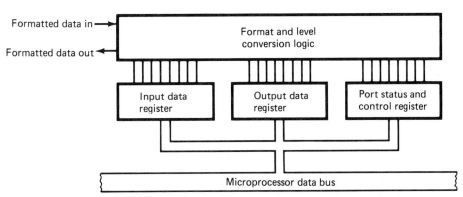

Fig. 7-4 I/O interface logic. The logic which connects the outside world to the microprocessor's I/O port. This is the logic which makes one kind of I/O port different from another. The I/O interface logic is connected to the outside world on one side and to the I/O registers on the other.

tion is called a *programmed data transfer*. Later we will learn of another type of transfer called direct memory access (DMA). Both the programmed data transfer and the DMA use the I/O port to let the microprocessor communicate with the outside world. A programmed data transfer is diagrammed in Fig. 7-5.

The I/O data can come from many different places and can go to many different places. Where the data comes from and where the data goes to depend on the function the microprocessor-based system is to perform. The two basic kinds of I/O data exchange are between the microprocessor-based system and a human and between the microprocessor-based system and another machine.

Operations between a microprocessor-based system and a human are slow. They may be alphanumeric data, commands, or status. Operations between a microprocessor-based system and another machine tend to be much faster. They too may be either alphanumeric data or commands, but they can use many coded transmissions which a human would not understand. For example, think of how slowly data enters a computer as you type at its keyboard. On the other hand, think of how fast the data must be sent to keep up with a printer which types the data at a speed of 120 characters per second.

Often you will hear the term *interface*. When someone speaks of the microprocessor's interface, they are talking in a broad way about all of the microprocessor's circuits which are used to let the microprocessor communicate with an external device. The microprocessor's serial interface, for example, describes the microprocessor's I/O port, control logic, data format conversion logic, connector, and maybe even

cables which are used to allow the microprocessor-based system to connect to a printer, modem, or other similar peripheral.

Self-Test

Answer the following questions.

1. The fundamental purpose of an I/O port is to
   a. Send data to an external device
   b. Receive data from an external device
   c. Allow the microprocessor-based system to have a purpose
   d. Allow the microprocessor to create data on its own
2. The abbreviation I/O stands for
   a. Inside/outside
   b. Internal/outside
   c. Input/output
   d. Intermediate/occasional
3. A simple digital alarm clock is built using a microprocessor. Which of the following is not an input?
   a. The AM-PM indicator
   b. The 60-Hz power line signal which is counted to give time
   c. The switch which turns the alarm function on or off
   d. The "snooze" switch
4. In the above digital alarm clock example, one of the I/O interfaces is a microprocessor-to-machine interface, and the others are microprocessor-to-human interfaces. Which is the microprocessor-to-machine interface? Why?
5. Briefly explain why a microprocessor-based system needs to be interfaced to the outside world. Your answer should discuss the fact that a microprocessor-based system must have both input and output interfaces and why.

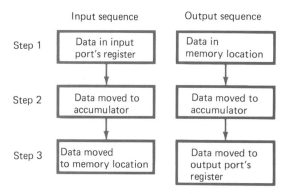

Fig. 7-5 A programmed data transfer. Moving data between a memory location and an I/O port. Note that there are two different flows, depending on the direction of the data. However, no matter what direction the data flows, it always passes through the accumulator.

7-2 CONNECTING THE I/O PORT TO THE MICROPROCESSOR

The simplest microprocessor system usually has more than one I/O port. This means that the microprocessor must have a way to tell which I/O port it is talking to. It does this by addressing the I/O port in much the same way that it addresses a memory location.

There are two ways the I/O port can be connected to the microprocessor so that it can be addressed. It can be memory-mapped or it can be I/O-mapped. In this section, we will look at both types of I/O ports.

If the I/O port is *memory-mapped,* the I/O registers act like read-write RAM at the addressed memory locations. Of course, the I/O ports are addressed (mapped) at memory locations which do not actually have any RAM.

Some microprocessors, the 6802 for example, only allow memory-mapped I/O. That is, the only way an I/O port can be addressed by the microprocessor is to place it at a memory address. However, all microprocessors can use this if the designer feels that memory-mapped I/O is the best way to address I/O in the particular system.

Memory-mapped I/O has advantages and disadvantages. For example, you will find that it allows the programmer to use the wide range of instruction types and memory addressing modes for I/O data transfers. When memory-mapped I/O is used, there is no programming difference between an I/O data transfer and a data transfer with a memory location.

A disadvantage, however, is that memory-mapped I/O reduces the number of memory locations available to the programmer. If an I/O port has an input register and an output register, you must give up at least two memory locations. In many cases, a large block of memory locations are assigned to I/O. When this happens, the entire block of memory locations is not available for ROM or RAM.

Some microprocessors have a special addressing mode called the *I/O mode.* When the programmer wants to make an I/O data transfer, a different set of the microprocessor's control lines is used. These I/O commands, like the memory commands, use the microprocessor's address bus and data bus, but there is one difference. These microprocessors have a special I/O control line or set of I/O control lines.

Control lines signal the external microprocessor support logic that the data transfer is to be with memory or with an I/O device. Usually, there is one line to tell the logic that the transfer is an I/O transfer and another that tells the logic that the transfer is a memory transfer. When the control lines signal a memory data transfer, the I/O support logic is disabled. Likewise, when the control lines signal an I/O transfer, the memory support logic is disabled.

Often the microprocessor does not have as many I/O locations as it has memory locations. For example, one common 8-bit microprocessor (the Zilog Z80) has 64K (65,536) memory locations. However, it only has 256 I/O locations. When an I/O operation happens, the Z80 uses the lower 8 lines of the 16 lines in its address bus. During a data transfer with an I/O location, the data is transferred over the 8-bit data bus just as it is during a data transfer with a memory location. Figure 7-6 on page 132 shows the two forms that microprocessors use to communicate with memory and I/O.

Self-Test

Answer the following questions.

6. Which of the following *could not* be a true statement?
   a. The Z80 can use memory-mapped I/O.
   b. The 6802 can only use memory-mapped I/O.
   c. The Z80 cannot use memory-mapped I/O.
   d. The 6802 does not have special I/O addressing.

7. An I/O port which can do both input and output data transfers uses a minimum of _____ I/O locations.
   a. 1   *c.* 3
   b. 2   *d.* 4

8. An advantage of memory-mapped I/O is that
   a. You do not use up available memory locations
   b. You only have 256 I/O addresses
   c. You have use of a wide range of data transfer instructions
   d. All of the above

9. An I/O data transfer using the Z80 microprocessor's input or output instruction must have an address lying in the range from 00H to FFH. The 6800 microprocessor can have I/O data transfers with addresses lying from 0000H to FFFFH. What is the difference between the way these two 8-bit microprocessors perform an I/O transfer?

7-3 POLLING AND INTERRUPTS

When the microprocessor is asked to communicate with an I/O port, we say that the microprocessor is *servicing* the I/O. The software usually has a special part of the program which is used to service I/O requests. For example, each time you type a character on a personal computer keyboard, a keyboard service routine is called. It transfers the character you typed from the keyboard I/O port into the CPU and then to a data buffer in memory.

When you have one or more I/O devices connected to a microprocessor system, any one of

Answer the following questions.

7-1. What do we mean by the term *I/O port*? Your answer should identify what we mean by the initials I and O and why we use the term *port*.

7-2. Why must a microprocessor have an input port and an output port?

7-3. We say that an I/O port is, in many ways, just like a memory location. How is it like a memory location, and how is it different from a memory location?

7-4. An data register at an I/O port is dual-ported. What does this mean? Why must the I/O port data register be dual-ported but a memory location not have to be dual-ported?

7-5. Frequently, you will find that there are logic circuits between the I/O port's data registers and the circuits in the outside world. What two major functions are performed by these logic circuits?

7-6. What is a programmed data transfer?

7-7. Briefly explain some of the different characteristics you would expect to find between I/O data transfer to humans and to machines.

7-8. Interfacing a microprocessor is the process of making the microprocessor _____ with another device?

7-9. An I/O port's connection to the microprocessor can be described as either _____ mapped or _____ mapped.

7-10. What is the major advantage of having an I/O port replace a memory location? What is a disadvantage?

7-11. Briefly describe how an I/O port is connected to the microprocessor if it does not simply replace RAM or ROM.

7-12. What do we mean when we say we are going to service an I/O port?

7-13. What is the major difference between servicing an I/O port with a polling routine and with an interrupt?

7-14. List the basic steps that a polling routine goes through to find the I/O ports which need service and to service them.

7-15. How can you change the order or priority of service given to a particular I/O port which is serviced by a polling routine?

7-16. What is the major function of an interrupt input on a microprocessor?

7-17. List the major steps which happen once a signal is received at a microprocessor's active interrupt input port.

7-18. Why is it fair to say that an interrupt is very much like a subroutine call?

7-19. What are three different ways a microprocessor can use to tell which external device caused an interrupt?

7-20. List the three different kinds of interrupt inputs found on most microprocessors and briefly explain the major features of each one.

7-1. In physics, there is a conservation of energy law which says that energy can be neither created nor destroyed; it can only change form. How can a conservation of data law explain why a microprocessor must have both input and output ports?

7-2. Often the dual-ported I/O register performs two functions: I/O data word storage and electrical level shifting. Can you think of some examples in everyday life when the microprocessor's I/O port is probably performing electrical level shifting as well as data buffering?

7-3. Programmed data transfer usually moves data through the accumulator as the data passes to or from the microprocessor's memory. As we learned earlier, the accumulator is often the microprocessor's most powerful register. Why do you think this might be an advantage for a lot of I/O transfers, especially those which are human interfaces?

7-4. Small microprocessor systems which do not need enough memory (ROM and RAM) to use the microprocessor's full memory address range often use memory-mapped I/O. Why do you think the designers do this?

7-5. Like many microprocessor processes, the polling and interrupt routines are also used in real life. Can you think of equivalent processes which are used when you are conducting a formal meeting?

7-6. Some hardware systems prioritize interrupt handling by how close the card causing the interrupt is to the CPU card on the system bus. The interrupt line passes through each card, and the card which causes the interrupt can block interrupts from cards further down the bus from reaching the processor. How is this somewhat like using the microprocessor's maskable interrupt? What do you suppose happens if an interrupt is received by a card closer to the CPU when the CPU is processing an interrupt from a card further from the CPU?

Answers to Self-Tests

1. *c*   2. *c*   3. *a*
4. *b*. The power line signal which is counted to give a time reference is a machine-generated signal. All the other inputs come from human actions.
5. A microprocessor processes data. It must receive the data from some source, and once the processing is complete, the microprocessor must output the data to either a human or another machine so the processed data can be used.
6. *c*   7. *b*   8. *c*
9. The Z80 only has 256 (FF) I/O addresses because the Z80 has special I/O control lines and I/O-mapped addresses. The 6802 does not have I/O-mapped I/O transfers and must use memory-mapped I/O, which gives the 6802 the entire range of 0 to 65,535 (0000 to FFFF) memory address locations for I/O address locations.
10. *c*   11. *a*   12. *c*   13. *a*
14. When the microprocessor receives an interrupt, it follows the following steps:
  a. The interrupt is received
  b. The current instruction is finished.

  c. The PC is put on the stack.
  d. The PC is loaded with the interrupt vector.
  e. The interrupt routine at this vector starts executing.
  f. The routine finishes with a return instruction.
  g. The return loads the PC from the stack.
  h. The interrupted program execution continues.
15. The interrupt vector is the name given to the memory location where the first instruction of the interrupt routine is stored.
16. The return instruction terminates the processing of the interrupt service routine and returns operation to the program step following the one being executed when the interrupt happened. The first program step in the program to be continued is pointed to by the PC, which is loaded with the value loaded on the stack when the interrupt occurred.
17. The microprocessor can tell which device caused the interrupt by
  a. Having a separate interrupt input for each device

  b. Having the interrupting device generate its own vector
  c. Polling all the possible devices to see which caused the interrupt
18. The three major different kinds of interrupts are the
  a. Reset
  b. Nonmaskable interrupt
  c. Maskable interrupt
19. The function of each interrupt type is:
  a. The reset interrupt causes the microprocessor to start over from the beginning. It is used for power-up initialization.
  b. The nonmaskable interrupt is a general-purpose interrupt input which is always active. If an input happens on this input, the microprocessor will respond to it.
  c. The maskable interrupt is also a general-purpose interrupt input, but the programmer can control when it is active and when it is turned off. It is an input for external devices which the programmer wants to be able control when they can request service.

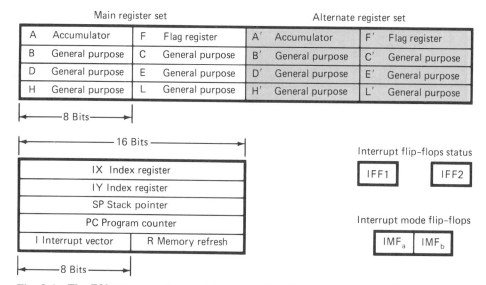

Main register set				Alternate register set			
A	Accumulator	F	Flag register	A′	Accumulator	F′	Flag register
B	General purpose	C	General purpose	B′	General purpose	C′	General purpose
D	General purpose	E	General purpose	D′	General purpose	E′	General purpose
H	General purpose	L	General purpose	H′	General purpose	L′	General purpose

←— 8 Bits —→

←——— 16 Bits ———→

IX Index register
IY Index register
SP Stack pointer
PC Program counter

I Interrupt vector	R Memory refresh

←— 8 Bits —→

Interrupt flip-flops status

| IFF1 | | IFF2 |

Interrupt mode flip-flops

| IMF_a | IMF_b |

Fig. 8-4 The Z80 programming model showing the alternate register set. The programmer can use these registers to hold the register operations from one program while the primary register set holds the operations from a second program.

7. The Z80 has _____ registers.
  a. 8-bit   c. 16-bit
  b. 24   d. All of the above
8. In general, you would expect the results of an arithmetic operation to appear in the
  a. Accumulator
  b. 8-bit C or D registers
  c. 16-bit HL register pair
  d. Memory
9. The B, C, D, E, H, and L registers are called general-purpose registers. They can be used as
  a. 8-bit registers
  b. Index registers
  c. 16-bit registers
  d. All of the above
10. The general-purpose registers can be addressed as BC, DE, and HL register pairs as well as single registers. This is done so that they can be used as
  a. 8-bit registers
  b. Index registers
  c. 16-bit registers
  d. All of the above
11. The purpose of the flag register is to
  a. Let the programmer change quickly between the main and alternate registers
  b. Store special results (zero, negative, carry, etc.) of different data operations
  c. Provide expanded memory addressing in cases where 64 kbytes is not enough
  d. Show the programmer when a data operation resulted in a nonzero byte

12. The alternate register set is provided to let the programmer keep
  a. Large programs in memory rather than on disk
  b. Register data for two programs in the microprocessor at one time
  c. More than one pointer to memory locations
  d. An alternate stack

8-4 THE ASSIGNED REGISTERS

The other two Z80 register groups have assigned functions. That is, each register in the group has a specific function to perform. It is not used for any other purpose. These two groups are the *special registers* and the *interrupt control flip-flops*.

The *special register group* is made up of the interrupt vector (I) register, the memory refresh register (R), two index registers (IX and IY), the stack pointer, and the program counter. The interrupt vector and memory refresh registers are 8-bit registers, and the other registers in this group are 16-bit registers. They are highlighted in Fig. 8-5.

The interrupt register stores the High byte of an interrupt vector. This is needed when the Z80 performs a certain kind of interrupt processing. The Low byte of the interrupt vector is provided by the external interrupting device.

earlier, the accumulator is often the microprocessor's most powerful register. Why do you think this might be an advantage for a lot of I/O transfers, especially those which are human interfaces?

7-4. Small microprocessor systems which do not need enough memory (ROM and RAM) to use the microprocessor's full memory address range often use memory-mapped I/O. Why do you think the designers do this?

7-5. Like many microprocessor processes, the polling and interrupt routines are also used in real life. Can you think of equivalent processes which are used when you are conducting a formal meeting?

7-6. Some hardware systems prioritize interrupt handling by how close the card causing the interrupt is to the CPU card on the system bus. The interrupt line passes through each card, and the card which causes the interrupt can block interrupts from cards further down the bus from reaching the processor. How is this somewhat like using the microprocessor's maskable interrupt? What do you suppose happens if an interrupt is received by a card closer to the CPU when the CPU is processing an interrupt from a card further from the CPU?

Answers to Self-Tests

1. *c*   2. *c*   3. *a*
4. *b*. The power line signal which is counted to give a time reference is a machine-generated signal. All the other inputs come from human actions.
5. A microprocessor processes data. It must receive the data from some source, and once the processing is complete, the microprocessor must output the data to either a human or another machine so the processed data can be used.
6. *c*   7. *b*   8. *c*
9. The Z80 only has 256 (FF) I/O addresses because the Z80 has special I/O control lines and I/O-mapped addresses. The 6802 does not have I/O-mapped I/O transfers and must use memory-mapped I/O, which gives the 6802 the entire range of 0 to 65,535 (0000 to FFFF) memory address locations for I/O address locations.
10. *c*   11. *a*   12. *c*   13. *a*
14. When the microprocessor receives an interrupt, it follows the following steps:
   a. The interrupt is received
   b. The current instruction is finished.

c. The PC is put on the stack.
d. The PC is loaded with the interrupt vector.
e. The interrupt routine at this vector starts executing.
f. The routine finishes with a return instruction.
g. The return loads the PC from the stack.
h. The interrupted program execution continues.
15. The interrupt vector is the name given to the memory location where the first instruction of the interrupt routine is stored.
16. The return instruction terminates the processing of the interrupt service routine and returns operation to the program step following the one being executed when the interrupt happened. The first program step in the program to be continued is pointed to by the PC, which is loaded with the value loaded on the stack when the interrupt occurred.
17. The microprocessor can tell which device caused the interrupt by
   a. Having a separate interrupt input for each device

b. Having the interrupting device generate its own vector
   c. Polling all the possible devices to see which caused the interrupt
18. The three major different kinds of interrupts are the
   a. Reset
   b. Nonmaskable interrupt
   c. Maskable interrupt
19. The function of each interrupt type is:
   a. The reset interrupt causes the microprocessor to start over from the beginning. It is used for power-up initialization.
   b. The nonmaskable interrupt is a general-purpose interrupt input which is always active. If an input happens on this input, the microprocessor will respond to it.
   c. The maskable interrupt is also a general-purpose interrupt input, but the programmer can control when it is active and when it is turned off. It is an input for external devices which the programmer wants to be able control when they can request service.

CHAPTER 8

Two 8-Bit Microprocessors: The Z80 and 6802

∎

CHAPTER OBJECTIVES

This chapter will help you to:

1. *Compare* the major features of two popular 8-bit microprocessors: the Z80 and the 6802.
2. *Identify* the registers of each microprocessor and explain their functions.
3. *Differentiate between* register-intensive and memory-intensive architectures.
4. *Recognize* the major mnemonics and op codes pertaining to the Z80 and 6802 microprocessors.
5. *Compare* the Z80 and 6802 instruction set operations.

───────────

There are two major 8-bit microprocessor architectures in use today—the Intel architecture and the Motorola architecture. The Z80 is one of the most popular and versatile versions of the Intel architecture, and the 6811 is the most current version of the Motorola architecture.

8-1 AN INTRODUCTION TO 8-BIT MICROPROCESSORS

The *8-bit microprocessor* is a second-generation microprocessor. This means that it is one of the second group of microprocessors developed. The 4-bit microprocessors were the first developed. As we learned in Chap. 1, there were a number of good reasons for the 8-bit microprocessor being the second-generation microprocessor.

One reason for building the next generation of microprocessors with 8-bit capability comes from the data capacity it gives the microprocessor. The 4-bit microprocessor is very good for working with binary-coded decimal numbers. These BCD numbers use four binary digits to express decimal values from 0 through 9. The 4-bit size was chosen because it fits a certain type of data very well.

In the same way, the 8-bit microprocessor was chosen to handle a certain type of data. By the early 1970s, the use of 8-bit data to represent the full alphanumeric character set was a very well accepted standard. This 8-bit data standard is called the *American Standard Code for Information Interchange* (ASCII, pronounced *ask-key*). Eight bits gives enough combinations for all of the alphanumeric characters, 36 additional general and scientific characters, and 32 more command (control) codes. There is still 1 bit left over for error checking.

The Intel 8080 8-bit microprocessor became the basis for personal computers which used the CP/M operating system. The CP/M-based microcomputers were replaced with 16-bit personal computers in the 1980s. These PCs use DOS. Microcomputers built around the 6800 never became popular. Microcomputers built around the *MOS Technology 6502* became very popular. Two of these, the Apple II and the Commodore 64, were developed in the late 1970s with mid-1970 microprocessors and remained in production into the last of the 1980s.

In the foregoing discussion, we see how three different microprocessors were born and lived. One was lost as a general-purpose device. Two microprocessors made deep inroads into the personal computer field, and two developed a strong position as the basis for commercial and industrial products.

What types of products in today's world are being built with 8-bit microprocessors if they are not being used for new personal computer designs? The list of product types is nearly endless. In the world of personal computing, for example, the 8-bit microprocessor is still being used to provide *intelligent control* of keyboards, printers, modems, disk controllers, and communications devices.

In the world of home electronics, 8-bit microprocessors are found in sophisticated cassette recorders for audio systems, TVs, VCRs, compact disc players, and even home telephones with dialers and other advanced features. In industrial applications, 8-bit microprocessors are the basis for *smart (intelligent) instruments*, remote data-gathering stations, machine tool controllers, and elevator systems.

In each case, the microprocessor is added to a familiar product. This makes the product smart or intelligent. For example, a simple digital multimeter displays the results of linear voltage, current, and resistance measurements. However, add a microprocessor to the simple multimeter and it can convert nonlinear voltage measurements taken from a thermocouple into temperature readings displayed in degrees Celsius or Fahrenheit. Also, it can respond to a remote data request from a microcomputer and supply its current status, current readings and the average of the past readings.

The rest of this chapter looks at two of the popular 8-bit microprocessors: the *Z80* and the *6802*. For each microprocessor, we will look at the programming model, its instruction set, its internal construction, and the microprocessor hardware. One of the important points for you to understand is that the microprocessor, while a very powerful device, cannot stand alone. The use of a microprocessor means that software must be developed and support hardware must be added to the microprocessor.

As we look at each programming model, we will build on the knowledge gained in

Chaps. 4 through 7. There you learned many of the fundamental building blocks which make up a microprocessor. In this chapter, you will see how those fundamental building blocks have been put together by specific manufacturers for specific microprocessors We will take our general knowledge of microprocessor architecture and use it to learn the special architecture of a specific microprocessor.

Self-Test

Answer the following questions.

1. The 8-bit microprocessor is a _____ generation product.
   a. First   *c.* Third
   b. Second   *d.* Fourth
2. One of the biggest advantages that the 8-bit microprocessor has over the 4-bit microprocessor is its ability to work with _____ data.
   a. ASCII   *c.* BCD
   b. ASCII and BCD   *d.* Digital
3. One of the popular uses for a number of 8-bit microprocessors in the late 1970s was
   a. VCR control   *c.* Microcomputers
   b. Laser disk control *d.* All of the above
4. The majority of applications for the 8-bit microprocessor today is for
   a. High-speed general-purpose computing
   b. Apple computers
   c. Consumer, commercial, and industrial controllers
   d. All of the above
5. When we add a microprocessor to a piece of consumer, commercial, or industrial equipment to give control flexibility, programmability, and data modification ability, we say the device is
   a. Smart
   b. Microprocessor-based
   c. Intelligent
   d. All of the above

8-2 A PROGRAMMING MODEL FOR THE Z80

The Z80 is really a second-and-one-half-generation product. It had its beginnings with the Intel 4004, which was the very first commercially available 4-bit microprocessor. The 4004 was improved and became the 4040, and the 4040 led to a second-generation microprocessor, the 8-bit 8080.

From page 140:

8-bit microprocessor

American Standard Code for Information Interchange (ASCII)

MOS Technology 6502

On this page:

Intelligent control

Smart (intelligent) instruments

Z80 microprocessor

6802 microprocessor

The improved 8080 became the 8085, while the engineers at Zilog took the basic 8080 model and created the Z80, which is a superset of the 8080. This means that the Z80 performs all of the 8080 functions as well as many more. For the Z80 to pass the test of being an 8080 superset, it must run without failure any programs written for the 8080 without knowledge of the Z80. The Z80 does pass this test and provides many additional functions as well.

The Z80 programming model is shown in Fig. 8-1. As you can see, the Z80 programming model is similar to the programming model introduced in Chap. 6. The Z80 has more registers than the programming model from Chap. 6. These make programming easier and are also used to control Z80 hardware functions.

The Z80 register set is made up of 24 different registers which can be broken into four different groups. We will look at each of these groups individually in the next two sections.

8-3 THE Z80 8-BIT REGISTERS

The first Z80 register group are the eight main 8-bit registers A, F, B, C, D, E, H, L. These are the accumulator (A), the flag register (F), and six general-purpose registers. They are highlighted in Fig. 8-2.

The *accumulator* (A) stores either the operand or the results of an ALU operation. That

is, the accumulator either contains the data you are going to work on or data which is the result of an ALU operation. As a general rule, you can assume that operations on data in the accumulator will cause the data in the accumulator to be replaced by the result.

The F register is called the Z80 *flag register*. It can also be called a status register, as we referred to it in Chap. 5. Each bit in the flag register is set to a logic 1 or cleared to a logic 0, depending on the results of an ALU operation. The flag bits and their meanings are shown in Fig. 8-3.

The registers B, C, D, E, H, and L can either be used as *general-purpose* 8-bit registers or as 16-bit registers when used in pairs. The pairing is B and C, D and E, and H and L. You decide to use these registers as individual 8-bit registers or as 16-bit register pairs by the instruction you choose to use.

The Z80 is a *register-intensive architecture*. It offers the programmer many different registers to temporarily store data in while other operations are being done. Other more memory-oriented designs make the programmer use memory locations as temporary storage. These registers are called the *main register set*.

The Z80 has another eight identical registers. These registers are called the *alternate register set*. As shown in the programming model (Fig. 8-4 on page 144), the registers in the secondary register set are given the names A′, F′, B′, C′, D′, E′, H′, and L′. The alternate register

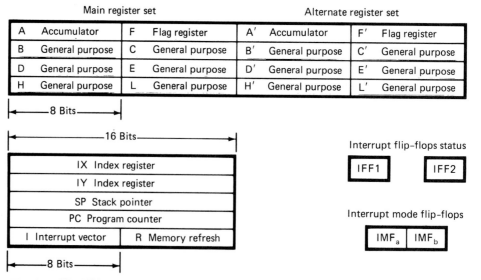

Fig. 8-1 **The Z80 programming model. The programming model shows you all the parts in the microprocessor that you as a programmer can change or use. Note that this means that parts like the ALU, which are shown on a block diagram but cannot be affected with programming, are left off.** *Adapted from Zilog Z80 data sheets.*

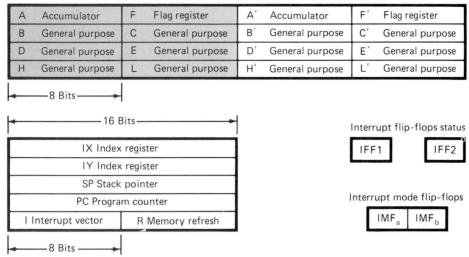

A	Accumulator	F	Flag register	A'	Accumulator	F'	Flag register
B	General purpose	C	General purpose	B'	General purpose	C'	General purpose
D	General purpose	E	General purpose	D'	General purpose	E'	General purpose
H	General purpose	L	General purpose	H'	General purpose	L'	General purpose

Main register set   Alternate register set

←—— 8 Bits ——→

←———— 16 Bits ————→

IX Index register
IY Index register
SP Stack pointer
PC Program counter
I Interrupt vector | R Memory refresh

←—— 8 Bits ——→

Interrupt flip-flops status

IFF1   IFF2

Interrupt mode flip-flops

IMF$_a$ | IMF$_b$

Fig. 8-2 The Z80 six general-purpose registers, its accumulator, and the flag register. The general-purpose registers are organized as three 8-bit pairs which can be used in either the 8-bit or 16-bit modes.

set is identical to the main one. Why and how are the alternate registers used?

The alternate register set lets a programmer quickly switch from executing one program to executing a second program. Each program can have its own set of temporary data stored in the accumulator, flags, and general-purpose registers. Without an alternate set, changing between two programs is a much more complex process. First, the programmer must store the contents of all the registers in memory. Second, the programmer must load the registers with the data needed to support the second program. These operations require multiple steps using most microprocessor instruction sets.

With the Z80, the programmer only needs to execute a single exchange instruction. The *exchange instruction* swaps the data in the main registers with the data in the secondary regis-

ters, and the programmer may start execution using the new program data. Because this exchange is performed with a single instruction, the Z80 offers a very powerful architecture for multiple programming efforts.

Self-Test

Answer the following questions.

6. The Z80 is an 8-bit microprocessor developed by Zilog. Its architecture is a _____ the Intel 8080/8085 architecture.
   a. Subset of
   b. Copy of
   c. Direct replacement for
   d. Superset of

S Sign	Z Zero	X	H Half-carry	X	P/V Parity overflow	N Add subtract	C Carry
7	6	5	4	3	2	1	0

S   = 1 if the MSB of the result is 1
Z   = 1 if the result of the operation is 0
X   = Don't care (not used)
H   = 1 if the add or subtract caused a carry or borrow from bit 4
X   = Don't care (not used)
P/V = 1 for odd parity or arithmetic overflow
N   = 1 if the previous operation was a subtract
C   = 1 if the operation produced a carry from the MSB

Fig. 8-3 The Z80 flag register. The flag bits are set (or cleared) by different data operations. You can use these flag bits to control the flow of a program by testing one or more of these flag bits after an operation.

Main register set				Alternate register set			
A	Accumulator	F	Flag register	A′	Accumulator	F′	Flag register
B	General purpose	C	General purpose	B′	General purpose	C′	General purpose
D	General purpose	E	General purpose	D′	General purpose	E′	General purpose
H	General purpose	L	General purpose	H′	General purpose	L′	General purpose

←— 8 Bits —→

←——— 16 Bits ———→

IX Index register
IY Index register
SP Stack pointer
PC Program counter

I Interrupt vector	R Memory refresh

←— 8 Bits —→

Interrupt flip-flops status

| IFF1 | | IFF2 |

Interrupt mode flip-flops

| IMF$_a$ | IMF$_b$ |

Fig. 8-4 The Z80 programming model showing the alternate register set. The programmer can use these registers to hold the register operations from one program while the primary register set holds the operations from a second program.

7. The Z80 has _____ registers.
   a. 8-bit   c. 16-bit
   b. 24   d. All of the above
8. In general, you would expect the results of an arithmetic operation to appear in the
   a. Accumulator
   b. 8-bit C or D registers
   c. 16-bit HL register pair
   d. Memory
9. The B, C, D, E, H, and L registers are called general-purpose registers. They can be used as
   a. 8-bit registers
   b. Index registers
   c. 16-bit registers
   d. All of the above
10. The general-purpose registers can be addressed as BC, DE, and HL register pairs as well as single registers. This is done so that they can be used as
    a. 8-bit registers
    b. Index registers
    c. 16-bit registers
    d. All of the above
11. The purpose of the flag register is to
    a. Let the programmer change quickly between the main and alternate registers
    b. Store special results (zero, negative, carry, etc.) of different data operations
    c. Provide expanded memory addressing in cases where 64 kbytes is not enough
    d. Show the programmer when a data operation resulted in a nonzero byte

12. The alternate register set is provided to let the programmer keep
    a. Large programs in memory rather than on disk
    b. Register data for two programs in the microprocessor at one time
    c. More than one pointer to memory locations
    d. An alternate stack

8-4 THE ASSIGNED REGISTERS

The other two Z80 register groups have assigned functions. That is, each register in the group has a specific function to perform. It is not used for any other purpose. These two groups are the *special registers* and the *interrupt control flip-flops*.

The *special register group* is made up of the interrupt vector (I) register, the memory refresh register (R), two index registers (IX and IY), the stack pointer, and the program counter. The interrupt vector and memory refresh registers are 8-bit registers, and the other registers in this group are 16-bit registers. They are highlighted in Fig. 8-5.

The interrupt register stores the High byte of an interrupt vector. This is needed when the Z80 performs a certain kind of interrupt processing. The Low byte of the interrupt vector is provided by the external interrupting device.

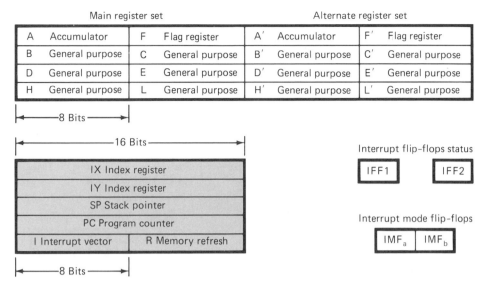

Fig. 8-5 The Z80 special register group. This group had four 16-bit registers and two 8-bit registers. The four 16-bit registers are the program counter, stack pointer, and two (IX and IY) index registers. The two 8-bit registers are the interrupt vector register (high-order 8-bits only) and the memory refresh register.

The programmer must load an 8-bit number into the interrupt register before it can be used.

The memory refresh register automatically increments and its contents are placed on the bus during each fetch operation. This is done so that the Z80 can keep dynamic RAM ICs refreshed if they are connected to the microprocessor. Dynamic RAM and its need to be refreshed is explained in Chap. 11. The programmer has no control over this register, but it is often included in Z80 programming models.

The Z80 provides two 16-bit registers for indexed addressing. These are known as the IX and IY registers. Both are identical in operation, and the programmer selects either one as needed. Two *index registers* are offered so that the programmer can use two different offsets without having to reload the index register. Either of the index registers is available at any time by using an instruction which refers to that register. The Z80 provides a large number of index register instructions. These allow the programmer to load the index registers with data or to put the index register data into other locations. The exact use of the two index registers depends on the programmer.

The Z80 has one 16-bit *stack pointer (SP)*. The Z80 instruction set provides the programmer with instructions to load the stack pointer from memory or another 16-bit register or register pair. There is also an instruction to save the current contents of the stack pointer on the stack. Because the Z80 only provides a single stack pointer, the programmer must perform all programming functions with a single stack or include programming steps to change the contents of the stack pointer each time a separate stack is to be used.

The Z80 *program counter (PC)* is also a dedicated 16-bit register. When the Z80 is reset, the program counter is initialized pointing to memory location zero. Therefore, all Z80 programs must start at memory location 0000H. As with any program counter, the contents may be changed by a number of special processes, such as the subroutine instructions and the interrupts.

The last register group is the *interrupt control flip-flops*. There are four of these special-purpose single-bit registers. The four flip-flops are IFF1, the interrupt enabled/disabled flip-flop; IFF2, the interrupt status storage flip-flop; and IMF_a and IMF_b, the interrupt mode flip-flops. See Fig. 8-6 on page 146.

The Z80 has two interrupts: the *nonmaskable interrupt (NMI)* and the *maskable interrupt (INT)*. The nonmaskable interrupt cannot be turned off; however, the maskable interrupt can be enabled or disabled by the programmer.

The Z80 interrupt service routines start at memory locations which depend on the type of interrupt being processed. When an interrupt occurs, the current PC value is placed on the stack and program execution begins at the memory location pointed to by the selected interrupt vector. When the interrupt routine processes a return instruction, the PC is loaded with the PC value from the stack, and the interrupted program execution continues.

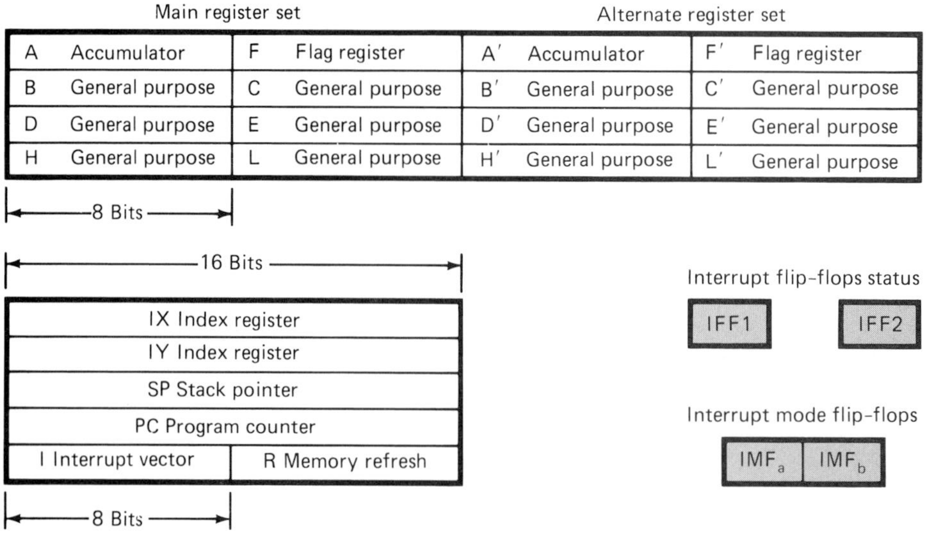

Fig. 8-6 The interrupt control flip-flops. These four single-bit registers are used to control and show the status of the Z80 maskable interrupt system. IFF1 shows and controls the On/Off status of the maskable interrupt, and IFF2 shows the prior status of IFF1 if the nonmaskable interrupt was activated and turned the maskable interrupt off. IMF_a and IMF_b are used to control the maskable interrupt mode.

The nonmaskable interrupt cannot be disabled. When the Z80 is interrupted by a signal at the NMI input, the program counter is set to 0066H. The processor then begins to execute program steps starting with the instruction stored at memory location 0066H. Usually the use of the NMI is reserved for unusual needs.

The interrupt capability at the INT input can be turned on or turned off using the EI and the DI (enable interrupt and disable interrupt) instructions. The current status of the interrupt is set in the IFF1 flip-flop. When the programmer *enables* the maskable interrupt, IFF1 is set to a *logic* 1. When the programmer disables the maskable interrupt, IFF1 is set to logic 0.

If a nonmaskable interrupt occurs, the maskable interrupt must be turned off (if it is on) while the nonmaskable interrupt is processed. The nonmaskable interrupt can be turned back on (if it was on) once the nonmaskable interrupt process is complete. The current value of IFF1 is stored in IFF2 while the nonmaskable interrupt process takes place. Once the return instruction for the nonmaskable interrupt process is executed, the content of IFF2 is transferred back to IFF1.

The maskable interrupt has three modes of operation. These modes (called mode 0, mode 1, and mode 2) are determined by the values set in two interrupt mode flip-flops IMF_a and IMF_b. The modes are shown in Fig. 8-7.

In response to a mode 0 interrupt, the Z80 goes into a special instruction fetch cycle. In this instruction fetch cycle, it waits for external hardware to supply an RST (restart) instruction. There are eight RST instructions: RST0 to RST7. The binary values of the RST number (0 to 7) are used as the A_3, A_4, and A_5 address bits. This means that the RST instructions start interrupt routines at memory locations 0000H, 0008H, 0010H, 0018H, 0020H, 0028H, 0030H, and 0038H. Mode 0 gives the Z80 the ability to start interrupt processing at one of eight different addresses. It is the Intel 8080- and 8085-compatible interrupt mode.

In interrupt mode 1, the Z80 has only one interrupt address, 0038H. There is no requirement for external hardware to provide an RST instruction. This is the simplest interrupt mode, but it is not compatible with the 8080.

Interrupt mode 2 is the most sophisticated

IMF_a	IMF_b	
0	0	Mode 0
0	1	Not used
1	0	Mode 1
1	1	Mode 2

Fig. 8-7 The three different modes for the maskable interrupt input (INT). Mode 0 is just like (emulates) the 8080/8085 interrupt system using eight RST instructions. Mode 1 is a simple interrupt system with one interrupt vector at memory location 0038H. Mode 2 is the Z80 interrupt system which allows the external device to supply the 8 low-order bits of the interrupt vector.

interrupt mode. In this mode, the Z80 can start the interrupt processing routine at any address. As you learned earlier, the upper 8 bits of the address are supplied by the interrupt register. The lower 8 bits are provided by the external hardware generating the interrupt. This process allows every interrupting hardware device to supply its own interrupt vector.

Self-Test

Answer the following questions.

13. The registers found in the special register group have
    a. 16 bits
    b. Dedicated functions
    c. 8 bits
    d. General-purpose functions
14. The interrupt vector register is loaded with an 8-bit number which the Z80 uses as the _____ when a mode 2 interrupt happens.
    a. High byte of the interrupt vector
    b. Low byte of the interrupt vector
    c. Program counter's preset value
    d. Memory refresh counter
15. The memory refresh register is
    a. Usually loaded with an 8-bit number by the programmer when the microprocessor is initialized
    b. Not loaded by the programmer because it cannot be
    c. Is not part of the 8080/8085 programming model
    d. Both b and c
16. The two Z80 index registers are
    a. Made up from two general-purpose registers working as a 16-bit pair
    b. Used as substitutes for the program counter when a swap occurs with the alternate register set
    c. Two 8-bit registers in the special register group used to provide offsets to the programmer, reducing the instruction by 1 byte
    d. Two 16-bit registers
17. The two Z80 index registers are
    a. Dedicated, one to the main register set (IX) and one to the alternate register set (IY)
    b. Dedicated, one to the main register set (IY) and one to the alternate register set (IX)
    c. Available at any time as the programmer chooses
    d. Not available when the Z80 is processing an interrupt
18. If the programmer wishes to build two different stacks in two different areas of Z80 memory,
    a. The stack pointer (SP) must be loaded each time the other stack is to be used
    b. The programmer switches between the two stack pointers in the special register set (SP1 and SP2)
    c. There is never any need to have two stacks
    d. Both a and c
19. The Z80 program counter is
    a. A dedicated 16-bit register in the special register set.
    b. Made up of an unused register pair.
    c. One of two registers in the special register set. The one being used depends on the use of the 8-bit registers.
    d. A dedicated 8-bit register pair in the special register set.
20. The Z80 interrupt control
    a. Comes from the unused general-purpose register pair HL
    b. Depends on the polarity of the signal applied to INT
    c. Can only be used when the maskable interrupt (INT) is disabled
    d. Uses four single bit registers IFF1, IFF2, IMF_a, and IMF_b
21. The Z80 nonmaskable interrupt
    a. Loads the program counter with 0066H
    b. Shuts off the maskable interrupt (INT) when it begins processing an interrupt
    c. Cannot be shut off by the programmer using IFF1
    d. All of the above
22. Mode 2 is the Z80's _____ interrupt processing mode.
    a. Simplest
    b. Second
    c. Most versatile and complex
    d. 8080-compatible
23. The only action taken when the Z80 processes an INT signal in mode 1 is that the
    a. Maskable interrupt is shut off if it is on, and the value is stored in IFF2 for use after INT is complete
    b. Program counter is loaded with 0000H
    c. Program counter is loaded with 0038H
    d. Program counter is loaded with PC + 0005H
24. Mode 0 is the _____ interrupt processing mode.
    a. Simplest
    b. Second
    c. Most versatile and complex
    d. 8080-compatible

8-5 THE Z80 BLOCK DIAGRAM

The Z80 has an ALU, registers, an external 8-bit data bus, and a 16-bit memory address bus as shown in Fig. 8-8. The 8-bit *data bus* is a *bidirectional bus*. It allows programmers to send data from the Z80 CPU to an external device and receive data from an external device.

Each data bit on the Z80 data bus (D_0 to D_7) can have one of three different electrical states: high, low, or high impedance. When the Z80 is receiving or transmitting data, the bus signals are either high or low, depending on the data being transmitted. When the Z80 bus is in the high-impedance state, the Z80 cannot drive or receive from the bus. In this state, an external device can take over the bus and supply its own signals. When a bus driver is in the high impedance mode, it is often said to be *tristated*. That is, it is in its third state (high impedance) rather than at a low output impedance at either the electrically high or the electrically low condition.

The *memory address bus* is a 16-bit bus made up of the memory address signals A_0 to A_{15}. This is a *unidirectional bus*. That is, the signals from the memory address bus only go out of the Z80; the Z80 does not receive memory address signals. Each bit in the memory address bus may have one of three states: high, low, or high impedance.

The Z80 buses are supported with a group of command and control signals which are covered in Sec. 8-7. These signals indicate the processor status to external devices and accept commands generated by external hardware.

Self-Test

Answer the following questions.

25. The ALU is a very powerful part of the architecture, and all instructions which modify and test Z80 data use the ALU. However, the ALU is usually only shown

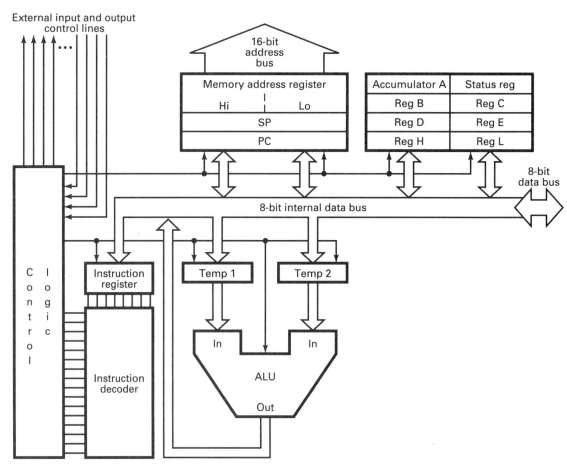

Fig. 8-8   The Z80 block diagram. This block diagram was used in Chap. 5, when we learned about microprocessor architecture. It shows the ALU and the buses in addition to the registers and is used to give the hardware designer a good picture of the device being connected to the outside world. *Adapted from Zilog Z80 data sheets.*

on the block diagram rather than on its programming model because

a. Programmers do not use the ALU
b. Programming cannot change what the ALU does
c. The programmer cannot store any data in the ALU
d. All of the above

26. The Z80 memory address bus is _____ and _____ . (Choose two of the four answers.)

a. A unidirectional 16-bit bus whose output lines are high, low, or tristated (high impedance)
b. An 8-bit bidirectional bus whose lines are high, low, or tristated (high impedance)
c. Used to transmit data to and from memory
d. Used to transmit address information to memory

27. The Z80 data bus is _____ and _____ . (Choose two of the four answers.)

a. A unidirectional 16-bit bus whose output lines are high, low, or tristated
b. An 8-bit bidirectional bus whose lines are high, low, or tristated
c. Used to transmit data to and from memory
d. Used to transmit address information to memory

28. If the buses are _____ , an external device can take over the memory or I/O systems connected to the Z80.

a. Transmitting an electrically low signal
b. Transmitting an electrically high signal
c. Placed in a high-impedance state
d. All of the above

Immediate The byte following the op code holds the data to be worked on. A 2-byte instruction.

Immediate extended The two bytes following the op code hold the data to be worked on. A 3-byte instruction.

Modified page zero The byte following the op code is an address for memory or I/O locations 0000 to 00FF, where the data to be worked on is held. A 2-byte instruction.

Relative The byte following the op code is added to the PC value (2's complement) to give the address for the data that is to be worked on. A 2-byte instruction.

Extended The two bytes following the op code are the address the instruction is to act on. A 3-byte instruction.

Indexed The byte following the op code is added to the indicated index register using unsigned arithmetic. The resulting 16-bit number is used as the address of the data the instruction is to act on. A 2-byte instruction.

Register The instruction includes the name of a register which holds the data the instruction is to act on. A 1- or 2-byte instruction depending on the instruction.

Register indirect The instruction includes the name of a register pair which points to the memory locations holding the data the instruction is to work on. A 1- or 2-byte instruction depending on the instruction.

Implied The op code tells what it is to work on. A 1-byte instruction.

Bit The bytes following the op code tell which bit of a memory location or register the instruction is to work on. A 2- or 4-byte instruction.

Fig. 8-9 A summary of the 10 Z80 addressing modes.

8-6 THE Z80 INSTRUCTION SET

The Z80 instruction set of 158 instructions includes the 78 instructions of the 8080. It is a very powerful instruction set. In the following paragraphs, we will look at the 11 different types of instructions available to the Z80 programmer. These 11 types are made much more powerful by the 10 different addressing modes summarized in Fig. 8-9 (above).

Tables 8-1 through 8-11 (at the end of this chapter, preceding the summary) show the complete set of instructions for each of these groups and the different ways they can be used with the addressing modes. Review the

instructions in the tables as you read the following paragraphs to be sure you understand the instructions in each group. The notes in the tables will help you to understand what the instruction does. To keep the tables simple, symbols are used to show how the flags will respond to executing the instruction: 1 = Set, 0 = Reset, ● = No Change, ↕ = Flag set by operation to 1 or 0 as appropriate, and X = Don't Care.

The load instructions (Table 8-1) copy data from register to register and between registers and memory locations. The Z80 8-bit load instructions come in three forms: register to register; register to memory and memory to register; and immediate. The Z80 16-bit load instructions (Table 8-2) include a fourth form which allows the register contents to be put on or taken from the stack.

The exchange instructions (Table 8-3) are used to swap data between the 16-bit register pairs. For example, the EX DE,HL instruction swaps the contents of the DE and HL register pairs. The EXX instruction swaps data between the main and auxiliary (') register sets.

The *block move instructions* move (copy) data from the memory location pointed to by the HL register pair (HL) to the memory location pointed to by the DE register pair (DE). After each byte of data is moved, both DE and HL are incremented (LDI or LDIR) or decremented (LDD or LDDR). BC is always decremented. There are two forms. LDI and LDD move a single byte of data. LDIR and LDDR repeat the transfer until BC = 0. The BC register pair has the number of bytes of data to be moved.

Two similar instructions allow the programmer to search a block of memory locations for a byte of data which matches the data in the accumulator. Again, the BC register pair is used as a counter and the HL register pair is an incrementing or decrementing memory location pointer.

The 8-bit arithmetic and logical instructions (Table 8-4) are used to modify data or to arithmetically or logically combine data. The Z80 has many forms of add and subtract, including increment and decrement. The logical instructions include AND, OR, and Exclusive OR.

The 16-bit arithmetic and logical instructions (Table 8-5) perform 16-bit adds and subtracts using the HL register pair and the BC, DE, or HL register pairs or the stack pointer. These instructions also perform increments or decrements on the 16-bit register pairs, the stack pointer, and the IX and IY index registers. These instructions help the programmer compute new addresses. They are not used for general-purpose arithmetic computations.

The *rotate and shift instructions* (Table 8-6) move data to the right or left. The Z80 provides both rotates through the carry flag and rotates without the carry flag.

The *general-purpose arithmetic* and *CPU control instructions* (Table 8-7) are the miscellaneous instructions. They include such features as the ability to complement the accumulator (1's complement or 2's complement), halt the processor, and enable or disable the interrupts.

The *bit set*, *reset*, and *test instructions* (Table 8-8) are powerful for working with external hardware. Frequently, the external hardware generates a byte of data, but each bit in the byte has a different meaning. The bit set, reset, and test instructions are used to work with each bit alone.

The Z80 provides two different kinds of instructions which allow the programmer to change the program counter address. These are the *jump* (Table 8-9) and *call instructions* (Table 8-10). The call instructions include the return instructions.

As you learned earlier, the jump instructions simply change the PC value. All earlier values for the PC are lost. There are two forms of the jump instruction. The absolute instruction simply loads the PC with the new value when the instruction is executed. The second type is the conditional jump instruction. The conditional jump instruction tests the flag register for certain conditions. If the condition is there, the jump happens. If the condition is not in the flag register, the jump does not happen.

The call instructions (Table 8-10) are like the jump instructions except the current value of the PC is placed on the stack when the call instruction is executed. This operation saves the current value of the program counter so that program execution may be resumed after the subroutine executes. The call instruction also has unconditional and conditional forms. If an unconditional call instruction is executed, the current value of the PC is saved. Program execution begins at the instruction stored in the memory location whose address is loaded into the PC by the call instruction. If a conditional call instruction is executed, the PC is only changed if the condition is met.

Both forms of call instruction use the *return (RET) instruction*. The return instruction continues executing the program suspended by the call instruction. Executing the return instruction causes the PC to be loaded from the stack, thus restoring the original PC value. Because the PC was pointing at the next instruction when the original call instruction was executed, the value taken from the stack points the PC to the instruction which comes after the call instruction. Return instructions in the Z80 have both conditional and unconditional forms.

The special interrupt instruction, the *RST instruction*, is also included in this group. When a mode 0 interrupt is received, the Z80 looks for an RST instruction to be placed on the data bus by external hardware. Once the RST instruction is received, the Z80 acts as if it had executed a call instruction to one of the eight RST addresses (0 to 7).

The *Z80 input and output instructions* (Table 8-11) move data between an external hardware device connected to the data bus and the Z80. The Z80 uses a separate 256-address I/O space. That is, the Z80 does not use memory-mapped I/O but has a special set of I/O instructions.

Self-Test

Answer the following question.

29. Complete the following table of Z80 instructions and fill in the flags using 1 = Set, 0 = Reset, ● = No Change, ↕ = Flag set by operation, and X = Don't care.

Table for Self-Test Question 29

Mnemonic	Symbolic operation	Flags					
		S	Z	R	P/V	N	C
a. LD (HL), r							
b.	$(nn) \leftarrow A$						
c. LD r, A							
d. LD IX, nn							
e.	$IY_H \leftarrow (nn + 1)$ $IY_L \leftarrow (nn)$						
f.	$(SP - 2) \leftarrow BC_L$ $(SP - 1) \leftarrow BC_H$ $SP \rightarrow SP - 2$						
g. LDI							
h.	$A - (HL)$ $HL \leftarrow HL + 1$ $BC \leftarrow BC - 1$						
i.	$A \leftarrow A + D + CY$						
j. INC(HL)							
k. CPL							
l.	$HL \leftarrow HL + BC$						
m. DEC IY							
n. RRA							
o. JP NZ,nn							
p.	$PC \leftarrow IX$						
q.	$PC_L \leftarrow (SP)$ $PC_H \leftarrow (SP + 1)$						

8-7 Z80 HARDWARE

In this section, we will look at the Z80 as an IC. First, we will look at the signals which go into and go out of the Z80. Then, we will look at some of the support ICs developed to work with the Z80.

Figure 8-10 shows the pin function and pinout diagrams for the Z80. The pin assignment diagram groups the different kinds of signals showing you what connections must be made to a Z80 before it can become a part of a working system. The pinout diagram shows where all of the different signals are connected to the 40-pin package.

Looking at Fig. 8-10, you see the connections shown as six groups of signals:

The 16-line address bus
The 8-line data bus
Six system control lines
Five CPU control lines

Two CPU bus control lines
Three supply lines

In the following paragraphs, we will look at each of these groups. *Note:* Many of the signals are shown with an overbar. As in any digital logic, this indicates that these signals are true (asserted) when they are at their lowest electrical level.

The *16-line address bus* (A_0 to A_{15}) lets the Z80 address up to 64 kbytes of memory. The lower eight lines are also used to address up to 256 I/O ports. As we learned earlier, the address bus is only an output, but it can be switched into a high-impedance state if some external device needs to drive the address bus.

The *8-line data bus* (D_0 to D_7) is an 8-bit bidirectional data communications path with memory or I/O devices. The data bus works with the address bus. First, the the external device is addressed, and then a data exchange takes place. The data bus can be switched into

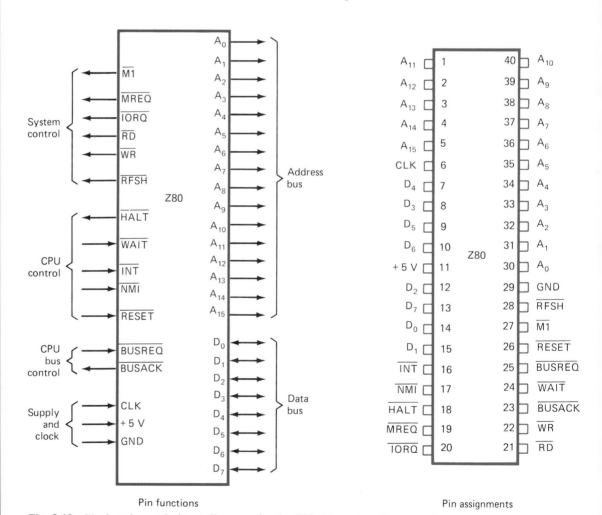

Pin functions

Pin assignments

Fig. 8-10 **Pin function and pinout diagrams for the Z80. These two diagrams show you the electrical and mechanical signal connections for the processor.** *Adapted from Zilog Z80 data sheets.*

a high-impedance state if an external device needs to drive the data bus.

The six *system control lines* tell memory and I/O devices when they can use the signals on the address and data buses. The $\overline{M1}$ signal is a timing signal which comes from the clock. The $\overline{M1}$ signal is used to tell external devices that the Z80 is fetching an instruction, thus beginning a machine cycle. The $\overline{MREQ}$ and $\overline{IORQ}$ lines tell external devices that the Z80 has valid address data on the address bus for memory ($\overline{MREQ}$) or for an I/O device ($\overline{IORQ}$). In the same way, the $\overline{RD}$ and $\overline{WR}$ tell external devices that the data on the data bus is valid ($\overline{WR}$) or that the Z80 is ready to read the data on the data bus ($\overline{RD}$). $\overline{RFSH}$ is the refresh signal generated by the Z80 to refresh dynamic RAMs. As we will learn in Chap. 11, dynamic RAMs must be refreshed regularly or they will lose their data.

The *CPU control lines* let the outside world tell the Z80 what mode it should operate in, and they allow the Z80 to tell the outside world when it is halted. If the Z80 executes a halt instruction, the $\overline{HALT}$ line becomes active, and program execution stops. The Z80 executes NOP instructions until an interrupt signal is received. $\overline{NMI}$ and $\overline{INT}$ are the two interrupt inputs. Once an active low signal is put at either one of these two inputs, the Z80 begins to process the inrupt as soon as the current instruction is done. $\overline{RESET}$ is used to initialize the Z80. When a reset signal is received, the program counter is set to zero, the maskable interrupt is turned off, and the interrupt vector is set to zero. The reset line must be held active for at least three clock cycles to give the reset operation time to complete. Once the reset line is released, the Z80 begins to execute the instruction starting at memory location 0000H.

The $\overline{WAIT}$ input lets a slow memory or I/O device tell the Z80 it needs more time. This allows a device which responds slowly to the Z80 address information to take the time it needs to get its data ready or to receive data. Often, memory systems which are built with slower memory ICs make the processor wait.

The two *CPU bus control lines* allows an external device to request the use of the address and data buses. Some very intelligent external devices can work directly with the memory, for example. This allows them to place data in the memory without going through the processor. A device with this capability places a request on the $\overline{BUSREQ}$ input. Once the Z80 receives a bus request signal, it finishes its current machine cycle, puts the address and data buses in the high-impedance state and acknowledges the request with a $\overline{BUSACK}$ signal. The buses are returned to Z80 control after the $\overline{BUSREQ}$ signal is taken away.

The Z80 operates from a 5-V power supply and draws between 150 and 250 mA. The power is supplied to the +5-V and GND pins.

The Z80 also requires an external clock that provides the timing for all Z80 operations. This signal is a square wave with a maximum frequency of 8 MHz for a Z80H, 6 MHz for a Z80B, 4 MHz for a Z80A, and 2.5 MHz for a Z80. The higher-frequency clocks let the Z80 operate faster but also cause it to use more power.

When designing a microprocessor-based system using the Z80 or other microprocessor, it is important to understand the processor timing diagrams. Figure 8-11 (page 154) shows the *timing diagram* for a Z80 performing a memory read cycle. These detailed diagrams show exactly when address information is valid, when data will be read, and what effect a wait request has on the memory read cycle. Such detailed timing diagrams are usually given by a microprocessor manufacturer for the fetch cycle, memory read and write cycles, I/O cycles, interrupt request and acknowledge cycles, bus request cycles, and any other cycles that microprocessor may have.

A number of *support ICs* have been developed to work with the Z80. In the following paragraphs, we will look at three support ICs which have become very popular: the *parallel input/output (PIO) controller*, the *counter-timer circuit (CTC)*, and the *serial input/output (SIO) controller*. Although the logic in any one of these ICs can be duplicated with SSI and MSI TTL parts, these ICs bring numerous functions into one easy-to-use IC.

The Z80 PIO is shown in Fig. 8-12 (page 155). The PIO gives the Z80 two *8-bit parallel ports* with hand-shaking lines. The Z80 PIO is designed to connect directly (interface) to the buses and control lines. All that is needed to use a PIO with the Z80 is address decoding logic. This logic selects the four addresses needed by the PIO from the 256 I/O addresses which the Z80 can generate.

The PIO has two 8-bit parallel ports with *handshaking lines* $\overline{ASTB}$ and ARDY (or $\overline{BSTB}$ and BRDY). The 8-bit parallel ports can output or input data. The programmer can set each *pin* as an input or output with programming. A signal on $\overline{ASTB}$ tells the PIO's A port to take

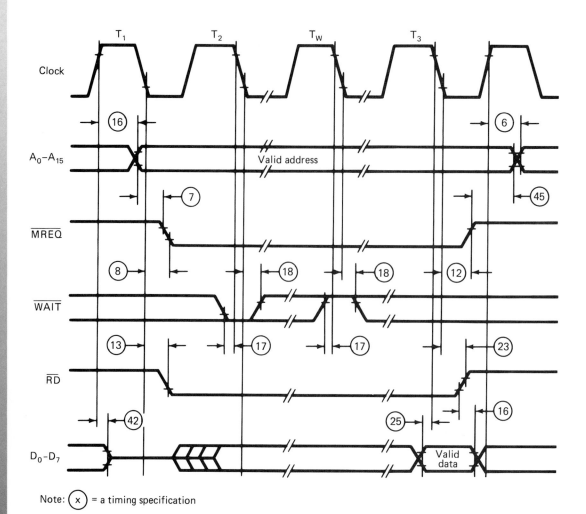

Note: (x) = a timing specification

Fig. 8-11 A Z80 memory read timing diagram. This diagram shows the time when the address bus information and the data bus data are valid. Note that all timing is given in relation to the Z80 clock signal. *Adapted from Zilog Z80 data sheets.*

data in. If the ports are used as outputs, the PIO places a signal on BRDY to indicate there is valid data on the B output port.

Before you can use the PIO to output or input data, you must send it control information so it knows how to operate. Because the PIO can work in many different ways, sending it the control information is somewhat complex. Selection of the control or data mode is done by a signal on the C/$\overline{\text{D}}$ input. Likewise, the A/$\overline{\text{B}}$ input is used to select operation with the A or B port. The $\overline{\text{A}}$ with C or $\overline{\text{D}}$ and B with C and $\overline{\text{D}}$ signals make up the PIO's four different I/O addresses.

In the control mode, you can send the PIO a series of bytes which tell it how to operate and what kind of interrupt signals to generate. For example, the PIO can generate an interrupt signal when a signal is received on its $\overline{\text{ASTB}}$ or $\overline{\text{BSTB}}$ handshake line, or it can generate an interrupt from any one of the input

bits. Additionally, the PIO can be programmed to send a previously loaded byte as the Low byte of the mode 2 interrupt vector.

The PIO has two special pins to let you daisy-chain an interrupt control signal. Three PIOs are shown in Fig. 8-13 with their interrupt enable (IE) signals wired in a daisy chain. All of the interrupts ($\overline{\text{INT}}$) are wired together. This is called wired OR. If PIO A OR B OR C generates an interrupt signal, the Z80 receives the interrupt.

How does the Z80 tell which I/O device caused the interrupt? The interrupt enable daisy-chain line controls which PIO interrupts the Z80. For example, if the PIO A generates an interrupt, it also outputs a signal on its interrupt enable output (IEO). This signal is passed to PIO B by the daisy chain. PIO B receives the signal on IEI. PIO B's interrupt output ($\overline{\text{INT}}$) is then disabled, and PIO B also outputs a signal on IEO. This PIO disables PIO

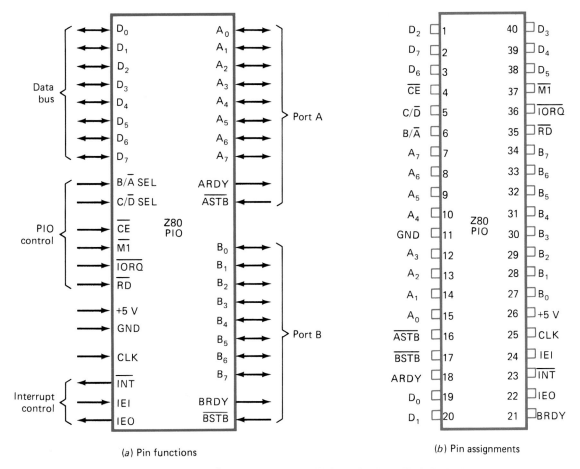

Data bus

D_0	A_0
D_1	A_1
D_2	A_2
D_3	A_3
D_4	A_4
D_5	A_5
D_6	A_6
D_7	A_7

Port A

PIO control

$B/\overline{A}$ SEL
$C/\overline{D}$ SEL
$\overline{CE}$
$\overline{M1}$
$\overline{IORQ}$
$\overline{RD}$
+5 V
GND
CLK

ARDY
$\overline{ASTB}$

Z80 PIO

B_0
B_1
B_2
B_3
B_4
B_5
B_6
B_7

Port B

Interrupt control

$\overline{INT}$
IEI
IEO

BRDY
$\overline{BSTB}$

(a) Pin functions

D_2	1	40	D_3
D_7	2	39	D_4
D_6	3	38	D_5
$\overline{CE}$	4	37	$\overline{M1}$
$C/\overline{D}$	5	36	$\overline{IORQ}$
$B/\overline{A}$	6	35	$\overline{RD}$
A_7	7	34	B_7
A_6	8	33	B_6
A_5	9	32	B_5
A_4	10	31	B_4
GND	11	30	B_3
A_3	12	29	B_2
A_2	13	28	B_1
A_1	14	27	B_0
A_0	15	26	+5 V
$\overline{ASTB}$	16	25	CLK
$\overline{BSTB}$	17	24	IEI
ARDY	18	23	$\overline{INT}$
D_0	19	22	IEO
D_1	20	21	BRDY

Z80 PIO

(b) Pin assignments

Fig. 8-12 The Z80 PIO (parallel input/output) controller. This device has all of the logic for two sophisticated 8-bit parallel I/O ports in one LSI circuit. The PIO gives the system advanced interrupt handling. *Adapted from Zilog Z80 data sheets.*

C's interrupt. Therefore, both PIO B and PIO C cannot generate interrupt signals.

If PIO B is generating an interrupt, it stops PIO C from generating an interrupt. However, if PIO A generates an interrupt while the PIO B interrupt is being processed, the PIO A interrupt takes over because the daisy chain gives PIO A a higher priority than PIO B.

Figure 8-14 (page 156) shows the pin functions for the Z80 CTC. Like the Z80 PIO, the CTC is a complex I/O device. It does a job which takes many SSI and MSI circuits to do.

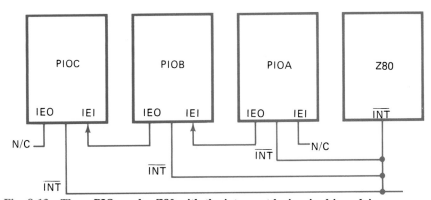

Fig. 8-13 Three PIOs and a Z80 with the interrupt logic wired in a daisy chain. The I/O device (PIO A) closest to the Z80 has the highest priority interrupt, and the I/O device farthest from the Z80 (PIO C) has the lowest priority.

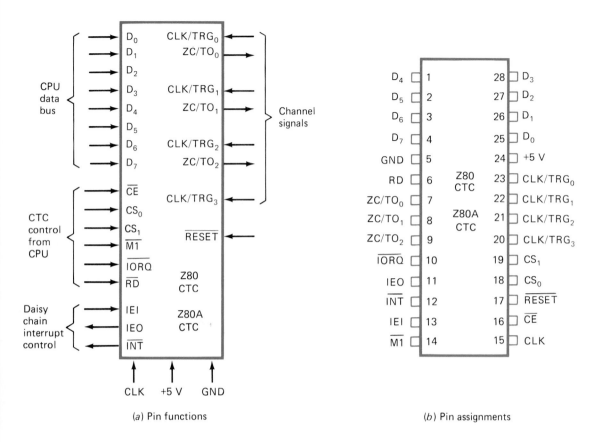

(a) Pin functions   (b) Pin assignments

Fig. 8-14   The Z80 CTC (counter-timer circuit). This support IC is used to let the Z80 system designer do counting and timing functions with a hardware device instead of doing these jobs with software, which ties up the processor so nothing else can be done. *Adapted from Zilog Z80 data sheets.*

The CTC has four independently *programmable counter-timers*. Hardware counters and timers are very common devices in microprocessor-based control products because any counting or timing which uses software keeps the processor busy when it could be doing other jobs.

Each of the four counter-timers can be programmed to divide by 1 to 255. They can work in one of two different modes. In the timer mode they cause an interrupt after a programmed amount of time has passed. In the counter mode they divide the frequency of the input signal by n, where n is 1 to 255. There is a prescaler (fixed divider) in front of each timer which divides the incoming signal by 1, 16, or 64. The counter can be used to generate low-frequency signals from the Z80 clock frequency.

The CTC interfaces with the Z80 in the same way that the PIO interfaces with the Z80. Each counter can cause an interrupt. The four counters become part of the daisy chain if daisy chain interrupt logic is used.

The Z80 SIO is shown in Fig. 8-15. The SIO converts parallel data signals from the Z80 data bus into serial signals. The SIO combines the equivalent logic of many different ICs including two USARTs (universal synchronous-asynchronous receiver-transmitters) into one IC. We will cover serial transmission devices in more detail in Chap. 13.

Like the CTC and the PIO, the SIO interfaces with the Z80 with nearly the same set of signals. The channel A and channel B signals input and output serial data and also provide the handshaking needed to transfer data with certain serial devices.

Self-Test

Answer the following questions.

30. When the Z80 $\overline{\text{MREQ}}$ line is asserted it indicates that the
    a. Address bus has valid address data for an I/O device

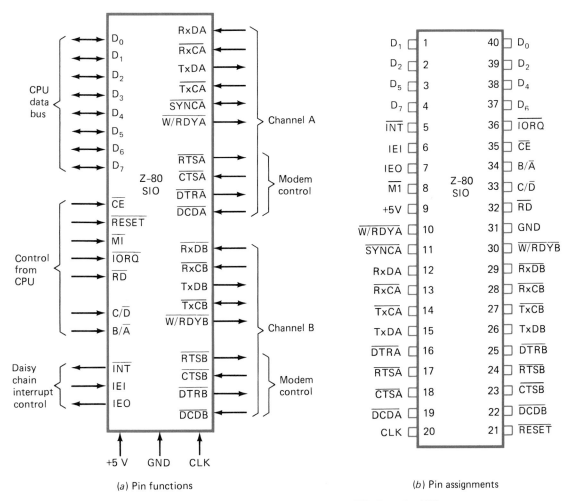

(a) Pin functions

(b) Pin assignments

Fig. 8-15 **The Z80 SIO (serial input/output) circuit. This support IC gives the Z80 system two serial ports which it can use to communicate with serial peripheral devices. If the SIO is not used, the function must be made with SSI, MSI, and LSI parts.** *Adapted from Zilog Z80 data sheets.*

    b. Address bus has valid address data for a memory location

    c. Data bus has valid data to send to the addressed memory location or I/O port

    d. The Z80 is ready to receive data over the data bus from an addressed I/O device or memory location

31. When the Z80 $\overline{\text{RD}}$ line is asserted it indicates that the

    a. Address bus has valid address data for an I/O device

    b. Address bus has valid address data for a memory location

    c. Data bus has valid data to send to the addressed memory location or I/O port

    d. Z80 is ready to receive data over the data bus from an addressed I/O device or memory location

32. When the Z80 $\overline{\text{IORQ}}$ line is asserted it indicates that the

    a. Address bus has valid address data for an I/O device

    b. Address bus has valid address data for a memory location

    c. Data bus has valid data to send to the addressed memory location or I/O port

    d. Z80 is ready to receive data over the data bus from an addressed I/O device or memory location

33. When the Z80 $\overline{\text{WR}}$ line is asserted it indicates that the

    a. Address bus has valid address data for an I/O device

    b. Address bus has valid address data for a memory location *c.* Data bus has valid data to send to the addressed memory location or I/O port

d. Z80 is ready to receive data over the data bus from an addressed I/O device or memory location

34. If you want the Z80 to stop executing the current program and execute NOP instructions until it receives an input on one of its interrupt lines, you would assert its _____ line.
   a. $\overline{\text{WAIT}}$ *c.* $\overline{\text{RESET}}$
   b. $\overline{\text{HALT}}$ *d.* $\overline{\text{NMI}}$

35. If you want to start the program counter executing at location 0000H, you would assert the Z80 _____ line.
   a. $\overline{\text{WAIT}}$
   b. $\overline{\text{HALT}}$
   c. $\overline{\text{RESET}}$
   d. $\overline{\text{NMI}}$

36. If you want to start the program counter executing at location 0066H, you would assert the Z80 _____ line.
   a. $\overline{\text{WAIT}}$ *c.* $\overline{\text{RESET}}$
   b. $\overline{\text{HALT}}$ *d.* $\overline{\text{NMI}}$

37. Briefly explain what happens when you assert the $\overline{\text{WAIT}}$ input.

38. Briefly explain what happens when you assert the $\overline{\text{BUSREQ}}$ input.

39. When it is operating, the Z80 feels somewhat warm to the touch. Does this make sense when you compute the power drawn by a Z80?

40. Why may a Z80B need to work with faster memory than does a Z80?

41. You are servicing a Z80-based product, and you find that the parallel I/O ports are implemented with two PIOs. Other logic you would expect to find would include
   a. An input data latch
   b. A TTL control register
   c. An address decoder
   d. All of the above

42. When an external device needs to tell a PIO that it has data to transfer into the PIO's A port it asserts the _____ line.
   a. $\overline{\text{A}}/\text{B}$
   b. $\overline{\text{INT}}$
   c. IEO
   d. $\overline{\text{ASTB}}$

43. The PIO and SIO each use four different I/O addresses because
   a. Each bit of each I/O port requires a separate address

b. The two I/O ports each have control and data addresses

c. A daisy chain requires a minimum of four addresses

d. Two addresses are needed to set the interrupt vector

44. If the interrupt outputs of a number of PIOs, SIOs, and CTCs are wired-ORed to the Z80 interrupt input, the priority of interrupt signals reaching the Z80 can be set by using the _____ signals on the I/O devices.
   a. $\overline{\text{INT}}$ and $\overline{\text{NMI}}$
   b. IEI and IEO
   c. $\text{C}/\overline{\text{D}}$ and $\overline{\text{A}}/\text{B}$
   d. All of the above

45. You are using a CTC to generate a low-frequency signal from the 4-MHz Z80A clock. You use two CTC dividers, one after the other, both set to their maximum divide ratio with no prescaling. The lowest output frequency available is
   a. 61.5 Hz
   b. 15.686 kHz
   c. 4.0 MHz
   d. 62,500 Hz

46. While you are servicing a Z80-based product you notice that there is an SIO as part of the system. From this information you would expect to find one or two
   a. Parallel data ports
   b. 64-kbyte memory banks
   c. High-speed disk drives
   d. Serial data ports

8-8 A PROGRAMMING MODEL FOR THE 6802

The MC6800 microprocessor was introduced by Motorola at nearly the same time that Intel introduced the 8080. Both were first implementations of an 8-bit microprocessor. In the late 1970s, Intel introduced an improved 8080 called the 8085. At nearly the same time, Motorola introduced an improved 6800 called the 6802. The programming model did not change for either of these improved products, but the hardware needed to make a system became much easier.

The programming model for the 6802 is shown in Fig. 8-16. The 6802 is much simpler

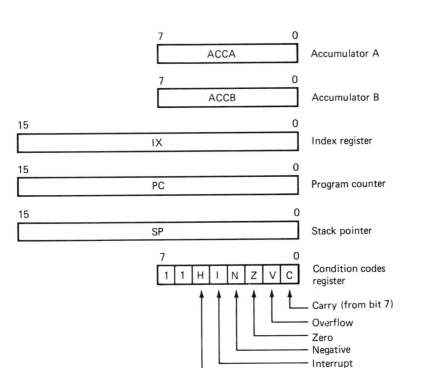

Fig. 8-16   **The programming model for the 6802 microprocessor. Compare this simple model to the Z80 programming model.** *Adapted from Motorola 6802 data sheets.*

than the Z80. There are no general-purpose registers, and there is only a single index register. There are, however, two accumulators. Both processors use 6 status bits in the status register. The 6802 uses memory-mapped I/O.

The programmer using the 6802 must make use of memory to perform many tasks that a programmer might do with registers when using an 8080 or Z80. For these reasons, the 6802 is said to be a *memory-intensive* architecture, whereas the 8080 and Z80 are called *register-intensive* architectures. The next few paragraphs review the 6802 register functions.

The 6802 accumulators are two identical 8-bit registers. Either of the two accumulators can receive the results of an operation. The source of data for an operation can come from either of the two accumulators or the data can come from a memory location. The accumulators are selected by the instruction used.

The 6802 condition code (status) register bit meanings are shown in Fig. 8-17 on page 160. These bits have nearly identical meanings as the same bits in the Z80 flag register. However, the 6802 does not have an add/subtract (N)

bit. Also, the 6802 shows the maskable interrupt status and does not have a parity bit.

The 6802 index register, stack pointer and program counter are special-purpose 16-bit registers. Each has a fixed function it always performs. Most all 6802 instructions can use the index register. The stack pointer allows the programmer to maintain a stack anywhere in the 6802 64-kyte memory space.

When the 6802 is reset, the program counter is loaded with the contents of memory locations FFFE and FFFF. That is, the address of the first 6802 program instruction is taken from the 2 *highest* bytes of addressable memory. *Note:* This is different from a Z80 operation. The Z80 reset causes the program counter to point to memory location 0000H, which contains the program's first instruction. That is, the Z80 reset vector is 0000H. The 6802 reset vector is contained in memory locations FFFE and FFFF.

Like the Z80, the 6802 has a nonmaskable interrupt and a maskable interrupt. The 6802 also has a software interrupt. The 6802 interrupt modes are much simpler to understand

1	1	H Half- carry	I Interrupt	N Negative	Z Zero	V Overflow	C Carry
7	6	5	4	3	2	1	0

```
1 = Permanent logic 1
1 = Permanent logic 1
H = 1 if the add or subtract caused a carry or borrow from bit 4
I  = 1 if the maskable interrupt (INT) is enabled (on)
N = 1 if the MSB of the result is 1
Z = 1 if the result of the operation is 0
V = 1 for arithmetic overflow
C = 1 if the operation produced a carry from the MSB
```

Fig. 8-17 **The 6802 condition code (status) register. Note that the 6802 condition code register includes the status of the maskable interrupt in addition to the status information in the Z80 status register, but does not have a parity bit.**

than the three Z80 modes. There is, however, one important difference.

When any interrupt signal is received, the 6802 first completes the current instruction. It then *stores the contents of all of its registers* (except the stack pointer) on the stack, an operation not performed by the Z80. Finally, it loads the program counter with the interrupt vector. All interrupt routines are ended by executing a return from interrupt (RTI) instruction. This instruction loads the 6802 registers from the stack, including the program counter, and restarts the interrupted program.

The nonmaskable interrupt loads the program counter with the interrupt vector in memory locations FFFC and FFFD, the next-to-last two locations in memory. Once again, you should note the difference between this operation and the Z80 NMI operation. The Z80 interrupt vector points to memory location 0066H in response to an NMI. The 6802 finds its NMI interrupt vector at memory locations FFFC and FFFD.

The 6802 maskable interrupt ($\overline{\text{IRQ}}$) is a simple interrupt with a single mode of operation. When the interrupt mask in the condition code register is cleared (I = logic 0), a signal at $\overline{\text{IRQ}}$ (active low) causes the interrupt process to start. Maskable interrupts get the interrupt vector from memory locations FFF8 and FFF9.

A special software interrupt is caused when the 6802 executes the *SWI (software interrupt) instruction*. Executing the SWI instruction causes the same operation as if a hardware interrupt has happened. The only difference is that the interrupt vector is taken from memory locations FFFA and FFFB.

Self-Test

Answer the following questions.

47. The 6802 is the same kind of change from the 6800 as the _____ is from the 8080.
    a. Z80
    b. 8086
    c. 8085
    d. 8051

48. The 6802 has _____ registers.
    a. 6
    b. 24
    c. 8-bit
    d. 16-bit

49. The 6802 has _____ architecture.
    a. A register-intensive
    b. A complex
    c. An interrupt-intensive
    d. A memory-intensive

50. The two 6802 accumulators are best described as
    a. The main and the alternate accumulators
    b. Accumulators matched to their index registers
    c. Two identical accumulators which are always available
    d. All of the above

51. In the 6802 the condition code register is the same as the Z80
    a. Main register set
    b. Double accumulators
    c. Alternate register set
    d. Flag register

52. The 6802 stack pointer, index register, and program counter are all 16-bit
    a. Specific-purpose registers

Branch to
subroutine (BSR)

Jump to
subroutine (JSR)

Return from
subroutine (RTS)

b. General-purpose registers
c. General-purpose register pairs
d. Specific-purpose register pairs

53. The 6802 interrupt system is different from the Z80 interrupt system because _____ and _____ . (Choose two of the four answers.)
   a. An interrupt starts being processed after the current instruction is complete
   b. Processing the interrupt saves all of the registers on the stack, not just the current program counter value
   c. The interrupt vector is taken from a specific memory location rather than being a specific memory location
   d. Both nonmaskable and maskable interrupts are available

54. When the programmer wants to end a 6802 interrupt routine, the _____ instruction is used.
   a. RTI   c. SWI
   b. HLT   d. Return

8-9 THE 6802 INSTRUCTION SET

The last section showed us the 6802 has a simpler architecture than does the Z80. As you expect, the 6802 instruction set is also simpler. The 6802 has 72 instructions used with seven different addressing modes. The addressing modes are shown in Fig. 8-18.

Tables 8-12 through 8-15 (at the end of this chapter, preceding the summary) are the 6802 instructions as Motorola shows them. The following paragraphs briefly review these instructions. Be sure to study the instructions in the table carefully, comparing them to the Z80 instructions and to the general instructions you studied in Chap. 6.

The 6802 accumulator and memory instructions (Table 8-12) are used to move and combine data arithmetically and logically. Note that most instructions have three forms. One acts on accumulator A, one on accumulator B, and one on a memory location. Each of the accumulator and memory instructions can use one or more of five different addressing modes as shown in the table.

The index register and stack manipulation instructions (Table 8-13) are 11 instructions which work with the stack pointer and index register.

The jump and branch instructions (Table 8-14) use a different set of addressing modes.

Accumulator Accumulator A or B is the object of the instruction. No other register or memory location is used. *Note:* This is shown in the implied column for the ASL, ASR, CLR COM, DEC, INC, LSR, NEG, ROL, ROR, and TST instructions.

Immediate The byte following the op code holds the data to be worked on except for LDS and LDX, which have two bytes of data. A 2- or 3-byte instruction.

Direct The byte following the op code is an address for memory locations 0000 to 00FF, where the data to be worked on is held. A 2-byte instruction.

Direct extended The two bytes following the op code are the address the instruction is to act on. A 3-byte instruction.

Indexed The byte following the op code is added to the index register using unsigned arithmetic. The resulting 16-bit number is used as the address of the data the instruction is to act on. A 2-byte instruction.

Implied The op code tells what it is to work on. A 1-byte instruction.

Relative The byte following the op code is added to the PC value (2's complement) to give the address for the data that is to be worked on. A 2-byte instruction.

Fig. 8-18   A summary of the 6802 addressing modes.

The immediate and direct modes are replaced by the relative mode. The branch instructions use relative addressing, and most test the condition code register to see when they should branch. Except for *BSR* (*branch to subroutine*), the branch instructions are the same as a jump or conditional jump instruction, except that they use relative addressing.

The 6802 jump instruction uses either indexed, extended, or relative addressing. The *jump to subroutine* (*JSR*) instruction is an unconditional subroutine call. Both the JSR and the BSR instructions are followed with an *RTS* (*return from subroutine*) instruction when the subroutine is complete. The NOP and interrupt instructions are included in this group.

The condition code register manipulation instructions (Table 8-15) let the programmer set or clear the condition code register carry, interrupt, and overflow bits. The condition code register contents can be loaded from the accumulator or loaded into the accumulator.

Although the 6802 instruction set is much simpler than the Z80's, there are many programmers who like the 6802 architecture better than the Z80 architecture. In many cases, they can write programs using the 6802 architecture and instruction set that perform just as well as a Z80 program. One reason that programmers like the 6802 is its simple architecture and very logical, easy-to-use instruction set.

Self-Test

Answer the following questions.

55. If you are using a 6802, you would use the _____ instruction and addressing to increment a memory location.
    a. INCA implied
    b. INCB implied
    c. INC extended
    d. None of the above

56. The ASRA instruction tells the 6802 to
    a. Perform an arithmetic shift left using extended addressing
    b. Perform an arithmetic shift right on accumulator A using implied addressing
    c. Perform an arithmetic shift left on accumulator A using extended addressing
    d. None of the above

57. The 6802 instruction which uses the fewest number of processor cycles to load the index register from memory location 000AH is
    a. LDX direct   c. INX direct
    b. LDX extended   d. DES extended

58. The JMP instruction which has the op code 6E tells the processor to load the program counter
    a. From memory location 6E
    b. With the contents of the following 2 bytes
    c. With the contents of the sum of the following byte plus the current value of the index register
    d. With the value 6E

59. The 6802 WAI instruction is a
    a. Two-byte instruction telling the processor to wait for a subroutine
    b. Single-byte instruction telling the processor to wait for an interrupt
    c. Single-byte instruction telling the processor to wait for a subroutine return
    d. Two-byte instruction telling the processor to wait for an interrupt

60. Executing a 6802 subtract with carry (SBCA) instruction changes the condition code register _____ bits.
    a. I, N, and Z
    b. Z, V, and C
    c. H, I, N, Z, V, and C
    d. N, Z, V, and C

8-10 THE 6802 HARDWARE

The 6802 block diagram is shown in Fig. 8-19. This diagram shows the ALU, registers, 8-bit data bus, 16-bit address bus, and the internal 128 bytes of memory. This memory can be enabled or disabled with an external signal. The lower 32 bytes can be powered from an external source (such as a battery). This gives 32 bytes of nonvolatile RAM as long as external power is applied. *Note:* A special version of the 6802 is available which does not have this RAM. It is called the 6808.

Figure 8-20 on page 164 shows the 6802 pin function and pinout diagrams. Next, we will look at these connections and the signals which must go into or come from them.

The 16-bit address bus allows the 6802 to access up to 64K memory locations. The internal 128 bytes of memory do not require any memory address decoding. However, if they are being used, address decoding logic must be used to disable any external memory at memory locations 0000H to 007FH. The 6802 address bus cannot be placed in a high-impedance state. If a high-impedance state is needed, it must be done with external logic.

The 8-bit data bus is a bidirectional data bus which is normally in the data output state. That is, the 6802 is normally outputting data unless it has been told to input data. The data bus can be placed in a high-impedance mode if needed.

The 6802 only has three *system control lines*. Enable is a signal generated from the internal clock. It is used to time external devices which need to work with the CPU. The Read/Write (R/W) line tells external memory that the 6802 is reading data from the device or writing data to the external device. The valid memory address (VMA) line tells external devices when the 6802 has valid address data on its address bus.

The four *CPU control lines* trigger changes in the 6802 operation. Asserting the memory ready (MR) line slows the processor down by one or more clock cycles. This is used by slow

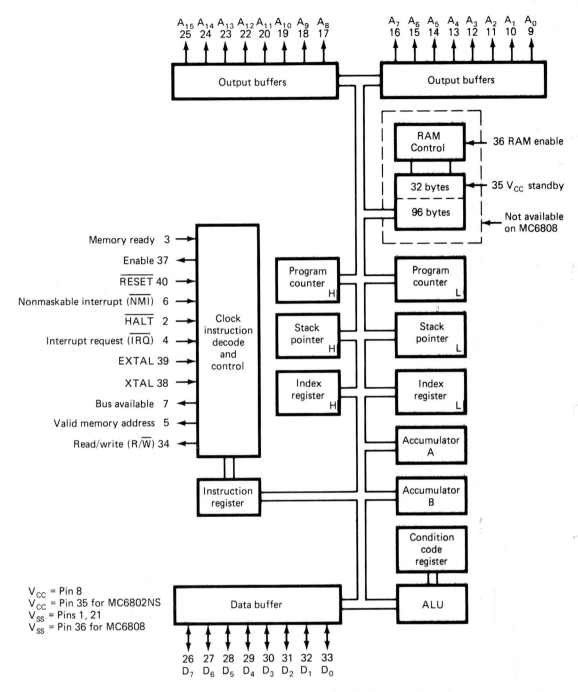

Fig. 8-19 **The 6802 block diagram. This block diagram is typically drawn to show the 6802 16-bit registers with the High and Low bytes separate; however, the programmer can only access these registers 16 bits at a time.** *Adapted from Motorola 6802 data sheets.*

devices which cannot transfer data as quickly as the 6802. This is the same as the Z80 wait state.

The $\overline{\text{RESET}}$ line initializes the processor and forces the program counter to be loaded from the contents of memory locations FFFE and FFFF. The $\overline{\text{NMI}}$ input is a nonmaskable interrupt, and the $\overline{\text{IRQ}}$ input is a maskable interrupt input.

The *bus control lines* allow an external device to take control of the 6802 data bus. The external device must work with external logic to take control of the address bus. A request for control on the $\overline{\text{HALT}}$ input is acknowledged on the bus available (BA) output.

The 6802 requires a 5-V supply capable of supplying 200 mA. Additionally, the 6802 re-

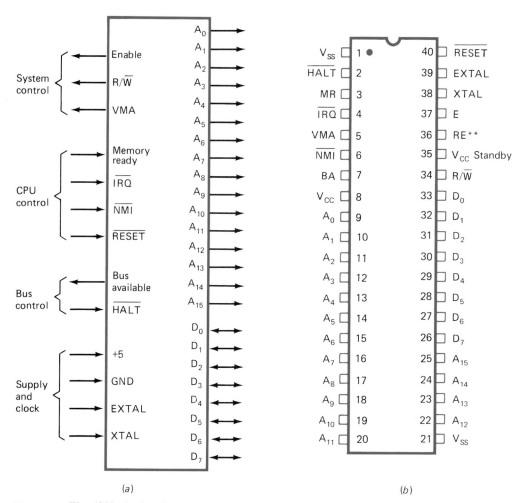

Fig. 8-20 **The 6802 pin function and pinout diagrams.** *Adapted from Motorola 6802 data sheets.*

quires an external quartz crystal to be connected between the EXTAL and XTAL pins. This quartz crystal sets the frequency of the internal oscillator which generates the clock signal for the processor. If an external clock signal is available, it can be connected to the EXTAL input.

Like the Z80, the 6800 family of processors has some LSI circuits designed specifically to provide functions needed by the 6800 processor to make a complete system with the fewest parts. In the following paragraphs, we will look at some of these different devices. You will see that these devices do many of the same jobs done by the Z80 devices.

The 6821 *PIA* (*parallel interface adapter*) is a single IC which gives the 6802 two 8-bit parallel I/O ports with handshaking. Figure 8-21 shows the pin assignment and pinout diagrams for the 6821. As you can see, this device is quite similar in function to the Z80 PIO.

The PIA connects to the external devices using the eight bidirectional data lines on each port and the two handshaking lines. All data lines can be programmed to act as either inputs or outputs. The handshaking lines CA1 and CB1 are inputs that cause the PIA to generate a system interrupt. The CA2 and CB2 handshaking lines can act as either inputs or outputs. In the input mode, they can be used to generate interrupts. In the output mode, they are used to tell an external device there is valid data on the eight data lines.

Each PIA port has a number of control registers. These control registers allow the programmer to define a direction for each bit on the port, to control the interrupt operation, and to find out which part of the PIA caused an interrupt signal. The different registers are accessed by different PIA addresses. These different addresses are connected to the RS0 and RS1 inputs.

The three chip select inputs (CS0, CS1, and $\overline{CS2}$) help the hardware designer reduce the amount of address decoding logic. The PIA is selected when CS0 and CS1 are high and $\overline{CS2}$

is low. For example, a very simple system only uses one PIA. The RS0 and RS1 lines are connected to the A_0 and A_1 lines. The $\overline{CS2}$ line is connected to A_2, CS1 is connected to A_3, and CS0 is connected to the output of a NOR gate, which has 12 inputs connected to A_4 to A_{15} and one input connected to VMA. These connections are shown in Fig. 8-22 on page 166.

In operation, the PIA is selected when address lines A_{15} to A_4 are low, A_3 is high, A_2 is low, and VMA is high. This means that the PIA is selected by addresses 0008H, 0009H, 000AH, and 000BH. The A_1 and A_0 lines are connected to RS0 and RS1. The four addresses select the different PIA registers for control and data transfer.

The enable input gives the PIA a clock signal so that its data transfers can be synchronized with the processor. The $\overline{RESET}$ is used to make sure the PIA is initialized along with the processor. The $R/\overline{W}$ line is used by the processor to tell the PIA in which direction the data is going.

A *programmable timer module (PTM)* gives a 6802 system access to three 16-bit timers. The 6840 PTM is shown in Fig. 8-23 on page 167. Like the CTC, the PTM is used in 6802 system designs, so timing does not have to be done with software routines. Each 16-bit timer is supported by a control register and a status register. The programmer can address these registers with the RS0, RS1, and RS2 inputs. The PTM is addressed and selected just like the PIA.

Like the PIO, the PTM can be programmed to generate an interrupt when the timer completes its count. The programmer can then examine the PTM's status registers to see which timer caused the interrupt.

Because serial communications are so common for microprocessor-based systems, the 6802 is also supported by a special LSI circuit which provides the system with a serial port. This part is called the 6850 *ACIA (asynchronous communications interface adapter)*. The 6850 is shown in Fig. 8-24 on page 168. The

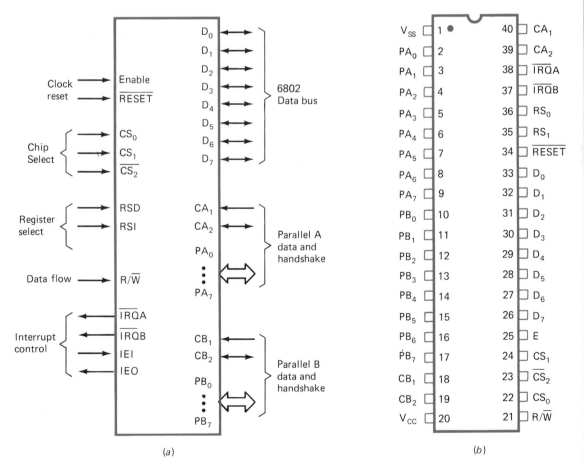

(a)

(b)

Fig. 8-21 The 6821 parallel interface adapter. This device gives the 6802 two 8-bit parallel I/O ports with handshaking and complete 6802 bus compatibility. The only external logic required is address decoding logic. *Adapted from Motorola 6802 data sheets.*

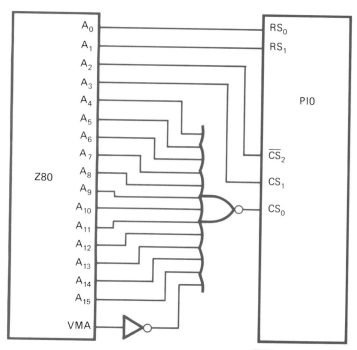

Fig. 8-22 Using the chip select (CS0, CS1, and $\overline{CS2}$) lines to build a minimum logic address decoder to select the four PIA registers for data and control.

6850 interfaces to the 6802 using the same control and data lines as the other I/O parts. The ACIA connections to the external world include serial transmit and receive lines as well as handshaking lines. These handshaking lines allow the ACIA to work with devices which use serial data and a few additional handshaking lines to control the data flow.

Self-Test

Answer the following questions.

61. A major hardware difference between the Z80 and the 6802 is
    a. The use of an internal 8-bit data bus
    b. The internal 128-byte memory
    c. The use of an accumulator
    d. None of these. Both processors have all of the above hardware features
62. If you are using the internal 128-byte memory, you must
    a. Disable any external ROM at locations FF00 to FFFF
    b. Enable any external RAM between locations 0000 and 007F
    c. Disable any external memory between locations FF00 and FFFF
    d. Disable any external memory between locations 0000 and 007F

63. The 6802 8-bit data bus
    a. Is bidirectional
    b. Can be placed in a high-impedance mode
    c. Only carries data from memory locations 0000 to FFFF
    d. All of the above
64. The 6802 _____ performs the same function as the Z80 $\overline{WAIT}$ input.
    a. Memory ready (MR)
    b. Enable
    c. $\overline{HALT}$
    d. Valid memory address (VMA)
65. The purpose of the EXTAL and XTAL connections and the circuits inside the 6802 which go with these connections is to
    a. Let the user put a quartz crystal between these two pins to set the frequency of the 6802 internal clock
    b. Allow the user to drive the 6802 with an external timing signal (through the EXTAL connection) if needed
    c. Reduce the number of support circuits needed to make a microprocessor system by getting rid of the need for an external oscillator
    d. All of the above
66. The 6802 support circuit called the PIA is very much like the Z80 support circuit called the PIO. The major difference between these two parts is

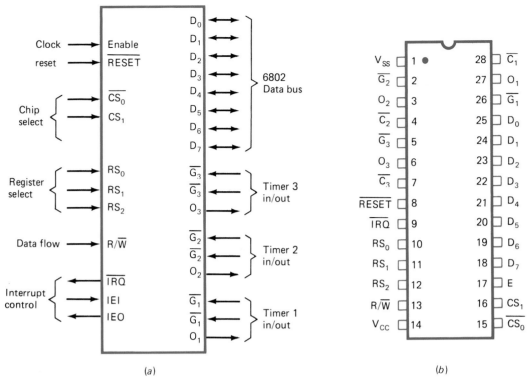

Fig. 8-23 **The 6840 programmable time module (PTM). This support IC gives the 6802 three 16-bit programmable counter-timers which can be used to move counting and timing functions from the 6802 to an external device.** *Adapted from Motorola 6802 data sheets.*

a. The kinds of bus control signals they are built to work with
b. How many parallel bits of information can be connected to the external ports
c. That the PIO can generate interrupts and the PIA cannot
d. All of the above

67. A programmer uses _____ to select the different control and data registers on a 6802 support circuit like the PIA or ACIA.
a. Information on the 8-bit data bus
b. Different addresses and the RSx (x = 0, 1, 2, etc.) inputs
c. The chip select (CSx) inputs
d. The interrupt output

68. A 6802 support circuit uses the _____ input to synchronize with the processor and to provide a timing signal.
a. Reset
b. VMA
c. $\overline{\text{HALT}}$
d. Enable

69. The PTM support circuit has
a. Three programmable 16-bit timers
b. The capability to interrupt the processor after a preprogramed amount of time has gone by
c. The ability to count the number of events which happen
d. All of the above

70. The major difference between the SIO and the ACIA is
a. The number of serial ports in the two parts
b. The use of control registers in the ACIA
c. That the SIO may be either memory mapped or I/O mapped
d. One is a serial port and the other is a parallel port

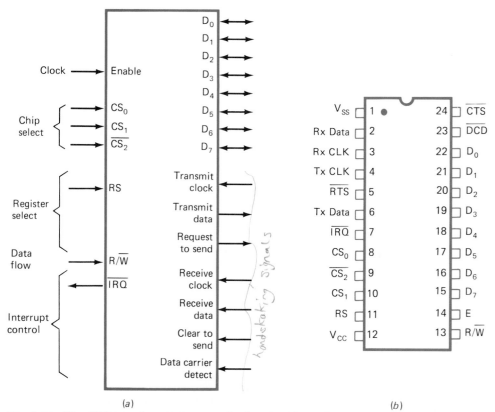

Fig. 8-24 **The 6850 asynchronous communications interface adapter (ACIA). The ACIA is used to add a serial data port to the 6802 with a minimum number of parts.** *Adapted from Motorola 6802 data sheets.*

Table 8-1 Z80 8-Bit Load Instructions

Mnemonic	Symbolic Operation	S	Z		H		P/V	N	C	Comments	
LD r, r′	r ← r′	•	•	X	•	X	•	•	•	r, r′	Reg.
LD r, n	r ← n	•	•	X	•	X	•	•	•	000	B
										001	C
LD r, (HL)	r ← (HL)	•	•	X	•	X	•	•	•	010	D
LD r, (IX + d)	r ← (IX + d)	•	•	X	•	X	•	•	•	011	E
										100	H
										101	L
LD r, (IY + d)	r ← (IY + d)	•	•	X	•	X	•	•	•	111	A
LD (HL), r	(HL) ← r	•	•	X	•	X	•	•	•		
LD (IX + d), r	(IX + d), r	•	•	X	•	X	•	•	•		
LD (IY + d), r	(IY + d) ← r	•	•	X	•	X	•	•	•		
LD (HL), n	(HL) ← n	•	•	X	•	X	•	•	•		
LD (IX + d), n	(IX + d) ← n	•	•	X	•	X	•	•	•		
LD (IY + d), n	(IY + d) ← n	•	•	X	•	X	•	•	•		
LD A, (BC)	A ← (BC)	•	•	X	•	X	•	•	•		
LD A, (DE)	A ← (DE)	•	•	X	•	X	•	•	•		
LD A, (nn)	A ← (nn)	•	•	X	•	X	•	•	•		
LD (BC), A	(BC) ← A	•	•	X	•	X	•	•	•		
LD (DE), A	(DE) ← A	•	•	X	•	X	•	•	•		
LD (nn), A	(nn) ← A	•	•	X	•	X	•	•	•		
LD A, I	A ← I	↕	↕	X	0	X	IFF	0	•		
LD A, R	A ← R	↕	↕	X	0	X	IFF	0	•		
LD I, A	I ← A	•	•	X	•	X	•	•	•		
LD R, A	R ← A	•	•	X	•	X	•	•	•		
n	8-bit value in range < 0, 255 >.										

Adapted from Zilog Z80 data sheets.

Table 8-2 Z80 16-Bit Load Instructions

Mnemonic	Symbolic Operation	S	Z		H		P/V	N	C	Comments	
					Flags						
LD dd, nn	dd ← nn	•	•	X	•	X	•	•	•	dd	Pair
										00	BC
										01	DE
LD IX, nn	IX ← nn	•	•	X	•	X	•	•	•	10	HL
										11	SP
LD IY, nn	IY ← nn	•	•	X	•	X	•	•	•		
LD HL, (nn)	H ← (nn + 1) L ← (nn)	•	•	X	•	X	•	•	•		
LD dd, (nn)	dd$_H$ ← (nn + 1) dd$_L$ ← (nn)	•	•	X	•	X	•	•	•		
LD IX, (nn)	IX$_H$ ← (nn + 1) IX$_L$ ← (nn)	•	•	X	•	X	•	•	•		
LD IY, (nn)	IY$_H$ ← (nn + 1) IY$_L$ ← (nn)	•	•	X	•	X	•	•	•		
LD (nn), HL	(nn + 1) ← H (nn) ← L	•	•	X	•	X	•	•	•		
LD (nn), dd	(nn + 1) ← dd$_H$ (nn) ← dd$_L$	•	•	X	•	X	•	•	•		
LD (nn), IX	(nn + 1) ← IX$_H$ (nn) ← IX$_L$	•	•	X	•	X	•	•	•		
LD (nn), IY	(nn + 1) ← IY$_H$ (nn) ← IY$_L$	•	•	X	•	X	•	•	•		
LD SP, HL	SP ← HL	•	•	X	•	X	•	•	•		
LD SP, IX	SP ← IX	•	•	X	•	X	•	•	•		
LD SP, IY	SP ← IY	•	•	X	•	X	•	•	•	qq	Pair
PUSH qq	(SP − 2) ← qq$_L$ (SP − 1) ← qq$_H$ SP → SP − 2	•	•	X	•	X	•	•	•	00 01 10	BC DE HL
PUSH IX	(SP − 2) ← IX$_L$ (SP − 1) ← IX$_H$ SP → SP − 2	•	•	X	•	X	•	•	•	11	AF
PUSH IY	(SP − 2) ← IYL (SP − 1) ← IY$_H$ SP → SP − 2	•	•	X	•	X	•	•	•		
POP qq	qq$_H$ ← (SP + 1) qq$_L$ ← (SP) SP → SP + 2	•	•	X	•	X	•	•	•		
POP IX	IX$_H$ ← (SP + 1) IX$_L$ ← (SP) SP → SP + 2	•	•	X	•	X	•	•	•		
POP IY	IY$_H$ ← (SP + 1) IY$_L$ ← (SP) SP → SP + 2	•	•	X	•	X	•	•	•		

Adapted from Zilog Z80 data sheets.

Mnemonic	Symbolic Operations	S	Z		H		P/V	N	C	Comments
EX DE, HL	DE ↔ HL	•	•	X	•	X	•	•	•	
EX AF, AF′	AF ↔ AF′	•	•	X	•	X	•	•	•	
EXX	BC ↔ BC′ DE ↔ DE′ HL ↔ HL′	•	•	X	•	X	•	•	•	
EX (SP), HL	H ↔ (SP + 1) L ↔ (SP)	•	•	X	•	X	•	•	•	
EX (SP), IX	IX_H ↔ (SP + 1) IX_L ↔ (SP)	•	•	X	•	X	•	•	•	
EX (SP), IY	IY_H ↔ (SP + 1) IY_L ↔ (SP)	•	•	X	•	X	•	•	•	
LDI	(DE) ← (HL) DE ← DE + 1 HL ← HL + 1 BC ← BC − 1	•	•	X	0	X	① ↕	0	•	
LDIR	(DE) ← (HL) DE ← DE + 1 HL ← HL + 1 BC ← BC − 1 Repeat until BC = 0	•	•	X	0	X	① 0	0	•	
LDD	(DE) ← (HL) DE ← DE − 1 HL ← HL − 1 BC ← BC − 1	•	•	X	0	X	① ↕	0	•	
LDDR	(DE) ← (HL) DE ← DE − 1 HL ← HL − 1 BC ← BC − 1 Repeat until BC = 0	•	•	X	0	X	② 0	0	•	If BC ≠ 0 If BC = 0
CPI	A − (HL) HL ← HL + 1 BC ← BC − 1	↕	③ ↕	X	↕	X	① ↕	1	•	

CPIR	A − (HL)
	
	HL ← HL + 1
	BC ← BC − 1
	Repeat until
	  A = (HL) or
	  BC = 0
		↕	③ ↕	X	↕	X	① ↕	1	•	If BC ≠ 0 and
										  A ≠ (HL)
										If BC = 0 or
										  A = (HL)

CPD	A − (HL) HL ← HL − 1 BC ← BC − 1	↕	③ ↕	X	↕	X	① ↕	1	•	

CPDR	A − (HL)
	
	HL ← HL − 1
	BC ← BC − 1
	Repeat until
	  A = (HL) or
	  BC = 0
		↕	③ ↕	X	↕	X	① ↕	1	•	If BC ≠ 0 and
										  A ≠ (HL)
										If BC = 0 or
										  A = (HL)

① P/V flag is 0 if the result of BC − 1 = 0, otherwise P/V = 1.
② P/V flag is 0 at completion of instruction only.
③ Z flag is 1 if A = (HL), otherwise Z = 0.

Adapted from Zilog Z80 data sheets.

Table 8-4 Z80 8-Bit Arithmetic and Logical Instructions

Mnemonic	Symbolic Operation	S	Z		H		P/V	N	C	Comments
ADD A, r	A ← A + r	↕	↕	X	↕	X	V	0	↕	r Reg.
ADD A, n	A ← A + n	↕	↕	X	↕	X	V	0	↕	000 B
										001 C
										010 D
ADD A, (HL)	A ← A + (HL)	↕	↕	X	↕	X	V	0	↕	011 E
ADD A, (IX + d)	A ← A + (IX + d)	↕	↕	X	↕	X	V	0	↕	100 H
										101 L
										111 A
ADD A, (IY + d)	A ← A + (IY + d)	↕	↕	X	↕	X	V	0	↕	
ADC A, s	A ← A + s + CY	↕	↕	X	↕	X	V	0	↕	s is any of r, n,
SUB s	A ← A − s	↕	↕	X	↕	X	V	1	↕	(HL), (IX + d),
SBC A, s	A ← A − s − CY	↕	↕	X	↕	X	V	1	↕	(IY + d) as shown
AND s	A ← A ∧ s	↕	↕	X	1	X	P	0	0	for ADD instruction.
OR s	A ← A ∨ s	↕	↕	X	0	X	P	0	0	
XOR s	A ← A ⊕ s	↕	↕	X	0	X	P	0	0	
CP s	A − s	↕	↕	X	↕	X	V	1	↕	
INC r	r ← r + 1	↕	↕	X	↕	X	V	0	•	
INC (HL)	(HL) ← (HL) + 1	↕	↕	X	↕	X	V	0	•	
INC (IX + d)	(IX + d) ← (IX + d) + 1	↕	↕	X	↕	X	V	0	•	
INC (IY + d)	(IY + d) ← (IY + d) + 1	↕	↕	X	↕	X	V	0	•	
DEC m	m ← m − 1	↕	↕	X	↕	X	V	1	•	m is any of r, (HL), (IX + d), (IY + d) as shown for INC. DEC same format and states as INC.
n	8-bit value in range < 0,255 >.									
d	8-bit value in range < 0,255 >.									

Adapted from Zilog Z80 data sheets.

Table 8-5 Z80 16-Bit Arithmetic and Logical Instructions

Mnemonic	Symbolic Operation	S	Z		H		P/V	N	C	Comments
ADD HL, ss	HL ← HL + ss	•	•	X	X	X	•	0	↕	ss: 00 BC, 01 DE, 10 HL, 11 SP
ADC HL, ss	HL ← HL + ss + CY	↕	↕	X	X	X	V	0	↕	
SBC HL, ss	HL ← HL − ss − CY	↕	↕	X	X	X	V	1	↕	
ADD IX, pp	IX ← IX + pp	•	•	X	X	X	•	0	↕	pp: 00 BC, 01 DE, 10 IX, 11 SP
ADD IY, rr	IY ← IY + rr	•	•	X	X	X	•	0	↕	rr: 00 BC, 01 DE, 10 IY, 11 SP
INC ss	ss ← ss + 1	•	•	X	•	X	•	•	•	
INC IX	IX ← IX + 1	•	•	X	•	X	•	•	•	
INC IY	IY ← IY + 1	•	•	X	•	X	•	•	•	
DEC ss	ss ← ss − 1	•	•	X	•	X	•	•	•	
DEC IX	IX ← IX − 1	•	•	X	•	X	•	•	•	
DEC IY	IY ← IY − 1	•	•	X	•	X	•	•	•	

Adapted from Zilog Z80 data sheets.

Table 8-6 Z80 Rotate and Shift Instructions

Mnemonic	Symbolic Operation	S	Z		H		P/V	N	C	Comments
RLCA	CY ← [7 ← 0] ← , A	•	•	X	0	X	•	0	↕	Rotate left circular accumulator.
RLA	CY ← [7 ← 0] ← , A	•	•	X	0	X	•	0	↕	Rotate left accumulator.
RRCA	[7 → 0] → CY, A	•	•	X	0	X	•	0	↕	Rotate right circular accumulator.
RRA	[7 → 0] → CY, A	•	•	X	0	X	•	0	↕	Rotate right accumulator.
RLC r		↕	↕	X	0	X	P	0	↕	Rotate left circular register r.
RLC (HL)		↕	↕	X	0	X	P	0	↕	r	Reg.
RLC (IX + d)	CY ← [7 ← 0] ← r,(HL),(IX + d),(IY + d)	↕	↕	X	0	X	P	0	↕	
RLC (IY + d)		↕	↕	X	0	X	P	0	↕	
RLm	CY ← [7 ← 0] ← m ≡ r,(HL),(IX + d),(IY + d)	↕	↕	X	0	X	P	0	↕	Instruction format and states are as shown for RLC's.
RRC m	[7 → 0] → CY m ≡ r,(HL),(IX + d),(IY + d)	↕	↕	X	0	X	P	0	↕	
RR m	[7 → 0] → CY m ≡ r,(HL),(IX + d),(IY + d)	↕	↕	X	0	X	P	0	↕	
SLA m	CY ← [7 ← 0] ← 0 m ≡ r,(HL),(IX + d),(IY + d)	↕	↕	X	0	X	P	0	↕	
SRA m	[7 → 0] → CY m ≡ r,(HL),(IX + d), (IY + d)	↕	↕	X	0	X	P	0	↕	
SRL m	0 → [7 → 0] → CY m ≡ r,(HL),(IX + d),(IY + d)	↕	↕	X	0	X	P	0	↕	
RLD	[7-4][3-0] [7-4][3-0] A (HL)	↕	↕	X	0	X	P	0	•	Rotate digit left and right between the accumulator and location (HL).
RRD	[7-4][3-0] [7-4][3-0] A (HL)	↕	↕	X	0	X	P	0	•	The content of the upper half of the accumulator is unaffected.

Register encoding (for RLC's):

r	Reg.
000	B
001	C
010	D
011	E
100	H
101	L
111	A

d   8-bit value in range < 0,255 >.

Adapted from Zilog Z80 data sheets.

Table 8-7   Z80 General-Purpose Arithmetic and Control Instructions

Mnemonic	Symbolic Operation	S	Z		H		P/V	N	C	Comments
DAA	Converts acc. content into packed BCD following add or subtract with packed BCD operands.	↕	↕	X	↕	X	P	•	↕	Decimal adjust accumulator.
CPL	$A \leftarrow \overline{A}$	•	•	X	1	X	•	1	•	Complement accumulator (one's complement).
NEG	$A \leftarrow 0 - A$	↕	↕	X	↕	X	V	1	↕	Negate acc. (two's complement).
CCF	$CY \leftarrow \overline{CY}$	•	•	X	X	X	•	0	↕	Complement carry flag.
SCF	$CY \leftarrow 1$	•	•	X	0	X	•	0	1	Set carry flag.
NOP	No operation	•	•	X	•	X	•	•	•	
HALT	CPU halted	•	•	X	•	X	•	•	•	
DI	$IFF \leftarrow 0$	•	•	X	•	X	•	•	•	
EI	$IFF \leftarrow 1$	•	•	X	•	X	•	•	•	
IM 0	Set interrupt mode 0	•	•	X	•	X	•	•	•	
IM 1	Set interrupt mode 1	•	•	X	•	X	•	•	•	
IM 2	Set interrupt mode 2	•	•	X	•	X	•	•	•	

Adapted from Zilog Z80 data sheets.

Table 8-8   Z80 Bit Set, Reset, and Test Instructions

Mnemonic	Symbolic Operation	S	Z		H		P/V	N	C	Comments
BIT b, r	$Z \leftarrow r_b$	X	↕	X	1	X	X	0	•	
BIT b, (HL)	$Z \leftarrow \overline{(HL)}_b$	X	↕	X	1	X	X	0	•	
BIT b, $\overline{(IX + d)}_b$	$Z \leftarrow \overline{(IX + d)}_b$	X	↕	X	1	X	X	0	•	
BIT b, $\overline{(IY + d)}_b$	$Z \leftarrow \overline{(IY + d)}_b$	X	↕	X	1	X	X	0	•	
SET b, r	$r_b \leftarrow 1$	•	•	X	•	X	•	•	•	
SET b, (HL)	$(HL)_b \leftarrow 1$	•	•	X	•	X	•	•	•	
SET b, (IX + d)	$(IX + d)_b \leftarrow 1$	•	•	X	•	X	•	•	•	
SET b, (IY + d)	$(IY + d)_b \leftarrow 1$	•	•	X	•	X	•	•	•	
RES b, m	$m_b \leftarrow 0$ $m \equiv r, (HL),$ $(IX + d),$ $(IY + d)$	•	•	X	•	X	•	•	•	
d	8-bit value in range < 0, 255 >.									

Comments column (Tables reference):

r	Reg.
000	B
001	C
010	D
011	E
100	H
101	L
111	A

b	Bit Tested
000	0
001	1
010	2
011	3
100	4
101	5
110	6
111	7

Adapted from Zilog Z80 data sheets.

Table 8-9   Z80 Jump Instructions

Mnemonic	Symbolic Operation	S	Z ①		H		P/V	N	C	Comments
JP nn	PC ← nn	•	•	X	•	X	•	•	•	
JP cc, nn	If condition cc is true PC ← nn, otherwise continue	•	•	X	•	X	•	•	•	
JR e	PC ← PC + e	•	•	X	•	X	•	•	•	
JR C, e	If C = 0, continue If C = 1, PC ← PC + e	•	•	X	•	X	•	•	•	If condition not met. If condition is met.
JR NC, e	If C = 1, continue If C = 0, PC ← PC + e	•	•	X	•	X	○	•	•	If condition not met. If condition is met.
JP Z, e	If Z = 0, continue If Z = 1, PC ← PC + e	•	•	X	•	X	•	•	•	If condition not met. If condition is met.
JR NZ, e	If Z = 1, continue If Z = 0, PC ← PC + e	•	•	X	•	X	•	•	•	If condition not met. If condition is met.
JP (HL)	PC ← HL	•	•	X	•	X	•	•	•	
JP (IX)	PC ← IX	•	•	X	•	X	•	•	•	
JP (IY)	PC ← IY	•	•	X	•	X	•	•	•	
DJNZ, e	B ← B − 1 If B = 0, continue If B ≠ 0, PC ← PC + e	•	•	X	•	X	•	•	•	If B = 0. If B ≠ 0.

cc Condition
000 NZ non-zero
001 Z zero
010 NC non-carry
011 C carry
100 PO parity odd
101 PE parity even
110 P sign positive
111 M sign negative

e represents the extension in the relative addressing mode.
e is a signed two's complement number in the range < −126, 129 >.
e − 2 in the opcode provides an effective address of pc + e as PC is incremented
    by 2 prior to the addition of e.

Adapted from Zilog Z80 data sheets.

Table 8-10   Call Instructions

Mnemonic	Symbolic Operation	S	Z		H		P/V	N	C	Comments
CALL nn	(SP − 1) ← PC$_H$ (SP − 2) ← PC$_L$ PC ← nn	•	•	X	•	X	•	•	•	
CALL cc, nn	If condition cc is false, continue; otherwise same as CALL nn	•	•	X	•	X	•	•	•	If cc is false. If cc is true.
RET	PC$_L$ ← (SP) PC$_H$ ← (SP + 1)	•	•	•	•	X	•	•	•	
RET cc	If condition cc is false, continue; otherwise same as RET	•	•	X	•	X	•	•	•	If cc is false. If cc is true. cc Condition 000 NZ non-zero 001 Z zero 010 NC non-carry 011 C carry
RETI	Return from interrupt	•	•	X	•	X	•	•	•	100 PO parity odd
RETN	Return from non-maskable interrupt	•	•	X	•	X	•	•	•	101 PE parity even 110 P sign positive 111 M sign negative
RST p	(SP − 1) ← PC$_H$ (SP − 2) ← PC$_L$ PC$_H$ ← 0 PC$_L$ ← p	•	•	X	•	X	•	•	•	t p 000 00H 001 08H 010 10H 011 18H 100 20H 101 28H 110 30H 111 38H
nn	16-bit value in range < 0, 65535 >.									

Adapted from Zilog Z80 data sheets.

Table 8-11 Z80 Input/Output Instructions

Mnemonic	Symbolic Operation	S	Z		H		P/V	N	C	Comments
IN A, (n)	A ← (n)	•	•	X	•	X	•	•	•	n to $A_0 \sim A_7$ Acc. to $A_8 \sim A_{15}$
IN r, (C)	r ← (C) If r = 110, only the flags will be affected	↕	↕	X	↕	X	P	0	•	C to $A_0 \sim A_7$ B to $A_8 \sim A_{15}$
			①							
INI	(HL) ← (C) B ← B − 1 HL ← HL + 1	X	↕	X	X	X	X	1	X	C to $A_0 \sim A_7$ B to $A_8 \sim A_{15}$
			②							
INIR	(HL) ← (C) B ← B − 1 HL ← HL + 1 Repeat until B = 0	X	1	X	X	X	X	1	X	C to $A_0 \sim A_7$ B to $A_8 \sim A_{15}$
			①							
IND	(HL) ← (C) B ← B − 1 HL ← HL − 1	X	↕	X	X	X	X	1	X	C to $A_0 \sim A_7$ B to $A_8 \sim A_{15}$
			②							
INDR	(HL) ← (C) B ← B − 1 HL ← HL − 1 Repeat until B = 0	X	1	X	X	X	X	1	X	C to $A_0 \sim A_7$ B to $A_8 \sim A_{15}$
OUT (n), A	(n) ← A	•	•	X	•	X	•	•	•	n to $A_0 \sim A_7$ Acc. to $A_8 \sim A_{15}$
OUT (C), r	(C) ← r	•	•	X	•	X	•	•	•	C to $A_0 \sim A_7$ B to $A_8 \sim A_{15}$
			①							
OUTI	(C) ← (HL) B ← B − 1 HL ← HL + 1	X	↕	X	X	X	X	1	X	C to $A_0 \sim A_7$ B to $A_8 \sim A_{15}$
			②							
OTIR	(C) ← (HL) B ← B − 1 HL ← HL + 1 Repeat until B = 0	X	1	X	X	X	X	1	X	C to $A_0 \sim A_7$ B to $A_8 \sim A_{15}$
			①							
OUTD	(C) ← (HL) B ← B − 1 HL ← HL − 1	X	↕	X	X	X	X	1	X	C to $A_0 \sim A_7$ B to $A_8 \sim A_{15}$
			②							
OTDR	(C) ← (HL) B ← B − 1 HL ← HL − 1 Repeat until B = 0	X	1	X	X	X	X	1	X	C to $A_0 \sim A_7$ B to $A_8 \sim A_{15}$

① If the result of B−1 is zero the Z flag is set, otherwise it is reset.
② Z flag is set upon instruction completions only.
n   8-bit value in range < 0,255 >.

Adapted from Zilog Z80 data sheets.

Table 8-12 6802 Accumulator and Memory Instructions

Operations	Mnemonic	Immed OP	~	#	Direct OP	~	#	Index OP	~	#	Extend OP	~	#	Implied OP	~	#	Boolean/Arithmetic Operation (All register labels refer to contents)	H (5)	I (4)	N (3)	Z (2)	V (1)	C (0)
Add	ADDA	8B	2	2	9B	3	2	AB	5	2	BB	4	3				A + M · A	↕	•	↕	↕	↕	↕
	ADDB	CB	2	2	DB	3	2	EB	5	2	FB	4	3				B + M · B	↕	•	↕	↕	↕	↕
Add Acmltrs	ABA													1B	2	1	A + B · A	↕	•	↕	↕	↕	↕
Add with Carry	ADCA	89	2	2	99	3	2	A9	5	2	B9	4	3				A + M + C · A	↕	•	↕	↕	↕	↕
	ADCB	C9	2	2	D9	3	2	E9	5	2	F9	4	3				B + M + C · B	↕	•	↕	↕	↕	↕
and	ANDA	84	2	2	94	3	2	A4	5	2	B4	4	3				A · M · A	•	•	↕	↕	R	•
	ANDB	C4	2	2	D4	3	2	E4	5	2	F4	4	3				B · M · B	•	•	↕	↕	R	•
Bit Test	BITA	85	2	2	95	3	2	A5	5	2	B5	4	3				A · M	•	•	↕	↕	R	•
	BITB	C5	2	2	D5	3	2	E5	5	2	F5	4	3				B · M	•	•	↕	↕	R	•
Clear	CLR							6F	7	2	7F	6	3				00 · M	•	•	R	S	R	R
	CLRA													4F	2	1	00 · A	•	•	R	S	R	R
	CLRB													5F	2	1	00 · B	•	•	R	S	R	R
Compare	CMPA	81	2	2	91	3	2	A1	5	2	B1	4	3				A − M	•	•	↕	↕	↕	↕
	CMPB	C1	2	2	D1	3	2	E1	5	2	F1	4	3				B − M	•	•	↕	↕	↕	↕
Compare Acmltrs	CBA													11	2	1	A − B	•	•	↕	↕	↕	↕
Complement, 1s	COM							63	7	2	73	6	3				M̄ · M	•	•	↕	↕	R	S
	COMA													43	2	1)	Ā · A	•	•	↕	↕	R	S
	COMB													53	2	1	B̄ · B	•	•	↕	↕	R	S
Complement, 2s	NEG							60	7	2	70	6	3				00 − M · M	•	•	↕	↕	①	②
(Negate)	NEGA													40	2	1	00 − A · A	•	•	↕	↕	①	②
	NEGB													50	2	1	00 − B · B	•	•	↕	↕	①	②
Decimal Adjust, A	DAA													19	2	1	Converts Binary Add. of BCD Characters into BCD Format	•	•	↕	↕	↕	③
Decrement	DEC							6A	7	2	7A	6	3				M − 1 · M	•	•	↕	↕	④	•
	DECA													4A	2	1	A − 1 · A	•	•	↕	↕	④	•
	DECB													5A	2	1	B − 1 · B	•	•	↕	↕	④	•
Exclusive OR	EORA	88	2	2	98	3	2	A8	5	2	B8	4	3				A ⊕ M · A	•	•	↕	↕	R	•
	EORB	C8	2	2	D8	3	2	E8	5	2	F8	4	3				B ⊕ M · B	•	•	↕	↕	R	•
Increment	INC							6C	7	2	7C	6	3				M + 1 · M	•	•	↕	↕	⑤	•
	INCA													4C	2	1	A + 1 · A	•	•	↕	↕	⑤	•
	INCB													5C	2	1	B + 1 · B	•	•	↕	↕	⑤	•
Load Acmltr	LDAA	86	2	2	96	3	2	A6	5	2	B6	4	3				M · A	•	•	↕	↕	R	•
	LDAB	C6	2	2	D6	3	2	E6	5	2	F6	4	3				M · B	•	•	↕	↕	R	•
Or, Inclusive	ORAA	8A	2	2	9A	3	2	AA	5	2	BA	4	3				A + M · A	•	•	↕	↕	R	•
	ORAB	CA	2	2	DA	3	2	EA	5	2	FA	4	3				B + M · B	•	•	↕	↕	R	•
Push Data	PSHA													36	4	1	A · M$_{SP}$, SP 1 · SP	•	•	•	•	•	•
	PSHB													37	4	1	B · M$_{SP}$, SP 1 · SP	•	•	•	•	•	•
Pull Data	PULA													32	4	1	SP + 1 · SP, M$_{SP}$ · A	•	•	•	•	•	•
	PULB													33	4	1	SP + 1 · SP, M$_{SP}$ · B	•	•	•	•	•	•
Rotate Left	ROL							69	7	2	79	6	3				M	•	•	↕	↕	⑥	⑥
	ROLA													49	2	1	A	•	•	↕	↕	⑥	⑥
	ROLB													59	2	1	B	•	•	↕	↕	⑥	⑥
Rotate Right	ROR							66	7	2	76	6	3				M	•	•	↕	↕	⑥	⑥
	RORA													46	2	1	A	•	•	↕	↕	⑥	⑥
	RORB													56	2	1	B	•	•	↕	↕	⑥	⑥
Shift Left, Arithmetic	ASL							68	7	2	78	6	3				M	•	•	↕	↕	⑥	⑥
	ASLA													48	2	1	A	•	•	↕	↕	⑥	⑥
	ASLB													58	2	1	B	•	•	↕	↕	⑥	⑥
Shift Right, Arithmetic	ASR							67	7	2	77	6	3				M	•	•	↕	↕	⑥	⑥
	ASRA													47	2	1	A	•	•	↕	↕	⑥	⑥
	ASRB													57	2	1	B	•	•	↕	↕	⑥	⑥
Shift Right, Logic	LSR							64	7	2	74	6	3				M	•	•	R	↕	⑥	⑥
	LSRA													44	2	1	A	•	•	R	↕	⑥	⑥
	LSRB													54	2	1	B	•	•	R	↕	⑥	⑥
Store Acmltr	STAA				97	4	2	A7	6	2	B7	5	3				A · M	•	•	↕	↕	R	•
	STAB				D7	4	2	E7	6	2	F7	5	3				B · M	•	•	↕	↕	R	•
Subtract	SUBA	80	2	2	90	3	2	A0	5	2	B0	4	3				A − M · A	•	•	↕	↕	↕	↕
	SUBB	C0	2	2	D0	3	2	E0	5	2	F0	4	3				B − M · B	•	•	↕	↕	↕	↕
Subtract Acmltrs.	SBA													10	2	1	A − B · A	•	•	↕	↕	↕	↕
Subtr. with Carry	SBCA	82	2	2	92	3	2	A2	5	2	B2	4	3				A − M − C · A	•	•	↕	↕	↕	↕
	SBCB	C2	2	2	D2	3	2	E2	5	2	F2	4	3				B − M − C · B	•	•	↕	↕	↕	↕
Transfer Acmltrs.	TAB													16	2	1	A · B	•	•	↕	↕	R	•
	TBA													17	2	1	B · A	•	•	↕	↕	R	•
Test, Zero or Minus	TST							6D	7	2	7D	6	3				M − 00	•	•	↕	↕	R	R
	TSTA													4D	2	1	A − 00	•	•	↕	↕	R	R
	TSTB													5D	2	1	B − 00	•	•	↕	↕	R	R

Legend:
OP   Operation Code (Hexadecimal)
~   Number of CPU Cycles
#   Number of Program Bytes
+   Arithmetic Plus
−   Arithmetic Minus
·   Boolean AND
M$_{SP}$ Contents of memory location pointed to be Stack Pointer

+   Boolean Inclusive OR
⊕ Boolean Exclusive OR
M̄   Complement of M
→   Transfer Into
0   Bit = Zero
00   Byte = Zero

Condition Code Symbols:

H   Half-carry from bit 3
I   Interrupt mask
N   Negative (sign bit)
Z   Zero (byte)
V   Overflow, 2's complement
C   Carry from bit 7
R   Reset always
S   Set always
↕   Test and set if true, cleared otherwise
•   Not affected

Note − Accumulator addressing mode instructions are included in the column for IMPLIED addressing.

Adapted from Motorola 6802 data sheets.

Table 8-13　6802 Index Register and Stack Manipulation Instructions

Pointer Operations	Mnemonic	Immed OP	~	#	Direct OP	~	#	Index OP	~	#	Extnd OP	~	#	Implied OP	~	#	Boolean/Arithmetic Operation	5 H	4 I	3 N	2 Z	1 V	0 C
Compare Index Reg	CPX	8C	3	3	9C	4	2	AC	6	2	BC	5	3				$X_H - M, X_L - (M+1)$	•	•	⑦	I	⑧	•
Decrement Index Reg	DEX													09	4	1	$X - 1 \rightarrow X$	•	•	•	I	•	•
Decrement Stack Pntr	DES													34	4	1	$SP - 1 \rightarrow SP$	•	•	•	•	•	•
Increment Index Reg	INX													08	4	1	$X + 1 \rightarrow X$	•	•	•	I	•	•
Increment Stack Pntr	INS													31	4	1	$SP + 1 \rightarrow SP$	•	•	•	•	•	•
Load Index Reg	LDX	CE	3	3	DE	4	2	EE	6	2	FE	5	3				$M \rightarrow X_H, (M+1) \cdot X_L$	•	•	⑨	I	R	•
Load Stack Pntr	LDS	8E	3	3	9E	4	2	AE	6	2	BE	5	3				$M \rightarrow SP_H, (M+1) \rightarrow SP_L$	•	•	⑨	I	R	•
Store Index Reg	STX				DF	5	2	EF	7	2	FF	6	3				$X_H \cdot M, X_L \cdot (M+1)$	•	•	⑨	I	R	•
Store Stack Pntr	STS				9F	5	2	AF	7	2	BF	6	3				$SP_H \rightarrow M, SP_L \rightarrow (M+1)$	•	•	⑨	I	R	•
Indx Reg → Stack Pntr	TXS													35	4	1	$X - 1 \rightarrow SP$	•	•	•	•	•	•
Stack Pntr → Indx Reg	TSX													30	4	1	$SP + 1 \cdot X$	•	•	•	•	•	•

Adapted from Motorola 6802 data sheets.

Table 8-14　6802 Jump and Branch Instructions

Operations	Mnemonic	Relative OP	~	#	Index OP	~	#	Extend OP	~	#	Implied OP	~	#	Branch Test	5 H	4 I	3 N	2 Z	1 V	0 C
Branch Always	BRA	20	4	2										None	•	•	•	•	•	•
Branch If Carry Clear	BCC	24	4	2										C = 0	•	•	•	•	•	•
Branch If Carry Set	BCS	25	4	2										C = 1	•	•	•	•	•	•
Branch If = Zero	BEO	27	4	2										Z = 1	•	•	•	•	•	•
Branch If ⩾ Zero	BGE	2C	4	2										$N \oplus V = 0$	•	•	•	•	•	•
Branch If > Zero	BGT	2E	4	2										$Z + (N \oplus V) = 0$	•	•	•	•	•	•
Branch If Higher	BHI	22	4	2										C + Z = 0	•	•	•	•	•	•
Branch If ⩽ Zero	BLE	2F	4	2										$Z + (N \oplus V) = 1$	•	•	•	•	•	•
Branch If Lower Or Same	BLS	23	4	2										C + Z = 1	•	•	•	•	•	•
Branch If < Zero	BLT	2D	4	2										$N \oplus V = 1$	•	•	•	•	•	•
Branch If Minus	BMI	2B	4	2										N = 1	•	•	•	•	•	•
Branch If Not Equal Zero	BNE	26	4	2										Z = 0	•	•	•	•	•	•
Branch If Overflow Clear	BVC	28	4	2										V = 0	•	•	•	•	•	•
Branch If Overflow Set	BVS	29	4	2										V = 1	•	•	•	•	•	•
Branch If Plus	BPL	2A	4	2										N = 0	•	•	•	•	•	•
Branch To Subroutine	BSR	8D	8	2											•	•	•	•	•	•
Jump	JMP				GE	4	2	7E	3	3					•	•	•	•	•	•
Jump To Subroutine	JSR				AD	8	2	BD	9	3					•	•	•	•	•	•
No Operation	NOP										01	2	1		•	•	•	•	•	•
Return From Interrupt	RT1										3B	10	1							
Return From Subroutine	RTS										39	5	1		•	•	•	•	•	•
Software Interrupt	SWI										3F	12	1		•	•	•	•	•	•
Wait for Interrupt	WAI										3E	9	1		•		•	•	•	•

Adapted from Motorola 6802 data sheets.

Table 8-15　6802 Condition Code Register Manipulation Instructions

Operations	Mnemonic	Implied OP	~	#	Boolean operation	5 H	4 I	3 N	2 Z	1 V	0 C
Clear Carry	CLC	0C	2	1	$0 \cdot C$	•	•	•	•	•	R
Clear Interrupt Mask	CLI	0E	2	1	$0 \cdot I$	•	R	•	•	•	•
Clear Overflow	CLV	0A	2	1	$0 \rightarrow V$	•	•	•	•	R	•
Set Carry	SEC	0D	2	1	$I \cdot C$	•	•	•	•	•	S
Set Interrupt Mask	SEI	0F	2	1	$I \cdot I$	•	S	•	•	•	•
Set Overflow	SEV	08	2	1	$I \cdot V$	•	•	•	•	S	•
Acmltr A · CCR	TAP	06	2	1	$A \cdot CCR$						
CCR → Acmltr A	TPA	07	2	1	$CCR \cdot A$	•	•	•	•	•	•

Adapted from Motorola 6802 data sheets.

1. The 8-bit microprocessor is a second-generation microprocessor. Four-bit microprocessors were the first generation.

2. Today, 8-bit microprocessors are mostly used as controllers to make products intelligent. At one time they were very popular for personal computers.

3. The Zilog Z80 developed from the Intel 8080. Its instruction set is a superset of the 8080 instruction set.

4. The 6800 was developed by Motorola and is a very different architecture from the 8080 and Z80.

5. Both Z80 and 6800 architectures are still popular for control functions in consumer, commercial, and industrial equipment.

6. The Z80 8-bit main register set consists of an accumulator (A), a flag register (F), and six general-purpose registers (B, C, D, E, H, and L), which can also be used as three 16-bit register pairs. They are paired as BC, DE, and HL.

7. The Z80 has a register-intensive architecture. Its registers let the programmer perform many operations as register-to-register operations rather than accumulator-to-memory operations.

8. The Z80 flag register is the status register.

9. The Z80 secondary register set is another eight registers called A′, F′, B′, C′, D′, E′, H′, and L′. The data in these registers may be swapped with the data in the main register set. This lets the programmer quickly switch from executing one program to executing another program without saving the register contents in memory each time the other program is to be executed.

10. The Z80 special registers have assigned functions. These registers are the 8-bit interrupt vector register, the 8-bit memory refresh register, two 16-bit index registers, the 16-bit stack pointer, and the 16-bit program counter.

11. The interrupt control flip-flops are used to control and show the status of the Z80 interrupt system.

12. The Z80 has two interrupt inputs: the nonmaskable interrupt ($\overline{\text{NMI}}$) and the maskable interrupt ($\overline{\text{INT}}$).

13. Interrupt control flip-flops IFF1 and IFF2 are used to control the maskable interrupt. If IFF1 is set, INT is on. If IFF1 is cleared, INT is off (masked). IFF2 is used to temporarily store the IFF1 value if a nonmaskable interrupt occurs, since the nonmaskable interrupt turns the maskable interrupt off while it is being used.

14. A nonmaskable interrupt saves the current program counter value on the stack and loads the program counter with 0066H. Therefore, memory location 0066H must hold the first instruction of the NMI processing routine.

15. The maskable interrupt has three modes: mode 0, which is 8080 compatible; mode 1, a simple single-vector interrupt; and mode 2, an interrupt system which lets the interrupting device give the Z80 the lower 8-bits of the interrupt vector.

16. The maskable interrupt modes are controlled by interrupt flip-flops IMF_a and IMF_b.

17. In mode 0, an INT causes the processor to fetch a restart instruction word from the external device. The last 3 bits of this word (0 to 7) are used to complete interrupt vectors to eight different interrupt processing routines.

18. In mode 1, an INT gives the Z80 an interrupt vector to a single memory location for the first instruction of the interrupt processing routine.

19. In mode 2, the interrupt vector High byte comes from the interrupt vector register. The Low byte comes from the external device which caused the interrupt. It can be any 8-bit number.

20. The Z80 data bus is a bidirectional 8-bit bus used to move data between an addressed memory or I/O location and the processor.

21. The address bus is a unidirectional 16-bit bus which supplies 16-bit memory location addresses and 8-bit I/O addresses.

22. Both the address bus and the data bus can be placed in a high-impedance state so that an external device can drive the devices connected to these buses.

23. The Z80 has a very powerful instruction set with many different instruction types. These instruction types are made even more powerful by the Z80's many different addressing modes.

24. The load instructions move data between registers and memory locations.

25. The exchange instructions swap data between 16-bit register pairs.

26. The block move instructions let the programmer move data from one group of memory locations to another.

27. The search instructions let the programmer search a group of memory locations to find a byte of data which matches the byte of data in the accumulator.

28. The arithmetic and logical instructions let the programmer combine data in any register with any other register or memory location.

29. A special group of 16-bit arithmetic and logical instructions let the programmer perform operations on the 16-bit register pairs to aid in calculating memory addresses.

30. The Z80 has a group of bit instructions which let the programmer work on each individual bit in a word. These are very useful when processing data received from an external hardware device.

31. Both the jump and the call instructions have unconditional and conditional types. The conditional types check the indicated status register bit before the instruction is executed and only jump or call if the condition is true.

32. The return instruction is used to end a subroutine. The Z80 has both unconditional and conditional forms of the return instruction.

33. Most of the Z80's 40 pins are used for address and data information. The rest are used for power, ground, clock, and control.

34. The Z80 has separate lines to tell external devices that it is addressing a memory location or an I/O location and that it is reading or writing.

35. An external device can halt the Z80 so that it can take over its address and data bus.

36. The reset input clears the Z80 program counter so that the first instruction fetched is in memory location 0000H. A reset also turns off the maskable interrupt and sets the interrupt vector register to zero.

37. A slow external device can tell the Z80 to wait one or more clock cycles between the time it receives an address and the time it must work with the Z80 data bus.

38. The Z80 needs an external clock signal. This is a square wave with a maximum frequency of 2.5 to 10 MHz, depending on the Z80 version.

39. The Z80 is supported with a number of LSI devices which perform the I/O functions that would take many SSI and MSI TTL devices to perform.

40. The Z80 PIO has two 8-bit parallel ports. These can be used to transfer data into or out of the system. They include handshaking signals, a sophisticated interrupt management system, and compatibility with the Z80 data and control bus signals.

41. The Z80 PIO and other similar devices have a number of control registers which let the programmer set how the I/O device is to operate. These control registers are used to set the data direction for the I/O ports bit by bit, to set up how the interrupts are generated, and to match the interrupt output to the Z80 interrupt mode.

42. The Z80 support ICs have a special interrupt control daisy chain which lets the hardware designer give the different devices an interrupt priority, depending on how far they are from the Z80.

43. The Z80 CTC is a support IC which provides three programmable counter-timers. These counter-timers can be used to interrupt the Z80 system when a preprogrammed amount of time has passed or when a preprogrammed number of events have happened.

44. The Z80 SIO provides two parallel-to-serial and serial-to-parallel conversion circuits. These are used to let the system communicate with remote external devices such as terminals, printers, and modems.

45. The 6802 is an improved version of the 6800. It executes the same instruction set, but has an internal oscillator circuit for its clock and has improved control signals.

46. The 6802 uses memory-mapped I/O.

47. The 6802 has a very simple architecture with only two accumulators, one index register, one stack pointer, a condition code (status) register, and a program counter.

48. The 6802 is a memory-intensive architecture because it requires the programmer to perform most operations using memory rather than registers.

49. The 6802 has a nonmaskable interrupt and a maskable interrupt. Both are simple single-vector interrupts.

50. The 6802 takes its interrupt vectors from reserved memory locations instead of the interrupt vector pointing to a reserved memory location as the Z80 does.

51. The contents of all the 6802 registers are automatically saved on the stack when the 6802 responds to an interrupt.

52. The programmer can make an interrupt by executing the software interrupt instruction (SWI).
53. The reset and interrupt vectors are stored in the highest 6802 memory locations.
54. The 6802's simple architecture is supported with a simple instruction set. It has about half as many instructions as the Z80. For some instructions, the 6802 uses a relative addressing mode.
55. The 6802 data manipulation instructions typically have three forms: one to work with accumulator A, one to work with accumulator B, and one to work with an addressed memory location.
56. Branch instructions use the relative addressing mode. The instructions let the programmer make a test of the condition code register and branch to an address which is the sum of the instruction's second byte and the current program counter value.
57. The 6802 hardware architecture includes a 128-byte internal memory which can be used as a scratch-pad memory.
58. The 6802 addresses external devices as one of 65,536 memory locations. A valid address is shown by the VMA (valid memory address) control line, and the R/$\overline{\text{W}}$ line indicates the direction the data is to flow on the data bus.
59. The 6802 uses a signal called enable to give external devices timing information.
60. A slow external device can cause the 6802 to skip clock cycles between the address and data operations by using the memory ready line.
61. The 6802 reset line causes the processor to get its reset vector from memory locations FFFE and FFFF.
62. The 6802 has many support ICs. These give parallel I/O, serial I/O, timing, and other functions.
63. The 6802 support ICs are specially designed to work with the 6802 address, data, and control signals. They also provide special interrupt information.
64. The support ICs are programmed through a series of control registers which are at different addresses than the data exchanges.

CHAPTER REVIEW QUESTIONS

Answer the following questions.

8-1. The 8-bit microprocessors are thought of as _____ products.
  a. First-generation
  b. Second-generation
  c. Third-generation
  d. Older-generation

8-2. The major use for 8-bit microprocessors in new designs is
  a. In applications using a lot of BCD data
  b. In general-purpose personal computing applications
  c. For peripheral control (terminals, printers, modems, etc.) applications
  d. For general-purpose consumer, commercial, and industrial control and monitoring applications

8-3. The Zilog Z80 is an outgrowth of the _____ architecture.
  a. Intel 4004
  b. Intel 8080/8085
  c. Motorola 6800
  d. MOS Technology 6502

8-4. The Z80 has eight 8-bit registers which are made up of the accumulator, flag register, and
  a. Three 16-bit registers
  b. Six 8-bit general-purpose registers
  c. The A, F, B, C, D, E, H, and L registers
  d. Six 8-bit general-purpose registers which can also be used as three 16-bit register pairs

8-5. A microprocessor can have either a _____ or a _____ . The Z80 has a _____ . (Choose three of the four answers.)
   a. Memory-intensive architecture
   b. Memory-mapped I/O structure
   c. Register for managing I/O interrupts
   d. Register-intensive architecture

8-6. The Z80 status register is called its _____ register.
   a. Condition code
   b. Main
   c. Flag
   d. Alternate

8-7. The Z80 has an additional set of eight 8-bit registers. These are named A′, F′, B′, C′, D′, E′, H′, and L′ and are known as the _____ register set.
   a. Condition code
   b. Main
   c. Flag
   d. Alternate

8-8. The Z80 specific registers have assigned functions. One of these assigned functions is
   a. An index register
   b. A stack pointer
   c. A program counter
   d. All of the above

8-9. The Z80 has two kinds of interrupts. They are
   a. Controlled by the interrupt control flip-flops
   b. The nonmaskable interrupt (NMI)
   c. The maskable interrupt (INT)
   d. All of the above

8-10. When a NMI is received, the Z80 stores the current program counter value on the stack and loads the program counter
   a. With the contents of memory locations FFFC and FFFD.
   b. With the interrupt vector 0066H
   c. With the reset vector 0000H
   d. One of the above, depending on the kind of interrupt

8-11. The Z80 INT has three interrupt modes. Mode 0 is 8080-compatible, mode 1 is a simple single-vector interrupt, and mode 2
   a. Is used for devices which have slow response and need more time between the address and data operations
   b. Is only available for nonmaskable operations
   c. Lets the external device provide the interrupt vector Low byte
   d. Clears IFF1 and stores its present value in IFF2 until it is finished

8-12. The Z80 address and data buses
   a. Can be put in a high-impedance mode
   b. Are 8-bit bidirectional
   c. Are 16-bit unidirectional
   d. All of the above

8-13. The 158 different Z80 instructions are made even more powerful by
   a. Its 64-byte memory address space and its 256 I/O addresses
   b. The use of two index registers
   c. The alternate register set
   d. The 11 different Z80 addressing modes

8-14. Executing the Z80 LD A,(BC) instruction causes the
   a. Accumulator (A) to be loaded with the contents of the memory location pointed to by the BC register pair
   b. Contents of the accumulator (A) to be loaded into the memory location pointed to by the BC register pair
   c. Accumulator (A) to be loaded with the contents of the BC register pair Low byte
   d. Contents of the accumulator (A) to be loaded into the BC register pair Low byte

8-15. Executing the Z80 LD (nn),dd instruction causes the
   a. Memory location nn + 1 to be loaded with the High byte of the 16-bit number dd and the memory location nn to be loaded with the Low byte of the 16-bit number dd
   b. Memory location pointed to by the 16-bit value nn + 1 to be loaded with the High byte of the 16-bit number dd and the memory location pointed to by the 16-bit value nn to be loaded with the Low byte of the 16-bit number dd
   c. Memory location dd + 1 to be loaded with the High byte of the 16-bit number dd and the memory location dd to be loaded with the Low byte of the 16-bit number nn
   d. Memory location pointed to by the 16-bit value dd + 1 to be loaded with the High byte of the 16-bit number nn and the memory location pointed to by the 16-bit value dd to be loaded with the Low byte of the 16-bit number nn

8-16. The Z80 instruction EXX is used to
   a. Clear the EX register pair
   b. Set the EX register pair
   c. Swap the stack pointer with the index register
   d. Swap the main and alternate register sets

8-17. If you wish to search a block of 2048 memory locations for a byte of data set to E6 without stopping, you would use the Z80 _____ instruction.
   a. CPX
   b. CPIR
   c. DAA
   d. CPI

8-18. To increase the value of the stack pointer by 1, you would use the Z80 _____ instruction.
   a. INX ss
   b. INX (IX)
   c. INX IX
   d. INX (ss)

8-19. The Z80 RRCA instruction _____ (does/does not) rotate the accumulator data through the carry bit.

8-20. Data shifted out of the carry bit when executing a Z80 SLA m operation is
   a. Shifted into bit 0 of memory location m
   b. Shifted into bit 0 of the accumulator
   c. Shifted into bit 7 of the accumulator
   d. Lost

8-21. What happens in a Z80 if the IY register is set to 00A0, and you execute the instruction SET 6,(IY + 0004)?

8-22. What happens if you execute the Z80 instruction JR e?

8-23. What happens if you execute the Z80 instruction JP (IX)?

8-24. Briefly explain the events that happen when a Z80 executes the CALL cc, nn instruction. Your explanation should tell what cc and nn mean.

8-25. What is the difference between a RET and RET cc instruction in the Z80?

8-26. What happens when you have the Z80 execute the OUT (7F),A instruction?

8-27. Briefly explain the difference between the Z80 read and write control signals and the 6802 read and write control signal.

8-28. What functions are done by the biggest users of pins on both the Z80 and the 6802?

8-29. If an external device wants to take over the Z80 address and data buses, it asserts the Z80 _____ control line.

8-30. If you assert the Z80 $\overline{\text{RESET}}$ line, you expect the next instruction to be taken from memory location _____ .

8-31. The maximum frequency clock signal that can be used with a Z80 depends on _____ .

8-32. Briefly explain what function is performed by the Z80 PIO and the 6802 PIA.

8-33. What part of the PIO or PIA do you use to program how its interrupt will operate?

8-34. What functions do the IEI and IEO connections to a Z80 support IC perform for the interrupt system?

8-35. Briefly explain the functions performed by the Z80 CTC or the 6802 PTM support ICs.

8-36. Briefly explain the functions performed by the Z80 SIO or the 6802 ACIA support ICs.

8-37. Why do we say that the 6802 has memory-mapped I/O?

8-38. What is the major difference between the Z80 interrupt vectors and the 6802 interrupt vectors?

8-39. Why does the typical 6802 instruction have three forms?

8-40. Complete the following table of 6802 instructions and fill in the flags using S = set always, R = reset always, $\updownarrow$ = test and set, cleared otherwise, and ● = Not affected.

	Operations	Mnemonic	Boolean / arithmetic	H I N Z V C
a.		ADD A		
b.			$00 \rightarrow M$	
c.		NEG A		
d.		DEC		
e.			$A \cap M \rightarrow M$	
f.			$A + 1 \rightarrow A$	
g.		ASR		
h.		LSRB		
i.		SUBB		
j.			$X - 1 \rightarrow X$	
k.			$M \rightarrow X_H$ $M + 1 \rightarrow X_L$	
l.		TXS		
m.			$0 \rightarrow C$	
n.		SEC		

CRITICAL THINKING QUESTIONS

8-1. The Intel and Motorola 8-bit architectures are the major general-purpose microprocessor designs in use today. There are, however, quite a number of other 8-bit microprocessor designs in use. Why do you think this is so?

8-2. Often the execution of a program to service an interrupt requires the use of many registers of a microprocessor. The Z80 has a register exchange instruction which lets the programmer access a fresh set of registers with one instruction. The 6802 makes all registers available by saving their current contents on the stack. Can you think of any disadvantages there might be with the 6802 instead of the Z80 if your program must service interrupts?

8-3. The Z80 has three different interrupt processing modes. Mode 1 uses much less external hardware than mode 2. What is the point in having mode 0?

8.4. When you compare the Z80 and 6802 instruction sets, you see that the Z80 has many more instructions. Since the two have relatively the same level of processing power, why does the Z80 have so many more instructions?

8-5. Although the Z80 is a register-intensive architecture and the 6802 is a memory-intensive architecture, they really have a lot in common. What are some of their common attributes?

Answers to Self-Tests

1. *b*	10. *c*	19. *a*
2. *a*	11. *b*	20. *d*
3. *c*	12. *b*	21. *d*
4. *c*	13. *b*	22. *c*
5. *d*	14. *a*	23. *c*
6. *d*	15. *d*	24. *d*
7. *d*	16. *d*	25. *b*
8. *a*	17. *c*	26. *a* and *d*
9. *a*	18. *a*	27. *b* and *c*
		28. *c*

29.

	Mnemonic	Symbolic operation	S	Z	R	P/V	N	C
a.	LD (HL), r	$(HL) \leftarrow r$	●	●	●	●	●	●
b.	LD (nn), A	$(nn) \leftarrow A$	●	●	●	●	●	●
c.	LD r, A	$r \leftarrow A$	●	●	●	●	●	●
d.	LD IX, nn	$IX \leftarrow nn$	●	●	●	●	●	●
e.	LD IY,(nn)	$IY \leftarrow (nn + 1)$	●	●	●	●	●	●
		$IY \leftarrow (nn)$						
f.	PUSH BC	$(SP - 2) \leftarrow BC_L$	●	●	●	●	●	●
		$(SP - 1) \leftarrow BC_H$						
		$SP \rightarrow SP - 2$						

			S	Z	R	P/V	N	C
g.	LDI	$(DE) \leftarrow (HL)$	●	●	0	1	0	●
		$DE \leftarrow DE + 1$						
		$HL \leftarrow HL + 1$						
		$BC \leftarrow BC - 1$						
h.	CPI	$A - (HL)$	↕	↕	↕	↕	1	●
		$HL \leftarrow HL + 1$						
		$BC \leftarrow BC - 1$						
i.	ADC A,D	$A \leftarrow A + D + CY$	↕	↕	↕	V	O	↕
j.	INC(HL)	$(HL) \leftarrow (HL) + 1$	↕	↕	↕	V	O	●
k.	CPL	$A \leftarrow L$	●	●	1	●	1	●
l.	ADD HL, BC	$HL \leftarrow HL + BC$	●	●	X	●	0	1

CY	AC	F0	RS1	RS0	OV	—	P

CY	PSW. 7	Carry flag.
AC	PSW. 6	Auxiliary carry flag.
F0	PSW. 5	Flag 0 available to the user for use as a general-purpose indicator.
RS1	PSW. 4	Register bank selector bit 1.
RS0	PSW. 3	Register bank selector bit 0.
OV	PSW. 2	Overflow flag.
——	PSW. 1	Reserved.
P	PSW. 0	Parity flag. Set/cleared by hardware each instruction cycle to indicate an odd/even number of logical l's in the accumulator.

Fig. 9-7   The 8051 program status word (PSW). This 8-bit, bit-addressable register is the 8051 status register. The 7 active bits can be set or cleared by a program instruction or by the result of an arithmetic or logical instruction. *Adapted from Intel 8051 data sheets.*

If the programmer does not want the stack to start at memory location 08H, the initial stack pointer value must be changed before using the stack.

The *data pointer* (DPTR) is a 16-bit register. It is made up of a High byte (DPH) and a Low byte (DPL). The data pointer holds a 16-bit address used when addressing external memory. The 8051 instructions let you work with the data pointer as a 16-bit register or as two separate 8-bit registers. This gives the programmer maximum flexibility. If needed, a single- or double-byte instruction can operate on either the High or Low byte. If the entire data pointer value must be reloaded, the programmer can use a 3-byte instruction.

The *power control register* is shown in Fig. 9-8. Two bits of this register, PD (power down) and IDL (idle), are only used in CMOS versions of the 8051. These bits place the single-chip microprocessor in special modes which conserve power until you wake up the micro-

processor. The two general-purpose flag bits (GF1 and GF0) can be used by the programmer to store a single bit of information. The SMOD bit provides a special serial-interface control function. We will look at SMOD in greater detail when the serial port is discussed.

The other 8051 special-function registers are used to control the 8051 I/O ports, serial transmitter-receiver, its counter-timers, and its interrupt system. We will cover the operation of these registers in the sections on the 8051 I/O and interrupts.

Self-Test

Answer the following questions.

13. The first 32 internal memory locations in an 8051 can be addressed as
    a. Read-write bytes of data
    b. 8-bit registers
    c. Data memory
    d. All of the above

SMOD	—	—	—	GF1	GF0	PD	IDL

SMOD	Double baud rate bit. When timer 1 is used to generate baud rate and SMOD = 1, the baud rate is doubled when the serial port is used in modes 1, 2, or 3.
GF1	General purpose flag bit.
GF0	General purpose flag bit.
PD	Power down bit. Setting this bit activates power down operation in the 80C51.
IDL	Idle mode bit. Setting this bit activates idle mode operation in the 80C51.

Fig. 9-8   The power control register (PCON). The PD and IDL bits are only used in CMOS versions of the 8051 which can be powered-down or placed in an idle state to conserve power. Setting SMOD doubles the serial port baud rate. GF1 and GF0 are general-purpose flags. *Adapted from Intel 8051 data sheets.*

14. The 8051 uses register addressing for some memory locations because
    a. It is easier for a programmer to think of them as registers than memory locations
    b. They actually are registers, not read-write memory locations
    c. This lets the 8051 use single-byte instructions which reference these memory locations
    d. It is the only way to enable bit addressing

15. Bit addressing is a very important feature when the 8051 is
    a. Being used as a microcontroller
    b. Used with an external 64 kbytes of program ROM
    c. Being used to make a general-purpose microcomputer
    d. Used with expanded RAM

16. Registers R0 to R7 can be found in
    a. Bank 0, 1, 2, or 3
    b. Memory locations 80 to F8
    c. Bank 0 only
    d. All of the above

17. The 8051 has special flags to
    a. Tell which register bank is currently selected
    b. Use with register mode addressing
    c. Add to the register address so that the 8051 can select the correct memory location
    d. All of the above

18. When the 8051 stack is initialized, it uses memory location 08H. This is also the memory location used
    a. To change the bank selection
    b. To store program data
    c. To address external RAM
    d. For the first register (R0) in the second bank (bank 1)

19. The 8051 instructions allow bit addressing for
    a. All memory locations in the internal 128 bytes of RAM
    b. All of the first 128 internal and external memory locations
    c. 16 special memory locations (20H to 30H)
    d. The last 80 memory locations of internal memory

20. _____ are usually described as scratch-pad memory locations.
    a. All memory locations in the internal 128 bytes of RAM
    b. All of the first 128 internal and external memory locations
    c. 16 special memory locations (20H to 30H)
    d. The last 80 memory locations of internal memory

21. Each 8051 special-function register
    a. Is assigned a single job to do for the 8051
    b. Can be addressed as a register or as a memory location
    c. May not be used as general-purpose memory location or register
    d. All of the above

22. The SFRs are
    a. Located in external memory space
    b. Addressed as internal memory above the 128-byte internal RAM
    c. Part of the internal 128 bytes of memory
    d. Either specific- or general-purpose, depending on the program

23. The 8051 stack pointer is different from general-purpose microprocessor stack pointers because it is _____ and _____ . (Choose two of the four answers.)
    a. An 8-bit register and therefore the stack can only be in the first 128 bytes of internal RAM
    b. An 8-bit register and therefore the stack can only be in the last 128 bytes of external RAM
    c. An incrementing register
    d. A decrementing register

9-4 THE 8051 I/O PORTS

The 8051 architectural block diagram shows four 8-bit bidirectional I/O ports. However, these are more than simple parallel I/O ports. Each has special characteristics, and three of the ports can be used in several different ways. The following paragraphs look at each I/O port to help you understand how they help the 8051 do many different jobs.

Figure 9-9 on page 200 shows an equivalent logic diagram for one bit of each port. You can see the electrical and logical implementation of each of the four ports (P0–P3) is just a little different. The *port latch* allows you to store data going out of the port or coming into the port. The latch can be set by data on the microprocessor data bus or at the port pins. Also, the latch can place data on the microprocessor data bus or send it to the port pin.

As you can see from Fig. 9-9(*a*) and (*c*), the bits on port 0 and port 2 have controlled pull-ups.

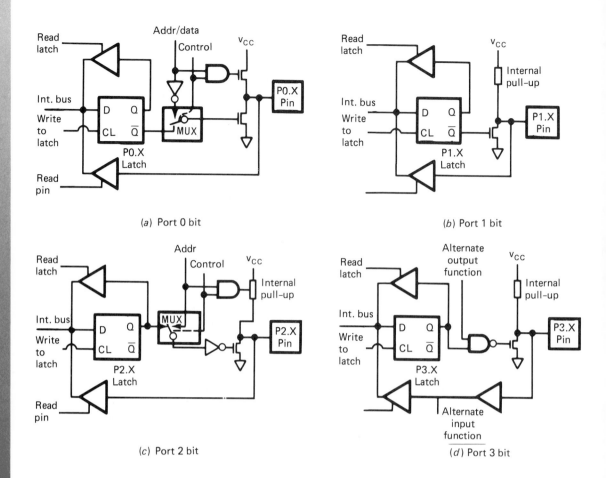

(a) Port 0 bit

(b) Port 1 bit

(c) Port 2 bit

(d) Port 3 bit

Fig. 9-9 An equivalent logic diagram for one bit of each I/O port. Because of alternate use, each port is configured somewhat differently from the others. *Adapted from Intel 8051 data sheets.*

These pull-ups are active or disconnected, depending on the port mode.

If external memory (either program memory or data memory) is connected to the 8051, port 0 and port 2 address this memory. Port 0 is also used to exchange data with this memory, meaning it performs two functions. First, it must output the Low byte of the 16-bit memory address. Then, it must either input data from or output data to the addressed memory location.

How can a single port serve both functions? As you know, a microprocessor addresses a memory location during the first part of an instruction cycle and exchanges data with a memory location during the second part of the instruction cycle. Because the addressing and data-exchange operations happen at different times, we can use a *multiplexed address-data bus*.

Simply, this means that some bus signals perform different functions at different times. In this case, we first place address information on the upper and lower 8 bits of the address bus. The lower 8 bits are latched (stored) at the external memory. During the second half of the instruction cycle, the lower 8 bits become the data bus.

Now a data exchange can take place with that memory location. A microprocessor with a multiplexed address-data bus has a special control output called the *address latch enable* (ALE). A signal on the ALE output tells an external address latch that the data on the lower 8 bits of the address bus are a valid memory location address. Once these 8 bits of data are latched at the memory, the lower 8 bits change from address bus lines to data bus lines. The data at port 2 do not change, so it is not necessary to latch the high-order address bits.

A multiplexed address-data bus is used on the Intel 8085 microprocessor as well as on the 8051 single-chip microprocessor. This technique reduces the total number of pins needed for the address and data buses.

Table 9-4 *continued*

Mnemonic	Instruction code D_7 D_6 D_5 D_4 D_3 D_2 D_1 D_0	Hexa-decimal	Explanation
CJNE Rn, #data, rel	1 0 1 1 1 n_2 n_1 n_0 d_7 d_6 d_5 d_4 d_3 d_2 d_1 d_0 r_7 r_6 r_5 r_4 r_3 r_2 r_1 r_0	B8 ~ BF Byte 2 Byte 3	(PC) ← (PC) + 3 IF #data < (Rn) THEN (PC) ← (PC) + rel and (C) ← 0 OR IF #data > (Rn) THEN (PC) ← (PC) + rel and (C) ← 1
CJNE @Ri, #data, rel	1 0 1 1 0 1 1 i d_7 d_6 d_5 d_4 d_3 d_2 d_1 d_0 r_7 r_6 r_5 r_4 r_3 r_2 r_1 r_0	B6 ~ B7 Byte 2 Byte 3	(PC) ← (PC) + 3 IF #data < ((Ri)) THEN (PC) ← (PC) + rel and (C) ← 0 OR IF #data > ((Ri)) THEN (PC) ← (PC) + rel and (C) ← 1
DJNZ Rn, rel	1 1 0 1 1 n_2 n_1 n_0 r_7 r_6 r_5 r_4 r_3 r_2 r_1 r_0	D8 ~ DF Byte 2	(PC) ← (PC) + 2 (Rn) ← (Rn) − 1 IF (Rn) ≠ 0 THEN (PC) ← (PC) + rel
DJNZ direct, rel	1 1 0 1 0 1 0 1 a_7 a_6 a_5 a_4 a_3 a_2 a_1 a_0 r_7 r_6 r_5 r_4 r_3 r_2 r_1 r_0	D5 Byte 2 Byte 3	(PC) ← (PC) + 3 (direct) ← (direct) − 1 IF (direct) ≠ 0 THEN (PC) ← (PC) + rel
NOP	0 0 0 0 0 0 0 0	00	(PC) ← (PC) + 1

Adapted from OKI Semiconductor MSM80C59 data sheets.

Table 9-5   8051 Bit Oriented Instructions

Mnemonic	Instruction code D_7 D_6 D_5 D_4 D_3 D_2 D_1 D_0	Hexa-decimal	Explanation
CLR C	1 1 0 0 0 0 1 1	C3	(C) ← 0
CLR bit	1 1 0 0 0 0 1 0 b_7 b_6 b_5 b_4 b_3 b_2 b_1 b_0	C2 Byte 2	(bit) ← 0
SETB C	1 1 0 1 0 0 1 1	D3	(C) ← 1
SETB bit	1 1 0 1 0 0 1 0 b_7 b_6 b_5 b_4 b_3 b_2 b_1 b_0	D2 Byte 2	(bit) ← 1
CPL C	1 0 1 1 0 0 1 1	B3	(C) ← $(\overline{C})$
CPL bit	1 0 1 1 0 0 1 0 b_7 b_6 b_5 b_4 b_3 b_2 b_1 b_0	B2 Byte 2	(bit) ← $(\overline{bit})$
ANL C, bit	1 0 0 0 0 0 1 0 b_7 b_6 b_5 b_4 b_3 b_2 b_1 b_0	82 Byte 2	(C) ← (C) AND bit
ANL C,/bit	1 0 1 1 0 0 0 0 b_7 b_6 b_5 b_4 b_3 b_2 b_1 b_0	B0 Byte 2	(C) ← (C) AND $(\overline{bit})$
ORL C, bit	0 1 1 1 0 0 1 0 b_7 b_6 b_5 b_4 b_3 b_2 b_1 b_0	72 Byte 2	(C) ← (C) OR (bit)
ORL C,/bit	1 0 1 0 0 0 0 0 b_7 b_6 b_5 b_4 b_3 b_2 b_1 b_0	A0 Byte 2	(C) ← (C) OR $(\overline{bit})$
MOV C, bit	1 0 1 0 0 0 1 0 b_7 b_6 b_5 b_4 b_3 b_2 b_1 b_0	A2 Byte 2	(C) ← (bit)
MOV bit, C	1 0 0 1 0 0 1 0 b_7 b_6 b_5 b_4 b_3 b_2 b_1 b_0	92 Byte 2	(bit) ← (C)

Adapted from OKI Semiconductor MSM80C59 data sheets.

When you compare the 8051 instruction set to the instruction set of a general-purpose 8-bit microprocessor (the Z80, for example), you find that in some ways it has more power and in others it has some real limitations. Many compromises have been made to allow it to operate as a microcontroller. In addition, many special instructions have been generated to allow the 8051 to operate, for example, in the bit mode as well as the byte mode (Table 9-5).

The 8051 limitations can be significant. For example, the 8051 has a very limited stack. As you know, the stack pointer can only address memory locations within the 8051 internal memory. If you need a larger stack, one must be created in external memory. However, this means multiple instructions are required each time you push information onto or pop information from this external stack.

Self-Test

Answer the following questions.

44. You could best describe the function of the 8051 MOV instructions as
    a. Moving data from one location to another
    b. Copying data from one location into another
    c. Exchanging data between two locations
    d. All of the above mean the same thing
45. The 8051 special instruction XCHD
    a. Exchanges data between the accumulator and the source
    b. Exchanges data between the accumulator and register D
    c. Exchanges the accumulator low-order nibble with the source low-order nibble
    d. All of the above, depending on the instruction addressing
46. All instructions which move data between the external data memory and the accumulator use
    a. Direct addressing
    b. Some form of indirect addressing
    c. A byte of program data
    d. All of the above
47. The MOVC instruction only has two forms (one for each type of addressing), whereas the MOVX has two forms for each type of addressing. This is because
    a. The MOVX moves twice as much data because it must read data from the data memory space, and this is twice as hard to address as the program memory space
    b. The MOVC only needs to move bytes. MOVX may be required to move nibbles, which are twice as long as bytes
    c. The MOVX either reads from or writes to data memory for each addressing

mode. The MOVC cannot write to program memory, so it does not need half of the instruction types
    d. Wrong—MOVC also has four forms
48. The only 3-byte MOV instruction is used to
    a. Move indirectly addressed RAM data to the accumulator
    b. Move directly addressed RAM data into a register
    c. Move a direct byte into the accumulator
    d. Load the data pointer with a 2-byte constant
49. The 8051 arithmetic instructions include the _____ instruction, which is not included in the Z80 arithmetic instructions.
    a. Add with carry
    b. Increment
    c. Multiply
    d. Decrement with carry
50. The 8051 swap instruction is used to
    a. Exchange the accumulator and register B
    b. Set the current bank for registers R0 to R7
    c. Rotate the accumulator left four places to swap nibbles
    d. The 8051 does not have a swap instruction; this is a Z80 instruction
51. The 8051 two-operand logical instructions logically combine the data in the two source operands and leave the result in the
    a. Accumulator
    b. R0 or R2 (current bank)
    c. Register B
    d. First operand
52. The ACALL and AJMP instructions make use of the 8051
    a. Addressing mode, which uses signed arithmetic and the current program counter value
    b. 11-bit absolute addressing mode to reach addresses up to 2 kbytes ahead of the current program counter value
    c. Full 64-kbyte memory addressing
    d. Direct addressing in internal data memory
53. The 8051 jump instructions are somewhat limited when compared to the Z80 jump instructions because
    a. There are no conditional jumps
    b. The conditional jumps only use relative addressing
    c. The 8051 jump instruction cannot test for a zero accumulator
    d. The two have just about identical sets of jump instructions

54. The 8051 CJNE and DJNZ instructions are very efficient because they
    a. Combine a test and a jump in one instruction
    b. Are single-byte instructions
    c. Use direct addressing
    d. Work within 2 kbytes of the current program counter value
55. The only difference between an 8051 I/O instruction and any other instruction is that
    a. The I/O instructions only have the accumulator as a destination
    b. The 8051 has no DMA facility
    c. That the 8051 I/O ports can be used for memory addressing
    d. There is no difference. There are no I/O-specific instructions

9-7 OTHER MICROCOMPUTERS IN THE 8051 FAMILY

In this section, we take a look at some other microcontrollers in the Intel 80XX family. For most of these you will find the microprocessor architecture is quite similar. However, the rest of the microcontroller is where the differences are found.

The 8052 is a simple expansion of the 8051. There are only two areas with specification differences. First, the 8052 has 8 kbytes of onboard ROM and 256 bytes of onboard RAM. The 8052 allows programmers to write larger programs and programs which use more variable data. As expected, the 8052 costs more than the 8051. Thus it is only used when extra ROM and RAM are required. Second, the 8052 has one extra 16-bit counter-timer. This counter-timer (TM2) gives programmers added flexibility, especially when one of the other two is used to control the serial port baud rate.

Because the 8051 and 8052 have onboard ROM, they must be programmed by the device manufacturer while the part is being made. This means the 8051 and 8052 are normally only used in high-volume applications which can afford the cost of this semicustom operation. For low-volume and prototyping applications there are special versions called the 8751 and 8752. These replace the onboard ROM with onboard EPROM. Using the EPROM, the user can erase program memory with an ultraviolet light and reprogram the memory with externally applied electrical signals.

Two alternate versions of the 8051 and 8052 are the 8031 and 8032. These devices do not have any onboard ROM (the onboard ROM may be defective). You must use an external ROM (or EPROM) for program memory. These are excellent devices for prototyping or low-volume products. An 8031 and an external 2764 8K EPROM have the same function as an 8751. However, the combination of an 8031 and a 2764 8K EPROM is significantly cheaper than an 8751, although it takes twice as much printed circuit board space and parallel I/O ports must be used to address the external program memory.

A third form of the 8052 is the 8052 AH-BASIC. This special 8052 has the BASIC programming language in ROM. Using BASIC instructions, a programmer can write instructions for this 8052 using the high-level language BASIC rather than assembly language.

As you learned earlier, the 8051 is a second-generation microcontroller. The first Intel microcontroller was the 8048. An architectural block diagram of it is shown in Fig. 9-20 on page 218. As you can see by examining this diagram, there are many similar characteristics between the 8048 and the 8051. But, the two microcontrollers do have significantly different architectures.

The 8048, 8049, and 8050 all have identical architectures with the exception of memory size. In each case, the memory doubles. The 8048 supports a 1K internal ROM or EPROM (8748). As you can see from the program memory map in Fig. 9-21 on page 219, the 8048 supports only 1 kbyte of internal memory, but the 8049 supports 2 kbytes and the 8050 supports 4 kbytes.

Figure 9-22 on page 219 shows a program memory map for these three microprocessors. You can see that the 8048 has 64 bytes of internal RAM, including 32 bytes which act as registers. The 8049 and 8050 use these same 32 bytes of register/memory locations and have a total of 128 and 256 bytes of RAM respectively.

The 8048, 8049, and 8050 are very popular single-chip microprocessors because they are very inexpensive. In volume, the 8048 can be obtained for less than a few dollars. In applications which do not require a serial port or in which a simple serial port can be generated from one parallel port bit, this is an extremely powerful controller device. Again, it lets you bring the power of a microprocessor to a very low-cost application.

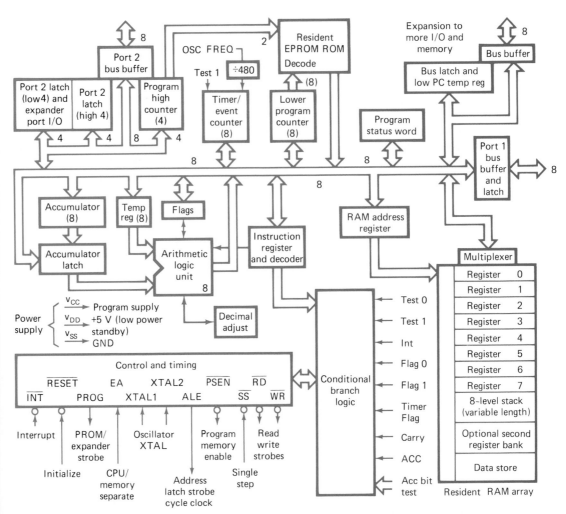

Fig. 9-20 **The architectural block diagram for the Intel 8048, 8049, and 8050. Note the 8748, 8749, and 8750 have identical architectural block diagrams, but the ROM is an EPROM.** *Adapted from Intel 8049 data sheets.*

Self-Test

Answer the following questions.

56. The major difference between the 8051 and the 8052 is
   a. The amount of RAM and ROM
   b. The lack of internal ROM
   c. No serial port
   d. There is no major difference except speed and cost

57. If someone has an 8751 single-chip microprocessor that has just been programmed, you know this is the part number for
   a. The low-cost model without a serial port
   b. The model which must use external ROM
   c. The model with a built-in EPROM
   d. A model with no internal RAM

58. An 8051 with a defective ROM could be used as an 8031, which is the part number for
   a. The low-cost model without a serial port
   b. The model which must use external ROM
   c. The model with a built-in EPROM
   d. A model with no internal RAM

59. You are servicing a product which uses an 8048. You find out that this single-chip microprocessor has a 1K ROM and only 64 bytes of RAM. This is the part number for
   a. The low-cost model without a serial port
   b. The model which must use external ROM
   c. The model with a built-in EPROM
   d. A model with no internal RAM

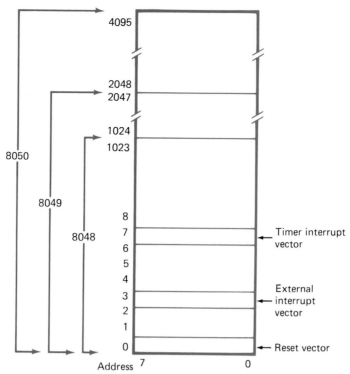

Fig. 9-21 **A memory map of program memory space for the Intel 8048, 8049, and 8050.** *Adapted from Intel 8051 data sheets.*

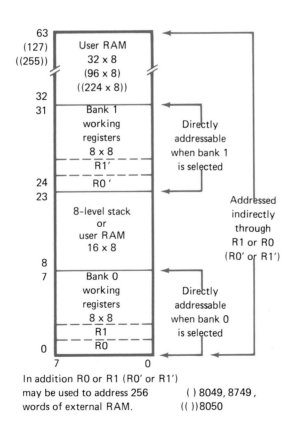

In addition R0 or R1 (R0' or R1')
may be used to address 256 ( ) 8049, 8749,
words of external RAM. (( ))8050

Fig. 9-22 **A memory map of data memory space for the Intel 8048, 8049, and 8051.** *Adapted from Intel 8051 data sheets*

SUMMARY

1. The microcontroller was developed to meet a market need. It gave the product designer a low-cost microprocessor system to put into low-cost products.

2. The microcontroller includes a CPU, ROM or EPROM, RAM, parallel I/O ports, timers, a clock (oscillator), and possibly a serial port.

3. A low-cost system built with a microcontroller does not have all of the functions and memory of a full microprocessor system. However, it is built with fewer ICs and costs three to five times less.

4. The microprocessor is a very good way to implement the control and processing functions for many products. The microcontroller is used to give intelligence to many consumer products and appliances as well as industrial products.

5. The Intel 8051 is a second-generation microcontroller. The first-generation Intel microcontrollers are the 8048 and 8049.

6. Major 8051 features include 4 kbytes of ROM, 128 bytes of RAM, four 8-bit parallel I/O ports, two 16-bit timers, and a full serial port.

7. Two of the 8051 parallel I/O ports can be used to address external memory.

8. External connections to an 8051 are simple. They are made up of the 32 I/O port lines, two power lines, four special timing and control lines, and two lines for a crystal.

9. The 8051 addresses two memory spaces: the program memory and the data memory.

10. The program memory space is a read-only memory space. It is used for storing program instructions. The processor cannot write to these memory locations. When the microprocessor fetches an instruction from one of these memory locations, the instruction is placed in the instruction register.

11. The data memory space is read-write memory space. The microprocessor uses this memory space to store variable data used by the program. It cannot execute from these memory locations.

12. The 8051 ROM is in program memory space, and the RAM is in data memory space.

13. The 8051 4-kbyte internal ROM can be expanded by adding as much as an additional 60 kbytes of external program memory. You can also assert the 8051 $\overline{EA}$ input, which forces the 8051 to use all external addressing for program memory. With $\overline{EA}$ asserted, you can have up to 64 kbytes of external program memory.

14. The 8051 internal 128-byte RAM can be supplemented with external data memory. You can add up to 64 kbytes of external data memory. This is in a separate memory space from the internal data memory, so you can have a maximum of 64K + 128 bytes of data memory.

15. If needed, you can overlap the 8051 program and data memory spaces. This allows you to have a read-write program memory, but reduces the total external memory from 128 kbytes to 64 kbytes.

16. The first 32 internal data memory locations are often called registers because they can be addressed using single-byte register instructions. There are four banks of eight registers per bank. The registers are R0 to R7, and the banks are bank 0 to bank 3.

17. The programmer must select the bank to be used before one of the eight registers (R0–R7) is addressed. The bank selection stays until the programmer changes it or the processor is reset.

18. The 16 bytes of RAM following the registers are bit-addressable. This means that special instructions allow you to read from or write to each bit of these 16 bytes. You can also address these memory locations by using the usual byte addresses.

19. The last 80 bytes of 8051 internal RAM are called scratch-pad memory locations. They are used for general-purpose read-write data storage.

20. The 8051 has 22 special-function registers (SFRs). The SFRs can be addressed as registers or as memory locations. However, each SFR has a dedicated function and can only be used for that function.

21. The microprocessor SFRs are the accumulator, the B register (for general-purpose storage), the program status word, the stack pointer, the data pointer and the power control register. The other SFRs are used for I/O control and data. Each SFR is an 8-bit register, except the data pointer, which is an 8-bit register pair.

22. The 8051 stack pointer is initialized so that the first variable written to the stack uses memory location 08. This is R0 in bank 1.

If the programmer does not want to use the bank 1 register memory locations for the stack, the stack pointer must be reinitialized to point at some other internal memory location. The stack pointer cannot point to external memory.

23. The stack is built up in memory. That is, as you write data to the stack, the stack addresses increase rather than decrease, as is done on most conventional 8-bit microprocessors.

24. The 8051 data pointer is used to address external data memory.

25. The power control register is used to control special power-down features in CMOS versions of the 8051.

26. Most of the 8051 parallel I/O ports have a dual function. That is, they can perform another I/O-related function, or they can act as an 8-bit bidirectional I/O port.

27. If the 8051 system uses external memory, ports 0 and 2 are used as a multiplexed address/data bus. Port 2 gives the high-order 8 bits of the address. Port 0 first gives the low-order 8 bits of the address and then changes into a bidirectional 8-bit data bus.

28. Port 3 can be an 8-bit parallel I/O port, or it can be used to give the 8051 six special inputs: two serial port (data in and data out) lines; two external interrupt inputs; and two external inputs to the 16-bit counter-timers.

29. The ports are in the read (input) or the write (output) mode, depending on the instruction used with the port.

30. The 8051 serial interface is a full-duplex, programmable, and buffered UART. It can receive another character before the first one is taken from the UART and can transmit a character while it is receiving one.

31. The serial interface is controlled by a special register. This register can program the number of data bits and stop bits, the source of the baud-rate clock, and the type of signal which interrupts the processor.

32. The 8051 has two 16-bit counter-timer registers. These can be used as counters or timers. In the counter mode, the registers increment once for each pulse at the external timer inputs (port 3). In the timer mode, the registers increment once for each machine cycle.

33. The counter-timer modes are controlled by function registers. The different modes give the counter-timers more flexibility.

34. Both the serial interface and the timers are complex single-chip microprocessor functions. They add a great deal to the device flexibility, but they require a great deal of detailed study before they can be used to their maximum advantage.

35. There are five different interrupt sources for an 8051: two external interrupts, two counter-timer interrupts, and a serial interface interrupt.

36. The external interrupts can either be transition-activated or level-activated.

37. A transition-activated interrupt responds to the trailing edge of the interrupt signal and is cleared when the 8051 vectors to the interrupt service routine.

38. A level-activated interrupt responds to a logic 0 interrupting signal and is cleared when the external hardware removes the interrupting signal.

39. The counter-timer interrupts are caused by a 16-bit register overflow (all bits go from logic 1s to logic 0s).

40. Either receiving a character in the serial interface receive buffer or sending a character out of the transmit buffer causes a serial interrupt. The software must read the serial interrupt status flags to see if the interrupt was caused by a receive or transmit operation.

41. The 8051 interrupt enable register lets the programmer turn all of the interrupts on or off or to selectively turn each interrupt source on or off.

42. The interrupt priority register lets the programmer give each interrupt source a high- or low-priority status.

43. Each 8051 interrupt source has a simple interrupt vector with only one assigned memory location.

44. The 8051 has single-, double-, and triple-byte instructions. Most instructions are single-byte instructions, because the 8051 instruction set is designed to generate efficient object code to fit in the limited program memory space.

45. A special group of the 8051 instructions lets the programmer perform operations on bit addresses instead of byte or multiple-byte addresses. These allow the programmer to work with data generated from external hardware in control systems.

46. The data transfer instructions copy data from a source to a destination without affecting the status register flag bits.

47. The 8051 mathematical operations include add, subtract, multiply, and divide. The multiply and divide operations are performed between the accumulator and the B register. The 16-bit result is left in the accumulator and the B register.

48. The 8051 logical operations include AND, OR, Exclusive OR, and the rotate accumulator instructions. A special rotate accumulator instruction lets you swap the 4 high-order bits with the 4 low-order bits.

49. The control-transfer instructions include the call and jump instructions. A number of these only use relative addressing. The addresses are relative to either the program counter or the data pointer.

50. The 8052 is an 8051 with double the amount of ROM and RAM; it also has an extra 16-bit counter-timer.

51. The 8031 and 8032 are versions of the 8051 and 8052 that do not have any internal ROM. They are used for development or for applications where all program memory must be external.

52. The 8048 and 8049 are simpler microcontrollers. They do not have a serial interface and are limited to a single timer. They are very low cost microcontrollers.

53. The 8748, 8749, 8751, and 8752 are microcontrollers with EPROMs instead of ROMs for program memory.

CHAPTER REVIEW QUESTIONS

Answer the following questions.

9-1. The microcontroller was developed
  a. To allow faster CPU-to-I/O communications with a shorter bus
  b. To give product designers a low-cost way to implement product designs using a microprocessor system
  c. Prior to the introduction of the 8-bit microprocessor
  d. With over 64K of address space

9-2. The microcontroller does not include
  a. RAM
  b. ROM
  c. An arithmetic coprocessor
  d. I/O ports and devices

9-3. _____ is characteristic of systems built with a single-chip microprocessor.
  a. Low cost relative to discrete microprocessor systems
  b. A low IC count
  c. Limited ROM and RAM
  d. These are all characteristics of microcontrollers

9-4. Many consumer and commercial products need the intelligence which can be added by a microprocessor. A microcontroller is used because of its
  a. Low cost relative to a discrete microprocessor system
  b. Low IC count
  c. Extensive ROM and RAM memory space
  d. For all of the above reasons

9-5. Which of the following is not a characteristic of the 8051?
  a. 4 kbytes of internal ROM
  b. 256 bytes of internal RAM
  c. Four parallel bidirectional I/O ports
  d. A full serial port

9-6. The 8051 addresses two different memory spaces. These are the _____ memory space and the _____ memory space. (Choose two of the four answers.)
  a. Data
  b. Register
  c. Counter-timer
  d. Program

9-7. _____ memory space is read only and is used to store instructions.
  a. Data
  b. Register
  c. Counter-timer
  d. Program

9-8. _____ memory space is read-write and is used for variable storage.
   a. Data   *c.* Counter-timer
   b. Register   *d.* Program

9-9. The 8051 internal ROM is _____ memory space, and the internal RAM is _____ memory space. (Choose two of the four answers.)
   a. Data   *c.* Counter-timer
   b. Register   *d.* Program

9-10. The 8051 memory can be expanded by adding up to
   a. 60 kbytes of external ROM in addition to the internal 4 kbytes of ROM
   b. 64 kbytes of completely external ROM if $\overline{EA}$ is asserted to disable the internal ROM between memory locations 0000 and 0FFF
   c. 64 kbytes of external RAM. The added RAM address space does not overlap the internal RAM address space.
   d. All of the above

9-11. The first 32 internal RAM locations are addressable as _____ or _____ . (Choose two of the four answers.)
   a. Program instructions   *c.* Registers
   b. Banks   *d.* Memory locations

9-12. Each of the 8051 banks has _____ registers.
   a. 1   *c.* 4
   b. 2   *d.* 8

9-13. The 8051 has 128 bit-addressable memory locations in _____ memory space.
   a. External program   *c.* External data
   b. Internal data   *d.* Internal program

9-14. The 8051 special function registers can be addressed as either _____ or as _____ , but they must be used for their dedicated function, not for general-purpose storage. (Choose two of the four answers.)
   a. Data locations   *c.* Counter-timers
   b. Registers   *d.* Memory locations

9-15. The 8051 stack pointer is different from conventional microprocessors because it
   a. Can address the full 64 kbytes of external data memory
   b. Is only 8 bits long and decrements each time it stores data on the stack
   c. Can only address external memory because it is 8 bits long
   d. Is only 8 bits long, addresses internal memory, and increments each time it stores data on the stack

9-16. If the 8051 programmer wishes to address external data memory, the _____ must be initialized to point at external data memory locations, since they can only be indirectly addressed.
   a. Stack pointer   *c.* Program counter
   b. Stack   *d.* Data pointer

9-17. Three of the 8051 parallel I/O ports can
   a. Be used to address external memory locations
   b. Have more than one function
   c. Perform serial I/O interface functions
   d. All of the above

9-18. The 8051 uses a multiplexed address data bus when working with external memory. This means that
   a. The external memory must have a latch to store the lower 8 address bits
   b. One of the parallel I/O ports is used for addressing and for data transfers
   c. The 8051 uses two general-purpose bidirectional I/O ports for this
   d. All of the above

9-19. We say that the 8051 has a fully buffered duplex serial interface. This means that
   a. The serial port can receive data while sending data

b. The serial port can hold at least two received data words at one time

c. There is less chance of a framing error due to overrun

d. All of the above statements are true of the 8051 serial interface

9-20. The 8051 counter-timer registers increment one count for each externally received pulse when they are in the _____ mode and one count for each machine cycle when they are in the _____ mode. (Choose two of the four answers.)

a. Event
b. 16-bit
c. Counter
d. Timer

9-21. The 8051 can be interrupted by

a. One of two external signals

b. The counter-timers rolling over

c. A received or transmitted serial character

d. All of the above, if the proper interrupts are enabled

9-22. By using the interrupt enable register and the interrupt priority register, you can always

a. Make any one interrupt source a higher priority than any other interrupt source

b. Ensure the serial interrupt vectors to 0008

c. Completely rearrange the entire interrupt priority structure

d. None of the above are changed by these two registers

9-23. A transition-activated interrupt is set by the trailing edge of the external pulse and cleared when the

a. Signal at the interrupt input returns to logic 0

b. Signal at the interrupt input returns to logic 1

c. Processor vectors to the interrupt service routine

d. Processor completes the interrupt service routine

9-24. A level-activated interrupt is set by a logic 0 at the external interrupt input and is cleared when the

a. Signal at the interrupt input returns to logic 0

b. Signal at the interrupt input returns to logic 1

c. Processor vectors to the interrupt service routine

d. Processor completes the interrupt service routine

9-25. When one of the 16-bit counter-timer registers rolls over, the counter timer function can be set so that

a. The register is preset with a preprogrammed number

b. An interrupt happens

c. The register continues accepting counts starting with 0000

d. All of the above can be set by selecting different modes

9-26. The 8051 has a large number of single-byte instructions which

a. Perform routine housekeeping functions

b. Implement the arithmetic and logic instructions

c. Allow efficient object code to be written

d. There are only a few single-byte 8051 instructions

9-27. The 8051 bit address instructions

a. Are evidence that the 8051 is designed for controller applications

b. Eliminate masking and rotate instruction sequences to access a bit of data in a byte

c. Only make up a part of the 8051 instruction set

d. All of the above

9-28. Two 8051 arithmetic instructions which are not found on the Z80 or the 6802 are _____ and _____ . (Choose two of the four answers.)

a. Divide
b. Add
c. Multiply
d. Subtract

9-29. The 8052 has twice as much internal ROM and RAM as the 8051. How does this change the 8052 instruction set when compared to the 8051 instruction set? Why?

9-30. How can either the 8048 or 8049 microcontrollers be used if they do not have a serial interface?

9-31. What kinds of applications are there for the 8031 microcontroller?

9-32. What is an 8751 microcontroller?

CRITICAL THINKING QUESTIONS

9-1. Manufacturing labor (time) and product size are becoming two very critical areas in new product designs. How does the microcontroller fit into these areas?

9-2. Troubleshooting products built with microcontrollers can be easier than troubleshooting systems built with microprocessors because the data bus is usually entirely inside of the microcontroller and therefore not subject to external damage. On the other hand, there are some ways a microcontroller-based product is more difficult to service. What might those ways be?

9-3. The microcontroller is not used as the basis for a general-purpose microcomputer (a PC) because it has limited memory and I/O. What technology do you think is keeping the use of microcontrollers out of general-purpose computing?

9-4. The 8051 has two different kinds of memory space—read only for instructions and read/write for data storage. Why do you think this makes more sense for a device that is used in controller applications than it does for a device used in general computing applications?

9-5. Why do you think a developer would choose an 8751 over an 8031 and an external EPROM?

Answers to Self-Tests

1. *a*   2. *c*   3. *b*   4. *d*
5. **a.** Not a target. This is a large general-purpose device.
   b. A target. A good controller function with limited requirements.
   c. A good controller target with limited functions and low-cost objectives.
   d. A poor target. The need for expandability is not something microcontrollers do well.
   e. A good way to get rid of lots of logic with one IC.
   f. A good controller and processing application. The DMM costs must be kept at a minimum.
   g. A good consumer product target. Televisions are very cost-conscious products.

6. *b*   7. *d*   8. *c*   9. *d*
10. *b* and *c*
11. *d*   12. *a*   13. *d*   14. *c*
15. *a*   16. *a*   17. *a*   18. *d*
19. *c*   20. *d*   21. *d*   22. *b*
23. *a* and *c*
24. *b* and *c*
25. *b, c,* and *d*
26. *d*   27. *c*   28. *a*   29. *b*
30. *c*
31. To make the 8051 transmit data with even parity, the 8-bit word which is sent to the serial interface must have the parity computed and added before the character gets to the port. This can be done with a few instructions working on the character while it is in the accumulator.
32. *b*   33. *b*
34. *b* and *c*
35. *c*   36. *d*   37. *d*   38. *b*
39. *d*   40. *b*   41. *d*

42. The interrupt priority register has one bit for each 8051 interrupt source. By setting this bit, we set the interrupt source to priority level 1. If the bit is cleared, the interrupt source has priority level 0. All of the level 1 interrupt sources have a higher priority than the level 0 interrupt sources. This means that setting the $\overline{\text{INT1}}$ bit and clearing the $\overline{\text{INT0}}$ bit gives INT1 the highest interrupt priority.

43. *b*   44. *b*   45. *c*   46. *b*
47. *c*   48. *d*   49. *c*   50. *c*
51. *d*   52. *b*   53. *b*   54. *a*
55. *d*   56. *a*   57. *c*   58. *b*
59. *a*

CHAPTER 10

Two Advanced Microprocessors

■

CHAPTER OBJECTIVES

This chapter will help you to:

1. *Describe* the typical applications for 16- and 32-bit microprocessors.
2. *Compare* the two major architectures used in 16- and 32-bit microprocessors.
3. *Identify* the registers found in the Intel and Motorola architectures and *describe* their uses.
4. *Compare* the Intel and Motorola addressing modes.
5. *Recognize* the principal characteristics of the instruction sets for the Intel and Motorola microprocessors.
6. *Describe* the typical hardware configuration and I/O control lines for the Intel and Motorola families.

───────────

There are two popular architectures in use today for microprocessors which process 16- or 32-bit words. They are known by the names of the companies that developed them: Intel and Motorola. Today there are many versions of both architectures, but once you understand the basic difference between the two architectures, you will understand all the family members. The major use of these advanced microprocessors is to build microcomputers. However, they are becoming more and more common in other dedicated applications, just as the 8-bit microprocessors are found in many common applications. In this text, we use the term *advanced microprocessor* to refer to the different families of 16- and 32-bit microprocessors.

■

10-1 AN INTRODUCTION TO THE ADVANCED MICROPROCESSORS

In the world of digital electronics, it is natural that each new generation in a product area doubles in size. Microprocessors are no exception. After the 4-bit microprocessor came the 8-bit microprocessor, and after the 8-bit microprocessor the next logical architecture was the 16-bit microprocessor. Both of these steps doubled the word length and included major changes in the microprocessor's architecture.

The next step was the 32-bit microprocessor. However, growth from 16 bits to 32 bits did not bring entirely new architectures, as did

the change from 4 bits to 8 bits or the change from 8 bits to 16 bits. In fact, the commonality between the architecture of each new generation and the previous generation is very high. This permits programs from earlier microprocessors to run (usually a lot faster) on the new-generation microprocessor. For this reason, we cover the 16- and 32-bit microprocessors in a single chapter under the title *Two Advanced Microprocessors*.

Two families of advanced microprocessors are of particular interest. These are the Intel and Motorola families. Although they are not the only developers of advanced (16-, 32-, and even some 64-bit) microprocessors, the Intel and Motorola advanced microprocessors have become the most popular—especially for creating personal computers.

There are other advanced microprocessor families, but none of them enjoys the popularity of the Motorola and Intel families. Once you understand the basic concepts behind these two microprocessors, you will be able to quickly learn other advanced microprocessors if you need to.

The advanced microprocessors are very powerful computing devices. Typically, products which use these microprocessors are configured as *general-purpose computers*. What do we mean by "configured as general-purpose computers"? We mean that the basic system architecture is like the general-purpose computing system architecture in Fig. 10-1. In this diagram, the CPU is connected to a bus along with memory, mass storage, general-purpose I/O, and a user interface. The user interface includes an alphanumeric and/or graphic display system and data entry system. This is the architecture of a general-purpose computer. The general-purpose architecture lets you use a wide variety of general-purpose software to make the unit do the job you need to perform.

Not every advanced microprocessor application is configured as a general-purpose computer. However, the hardware to make an advanced microprocessor system is more expensive than the hardware needed to make an 8-bit microprocessor-based system. Therefore, most advanced microprocessor-based systems either use a general-purpose computer architecture or are built from general-purpose computer subassemblies such as CPU, RAM, I/O, or mass storage cards. This lets the designers make use of low-cost microcomputer assemblies.

Dedicated systems with advanced microprocessors typically need a great deal of computing power. For example, the 32-bit microprocessors are very popular for graphic display systems. Presenting data in graphic form (using figures, lines, drawings, and other graphics techniques) requires many complex mathematical operations. For example, a graphic presentation can be described as a multidimensional array with x and y dimensions describing the different points of the display. The z axis is used to define attributes (characteristics) such as color and intensity. To manipulate these images, you must quickly process multidimensional arrays.

Frequently, advanced microprocessors are supported by a *numeric coprocessor*. The numeric coprocessor, like the 8-bit coprocessor, is a separate, dedicated CPU. It performs

From page 226:

Advanced microprocessors

On this page:

General-purpose computers

Numeric coprocessor

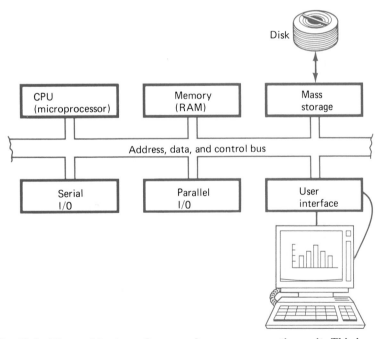

Fig. 10-1 **The architecture of a general-purpose computing unit. This is the architecture most commonly used by products which include advanced microprocessors as part of their system. Note: The areas of user interface and mass storage are the ones most likely to be different from one system to the next.**

such arithmetic functions as add, subtract, multiply, divide, and trigonometric functions on two variables. It returns a result much faster than a general-purpose CPU can. The general-purpose CPU may also be busy handling a display, managing mass storage devices, and performing other monitoring and maintenance functions while the coprocessor is working.

The following sections of this chapter examine the Intel and Motorola advanced microprocessors in more detail. We will look at the programming models and provide an overview of their instruction sets and their hardware. It is important that you understand that they are very complex devices. We cannot cover them in the same depth we gave 8-bit microprocessors. On the other hand, you may never find these microprocessors used in situations where you are required to work at the depth you may work with the 8-bit systems.

Often products using advanced microprocessors also require sophisticated operating and applications software. Typically, these are very complex systems, and unless you understand the software to some level of detail, you will not be able to understand the system itself. All of this means that, to understand an advanced microprocessor-based system, you may require more system knowledge than hardware knowledge. A brief overview of some of these software issues is presented in the next few paragraphs. This is done so that you can see why the advanced microprocessors are designed with complex capabilities.

One of the common applications for advanced microprocessors is with multiuser or multitasking operating systems. A multiuser operating system allows more than one user to *appear* to use the CPU at one time. Likewise, a multitasking operating system allows more than one task to appear to use the CPU at one time.

Advanced microprocessors, like the simpler microprocessors, can really execute only one instruction at a time and therefore deal with only one user or one task at a time. *However, systems using advanced microprocessors often have special software which switches the advanced microprocessor between software which is performing computations for multiple users or multiple tasks.*

When a microprocessor is used in this mode, it is important to make sure that the individual users or tasks do not have access to the programs performing the switching between

users or tasks. The software which is managing the different users or tasks often has access to certain processor instructions to which the individual users or tasks do not have access. This helps prevent a programming error occurring in one of the user programs or in one of the task programs from causing the whole system to crash.

To do this, the advanced microprocessors offer a special way of operating which is called the *supervisory, protected,* or *privileged mode.* When the advanced microprocessor is operating in this special mode, it may have access to a number of special instructions, additional registers, and other features of the advanced microprocessor. Frequently, you will see diagrams such as the one shown in Fig. 10-2, which shows how different privilege levels are granted to different programs depending on how the programs are written. This diagram is often referred to as an *onionskin diagram* because the privileged layers look like the layers of an onion.

Another problem which systems utilizing advanced microprocessors often face is *data protection.* Protection techniques are used to keep two users or tasks from accessing the same data at the same time. For example, think of a system in which an advanced microprocessor is managing a multitasking system on a factory floor. In this example, two tasks need information about the available quantity of a certain part so that they can determine how the factory process will continue. If there is only one part available, the advanced microprocessor's protection techniques are used to ensure that both tasks cannot be told that the part is available. Again, special privilege levels are afforded certain data to ensure that the correct data goes to each task.

The software which is developed to carry out the complex problems to be solved by systems using advanced microprocessors is often very large and requires access to large quantities of data. *Virtual memory techniques* allow a programmer to develop programs (or groups of programs) which are much larger than the amount of physical memory actually available to the processor. The additional memory space is really on a disk, but through sophisticated memory addressing techniques, the memory on the disk is made to look as if it is in main memory. The ability to work with virtual memory can be achieved by a combination of special instructions and special hardware and software. Advanced micropro-

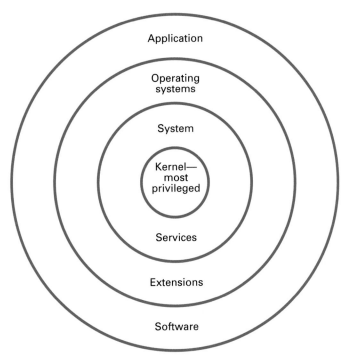

Fig. 10-2 An onionskin diagram showing the different privileged layers for a multiuser or multitasking operating system. The most privileged software is the kernel. Usually this is the heart of the operating system. Moving outward, each layer has fewer and fewer privileges.

cessors often include special instructions and internal hardware which allow a programmer to write software without knowing or caring how much memory is really available. The virtual memory system will ensure that the programs work properly.

Self-Test

Answer the following questions.

1. The first advanced microprocessors were
   a. Simple expansions of 8-bit predecessors
   b. Integrated circuit implementations of current 16-bit microcomputers
   c. New 16-bit architectures developed specifically for implementation as 16-bit microprocessors
   d. Developed before the 8-bit microprocessor came into being

2. The most common use for advanced microprocessors is as the CPU for
   a. Signal processing systems
   b. High-powered graphics processors for computer-aided engineering workstations and other similar products

   c. Personal computers
   d. All of the above

3. The most recent advanced microprocessors
   a. Are simple expansions of their 8-bit predecessors
   b. Are compatible expansions of the first 16-bit microprocessors
   c. Are new architectures, developed specifically for implementation as new advanced microprocessors
   d. Were developed before the 8-bit microprocessor came into being

4. Common applications which use advanced microprocessors include
   a. The Apple Macintosh (68XXX)
   b. The PC (X86 family)
   c. Engineering workstations (486, Pentium, and 68XXX)
   d. All of the above

5. One reason advanced microprocessors are not typically found as dedicated controllers is
   a. Relatively high-cost components are needed to make up a basic system
   b. A 32-bit address bus allows the microprocessor to address a great deal more memory than a 16-bit address bus

c. Expanding the microprocessor from 8 to 16 bits is one of the easy ways to double its power

d. All of the above

6. A key part of a general-purpose computing system is

a. I/O   d. Mass storage

b. Memory   e. User interface

c. The CPU   f. All of the above

10-2 AN INTRODUCTION TO THE INTEL X86 FAMILY OF ADVANCED MICROPROCESSORS

In this section, we review what is probably today's most popular family of advanced microprocessors, the *Intel X86* family of advanced microprocessors. This family was chosen for inclusion in this chapter because you will find one of its members, or a clone, at the heart of every IBM-compatible personal computer (PC). Although advanced microprocessors are new, you will find that your knowledge of 8-bit microprocessors, specifically the Z80, will make it easy to understand the X86 family of advanced microprocessors.

During the late 1970s, Intel introduced the 8088 and 8086. Internally, these two 16-bit microprocessors have identical architectures. The difference between the two is the size of the external data bus. The 8088 has an 8-bit external data bus, and the 8086 uses a 16-bit external data bus.

In 1981 IBM introduced a PC based on the Intel 8088. Since then, many manufacturers have developed PCs, and the 8088 has evolved. However, all but the most recent IBM-compatible PCs are based on microprocessors which use this Intel architecture.

The 8088 and 8086 evolved into faster versions, versions with wider external data buses, versions with greater memory addressing space, versions with advanced computing functions, and versions which process 32-bit data words. Because these versions also have a very similar architecture, they have a common numbering system. The 80286 followed the original introductions, and it was soon followed by the 80386 and then the Intel 486. The latest introduction broke the numbering rule: It is simply known as the Pentium. Often people speak of these microprocessors by their last three digits (e.g., 286, 386, 486), or simply as the X86 family.

Table 10-1 gives you an overview of the X86 family of advanced microprocessors and some key attributes associated with each member of the family. As you can see, as the new processors are introduced, the X86 family of advanced microprocessors grows in functions and features.

The 8086 was actually the first Intel member of the X86 family introduced to the market. It was followed within a few months by the 8088. As you can see in Table 10-1, the only

Table 10-1 Major Attributes of the X86 Processors

Attribute	8088	8086	286	386SX	386DX	486SX	486DX	Pentium
Data bus	8 bits	16 bits	16 bits	16 bits	32 bits	32 bits	32 bits	64 bits
Address bus	20 bits	20 bits	24 bits	32 bits	32 bits	32 bits	32 bits	32 bits
Operating speed (MHz)	5, 8	5, 8, 10	6, 8, 10, 12.5, 16, 20	16, 20, 25, 33	16, 20, 25, 33, 40, 50	25, 33, 50	25, 33, 50	50, 60, 66, 100
Instruction cache				16 bytes	16 bytes	32 bytes	32 bytes	8 kbytes
Data cache				256 bytes	256 bytes	8 kbytes	8 kbytes	8 kbytes
Math coprocessor	External 8087	External 8087	External 80287	External 80387	External 80387	External 80387	Internal	Internal
Memory management	External unit	External unit	Internal MMU	Internal MMU	Internal MMU	Internal MMU	Internal MMU	Internal MMU
Physical memory addressed	1 Mbyte	1 Mbyte	16 Mbytes	4 Gbytes	4 Gbytes	4 Gbytes	4 Gbytes	4 Gbytes
Internal data word size	16 bits	16 bits	16 bits	32 bits	16 bits	32 bits	32 bits	32 bits
Introduction date	1978	1978	1982	1985	1985	1989	1991	1993

difference between these two 16-bit microprocessors is that the 8088 uses an 8-bit data bus and the 8086 uses a 16-bit data bus. When it came time to pick a 16-bit microprocessor for the first IBM PC, the Intel 8088 was chosen because, among other reasons, its 8-bit data bus made it easy to use the low-cost 8-bit peripheral devices built for use with the 8080/8085 and other 8-bit microprocessors.

The next major introduction was the Intel 80286. Today it is simply called the *286*. IBM introduced the PC/AT (personal computer/advanced technology) version of its PC using the 286 in 1984. One of the features introduced with the 286 was real and protected modes of operation. In the real mode, the processor can address only 1 Mbyte of memory, whereas in the protected mode it can address 16 Mbytes. Another new feature was the ability to work with up to 1 Gbyte of virtual memory, and yet another feature added hardware multitasking.

The 8088, 8086, and 286 microprocessor ICs are also built by a number of integrated circuit (IC) manufacturers other than Intel. Many of them are made under license to Intel and are therefore identical to the Intel models. Some were made by reverse-engineering the processors and therefore are *clones* rather than identical copies. Occasionally, you may find some differences in operation between an Intel processor and a clone.

The 80386 was a major next step in the X86 family. As you can see in Table 10-1, the 386 is a full 32-bit microprocessor. It has a 32-bit data bus and a 32-bit address bus, and it uses 32-bit internal registers. The base 386 internal architecture is, in many ways, very much like the 8088, 8086, and 286 architectures. The major difference in the base architecture is that there are a few more registers and some register sets are now 32 bit rather than 16 bit.

Because there were many 16-bit peripheral devices on the market, Intel introduced a special version of the 386 called the *386SX*. It uses a 16-bit data bus. The original 80386 processor was renamed the *386DX*. In addition to 32-bit processing, the 386 microprocessors offered advanced virtual memory, advanced protected mode, and higher speeds.

Another variation of the 386DX is the 386SL. This is a low-power version. The 386SL can operate on either 3 or 5 V dc and has special power-management circuits which allow the processor to shut down when not being used. Other versions of the 386 are the 386DX2 and 386DX4. The DX2 version doubles the internal clock speed so, for example, a 33-MHz DX2 processor actually operates at 66 MHz internally, thus speeding up many calculations and operations. The I/O, however continues to operate at the lower speed, so there are no external interface problems if a 386DX2 is substituted for a 386DX. The 386DX4 triples the internal clock speed.

The 486 is basically a large integrated circuit which contains a fast 386 processor, a math coprocessor, a memory management unit (MMU), and an 8-kbyte cache memory. It is most frequently called the 486 or the i486. Like the 386, there are DX and SX versions. However, all 486 processors have a 32-bit data bus. As Table 10-1 shows, the SX version does not have the on-chip numeric coprocessor. The 486 achieves its high-speed operation from faster clock speeds, an internal pipelined architecture, and the use of *reduced instruction set computing* (*RISC*) to speed up the internal microcode. Like the 386, the 486 has SL, DX2, and DX4 versions which are low-power, clock-doubling, and clock-tripling versions respectively.

Neither the 386 nor the 486 was ever licensed to another IC manufacturer by Intel. However, there are a number of IC manufacturers with their own versions of the 386 and 486 processors. For all practical purposes, they are identical to the Intel equivalent; however, you should always keep in mind that they are clones, not licensed identical copies.

The Pentium is the next Intel member of the X86 family. Although the 586 nomenclature was expected, for marketing reasons, Intel elected to name this device *Pentium*. The Pentium introduces a number of new features. The use of superscalar architecture incorporates a dual-pipelined processor which lets the Pentium process more than one instruction per clock cycle. The addition of both data and code caches on chip is also a feature designed to improve processing speed.

A new advanced computing technique used in the Pentium is called *branch prediction*. Using branch prediction, the Pentium makes an educated guess where the next instruction following a conditional instruction will be. This prevents the instruction cache from running dry during conditional instructions. The Pentium has a 64-bit data bus. This means that it can perform data transfers with an external device (memory, for example) twice as fast as a processor with a 32-bit data bus.

286

Clones

386SX

386DX

Reduced instruction set computing (RISC)

Pentium

Branch prediction

Self-Test

Answer the following questions.

7. Internally, the X86 family of advanced microprocessors has a(n) _____ architecture.

   a. 4-bit   f. 4-bit or 8-bit
   b. 8-bit   g. 8-bit or 16-bit
   c. 16-bit   h. 16-bit or 32-bit
   d. 32-bit   i. 32-bit or 64-bit
   e. 64-bit

8. The first microprocessor in the X86 family to support an on-chip instruction cache is

   a. 8088   e. 386DX
   b. 8086   f. 486SX
   c. 286   g. 486DX
   d. 386SX   h. Pentium

9. The maximum physical memory space which can be addressed by the 286 is

   a. 640 kbytes   d. 4 Gbytes
   b. 1 Mbyte   e. All of the above
   c. 16 Mbytes

10. An internal memory management unit (MMU) first appeared on the

    a. 8088   e. 386DX
    b. 8086   f. 486SX
    c. 286   g. 486DX
    d. 386SX   h. Pentium

11. On-chip floating point arithmetic units first appeared on the

    a. 8088   d. 386
    b. 8086   e. 486
    c. 286   f. Pentium

12. If the processor you are using does not have enough physical memory for the program that is being used, you can use _____ to make the program believe that the processor has enough main memory for the program.

    a. An on-chip floating point processor
    b. Virtual memory capability
    c. Cache memory
    d. Any of the above

13. The first X86 advanced microprocessor to use full 32-bit data words was the

    a. 8088   d. 386
    b. 8086   e. 486
    c. 286   f. Pentium

10-3 A PROGRAMMING MODEL FOR THE X86 FAMILY OF ADVANCED MICROPROCESSORS

The 8088 and 8086 microprocessors were the first advanced microprocessors introduced by Intel. These two microprocessors have their own unique architecture; it is not a copy of a microcomputer architecture, nor is it an extension of an 8-bit microprocessor architecture. The 8086 is different from the 8088 because it is a true 16-bit microprocessor; that is, it has a full 16-bit data bus. Otherwise, the two microprocessors are the same. Both the 8088 and the 8086 internal architectures use 16-bit buses and registers.

The 8088 and the 8086 define the base programming model for the entire X86 family of advanced microprocessors. The newer members of the X86 family of advanced microprocessors have greater computing power because they are faster, they use 32-bit registers instead of the 16-bit registers used in the earlier (8088, 8086, and 286) advanced microprocessors, and they have advanced addressing techniques.

Once you understand the basic programming model for the 8088 and 8086 processors, you will be able to understand the improvements made with the newer models.

Figure 10-3 shows the base programming model for the X86 advanced microprocessors. Fig. 10-3(*a*) shows the register set used in the 8088, 8086, and 286 processors. Figure 10-3(*b*) shows the register set used in the 386 and 486 processors. As you can see, the general programming model for all the X86 advanced microprocessors is very much the same. The difference is in register length, extra data segment registers, and added features.

The base programming model is made up of three register groups. The first set contains eight general-purpose registers called the A, B, C, D, SI (source index), DI (destination index), SP (stack pointer), and BP (base pointer) registers. Depending on the specific processor, these are either 16- or 32-bit registers.

The lower two bytes of the A, B, C, and D registers are broken into low and high bytes called the AL, BL, CL, and DL and AH, BH, CH, and DH respectively. The full 16 bits are referred to as AX, BX, CX, and DX, where the X stands for *eXtended*. The SI, DI, BP, and SP registers are always treated as 16-bit registers. These four registers are called the *pointer registers* and the *index registers*, because they are used to point to locations within a segment. Figure 10-4 on page 234 shows how the stack pointer, for example, is used to point to the current stack location in memory.

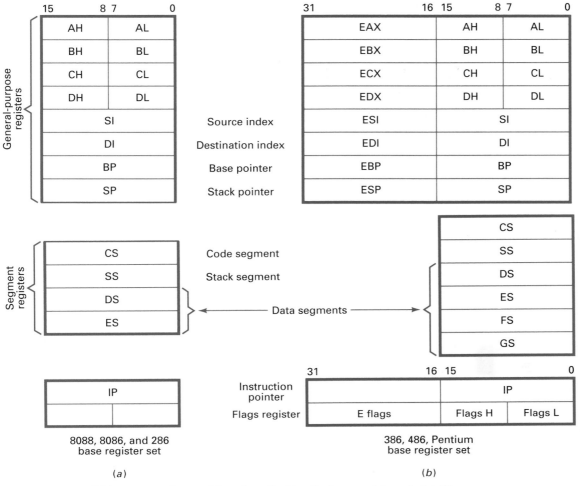

Fig. 10-3   The X86 programming model as described by the base register set. (*a*) The base register set for the 16-bit versions of the X86 family. (*b*) The base register set for the 32-bit versions of the X86 family. Note the addition of two new segment registers. Also note that the segment registers remain 16-bit registers.

When the microprocessor uses 32-bit registers, the eight general-purpose registers are called the EAX, EBX, ECX, EDX, ESI, EDI, ESP, and EBP registers. The E tells us that these registers have extended length. Each register can be addressed in 1-, 8-, 16-, or 32-bit (if available) modes. The arithmetic logic unit (ALU) works with these registers to give the X86 microprocessors their computation and data movement capability.

The second set of registers are the *segment registers*. This set of registers consists of the code segment (CS) and stack segment (SS) registers and either two or four data segment registers. The data segment registers are called DS, ES, and, if they are used, FS and GS. These are 16-bit registers for all processors. These X86 registers manage operations with external memory. Address computation and data movements are performed here. A second

ALU is dedicated to processing memory address data which works with the segment registers.

The third set of registers includes the instruction pointer (program counter) and the flags register. The *instruction pointer* is either a 16- or a 32-bit (for the 386 and newer processors) register. The programming model also includes the flags register. Special results from data operations working on data in the registers and memory locations are stored in the flag registers. The X86 flag register is shown in Fig. 10-5 on page 234. The first five bits are identical to the flag bits in the 8085 8-bit microprocessor. The flags in bits 6–11 were introduced with the 8088/8086. The flags in bits 12–14 were introduced with the 286. The flags in bits 16 and 17 were introduced with the 386, and the flag in bit 18 was introduced with the 486.

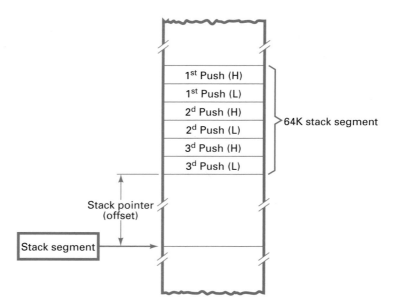

Fig. 10-4 **Building the X86 stack. The stack segment register is used to define the base address for the stack, and the stack builds down from the memory location pointed to by the sum of the offset stack segment register (stack segment base address) and the stack pointer value, which is used as an offset.**

These three register groups and their support circuits let the X86 processors manage memory operations and compute at the same time. When the processor includes a math coprocessor, additional computation may take place there as well.

To give the X86 processors even greater speed, the processors include various levels of instruction *prefetching* and *data caching*. All X86 processors include some instruction prefetching (often called *code caching*). The newer processors have significant amounts of internal memory in which they can cache data as well as code.

The next few instructions are prefetched whenever the address and data buses are available. The additional instructions are fetched and stored in the processor's *prefetch queue* or *code cache*. These are simply on-processor memory systems which store from six instructions up to 8 kbytes of instructions depending on the processor version.

These prefetched instructions are immediately available for processing once the current instruction is complete; therefore, when executing this sofware, the processor does not wait for a fetch cycle. Prefetching can significantly speed up execution time. This process may not work for every instruction because software does not always execute one instruction after the other. For example, branching may cause the next instruction to come from

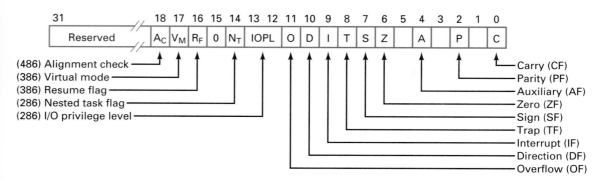

Fig. 10-5 **The X86 flag (status) register. Note that the first 5 bits are identical to those in the 8085//Z80. As the X86 processor family grew, new flag bits were added to accommodate the additional processor features.**

a very different location in memory. However, the greater the number of instructions prefetched, the greater the likelihood the next instruction to be executed is in the prefetched queue. The most advanced members of the X86 family have special logic which analyzes the prefetched instructions and attempts to anticipate branching and other such changes so that the correct instructions are prefetched.

The more advanced X86 processors also store frequently used data in a memory which is on board the processor. Again, having this data right at hand, in a data cache, avoids the need for an external memory access cycle and therefore speeds up processing time.

Self-Test

Answer the following questions.

14. The main difference between the older and newer versions of the X86 advanced microprocessors is that the general-purpose registers in the newer processors are
    a. 4-bit   d. 32-bit
    b. 8-bit   e. 64-bit
    c. 16-bit

15. You will find data segment registers called the FS and the GS registers on the _____ microprocessor.
    a. 8088   d. 386
    b. 8086   e. All of the above
    c. 286

16. The length of the 8088 instruction pointer is
    a. 4 bits
    b. 8 bits
    c. 16 bits
    d. 32 bits
    e. Variable depending on use and memory

17. A separate ALU is used with the _____ registers to perform memory location calculations.
    a. Segment
    b. Flag
    c. Instruction pointer
    d. General-purpose
    e. All of the above

18. A prefetch queue or code cache first appeared on the
    a. 8088   e. 386DX
    b. 8086   f. 486SX
    c. 286   g. 486DX
    d. 386SX   h. All of the above

19. Some of the newer versions of the X86 processors have an on-board memory which is used to store data as well as prefetched instructions. This is called
    a. An on-chip floating point processor
    b. Virtual memory capability
    c. Cache memory
    d. Any of the above

20. Logic which analyzes the instructions held in the code cache to anticipate what code will be needed after a branch was introduced with the
    a. 8088   e. 386DX
    b. 8086   f. 486SX
    c. 286   g. 486DX
    d. 386SX   h. Pentium

10-4 THE X86 ADDRESSING MODES

The X86 processors have many different addressing modes which give the X86 instructions a great deal of power. Once you understand the addressing modes, you can study the X86 instruction set.

Like other microprocessors, the X86 processors can address immediate data, I/O ports, registers, and memory locations. Because the range of potential memory locations is so great, addressing memory locations is the most complex addressing activity.

An X86 instruction performs an operation on one or two operands. The result is stored in the same location occupied by one of the original operands. If there are two operands, you can select which one gets the result. Each of the instruction operands can be a register or a memory location.

In two-operand instructions, the second operand can be a constant within the instruction itself; that is, it can be immediate data. These immediate instruction data are 2's complement numbers. The data transfer, arithmetic, and logical instructions can act on immediate data, registers, or memory locations and can work with words of varying length.

The X86 processors have both direct and indirect memory addressing. Their basic operation is shown in Fig. 10-6 on page 236.

If direct addressing is used, a displacement address is part of the instruction. Depending on the situation and the processor, this displacement address can be an 8-, 16-, or 32-bit word. In direct addressing, the displacement becomes the logical address, which is added to the shifted contents of the segment register (the segment base address) to give a physical

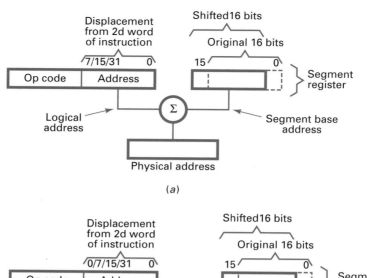

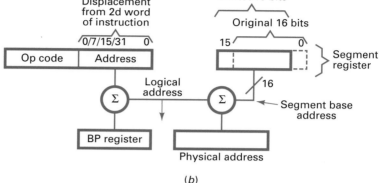

Fig. 10-6 (*a*) **Direct addressing in the X86 processors. The logical address from the second word of the instruction is added to the shifted 16-bit segment register value (the segment base address). The result of this addition is the physical address.** (*b*) **An indirect addressing operation in the X86 processors. In this figure, the content of the instruction second word is added to the content of the base pointer (BP) register. The result (the logical address) is added to the shifted value from the segment register (the segment base address). This final result becomes the physical address.**

memory address. Fig. 10-6(*a*) diagrams the direct addressing operation.

When you use indirect addressing, the instruction can have 0-, 8-, 16-, or 32-bit displacements. The logical address is made by adding a number of sources together. For example, you can address memory indirectly using the contents of a base register, using an index register, or using the sum of a base register and an index register. Figure 10-6(*b*) diagrams one of the many indirect addressing operations. In this example, the 16-bit displacement from the instruction is added to the contents of the base pointer register. That result (the logical address) is added to the offset contents of a segment register (the segment base address). The result of the addition is the physical address.

There are two important characteristics you must understand about the X86 memory addressing process. First, the X86 family of processors is able to address a wide range of

memory locations. The 8088/8086 addresses 1 Mbyte (2^{20}). The 286 addresses 16 Mbytes (2^{24}), and members after the 386 can address 4 Gbytes (2^{32}). Second, the X86 processors use only single-word operands for memory address instructions. This is one more reason why the processors are very fast.

Figure 10-7 is another view showing how the X86 family of processors calculates the address of a memory location. When the X86 processors address a memory location, the shifted value in one of the segment registers is added to a logical address. The logical address comes from a number of different places. For many instructions, the word following the instruction (called the *displacement*) becomes the logical address.

Often the X86 literature refers to this computed memory address as the *physical address*. Remember, the X86 processors have a separate ALU to perform this calculation, so the main ALU is free to perform data calculations

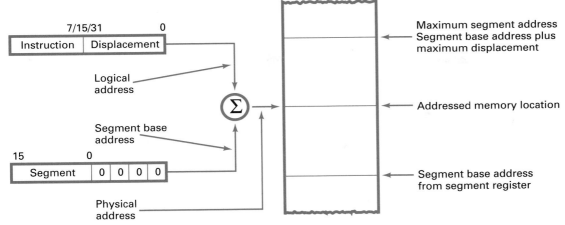

Fig. 10-7 Using the instruction displacement as the logical address and the content of the segment register to create the segment base address to, in turn, create a physical address. *Note:* The segment base address sets the lowest value in memory for the segment. The logical address defines how many memory locations above the segment base address can be in the segment.

at the same time as the address ALU is calculating a new memory address value.

As you can see, the segment base address is shifted four locations to the left; that is, it is multiplied by 10H (16_{10}). For the 8088, 8086, and 286 processors, the displacement and segment base addresses are 16-bit words. Therefore, there are only 64K base address locations ($2^{16} = 65,536 = 64K$). Because the base address is multiplied by 16, a segment base address starts on a 16-byte boundary. They start at memory location 0 and end at memory location 1,048,560 (16 bytes below the 1-Mbyte location).

The 16-bit displacement means that there are between 0 and 64K locations in any one segment. Between the segment base address and the displacement 1,048,576 memory locations can be addressed. Another way to look at this memory addressing scheme is to think of 64K segments, each of which can be up to 64 kbytes long.

Wherever possible, a programmer wants to keep all of the code for a particular application within one segment. In the past, this has been viewed as a restriction for the X86 processors because many programs need to exceed the 64-kbyte length. However, the newer X86 processors (386 and newer) no longer have the 64-kbyte segment restriction. We will learn more about this later in this section.

The displacement plus segment technique is a very fundamental memory addressing approach for all members of the X86 family. There are many variations which make it powerful and flexible. *However, all X86*

processors use the basic technique of adding a displacement to the segment base address.

The use of multiple segment registers by the X86 processors allows a programmer to divide memory space into multiple areas called *segments*. A programmer can use these different sections to store program instructions, data, and stack information. If needed, data can be put into multiple memory spaces. The number of spaces available depends on the number of data segment registers available. The older processors had two, and the newer processors have four. The segment registers are used to point to the base (lowest) address of its memory space.

The different segment types let you choose different kinds of addressing. The correct segment register is automatically chosen by the instruction being executed. Figure 10-8 on page 238 shows one way the different segment registers are used by the different kinds of instructions. The code segment (CS) register points to the base address of the currently executing program. The stack segment (SS) register points to the base of the stack. The data segment registers (DS, ES, FS, and GS) point to the base of the data areas.

When the X86 processor is performing an instruction fetch, the data in the code segment register is added to the logical address. Here, the logical address is the current instruction pointer value.

When you use the stack in an X86 microprocessor, the stack segment register is automatically added to the logical address. In this case,

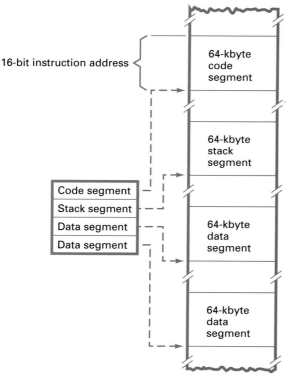

16-bit instruction address

64-kbyte code segment

64-kbyte stack segment

Code segment
Stack segment
Data segment
Data segment

64-kbyte data segment

64-kbyte data segment

Fig. 10-8   Using the X86 segment registers to select different parts of memory for code, stack, and data. *Note:* **These areas of memory can overlap if needed. Also, there can be as many data segments as the processor has data segment registers.**

the logical address is the contents of the stack pointer.

If you are referencing data, you can use any of the data segment registers. There are many different sources for data logical addresses. Logical address for data instructions is generated by each one of the different addressing modes.

The segment registers may point anywhere in memory. Therefore, all segments can use the same memory locations, overlapping memory locations, or totally separate memory locations. How the memory is used depends on how much memory is available and what the programmer wishes to do with the memory.

As you have learned, data from one of the segment registers is part of any X86 memory address operation. Comparing the X86 to one of the 8-bit microprocessors, you might think that all X86 memory addressing uses indirect addressing. However, this form of addressing is not considered to be indirect addressing in the X86 processors but, rather, is referred to as direct addressing. Remember, the X86 overcomes the slowness of conventional indirect addressing with a separate processor dedicated to addressing operations. The use of a separate ALU for memory calculations makes these calculations fast and independent of other calculations the processor is doing at the time.

Why were the X86 processors designed with this memory addressing, which seems to be more complicated than needed? One part of the answer is processing speed. Memory reference instructions use only a single operand for the address. Another part of the answer is that the segmentation of instructions, data, and the stack allow the programmer to develop more highly structured code. Because the X86 processors are so powerful, the designers knew that they would be used to execute very complex programs. Any help the processor architecture gives the programmer is a benefit.

As we have learned, all X86 memory reference instructions use a default segment register. That is, when you choose an addressing mode to go with an instruction, a segment register is automatically chosen. If you do not want to use the default segment register, you may use an alternative segment. To select a different segment register, add a segment *override prefix* in front of the instruction.

To understand the structure of an X86 instruction and the use of override prefixes, study Fig. 10-9. It shows the structure for two-

Fig. 10-9 (see page 239)   Building the X86 instruction word in the 8088, 8086, or 286. (*a*) **Each group begins with the optional segment override prefix. You start an instruction with this word if you do not want to use the default segment register. The segment register is taken from (*b*) the SEG table. The first (or second) word has the op code (*c*) and specifies (*d*) two operands, (*e*) a single operand with immediate data, or (*f*) a single operand. (*g*) The D, W, and S bits let you specify where the results will go (D = destination), if you are dealing with 8- or 16-bit data (W = width) and if you need the 8-bit data sign extended (S = sign). (*h*) Byte number 3 lets you choose the kind of addressing the instruction will use. If you set (*i*) MOD to 00, 01, or 10, you choose indirect addressing, and the R/M bits are used to choose how the effective address is to be calculated. The register used is selected by the REG bits according to the table (*l*). In the special case (*j*) where MOD = 00 and R/M = 110, you choose direct addressing and the address is contained in the (*k*) 2 displacement bytes. If you set MOD to (*l*) 11, you have specified a register as the address. The register is selected by the R/M value from the table.** *Adapted from Intel data sheets.*

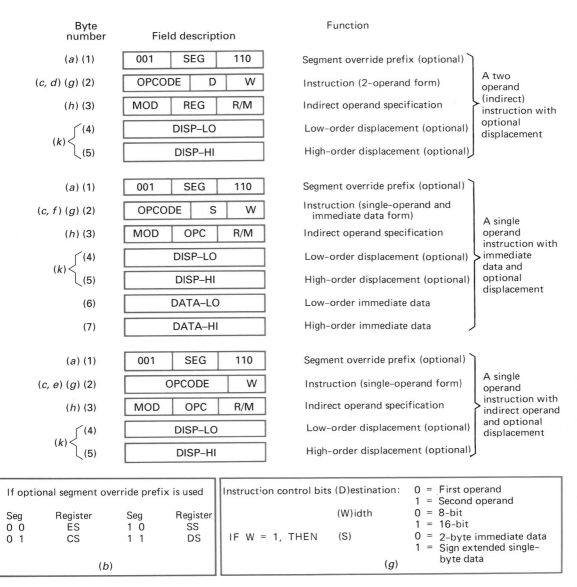

Byte number	Field description			Function	
(a) (1)	001	SEG	110	Segment override prefix (optional)	A two operand (indirect) instruction with optional displacement
(c, d) (g) (2)	OPCODE	D	W	Instruction (2-operand form)	
(h) (3)	MOD	REG	R/M	Indirect operand specification	
(k) (4)	DISP–LO			Low-order displacement (optional)	
(k) (5)	DISP–HI			High-order displacement (optional)	

Byte number	Field description			Function	
(a) (1)	001	SEG	110	Segment override prefix (optional)	A single operand instruction with immediate data and optional displacement
(c, f) (g) (2)	OPCODE	S	W	Instruction (single-operand and immediate data form)	
(h) (3)	MOD	OPC	R/M	Indirect operand specification	
(k) (4)	DISP–LO			Low-order displacement (optional)	
(k) (5)	DISP–HI			High-order displacement (optional)	
(6)	DATA–LO			Low-order immediate data	
(7)	DATA–HI			High-order immediate data	

Byte number	Field description			Function	
(a) (1)	001	SEG	110	Segment override prefix (optional)	A single operand instruction with indirect operand and optional displacement
(c, e) (g) (2)	OPCODE		W	Instruction (single-operand form)	
(h) (3)	MOD	OPC	R/M	Indirect operand specification	
(k) (4)	DISP–LO			Low-order displacement (optional)	
(k) (5)	DISP–HI			High-order displacement (optional)	

If optional segment override prefix is used

Seg	Register	Seg	Register
0 0	ES	1 0	SS
0 1	CS	1 1	DS

(b)

Instruction control bits (D)estination: 0 = First operand
                                   1 = Second operand
              (W)idth   0 = 8-bit
                                1 = 16-bit
IF W = 1, THEN   (S)   0 = 2-byte immediate data
                                1 = Sign extended single-byte data

(g)

First operand addressing

Memory addressing

Indirect addressing (e)

MOD = 00 No displacement
MOD = 01 Displacement 8-bits extended
MOD = 10 Displacement 16-bits

R/M effective address

000	(BX)	+	(SI)	+ DISP
001	(BX)	+	(DI)	+ DISP
010	(BP)	+	(SI)	+ DISP
011	(BP)	+	(DI)	+ DISP
100	(SI)	+	DISP	
101	(DI)	+	DISP	
110	(BP)	+	DISP	
111	(BX)	+	DISP	

Register addressing (l)

MOD = 11

R/M REG	W = 0	W = 1
000	AL	AX
001	CL	CX
010	DL	DX
011	BL	BX
100	AH	SP
101	CH	BP
110	DH	SI
111	BH	DI

Direct addressing (j)
MOD = 00 AND R/M = 110
R/M effective address
110 DISP

operand, single-operand with immediate data, and single-operand instructions. From this figure, you can see that the construction of an X86 instruction is complex, flexible, and powerful.

As you can see, the memory addressing method we have just learned about lets the programmer address any of the 1 million memory locations addressed by an 8088/8086 microprocessor. But how do the more advanced processors address the larger memory systems they support? The answer to this question is that they use advanced memory management techniques made available by the addition of an MMU in the 286 and newer processors.

Each of the newer X86 processors has two fundamental modes of operation. They are called the *real mode* and the *protected mode. In the real mode, all processors are limited to addressing 1 Mbyte of memory*. In this mode they work just like the 8088/8086 (thus it is often called the 8086 mode). In the protected mode, they are able to address many more memory locations (16 Mbytes for the 286 and 4 Gbytes for the 386 and newer processors).

What is the difference between the real and protected modes? First, the X86 processors are initialized in the real mode by the reset line. To put an X86 processor in the protected mode, the protection enable bit (bit 0) in the machine status register must be set to logic 1. The *machine status word (MSW)* is a processor control register which appears in the 286 and newer processors. There are four active bits in the MSV. Bit 0 enables the protected mode. Bits 1, 2, and 3 control and indicate the status of communications with an external numeric coprocessor.

Note: Once the protection enable bit is set, it can be cleared only by resetting the processor. This makes entering the protected mode a one-way street. You can get there, but there is no very practical way to get back (but probably you will not want to go back anyway!).

Second, once you enter the protected mode, use of the segment register data changes. In the real mode, the data in the segment register is multiplied by 16_{10} and the result is used as the segment base address. In the protected mode, the X86's MMU is used to allow the processor to address physical memory greater than 1 Mbyte and to allow the programmer to use virtual memory.

The introduction of the MMU means that there are more registers which now must be controlled. These registers are used to control how the MMU uses memory and which tasks have access to different memory segments.

In the protected mode, the address in the segment register points to a memory location called a *segment descriptor table*. Data in the segment descriptor table is used in place of the segment base address data to build the memory address.

As shown in Fig. 10-10, the upper 13 bits of the segment register data are used to select a particular segment descriptor. This is the segment descriptor address. The first three bits request a priority level (RPL_0 and RPL_1) and select either a global descriptor table or a local descriptor table. A global descriptor table points to information available for all programs (often the operating system software). A local descriptor table points to information available only to the particular program doing the addressing (usually a particular user or task). This gives the X86 programmer one more way to keep different addressing for different kinds of use and therefore to protect the data. Each segment descriptor table contains 8 bytes.

Figure 10-11 shows the data structure of a segment descriptor table. As you can see, the segment descriptor table consists of 8 bytes. These 8 bytes provide a 24-bit (286) or 32-bit (386 and higher) segment base address plus additional information. The first two bytes (plus a 4-bit extension from the next-to-last byte on the 386 and higher) indicate the segment limit. That is, they tell how many memory

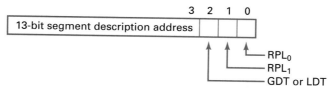

Fig. 10-10   Using the segment register to provide the segment descriptor address. *Note:* **Only the 13 most significant bits are used as the segment descriptor address. Bits 0, 1, and 2 have special uses.**

Segment base address A_{24}–A_{31}			Byte 7
G	16 or 32 bit	0	User bit	Segment limit D_{16}–D_{19}	Byte 6
P	Descriptor privilege level	Segment type	Type of descriptor	Byte 5
Segment base address A_{16}–A_{23}			Byte 4
Segment base address A_8–A_{15}			Byte 3
Segment base address A_0–A_7			Byte 2
Segment limit D_8–D_{15}			Byte 1
Segment limit D_0–D_7			Byte 0

P = Memory physically present or not
G = Segment size 1 Mbyte or 4 Gbytes

Fig. 10-11 The structure of the segment descriptor table. If byte 6 and 7 are all zeros, the segment descriptor is for a 286 processor. Byte 5 contains descriptor rights information such as type (global, local, etc.), segment type, the descriptor privilege level, and an indicator which tells whether the segment is in memory. If the segment is not in memory, it must be loaded before it can be used.

locations are in the segment. This information is used to make sure that the processor knows whether an out-of-bounds memory address is requested. In addition to the segment base and the segment limit, byte 6 (and part of byte 7 in the 386 and higher) is used to store information about how the segment is to be used. *Note:* If byte 6 and byte 7 are set to zero, the segment descriptor table has been built for the 286 processor.

When a memory location is addressed in the protected mode, the segment register contents point to the appropriate segment descriptor table. The data in the segment descriptor table are loaded into a special *segment descriptor register*. This register is part of the X86 MMU. There is one segment descriptor register for each segment register. As shown in Fig. 10-12, the contents of the segment descriptor register are combined with the instruction's logical address by the MMU to create a physical address.

Part of the information contained in the segment descriptor table for a particular segment tells the system whether this memory actually exists in the physical memory or

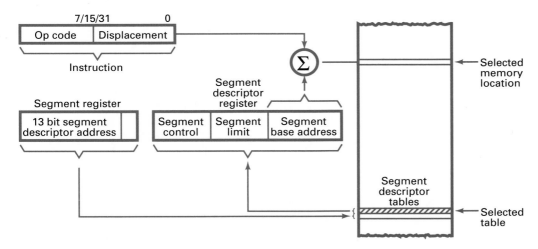

Fig. 10-12 Addressing memory using the segment descriptor tables. The segment register points to one of the segment descriptor tables. The content of the segment descriptor table is transferred into the segment descriptor register. The segment base address from the segment descriptor register is added to the logical address (in this case, the displacement from the instruction) to create the physical address which selects a memory location.

4-kbyte memory blocks

Task state segment (TSS)

whether this is virtual memory. If the memory does not really exist, then a subprogram is used to load the needed information from mass storage into the appropriate section of physical memory.

Segments for the 286 are limited to 64 kbytes. Segments for the 386 and newer processors can be either up to 1 Mbyte or up to 4 Gbytes long. If bit G (the granularity bit) in the segment descriptor table (see Fig. 10-11) is logic 0, the 20 limit bits define a maximum 1-Mbyte size. If bit G is set (logic 1), the 20 limit bits define the number of *4-kbyte memory blocks* which make up a segment. One million 4-kbyte segments fill the 4-Gbyte physical memory space.

As discussed earlier, the X86 processors offer four levels of privilege for any form of memory access. Again, these are selected by the first two bits of the segment register information. Figure 10-13 is an "onionskin" diagram which shows how the different privilege levels are often used. Once the privilege level is set, that section of code cannot access any information from a lower privilege level. For example, if the privilege level is set at PL1, it has access to data with privilege levels set at PL1, PL2, and PL3 but not at PL0. Different privilege levels are used to help the programmer when writing operating system software and software for a multiuser or multitasking environment.

One more register is added to the X86 processors with an MMU (286 and newer). This is the 16-bit task register (TR). The task register is used to address a special segment in memory called the *task state segment (TSS)*, which contains all the register information for a particular task. There is a TSS for each task the processor is running. When a new task is switched into operation, the contents of the appropriate TSS is used to initalize the processor. The size and the contents of the TSS depend on the processor.

Self-Test

Answer the following questions.

21. An X86 instruction performs an operation on one or two operands. The result is stored in
   a. One of the general-purpose registers
   b. The memory location to which the instruction points
   c. The same location as one of the original operands
   d. All of the above

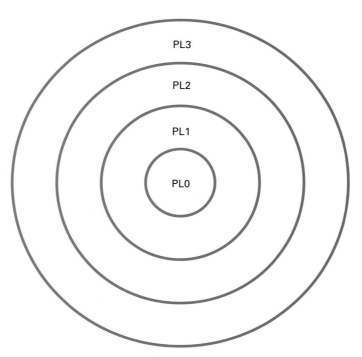

Fig. 10-13 **The X86 privilege levels. PL0 is the highest privilege; PL3 is the lowest privilege.**

22. The X86 advanced microprocessors use
    a. Nothing but indirect addressing (all memory reference instructions are calculed)
    b. Both direct and indirect addressing
    c. Only register indirect addressing
    d. All of the above
23. Either the displacement from an instruction or the displacement (which can be zero or nonexistent) plus the contents of other registers, which contain an index or an offset, are always added to the contents of one of the
    a. General-purpose registers
    b. Two operands in a two-operand instruction
    c. Segment registers
    d. All of the above
24. The segment register data is shifted 4 bits to the left (multiplied by 16) before it is used as the segment base address. This means that the X86 processors (when operating in the 8086 mode) have segments which start on 16-byte boundaries and there is a maximum of _____ of memory.
    a. 640 kbytes   c. 16 Mbytes
    b. 1 Mbyte   d. 4 Gbytes
25. The use of separate segment registers means that an X86 programmer can have separate sections of memory for
    a. Code   c. The stack pointer
    b. Data   d. All of the above
26. The newer X86 processors (286 and up) have real and protected modes. When the processor is running in the real mode, it acts like a(n)
    a. 8088   e. 386DX
    b. 8086   f. 486SX
    c. 286   g. 486DX
    d. 386SX   h. Pentium
27. When an X86 processor is running in the protected mode, the contents of the segment register are
    a. Combined with the displacement to generate the physical memory address
    b. Shifted 4 bits to the left (multiplied by 16_{10}) before being used
    c. Limited to 13 bits, which are used as the limit part of the segment descriptor table
    d. Broken into a 13-bit segment descriptor table address and a 3-bit control element, which sets the requested priority level and selects either a global or a local descriptor table

28. If the top 2 bytes of a segment descriptor table are set to logic 0s, this means that the segment descriptor table
    a. Was created by or for a 286 processor
    b. Was created by or for a 386/486 processor
    c. Has only two priority levels
    d. Has a segment limit which is 1 Mbyte long
    e. Has a segment limit which is 4 Gbytes long
29. If the G bit in the segment descriptor table is set to logic 1, this means that the segment descriptor table
    a. Was created by or for a 286 processor
    b. Was created by or for a 386/486 processor
    c. Has only two priority levels
    d. Has a segment limit which is 1 Mbyte long
    e. Has a segment limit which is 4 Gbytes long
30. Once the segment register addresses a segment descriptor table, the data in the segment descriptor table is loaded into a segment descriptor register. The _____ contents of the segment descriptor register is used as one element in calculating the physical memory address.
    a. Segment base address
    b. Segment limit
    c. Segment priority
    d. All of the above

10-5 THE X86 INSTRUCTION SET

The X86 processors have many different instruction types. As each new member of the X86 family is added to the line, a few more instructions are also added to the X86 instruction set. However, the processors are always *upwardly compatible*; that is, software written for an earlier version will always run on a later version. The newer processors have a special mode (the real mode) which allows them to operate as though they were an 8086.

Note: It is not true that software written for a new processor will necessarily run on an older processor. In fact, software written on the newer processor may use instructions which the older processors will not recognize, and therefore the software is *not downwardly compatible*.

In the previous section, we learned about the different addressing modes the X86 processors

Upwardly compatible processors

Downwardly compatible processors

use to let these instructions work within a wide range of memory space and registers. In this section, we will look at the instructions, but you must keep the addressing modes in mind. It is the combination of the wide variety of instructions and the wide variety of addressing modes which make the X86 processors so powerful.

A summary of the X86 instructions is given in Tables 10-2 through 10-10. These tables are organized by type of instruction. For each type, the table shows the instruction mnemonic, the instruction description, which of the X86 processors introduced the instruction to the X86 instruction set, and the instruction subtype (if the general instruction category has a number of different subtypes). *Note:* Once an instruction is introduced to the X86 instruction set, it is available to all newer processors. So, for example, the POPA instruction, introduced on the 286 processor, is not available for the 8088 or the 8086; it is available on the 286, 386, 486, and Pentium processors.

The following paragraphs provide general comments on different types of X86 instructions. As you read these paragraphs, review the instructions in the associated table. There are eight major types of X86 instructions: data transfer, arithmetic, logical and shift-rotate, string manipulation, bit manipulation, control transfer, high-level language support, and processor control instructions.

The purpose of the following review is to give a brief overview of the X86 instruction set. You cannot use this information to start programming the X86 processors. Programming these processors requires more detailed information than presented in this chapter. *Note:* The X86 processors can work with single-bit, nibble, byte, word, double-word, and extended double-word data. In the following discussions on the X86 instructions the general term *word* should be read to mean this full range of word lengths or whatever length is applicable to the instruction being discussed.

Tables 10-2 through 10-10 only summarize the X86 instructions. Depending on the addressing mode, each instruction can be used many different ways. This lets each instruction "explode" into many different instructions to do many different jobs.

Table 10-2 shows the four different subtypes of X86 data transfer instructions. There are six *general-purpose data transfer instructions:*

MOV, POP, POPA, PUSH, PUSHA, and XCHG. PUSH and POP change the stack pointer value by the word length being transferred and then transfer a word between the source operand and the stack. POPA and PUSHA are two extensions of the POP and PUSH instructions which were introduced with the 286 processor. POPA and PUSHA exchange *all* the register data with the stack. That is, you use POPA to retrieve the contents of all the X86 registers with a single instruction. XCHG exchanges a word in the source operand with a word in the destination operand. XCHG cannot be used to exchange data with the segment register.

There are three *accumulator-specific data transfer instructions:* IN, OUT, and XLAT. IN and OUT use two kinds of X86 I/O port addressing. Direct I/O addressing uses data in the low-order instruction byte to address an I/O port (0–255). Indirect I/O addressing uses an I/O port number stored in the DX register. This 16-bit register allows the X86 processors to access one of 65,536 different I/O ports.

XLAT is a table look-up instruction. The value in AL is an index into a 256-byte table, and BX holds the table base address. In operation, the table index value in AL is replaced by the value in the memory location addressed. This is shown in Fig. 10-14 on page 246. XLAT is a good example of the complex X86 addressing techniques which give the programmer quick access to special sections of memory. In this case, a single instruction word quickly accesses any one of 256 different memory locations in the special table. If the value in the BX register or in the data segment register is changed, a number of different 256-byte tables can be set up in memory.

There are 19 different kinds of *address-object data transfer instructions.* Many of these instructions were introduced with the 286 or newer processors. Many are used to load or store (put data in or retrieve data from) the new registers required to support the segment descriptor tables and the new F and G data segment registers.

BSWP (byte swap) is used in the 486 processor to reorder the 4 bytes in a 32-bit word; that is, the bytes are arranged in ascending or descending order within the 32-bit word.

LAR loads the access rights from a segment descriptor table into a general-purpose register. This allows the programmer to test the access rights prior to using the segment descriptor information.

Table 10-2   The X86 Data Transfer Instructions

Mnemonic	Description	8088/ 8086	286	386	486	Instruction Subtype
MOV	Move source to destination	X	X	X	X	General
POP	Pop source from stack	X	X	X	X	General
POPA	Pop all		X	X	X	General
PUSH	Push source onto stack	X	X	X	X	General
PUSHA	Push all		X	X	X	General
XCHG	Exchange source with destination	X	X	X	X	General
IN	Input to accumulator (AL or AX)	X	X	X	X	Accumulator-specific
OUT	Output from accumulator (AL or AX)	X	X	X	X	Accumulator-specific
XLAT	Table lookup to translate a byte	X	X	X	X	Accumulator-specific
BSWAP	Byte swap (reorders 4 bytes in 32-bit word)				X	Address-object
LAR	Load access rights		X	X	X	Address-object
LDS	Load pointer (segment in DS) (displacement in register)	X	X	X	X	Address-object
LEA	Load effective address in register	X	X	X	X	Address-object
LES	Load pointer (segment in ES) (displacement in register)	X	X	X	X	Address-object
LFS	Load pointer (segment in FS) (displacement in register)			X	X	Address-object
LGDT	Load global descriptor table register		X	X	X	Address-object
LGS	Load pointer (segment in GS) (displacement in register)			X	X	Address-object
LIDT	Load interrupt descriptor table register		X	X	X	Address-object
LLDT	Load local descriptor table register		X	X	X	Address-object
LMSW	Load machine status word		X	X	X	Address-object
LSL	Load segment limit		X	X	X	Address-object
LSS	Load pointer (segment in SS) (displacement in register)			X	X	Address-object
LTR	Load task register		X	X	X	Address-object
SGDT	Store global descriptor table		X	X	X	Address-object
SIDT	Store interrupt descriptor table		X	X	X	Address-object
SLDT	Store local descriptor table		X	X	X	Address-object
SMSW	Store machine status word		X	X	X	Address-object
STR	Store task register		X	X	X	Address-object
LAHF	Load AH from flags	X	X	X	X	Flag
POPF	Pop flags off stack	X	X	X	X	Flag
PUSHF	Push flags onto stack	X	X	X	X	Flag
SAHF	Store AH in flags	X	X	X	X	Flag

LEA transfers data from a memory location to a general, pointer, or index register, LDS, LES, LFS, LGS, and LSS each transfer a 32-bit double word (a displacement address and a segment address) from the source operand (memory) into a pair of destination registers. The segment address is transferred into the instruction specified register. For example, LES places the segment data in the ES register. The displacement address is transferred into one of the general, pointer, or index registers.

The LGDT, LIDT, and LLDT instructions, and their complements SGDT, SIGT, and SLGT, transfer the contents of a segment

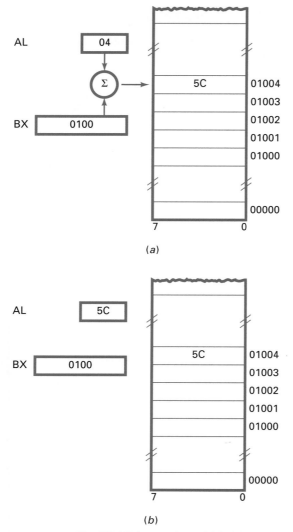

Fig. 10-14 The XLAT instruction which (*a*) uses the value in register AL to perform an offset look-up in the 256-byte table with a base value pointed to by register BX. (*b*) When the instruction is complete, the looked-up value (5C in this example) is put into register AL.

current machine status register contents into a general-purpose register. From there, the programmer can test the contents to determine the current MSW settings.

LSL (load segment limit) is used to copy the limit information from a particular segment descriptor table in memory into a general-purpose register. Once in the register, the information can be tested. The LTR and STR (load task register and store task register) instructions are used to load and store the task register. The task register is a special MMU register which points to the base of a stack (a segment) of information about each task running in the X86 environment.

There are four instructions used to transfer flag data. LAHF (load AH with flags) transfers the flag bits SF, ZF, AF, PF, and CF into the AH register. With the data in the AH register, you can, for example, use the rotate instruction to move specific bits into the carry, most significant bit, or other bits in the accumulator.

Only two data transfer instructions change settings in the flag register. These are the SAHF instruction, which stores the AH register in the flag register, and the POPF instruction, which places a byte from the stack into the flag register. SAHF (store AH into flags) allows you to store the contents of the accumulator in the flag register. When you do this, you must know what each bit means to the flag register. The original flag register contents are overwritten and therefore lost.

PUSHF (push flags) decrements the stack pointer register and transfers all flag bits onto the stack. This operation is reversed by POPF (pop flags), which transfers the indicated stack location into the flag register. The stack is always incremented or decremented for POPF and PUSHF instructions.

The X86 *arithmetic instructions* are broken down into the addition, subtraction, multiplication, and division subtypes. These instructions include the basic arithmetic operations ADD, SUB, MUL, and DIV and are summarized in Table 10-3. You can perform 8-, 16-, and 32-bit operations with most arithmetic instructions (32-bit instructions on 386 and newer processors). The multiplication and division instructions let you perform both signed and unsigned arithmetic (IMUL and MUL; IDIV and DIV).

In addition, the X86 processors allow you to perform arithmetic operations on unpacked decimal numbers. Figure 10-15 on page 248 shows how the X86 processors store packed

descriptor table (there are three types of segment descriptor tables—global, interrupt, and local) between the table and the corresponding segment descriptor registers in the X86 processor MMU.

Executing the LMSW (load machine status word) instruction with the bit 0 set places an X86 processor in the protected mode. LMSW is also used to set or clear the MSW bits 1–3. These bits indicate how numeric processor instructions are to be handled (in software if no coprocessor is present and by the coprocessor if it is installed). The complement to this instruction, SMSW (store machine status word), allows the programmer to copy the

Table 10-3   The X86 Arithmetic Instructions

Mnemonic	Description	8088/8086	286	386	486	Instruction Subtype
AAA	ASCII adjust for addition (unpacked BCD)	X	X	X	X	Addition
ADC	Add source to destination with carry	X	X	X	X	Addition
ADD	Add source to destination	X	X	X	X	Addition
DAA	Decimal adjust after addition (packed BCD)	X	X	X	X	Addition
INC	Increment operand by 1	X	X	X	X	Addition
AAS	ASCII adjust after subtraction (unpacked BCD)	X	X	X	X	Subtraction
CMP	Compare source to destination (no result except flag)	X	X	X	X	Subtraction
DAS	Decimal adjust after subtraction (packed BCD)	X	X	X	X	Subtraction
DEC	Decrement operand by 1	X	X	X	X	Subtraction
NEG	Negate (2's complement operand)	X	X	X	X	Subtraction
SBB	Subtract source from destination with borrow	X	X	X	X	Subtraction
SUB	Subtract source from destination	X	X	X	X	Subtraction
AAM	ASCII adjust after multiplication (unpacked BCD)	X	X	X	X	Multiplication
IMUL	Signed multiplication (source by destination)	X	X	X	X	Multiplication
MUL	Unsigned multiply (source by destination)	X	X	X	X	Multiplication
AAD	ASCII adjust before division (unpacked BCD)	X	X	X	X	Division
CBW	Convert byte to word	X	X	X	X	Division
CWD	Convert word to double word	X	X	X	X	Division
CWDE	Convert word to extended double word			X	X	Division
DIV	Unsigned division (accumulator by source)	X	X	X	X	Division
IDIV	Signed division (accumulator by source)	X	X	X	X	Division
MOVCX	Move with zero extend			X	X	Division
MOVSX	Move with sign extend			X	X	Division

and unpacked BCD numbers. Unpacked BCD numbers are found, for example, when you work with numbers derived from ASCII characters (the 4 most significant bits are set to zero on an ASCII character, which is a number). The packed BCD operations let you perform arithmetic operations on multiple BCD numbers packed into a single word.

The CBW (convert byte to word) instruction is used to convert a signed 8-bit number into a signed 16-bit number. Usually this is used prior to division. The CWD (convert word to double word) instruction performs the same function but converts a 16-bit signed word into a 32-bit signed word. Likewise, the CWDE instruction converts a 32-bit signed word into a 64-bit signed word. *Note:* The CBW and CWD/CWDE instructions are replaced in the 386 by the MOVSX and MOVCZ (move sign extended and move zero extended) instruc-

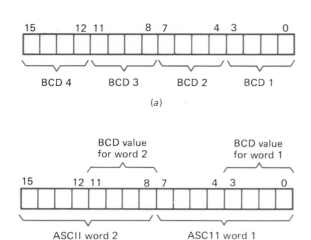

Fig. 10-15 **Storing BCD and ASCII characters in an X86 processor.** (*a*) Four BCD numbers are packed into a 16-bit word. (*b*) Two ASCII characters are packed into a 16-bit word. If these are numeric characters, the low-order nibble gives the BCD value for the number. Both BCD and ASCII characters can be processed using the X86 arithmetic instructions.

tions. These perform 8-bit to 16-bit, 16-bit to 32-bit, and 8-bit to 32-bit signed and unsigned extensions.

When executed, arithmetic operations affect (set or clear) six flag bits in the flag register. These flag bits are clear (CF), auxiliary carry (AF), zero (ZF), sign (SF), parity (PF), and overflow (OF).

The X86 processor *logical instructions* are summarized in Table 10-4(*a*). These can be either single- and double-operand instructions. The two-operand logical instructions include a TEST instruction. TEST performs the same logical operation as AND. However, only flags are affected—no result is returned.

The X86 *shift and rotate instructions* [Table 10-4(*b*)] are often shown as a subset of the logical instructions. The operation of the shift and rotate instructions is very much like the shift and rotate instructions for the 8-bit microprocessors. *Note:* Along with the introduction of the 32-bit word, the 386 introduced the double precision shift instruction.

Table 10-4(a)   The X86 Logical Instructions

Mnemonic	Description	8088/ 8086	286	386	486
AND	Logical AND (source and destination)	X	X	X	X
NOT	NOT operand (1's complement)	X	X	X	X
OR	Inclusive OR source and destination	X	X	X	X
TEST	Logical compare source and destination (no result but flags)	X	X	X	X
XOR	Exclusive OR source and destination	X	X	X	X

Table 10-4(b)   The X86 Shift and Rotate Instructions

Mnemonic	Description	8088/ 8086	286	386	486
RCL	Rotate through carry left	X	X	X	X
RCR	Rotate through carry right	X	X	X	X
ROL	Rotate left	X	X	X	X
ROR	Rotate right	X	X	X	X
SAL	Shift arithmetic left	X	X	X	X
SAR	Shift arithmetic right	X	X	X	X
SHL	Shift logical left	X	X	X	X
SHR	Shift logical right	X	X	X	X
SHLD	Double precision shift left			X	X
SHRD	Double precision shift right			X	X

String
manipulation
operations

Strings

Bit manipulation
instructions

Control transfer
instructions

Table 10-5   The X86 String Manipulation Instructions

Mnemonic	Description	8088/ 8086	286	386	486
INS	Input from port to string		X	X	X
OUTS	Output string data to port		X	X	X
CMPS	Compare source to destination string	X	X	X	X
LODS	Load string to AX or AL	X	X	X	X
MOVS	Move string source to destination	X	X	X	X
REP	Repeat while CX not equal zero	X	X	X	X
REPE	Repeat while CX zero	X	X	X	X
REPNZ	Repeat while CX not equal (not zero)	X	X	X	X
SCAS	Scan string for match with AL	X	X	X	X
STOS	Store AX or AL in string	X	X	X	X

The X86 *string manipulation operations* are like the Z80 block move, block search, and block compare instructions. These operations, summarized in Table 10-5, perform operations on a group of bytes or words called *strings*. The source and destination for two-operand instructions are pointed to by the source index and destination index registers. The X86 processors perform these string operations repeatedly if the instruction includes a repeat index. For repeat operation, the CX register holds a number which tells how many times the operation is to be repeated.

Very powerful X86 data manipulation operations can be created by combining repeated string operations with other operations. For example, the XLAT instruction performs a table lookup byte translation, which can be used to convert ASCII characters to EBCDIC characters. In this case, the ASCII character is an index into a table of 256 EBCDIC characters. When XLAT is combined with MOVS, a long string of ASCII data is converted into a long string of EBCDIC data. This complex conversion is done with just a few powerful X86 string instructions.

By including the SCAS instruction in this sequence, you can continuously check the string for the ASCII EOT (end of transmission) character. When you find EOT, it means that you are at the end of the ASCII file. You can use this to stop executing the instruction string. As you can well imagine, performing this kind of operation using an 8051, Z80, or 6802 requires many more instructions.

The *bit manipulation instructions* (Table 10-6) were introduced with the 386 processor. These instructions allow you to test or test and change a given bit in a given memory or register location. The TEST instruction sets or clears the carry bit so that it matches the status of the bit tested. Other instructions test and change the tested bit.

The *control transfer instructions* are broken down into the four subgroups shown in Table 10-7 on page 250. Like other microprocessor control transfer instructions these operations cause program execution to restart at a new

Table 10-6   The X86 Bit Manipulation Instructions

Mnemonic	Description	8088/ 8086	286	386	486
BSF	Bit scan forward			X	X
BSR	Bit scan reverse			X	X
BT	Bit test			X	X
BTC	Bit test and complement			X	X
BTR	Bit test and reset			X	X
BTS	Bit test and set			X	X

Table 10-7   The X86 Control Transfer Instructions

Mnemonic	Description	8088/ 8086	286	386	486	Instruction Subtype
SET (condition)*	Set conditional (sets byte to true or false based on condition)			X	X	Conditional
J (condition)*	Jump on condition	X	X	X	X	Conditional
JCXZ	Jump if CX equals zero	X	X	X	X	Iteration
JECXZ	Jump if ECX equals zero			X	X	Iteration
LOOP	Loop if CX does not equal zero	X	X	X	X	Iteration
LOOPE	Loop if CX does not equal zero and ZF = 1	X	X	X	X	Iteration
LOOPNE	Loop if CX does not equal zero and ZF does not equal zero	X	X	X	X	Iteration
INT	Interrupt	X	X	X	X	Interrupt
INTO	Interrupt on overflow	X	X	X	X	Interrupt
IRET	Interrupt return	X	X	X	X	Interrupt
CALL	Call procedure	X	X	X	X	Unconditional
JMP	Jump unconditionally	X	X	X	X	Unconditional
RET	Return from procedure	X	X	X	X	Unconditional

*The condition codes are:

O	Overflow
NO	No overflow
B/NAE	Below/not above or equal
NB/AE	Not below/above or equal
E/Z	Equal/zero
NE/NZ	Not equal/not zero
BE/NA	Below or equal/not above
S	Sign
NS	Not sign
P/PE	Parity/parity even
NP/PO	Not parity/parity odd
L/NGE	Less than/not greater or equal
NL/GE	Not less than/greater or equal
LE/NG	Less than or equal/greater than
NLE/G	Not less than or equal/greater than

memory location. With an X86 processor, this can be in a new code segment as well as a new location. It is important to remember that the X86 code segment is a fixed length depending on the processor. If you try to make a code segment longer than the segment length, the code will roll over the top of the segment to the bottom of the segment and begin to overwrite existing code in the lowest addresses.

The conditional transfer tests bits in the X86 processor flag register and then performs a relative jump. The relative jump is within −128 to +127 bytes of the current instruction pointer value when a 16-bit instruction is used. Greater range is available when a 32-bit instruction word is used. All conditional jump (JMP) instructions are one-word instructions, and so they can be executed very quickly.

Intrasegment jumps and calls (ones that start and end within the same code segment) use relative addressing. This allows the programmer to write position-independent code; that is, code which uses this relative addressing exclusively does not need to be modified just because you change its starting location in memory.

The unconditional transfers use two different kinds of addressing. The first kind transfers control within the current code segment. The second kind transfers control to a new code segment. Once you use JMP to transfer execution to a new code segment, it becomes he current code segment. The new execution address can be chosen by either direct or indirect addressing. If you use direct addressing, the new memory location is part of the instruction. If you use indirect addressing, you use any of the X86 memory addressing modes.

CALL and RET work together. When CALL executes, the next instruction's logical address is pushed onto the stack. If CALL switches execution to a new code segment, the next instruction code segment register value is pushed onto the stack first. This is followed by the instruction's logical address.

Once the stack operations are finished, execution begins at the new address specified by CALL. Execution RET reverses the process. Execution returns to the instruction pushed onto the stack. If a new code segment was called, executing RET starts execution in the old code segment at the old logical address. Using CALL and RET is different from using JMP because you can leave the current code segment and return.

The JMP instruction permanently transfers program execution to a new sequence of instructions. There are two forms of the JMP instruction: unconditional and conditional. The unconditional JMP instruction lets the programmer move program execution to a new address anywhere in the X86 processor's memory space. Figure 10-16(a) to (e) on page 252 shows five different types of unconditional jump statements. Each type lets you choose the length of the JMP instruction, depending on how far away the memory location to jump to is located. These five jumps can be divided into two intersegment jumps (jumps anywhere in the X86 memory address space) and three intrasegment jumps (jumps inside the current 64-kbyte code segment).

The 2-byte jump short (intrasegment direct) instruction restarts execution within −128 to +127 bytes of the current instruction. The 3-byte intrasegment direct instruction restarts execution anywhere within the 64 kbytes of the current code segment. A second 2-byte jump instruction (intrasegment indirect) restarts execution at the address stored in one of the X86 registers. These three operations are shown in Fig. 10-16(a) to (c).

The 5-byte jump far instruction (intersegment direct) gives two new byte values for the code segment register and two new byte values for the instruction pointer. This 5-byte instruction completely changes all registers which control the memory location from which the X86 processor executes instructions. A 2-byte jump far instruction (intersegment indirect) restarts execution at the address defined by the register specification contained in the second

byte. These three operations are shown in Fig. 10-16(d) and (e).

The X86 processors have a subtype of control transfer instructions we have not discussed: the *iteration control instructions*. They perform leading-decision and trailing-decision loop control. Different versions of these instructions are used to make a jump just before or just after the instructions in a loop are executed. The iteration control instructions also use single-byte relative addressing. This means that the destination address must be within −128 to +127 bytes of the initial instruction.

The iteration control instructions use the X86 CX register as a "count" register. At the start of the instruction, the CX register is loaded with the count value. The difference between each of the iteration instructions is in how they use the count register and the zero flag to control the decision process. Figure 10-17(a) to (d) on page 253 diagrams the operation of these iteration control instructions. Each instruction starts by decrementing the current value in CX. Then, tests are made on CX and, in some instructions, on the ZF flag. The change in instruction execution sequence is decided by these tests.

LOOP and JCXZ [Fig. 10-17(a) and (b)] test only CX. With the LOOP instruction, if CX is not zero, execution transfers to the new relative value. Otherwise, the instruction following LOOP is executed. With JCXZ, if CX is zero, control transfers to the new address. This instruction is used at the beginning of a loop to bypass the loop if the value in CX is zero from a previous operation. This instruction allows you to test the value of a flag before beginning a loop. The other instructions test the flag after the operation has passed through the loop.

LOOPZ and LOOPNZ [Fig. 10-17(c) and (d)] test both CX and ZF. LOOPZ makes a transfer happen if CX is not zero *and* the ZF flag is set. This creates a loop while not zero or loop while not equal instruction.

Like the 6802, the X86 processors let you create a software interrupt. All X86 interrupts transfer control after pushing the flag registers onto the stack and performing an indirect intersegment call. The X86 processors handle interrupts in two ways depending on the processor mode (real or protected). In the real mode, the interrupt call is an indirect intersegment call using an interrupt vector table located in absolute memory locations 0 through

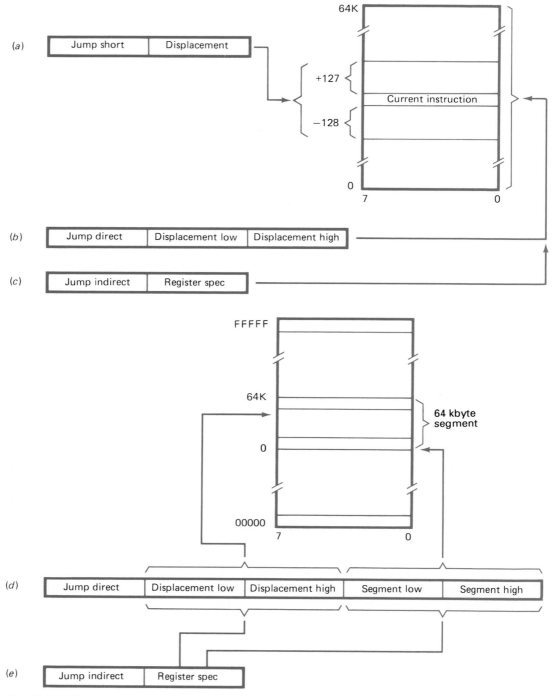

Fig. 10-16 Five unconditional jumps. (*a*) Jump to − 128 to + 127 bytes of the current instruction. (*b*) Jump to any point within the current 64-kbyte code segment. (*c*) Jump to any point within the current 64-kbyte code segment as pointed to by a register value. (*d*) Jump to any point within the memory space using a new segment register value and a new displacement value. (*e*) Jump to any point within the memory space using new segment register values and new displacement values pointed to by the registers in the instruction.

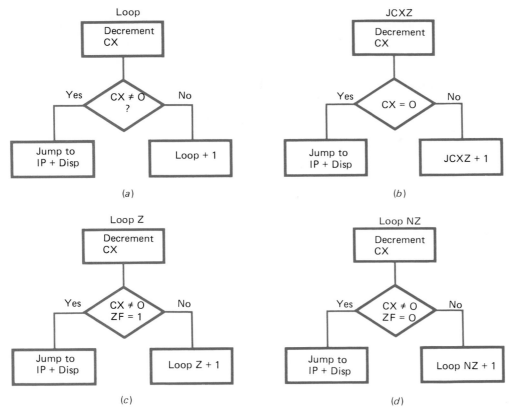

Fig. 10-17 Four iteration control instructions. (*a*) Looping continues until the CX register contains a nonzero value. (*b*) Looping continues until the CX register value is zero. (*c*) Looping continues until the CX register contains a nonzero value *and* the zero flag (ZF) is set. (*d*) Looping continues until the CX register value is zero *and* the zero flag (ZF) is cleared.

3FF. Each interrupt vector is made up of 4 bytes. The 4 bytes are used to provide new code segment register and instruction pointer values. In the protected mode, the interrupt call points to an interrupt descriptor table. The contents of this table are used to point to the interrupt processing routine. The X86 vector table has 256 different interrupt vectors. The X86 interrupt vectors are shown in Table 10-8 on page 254.

The two software interrupt instructions (INT and INTO) cause the same sequence of processor operations as an external interrupt. INT is a general-purpose software interrupt instruction. When executed, INT pushes the flag registers onto the stack, clears the TF and IF flags, and transfers program execution with an indirect call to one of the 256 vectors.

INTO is a special-purpose software interrupt instruction which executes only when the OF flag (trap on overflow) is set. When it executes, INTO pushes the flag registers onto the stack, clears the TF and IF flags, and then transfers

program execution by calling the fourth vector. If the trap on overflow flag is cleared, no software interrupt takes place.

Executing the IRET instruction (interrupt return instruction) transfers control to the return address saved by an interrupt operation. It restores the saved flag register, code segment register, and instruction pointer. Execution continues at the instruction following the instruction executing when the interrupt happened.

Table 10-9 on page 254 shows five instructions which are the *high-level language support instructions*. The BOUND instruction checks the contents of a register against the contents of any two memory locations and generates an exception (internal interrupt) if the contents of the register are outside the bounds specified. Normally, this is used to check the boundaries of a section of memory to ensure that the addressing is not out of bounds. CMPXCHG is a 486 instruction which is used to exchange and then compare two

Table 10-8   The X86 Interrupt Vector Assignment

Interrupt Vector Number	Interrupt Type	Instruction or Condition Which Can Cause Interrupt	First Processor to Use This Interrupt
0	Divide error	DIV or IDIV (division overflow)	8088/8086
1	Debug exception	Any instruction if the T (trap) flag is set	8088/8086
2	NMI interrupt	Logic 1 at the NMI input	8088/8086
3	1-byte interrupt	INT	8088/8086
4	Interrupt on overflow	The INTO checks the overflow bit	8088/8086
5	Array bounds check	The BOUND instruction checks for out-of-bounds data	286
6	Invalid op code	An instruction the processor does not recognize	286
7	Device not available	Either the ESC or WAIT instructions	286
8	Double fault	Two interrupts are caused by an instruction	286
9	Reserved		
10	Invalid TSS	Invalid TSS (segment) is addressed	286
11	Segment not present	Segment register instructions referencing a nonexistent segment	286
12	Stack fault	Stack references to nonexistent stack or outside of stack limit	286
13	General protection fault	Any memory reference which exceeds the limit, violates write-protect, privilege level, violates execute only, etc.	286
14	Page fault	Any memory access or code fetch	386
15	Reserved		
16	Floating point error	Floating point, system generates an error	486
17	Alignment check interrupt	Unaligned memory access	486
18–31	Reserved		
32–255	Interrupt	Interrupt vectors 32 to 255 as defined by user	8088/8086

16-bit words. ENTER and LEAVE are two instructions used to help high-level languages build the complex stacks they require. ENTER loads the information on the stack, and LEAVE removes the data from the stack. XADD (exchange and add) exchanges the high and low words in a 32-bit word and then performs an addition.

The last group of instructions is the *processor control instructions*, shown in Table 10-10. You can think of them as the X86 housekeeping instructions. For example, the seven flag oper-

Table 10-9   The X86 High-Level Language Support Instructions

Mnemonic	Description	8088/ 8086	286	386	486
BOUND	Check array bounds		X	X	X
CMPXCHG	Compare and exchange (reorder 16-bit data and compares)				X
ENTER	Make a stack frame		X	X	X
LEAVE	Procedure exit		X	X	X
XADD	Exchange and add				X

ations let the programmer set or clear individual flags in the flag register. Typically, these operations are used to initialize the flag register before some other operation.

The HLT instruction causes the X86 processor to halt. The halt state is cleared by hardware signals at the X86 RESET input or if an interrupt signal happens on an enabled external interrupt or on the X86 nonmaskable (NMI) interrupt.

The WAIT (processor wait) instruction causes the X86 processor to enter a wait state if the signal on its $\overline{\text{TEST}}$ pin is not asserted. The wait state is interrupted by an external interrupt. If this happens, the processor wait instruction is saved. Thus, when the interrupt return is executed, the X86 reenters the wait state. If $\overline{\text{TEST}}$ is asserted, the processor does not enter the wait state but continues executing in its normal sequence. The WAIT instruction allows the processor to synchronize external hardware.

ESC lets the X86 processor work with other processors such as the matching numeric coprocessor. When an X86 processor executes the ESC instruction, the instruction operand is fetched. This action puts the instruction on the X86 data bus. However, the X86 processor takes no other action with this information.

The other processors, such as the numeric coprocessor, can use (execute) this instruction. The X86 processor executes an NOP while the other processors are working with the data.

As each of the newer X86 processors introduced new functions, new instructions were provided to allow the programmer to work with the functions. These instructions are listed in Table 10-10.

As you can see, the X86 processors have an extremely complex instruction set. Their wide range of instructions is further complicated by a very wide range of addressing modes. However, both these qualities make it possible for the X86 processors carry out extremely complex computational processes.

Self-Test

Answer the following questions.

31. The X86 stack operations increment or decrement the stack pointer each time the stack is used. If you push or pop one of the data segment registers, you would expect the stack pointer to increment by
    a. 2 bytes   *c.* 8 bytes
    b. 4 bytes   *d.* 16 bytes

Table 10-10   The X86 Processor Control Instructions

Mnemonic	Description	8088/ 8086	286	386	486	Instruction Subtype
ESC	Escape	X	X	X	X	External sync
HLT	Halt	X	X	X	X	External sync
LOCK	Lock the bus	X	X	X	X	External sync
NOP	No operation	X	X	X	X	External sync
WAIT	Wait for test pin	X	X	X	X	External sync
CLC	Clear carry flag	X	X	X	X	Flag
CLD	Clear direction flag	X	X	X	X	Flag
CLI	Clear interrupt flag	X	X	X	X	Flag
CMC	Complement carry flag	X	X	X	X	Flag
STC	Set carry flag	X	X	X	X	Flag
STD	Set direction flag	X	X	X	X	Flag
STI	Set interrupt flag	X	X	X	X	Flag
ARPL	Adjust requested privilege level		X	X	X	General control
CLTS	Clear task-switched flag		X	X	X	General control
INVD	Invalid data cache				X	General control
INVLPG	Invalid TLB entry				X	General control
VERR	Verify read (permission to read segment)		X	X	X	General control
VERW	Verify write (permission to write segment)		X	X	X	General control
WBINVD	Write back and invalidate data cache				X	General control

32. One way to use two stacks and quickly switch between the two stack pointers is to use the _____ instruction.
    a. MOV   c. PUSH
    b. POP   d. XCHG

33. IN, OUT, and XLAT all have a common characteristic because
    a. They use the accumulator (AX or AL) for one part of their data transfer
    b. Each instruction is limited to 8-bit transfers
    c. All three instructions can work with either the accumulator or a memory location
    d. None of the above

34. The X86 arithmetic instructions are different from the instructions of other microprocessors we have studied because they let us perform arithmetic operations on
    a. Signed and unsigned binary numbers (multiplications)
    b. Bytes, words, and long words
    c. Unpacked (ASCII) and packed BCD
    d. All of the above

35. The X86 TEST instruction performs the same function as the _____ instruction, but there is no result other than setting flags.
    a. XOR   c. AND
    b. SUB   d. OR

36. You would expect that the SHLD instruction first appeared on the 386 processor because
    a. The 386 uses 32-bit segment registers
    b. The 386 uses 32-bit general-purpose registers
    c. There is no need to shift 32 times in a 286
    d. The SHLD and SHRD first appeared on the 286

37. The string manipulation instruction lets you
    a. Move a block of data from one set of memory locations to another set of memory locations
    b. Scan a block of data to see whether a byte or word matches a byte or word stored in the accumulator
    c. Store a constant byte or word value from the accumulator in all byte or word locations in a block of data
    d. All of the above

38. The REP (repeat) series of instructions let you
    a. Repeat an instruction indefinitely until an interrupt or a wait is received
    b. Fill memory with a constant value
    c. Perform a binary division using repeated subtraction
    d. Execute a sequence of instructions until decrementing the CX register causes a given condition

39. All the X86 jump instructions except JMP use
    a. Relative addressing
    b. Direct addressing
    c. Register indirect addressing
    d. Any of the above addressing plus any other kinds of addressing available to an X86 processor instruction

40. The CALL instruction can be used to call a subroutine or procedure within the _____ code segment.
    a. Same or different
    b. Same
    c. Different
    d. Data

41. An interrupt is much like calling a subroutine, except that the interrupt is caused by an external hardware input. Once the interrupt is happening and the interrupt service routine is executing, you must execute _____ to return to the program originally interrupted.
    a. A JMP instruction
    b. An IRET instruction
    c. A RET instruction
    d. Any of the above

42. The purpose of the X86 WAIT instruction and its accompanying $\overline{\text{TEST}}$ pin is
    a. To provide an alternative form of interrupt
    b. To allow slower-responding external devices to synchronize the processor
    c. To provide an input which is exclusively used by memory systems to slow the processor down to their speed
    d. All of the above

43. If you want an X86 processor to fetch an instruction and place it on its bus but do not want the instruction executed, you would use _____ with the memory location of the desired instruction as the operand for this instruction.
    a. STC   c. ESC
    b. NOP   d. WAIT

10-6 THE X86 HARDWARE

In this section, we will look at the X86 as an integrated circuit. We will look at the signals which go into and out of the X86 processors, and we will learn how they are used to build a working advanced microprocessor-based system. Finally, we will briefly look at the X86 processors in their most common usage—as the processors for PCs.

There are significant external hardware differences between the 8088/8086 and the newer X86 advanced microprocessors. Both the 8088 and the 8086 are available in 40-pin dual in-line packages (DIPs). The 286 and newer processors have too many connections to allow them to be packaged in a DIP; therefore they have been packaged so that they can have much higher external pin counts.

In order to fit into a 40-pin DIP, the 8088 and 8086 processors use a multiplexed address/data bus. All the newer processors have a high pin count package and therefore use separate pins for the address and data bus lines. In addition, the 8088 and 8086 have a minimum and maximum mode. These modes allow the processors to be used with minimum or maximum external hardware. This is of great use to a designer who wishes, for example, to use the power of X86 processing but needs to build a system with the minimum hardware possible. This feature was dropped from the 286 and from later versions of the X86 processor family.

Figure 10-18 on page 258 shows the pin assignment and pinout diagrams for the 8088 and 8086 processors. The pin assignment diagrams group the different kinds of signals to show what connections must be made to the processor before it can become part of a working microprocessor system. A pinout diagram shows where all the different signals are connected to the package. Because there are so many package variations, you need to refer to the manufacturer's data sheet to find the specific pinouts for the specific processor you are working with.

Figure 10-18(a) shows that the connections can be viewed as seven groups of signals

- 8-line multiplexed address/data bus
- 8-line high-order address lines (A8 to A15)
- 4-line high-order multiplexed addressed lines (A16 to A19) and status bus
- System control lines
- CPU control lines
- Bus control lines
- Clock and supply lines

In the following paragraphs, we will review each of these groups. *Note:* Many of the signals are shown with an overbar. As with other microprocessors, this tells you that these signals are TRUE (asserted) when they are at their lowest electrical level. Also, some 8088 and 8086 signals are shown with two different descriptions. The X86 processor operating mode (maximum or minimum) makes the difference in their description.

As mentioned earlier, the 8088 and 8086 processors can operate in the minimum mode or in the maximum mode. In the minimum mode, the signals let the microprocessor work with minimum external hardware. In the maximum mode, the processors use an external bus controller IC, which decodes the status lines ($\overline{S0}$, $\overline{S1}$, and $\overline{S2}$) and provides all the bus control signals. The 8088 and 8086 processors must be operated in the maximum mode when features such as the numeric coprocessor and special bus options are used. Virtually all applications where the 8088 is used as the processor for a PC are operated in the maximum mode.

The 8088 requires 20 address lines to access 1 Mbyte of memory. If 20 individual pins were to be dedicated to address lines, and 8 pins were dedicated to the data bus, only 12 additional lines would be available for all the other lines if a 40-pin package was used. This is not enough. For this reason, the 8088 and 8086, like the 8051 and 8085, use a multiplexed address/data bus.

The 8088 low-order address lines (A/D_0 to A/D_7) have two functions. During the address portion of a memory access cycle, they hold address information. The 8088/8086 ALE signal (a system control signal) tells the system hardware when to latch this information. During the data transfer portion of the memory address cycle, these lines are the 8088/8086 8-bit data bus. Address lines A_8 to A_{15} serve only a single function on the 8088. However, these lines also work as a multiplexed address/data bus on the 8086, which uses a 16-bit data bus instead of an 8-bit data bus.

The four highest address lines (A_{16} to A_{19}) are multiplexed during the last part of a memory address cycle (after the memory address information is latched) and during I/O operations. These lines are not used during I/O operations because the 8088 and 8086 use only 16 I/O address lines. During an I/O operation and during the last part of the processor cycle, these lines hold status information. At this time, they convert to the S6, S5, S4, and S3 status line.

The S6 status line is always low, and S5 shows the status of the interrupt enable flag.

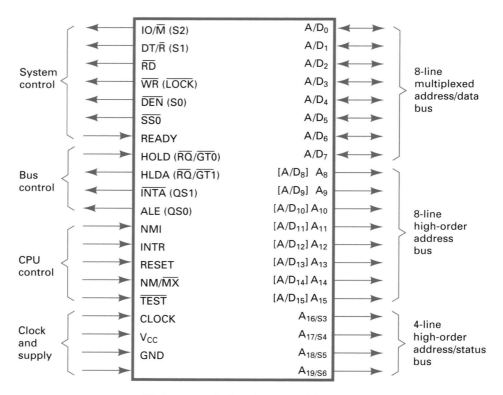

Minimum mode (maximum mode)

(a)

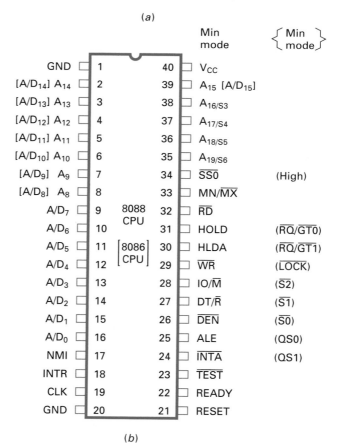

(b)

Fig. 10-18   The 8088 and 8086 (8086 differences noted in brackets) with their inputs and outputs grouped by function. (b) The 8088 and 8086 in a 40-pin DIP. *Note:* Two ground pins are used to give proper return for this device. *Adapted from Intel 8088 and 8086 data sheets.*

The combination of S4 and S3 can be decoded to show which memory segment is being addressed. Figure 10-19 shows how S3 and S4 are decoded. This information is used by some external devices to tell which segment register is accessing data.

The system and control signals tell memory and I/O devices when and how to use the signals on the memory address and data buses. IO/$\overline{\text{M}}$ tells external devices that the 8088/8086 is performing an I/O or memory transfer. The DT/$\overline{\text{R}}$ (data transmit/receive) signal is used in large hardware systems which need data bus transceivers. This signal controls the direction of data flow through external data bus transceivers.

$\overline{\text{RD}}$ is the read strobe. It indicates that the processor is performing memory or I/O reads. The signal is used to read devices connected to the 8088/8086 local bus. $\overline{\text{WR}}$ is the write strobe. It indicates that the processor is performing an I/O or memory write. When the 8088/8086 is used in maximum mode, this signal becomes $\overline{\text{LOCK}}$, which tells other devices on the system that they cannot gain control of the system bus. The $\overline{\text{LOCK}}$ signal is activated by a LOCK instruction prefix. It remains active until the completion of the next instruction.

The $\overline{\text{DEN}}$ (data enable) provides an output enable signal for minimum mode systems with a bus transceiver. In a maximum mode system, the $\overline{\text{DEN}}$, DT/$\overline{\text{R}}$, and IO/$\overline{\text{M}}$ lines become the $\overline{\text{S0}}$, $\overline{\text{S1}}$, and $\overline{\text{S2}}$ status lines. These three lines are used by a bus controller to generate all memory and I/O access and control signals. The $\overline{\text{SS0}}$ minimum mode status line is logically equivalent to the $\overline{\text{S0}}$ line used in the maximum mode. When decoded as shown in Fig. 10-20, these three lines provide more detailed information on the current bus cycles. In the maximum mode $\overline{\text{S0}}$ is always logic 1.

S4	S3	Register usage	
0	0	Alternate data	ES
0	1	Stack	SS
1	0	Code	CS
1	1	Data	DS

Fig. 10-19 **Decoding the S4 and S3 signals. This information, which is available during the data transfer part of the processor cycle, lets an external device know which segment register is being used to address memory.**

$\overline{\text{S2}}$	$\overline{\text{S1}}$	$\overline{\text{S0}}$	Status
0	0	0	Interrupt acknowledge
0	0	1	Read I/O port
0	1	0	Write I/O port
0	1	1	Halt
1	0	0	Code access
1	0	1	Read memory
1	1	0	Write memory
1	1	1	Passive

Fig. 10-20 **Decoding the $\overline{\text{S2}}$, $\overline{\text{S1}}$, and $\overline{\text{S0}}$ status lines in maximum mode. These three lines give an external device such as the 8288 bus controller eight different status conditions for the 8088/8086. The 8288 uses this information to properly manage external devices on the address and data buses.**

The last system control line is the READY input. External devices use READY to tell the 8088/8086 that they are ready to complete a data transfer. This signal can be used by slower devices to slow the 8088/8086 processor to their speed.

The CPU control lines are very similar to those used on the Z80. There are simply more of them, and they provide more sophisticated control. HOLD is an input which operates in conjunction with the HLDA (hold acknowledge) output. A logic 1 on hold indicates that another master is requesting the 8088/8086 local bus. The processor acknowledges on the HLDA lines and then tristates the local bus and control lines. In the maximum mode, HOLD and HLDA become the bus request-grant lines. They perform the same function with slightly different timing.

The $\overline{\text{INTA}}$ (interrupt acknowledge) output tells certain devices that the 8088/8086 (in minimum mode) is acknowledging an interrupt. As noted earlier, the ALE (address latch enable) is an 8088/8086 signal which tells external memory or I/O devices to latch the address information on the multiplexed address bus. In maximum mode $\overline{\text{INTA}}$ and ALE become two status lines (QS1 and QS0). These lines provide a two-bit status word which tells external devices the current status of the 8088/8086 instruction queue. This is important information for coprocessors and other devices which must work closely with the 8088/8086.

NMI is the 8088/8086 edge-triggered nonmaskable interrupt. A transition from logic 0 to logic 1 causes an interrupt at the end of the current instruction. The INTR (interrupt request) is a level (active high) triggered interrupt input. INTR is masked by setting the interrupt enable bit. When the 8088/8086 receives an external interrupt from NMI, it immediately begins processing the interrupt. If the interrupt is received on INTR, the 8088/8086 first checks to see whether the interrupt is masked by a logic 0 in the flag register IF (interrupt flag) flag.

When the 8088/8086 processes an external interrupt, it must receive the interrupt vector from the external device. To do this, it reads the data bus and uses this 8-bit value as a pointer into a 256-vector interrupt vector table (see Table 10-8). As the interrupt table shows, many more situations than having either the NMI or the INTA lines asserted can cause the start of interrupt processing.

For example, executing the BOUND instruction may cause the start of an interrupt procedure if the test performed by the BOUND instruction finds the tested data to be greater than the upper bound or less than the lower bound. This could be one way to test a memory address to see whether it exceeds the limit of a segment.

RESET causes the processor to stop all current operations. The MN/$\overline{\text{MX}}$ (minimum/maximum) input sets the processor's minimum/maximum mode. Usually this line is either permanently connected to a logic 0, putting the processor in the maximum mode, or to a logic 1, putting the processor in the minimum mode.

The $\overline{\text{WAIT}}$ (wait for test) instruction looks at the $\overline{\text{TEST}}$ input. If the $\overline{\text{TEST}}$ input is asserted, execution continues. Otherwise, the processor waits in an idle state.

The 8088/8086 clock signal provides basic timing for the processor and bus controller. This is a rectangular waveform with a 33 percent duty cycle. Both 5-MHz and 8-MHz versions of the 8088/8086 are available.

The X86 processors typically use +5V dc. However, there are 3-V dc versions as well. The processors have multiple power and ground pins. All of these must be connected in order to ensure a solid power and return path for the power supply.

A number of support ICs have been developed to work with the 8088/8086. In addition, the 8088 and 8086 processors work with a wide range of 8085 support ICs which use a similar multiplexed address/data bus.

As we discussed at the beginning of this section, the 286 and newer processors are

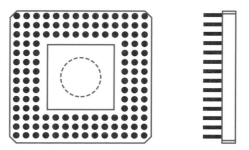

Fig. 10-21   A 114-pin pin grid array package used for advanced microprocessors and peripheral devices. *Note:* **Normally not all of the pins are dedicated to signals, because there are usually numerous V_{cc} and ground pins to ensure that current flow through any one pin is not too high and to distribute power to various parts of the integrated circuit.**

packaged quite differently from the 8088 and 8086. These packages have from 60 to nearly 300 pins depending on the specific package and the needs of the processor. Three packages are common. Figure 10-21 shows a typical pin grid array package, Fig. 10-22 shows a typical quad flat pack, and Fig. 10-23 shows a typical plastic leadless chip carrier (PLCC). The number of pins and the exact pinout depend on the processor (or peripheral part) you are working with. For example, the 286 is available in either a 68-pin pin grid array or a 68-pin quad flat pack. However, the 486DX2 requires a 168-pin pin grid array.

Table 10-11 shows some of the packages used for the different members of the X86 family. The pin grid array is designed to allow the device to be plugged in to a special pin grid array socket, and therefore it can be

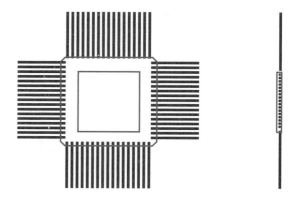

Fig. 10-22   A 132-pin quad flat pack. *Note:* **The side profile shows that the pins are formed to lie flat on printed circuit board pads. This is a surface mount device, and therefore the pins do not go through the printed circuit board as they do with a DIP.**

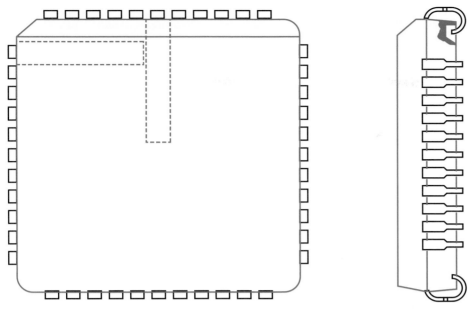

Fig. 10-23 **A 132-pin plastic leadless chip carrier.** *Note:* **The side profile shows that the pins are really just metal contacts on the side of the package. The PLCC package drops into a special socket designed to accept it.**

removed if necessary. Likewise, the PLCC is plugged into a special socket. The quad flat pack is designed to be soldered directly to the printed circuit board and is not generally considered to be a replaceable part. Quad flat packs can be removed and replaced with special tools, but they are certainly not field-serviceable.

Once the X86 processors broke the bounds of the DIP, the number of available pins went up significantly. This allowed significant changes in the external connections. First, the 286 and newer processors do not use a multiplexed address-data bus. This means that there are 40 pins on the 286 dedicated to the address and data lines and 64 pins on the 386DX and 486DX dedicated to the address and data lines. The Pentium dedicates 64 pins

to data alone and is packaged in a 273-pin pin grid array.

Table 10-12 on page 262 shows pin assignments for members of the X86 family. For reference, the first processor column shows the 8086 pins when the processor is used in the maximum mode. The following paragraphs describe some of the more unique pins found on these processors. As you can see from Table 10-12, many of the basic pins are common to all members of the X86 family. Common pins include such functions as address, data, RESET, interrupt (INTA and NMI), memory/IO select, and clock. You should note, however, that not all these lines operate quite the same. Their general function is the same, but there may be specific differences in timing, assertion level, and functional details.

Table 10-11 **The X86 Processor Packages and Pin Count**

Attribute	8088	8086	286	386SX	386DX	486DX	Pentium
Data lines	8	16	16	16	32	32	64
Address lines	12	4	24	32	32	32	32
Dual in-line package (DIP)	40 pin	40 pin	N/A	N/A	N/A	N/A	N/A
Pin grid array	68 pin	68 pin	68 pin	100 pin	132 pin	168 pin	273 pin
Quad flat pack and PLCC	44 pin	44 pin	68 pin	100 pin	132 pin		

Table 10-12   The X86 Processor Pin Assignment

Processor Signal Function	Input or Output	8086 (Max Mode)	286	386	486
Clock	Input	CLOCK	CLK	CLK2	CLK
Data bus	Input or output	A/D_0–A/D_{15}	D_0–D_{15}	D_0–D_{31}	D_0–D_{31}
Address bus	Output	$A_{16}/S3$–$A_{19}/S6$	A_0–A_{23}	A_2–A_{31} $\overline{BE0}$–$\overline{BE3}$	A_2–A_{31} $\overline{BE0}$–$\overline{BE3}$
Bus high byte enable	Output	$\overline{BHE}$	$\overline{BHE}$		
Bus cycle status	Output	$\overline{S0}$, $\overline{S1}$	$\overline{S0}$, $\overline{S1}$		
Memory I/O select	Output	$M/\overline{IO}$	$M/\overline{IO}$	$M/\overline{IO}$	$M/\overline{IO}$
Code/interrupt acknowledge	Output		$COD/\overline{INTA}$		
Bus lock	Output	$\overline{LOCK}$	$\overline{LOCK}$	$\overline{LOCK}$	$\overline{LOCK}$
Bus ready	Input	$\overline{READY}$	READY	READY	RDY
Bus hold request and	Input	$\overline{RQ/GT0}$	HOLD	HOLD	HOLD
bus hold acknowledge	Output	$\overline{RQ/GT1}$	HLDA	HLDA	HLDA
Maskable interrupt	Input	INTR	INTR	INTR	INTR
nonmaskable interrupt	Input	NMI	NMI	NMI	NMI
Processor extension operand	Input		$\overline{PEREQ}$	$\overline{PEREQ}$	
request and acknowledge	Output		$\overline{PEACK}$		
Processor extension busy	Input	$\overline{TEST}$	$\overline{BUSY}$	$\overline{BUSY}$	
and error	Input		$\overline{ERROR}$	$\overline{ERROR}$	
System reset	Input	RESET	RESET	RESET	RESET
Bus cycle write/read	Output	$\overline{RD}$		$W/\overline{R}$	$W/\overline{R}$
System status	Output	S0, S1, S2			
Queue status	Output	QS0, QS1			
Minimum/maximum mode	Input	$MN/\overline{MX}$			
System ground	Input	V_{SS}	V_{SS}	V_{SS}	V_{SS}
System power	Input	V_{CC}	V_{CC}	V_{CC}	V_{CC}
Address status	Output			$\overline{ADS}$	$\overline{ADS}$
Data/control	Output			$D/\overline{C}$	$D/\overline{C}$
Bus size 16	Input			$\overline{BS16}$	$\overline{BS16}$
Bus size 8	Input				$\overline{BS8}$
Next address	Input			$\overline{NA}$	
Even parity generation and detection	Input or output				DP0–DP3
Parity check	Output				$\overline{PCHK}$
Pseudo lock	Output				$\overline{PLOCK}$
Burst ready	Input				$\overline{BRDY}$
Burst last	Output				$\overline{BLAST}$
Internally generated bus request	Output				BREQ
Back off (float bus on next request)	Input				$\overline{BOFF}$
Address hold (allows access to address bus by another device	Input				$\overline{AHOLD}$
External address applied	Input				EADS
Cache enable (cache can fill from external data)	Input				$\overline{KEN}$
Flush (erase) cache	Input				$\overline{FLUSH}$
Page write through and	Output				PWT
page cache disable	Output				PCD
Floating point error	Output				$\overline{FERR}$
Ignore numeric error	Input				$\overline{IGNNE}$
Address bit 20 mask	Input				$\overline{A20M}$
Upgrade present	Input				UP
Test ports (for automatic	Input				TCK, TDI, TMS
device testing)	Output				TDO

The 386 and 486 introduced two lines to let an external device tell the processor the size of data it can supply. When the 486 $\overline{BS8}$ input is asserted, for example, the bus takes four cycles to input data from the peripheral because the processor knows that the data is not 32 bits wide but, rather, will come as four 8-bit bytes.

The 486 also introduced parity checking on each byte of input or output data. The four parity check lines (DP0 to DP3), for example, output signals to tell whether the parity is odd or even for each of the 4 bytes of input data. Most of the additional 486 lines, shown in the lower section of Table 10-12, indicate the status or allow some control of the cache memory and floating point unit. The 486 also provides four pins which allow a test system to gain access to the 486 to perform functional testing of the processor.

As discussed earlier, the 8088 can be used in either the maximum or the minimum mode. Two 8088 configurations are shown in Figs. 10-24 and 10-25 on pages 264 and 265. Although these configurations are for a specific member of the X86 family, you can extend them to cover newer processors. The main difference you will find between these configurations and configurations you might find using a 286 or 386 processor will be in the use of custom ICs. Most applications today use custom ICs rather than stock ICs to perform many of the interfacing functions. This is especially true as you begin to work with the newer PCs. Here you will find a few very complex VLSI components which, with a 386, for example, will create a complete PC. However, the functions you will find performed in these ICs are on a high level, very much like those performed in the following examples.

Figure 10-24 shows a very simple 8088-based controller. This is typical of a system you might find in a factory floor controller application. The program for this controller is held in a 2-kbyte EPROM, and the system uses a 256-byte scratch-pad RAM. The controller connects to external devices through four 8-bit parallel I/O ports. A memory map of the system is shown in Fig. 10-24(b).

The system uses a mix of 8088 and 8085 support ICs. The system clock comes from a 8284A clock generator. This 8088 support IC produces the proper square wave for the 8088 clock input. Two 8085 support devices are used: the 8156 and 8755A. The 8156 is a programmable 14-bit binary counter-timer with two programmable 8-bit I/O ports, one programmable 6-bit I/O port, and a 256-bit static RAM. The 8755 is a 2-kbyte EPROM with two 8-bit parallel I/O ports.

Although this system will not carry out complex functions because it has only limited RAM and EPROM, it is a good illustration of a minimum 8088-based system. Such a controller could be used, for example, to input 16-bit parallel data, perform a code conversion and error check, and output 16-bit parallel data. It can do all this at very high speeds. The 6-bit I/O lines can provide the necessary handshaking with the parallel ports interfaced by the four 8-bit data ports.

Figure 10-25 shows a block diagram of the 8088 system operating in the maximum mode. This system uses an 8284A clock generator to create the timing signals needed by the rest of the system. The 8288 bus controller decodes the $\overline{S0}$, $\overline{S1}$, and $\overline{S2}$ signals and generates the system control signals. Three 8282s latch the address lines to convert the multiplexed address-data bus into latched address lines for the memory and I/O. The 8286 octal transceiver buffers the 8088 data bus.

The bus between the 8088 and the support devices is the 8088 local bus. The bus on the other side of these devices is a fully buffered system bus, which supports ROM, RAM, parallel ports, mass storage controllers, serial communications controllers, support ICs, and other functions. There is no need for latch addressing or special data bus control on the fully buffered side of the bus transceivers. Such a configuration might be the heart of an early PC.

Figure 10-26 on page 266 shows how a few chips are used to create a PC today. Here a few VLSI parts are combined with a processor and memory (ROM and RAM) to create a full PC. They are very difficult to service if one of the VLSI circuits develops a fault. However, you should remember that these ICs still do no more than was done, or could be done, with many standard ICs.

Self-Test

Answer the following questions.

The 8088 and 8086 processors can be operated in either the minimum or the

(continued on page 266)

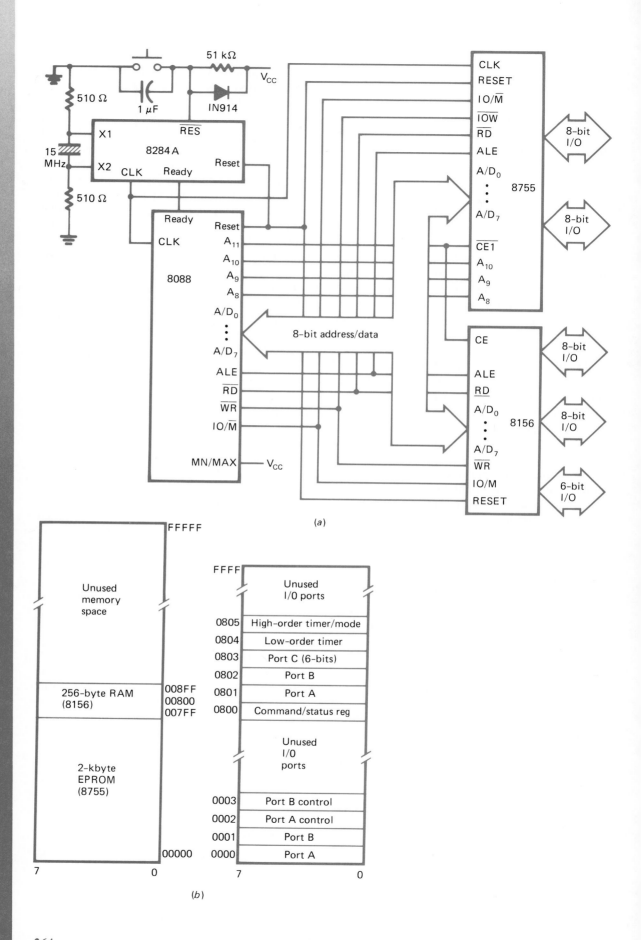

(a)

(b)

Fig. 10-24 (see page 264) A simple 8088-based controller. (*a*) The schematic diagram of a minimum mode system which communicates with external devices through 38 parallel I/O lines on the two I/O memory support ICs. The 8755 provides a 2-kbyte EPROM for the program storage, and the 8156 provides 256 bytes of RAM for system memory. Although this limited memory does not allow sophisticated calculations to be carried out, this system can perform very fast data transfers and conversions between its ports. (*b*) A memory map of the simple controller. *Note:* Because A_{12} to A_{20} are not decoded, memory addresses above 01000H will also be decided by the ROM or RAM.

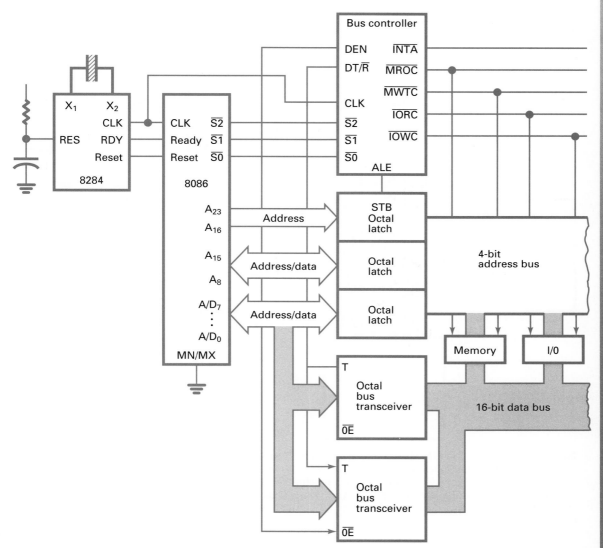

Fig. 10-25 A block diagram of a maximum mode 8086 system. Three 8282 octal latches buffer the processor from the system address bus, and two 8286 octal transceivers buffer the 8086 data bus from the 16-bit system data bus. Control signals from the 8288 tell the latches and bus transceiver when the address and data from the 8086 are valid as well as generating interrupt and memory-read-write signals.

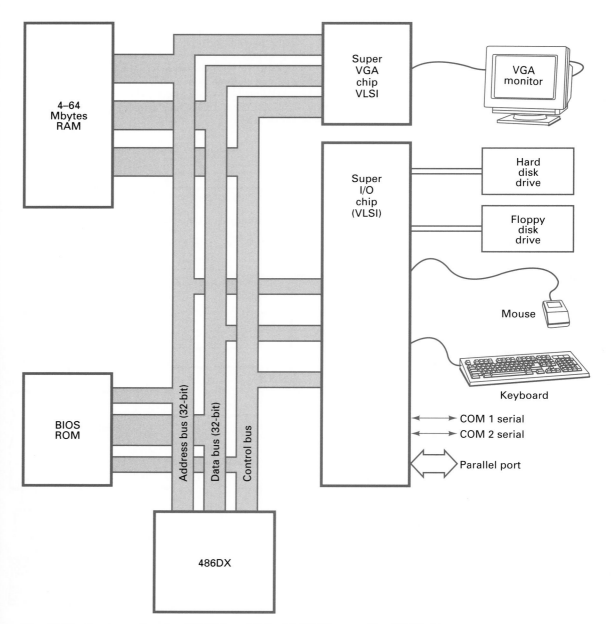

Fig. 10-26 Implementing the 486DX-based PC with VLSI parts. Two VLSI chips provide the human and machine interface. The super VGA chip creates an interface for a super VGA display. *Note:* **This chip may have on-chip video RAM, or it may need to be supplemented with additional video RAM. All I/O functions are handled through the super I/O chip, including disk control, serial and parallel ports, mouse interface, and keyboard.**

maximum modes. These modes are so named because the maximum mode
a. Lets you execute the maximum number of instructions
b. Lets you address the maximum number of memory locations (1 Mbyte)
c. Requires more support hardware than the minimum mode
d. All of the above

45. The 8088/8086 uses a multiplexed address and data bus because
a. 40 pins is a good size for the IC
b. Multiplexing is supported by 8085 support ICs
c. Multiplexing reduces the number of lines between the microprocessor and the auxiliary ICs
d. All of the above

46. The 8088 has a number of status lines. When decoded, the $\overline{S3}$ and $\overline{S4}$ status lines are used to tell an external device
  a. Which mode the microprocessor is running in
  b. The queue status
  c. The processor status
  d. Which segment register is in use

47. The purpose of the $\overline{LOCK}$ output is to
  a. Lock out any possible interrupts
  b. Keep any external device from using the bus
  c. Indicate that the processor is busy
  d. Show the processor mode

48. In a maximum mode system, the 8088, $\overline{S0}$, $\overline{S1}$, and $\overline{S2}$ signals are
  a. Decoded by an external bus controller
  b. Used to indicate the segment register in use
  c. Show the current status of the interrupt system
  d. Always low

49. The HOLD and HLDA (hold acknowledge) lines
  a. Control the segment register in use
  b. Are used by an external device which wishes to gain control of the X86 processor buses
  c. Are both outputs
  d. Are both inputs

50. The WAIT instruction causes the _____ input to be checked. If that input is not asserted, the processor waits another cycle.
  a. $\overline{TEST}$ or $\overline{BUSY}$ *c.* NMI
  b. $\overline{INTR}$ *d.* RESET

51. The 386 and 486 processors use the BIS16 input to tell the processor
  a. That the external device can output only 16-bit data
  b. To double-cycle the memory address sequence
  c. To cycle the data input four times
  d. All of the above

52. In Fig. 10-23 on page 261, the 8284 IC is used to
  a. Generate a RESET signal for the processor when the reset switch is pressed
  b. Initialize the processor with a RESET signal when the system is powered up
  c. Generate a CLOCK signal for the processor using a quartz crystal to set the clock frequency
  d. All of the above

10-7 AN INTRODUCTION TO THE MOTOROLA 68XXX FAMILY OF ADVANCED MICROPROCESSORS

In this section, we will look at another popular family of advanced microprocessors, the Motorola 68XXX family. It was chosen for inclusion in this chapter because you will find it in industrial and commercial work as well as in personal computers. Again, you will find that your knowledge of 8-bit microprocessors, specifically the 6802, and of the Intel X86 family of advanced microprocessors will make it easy to understand the 68XXX family of advanced microprocessors.

Late in 1976, Motorola introduced the first advanced microprocessor in integrated circuit form with an architecture developed specifically as a microprocessor. It was called the *Motorola MC68000*. Although the 68000 has architectural similarities to the 6800 family of 8-bit microprocessors and to the Digital Equipment Corporation PDP-11 family of 16-bit minicomputers, its architecture is very much its own. The MC68000 is found in a popular line of personal computers and is also popular for industrial and graphic-intense applications. It is the CPU used in Apple's Macintosh series of microcomputers, in a number of CAD/CAE (computer-aided design/computer-aided engineering) workstations, and in many advanced industrial control products.

Since its introduction, the 68000 evolved into faster versions, versions with wider external data buses, versions with greater memory addressing space, and versions with advanced computing functions. Internally, all versions of the 68000 have a 32-bit architecture. However, the external bus widths vary from 16 to 32 bits. Because these new versions have an architecture very similar to that of the original 68000, their numbering is also very similar. Some of the model numbers are 68000, 68010, 68020, 68030, 68040, and 68060, as well as variations on these numbers. People often speak of the 68XXX family of microprocessors. We will use that terminology throughout this chapter to refer to material which is common to all members of the 68XXX family.

Table 10-13 on page 268 gives an overview of the 68XXX family of advanced microprocessors and some key attributes associated with each member of the family. As you can see, as the newer processors are introduced, the

Table 10-13   Major Attributes of the 68XXX Processors

Attribute	68000	68010	68020	68030	68040	68060
Data bus	16 bits	16 bits	8, 16, or 32 bits	8, 16, or 32 bits	32 bits	32 bits
Address bus	24 bits	24 bits	32 bits	32 bits	32 bits	32 bits
Operating speed (MHz)	8, 10, 12, 16 (CMOS)	8, 10, 12	16, 20, 25, 33	16, 20, 25, 33, 40, 50	25, 40	50, 66
Memory word alignment	Word aligned	Word aligned	Instructions only	Instructions only	Instructions only	Instructions only
Instruction cache		3 words	256 bytes	256 bytes	4096 bytes	8 kbytes
Data cache				256 bytes	4096 bytes	8 kbytes
Floating point support	External processor	External processor	1–8 coprocessors	1–8 coprocessors	Internal FPU	Internal FPU
Memory management	Via external unit	Via external unit	MMU	Internal MMU	Internal MMU	Internal MMU
Physical memory addressed	16 Mbytes	16 Mbytes	4 Gbytes	4 Gbytes	4 Gbytes	4 Gbytes

68XXX family of advanced microprocessors grows in functions and features.

The 68000 is the original product. Although it has an internal 32-bit architecture, it uses a 16-bit data bus and a 24-bit memory address bus and therefore can address only 16 Mbytes of physical memory. Special instructions added to the 68010 give it virtual memory addressing capability. The 68020 introduced the 32-bit data bus and a coprocessor interface which supports up to eight coprocessors and added a 256-byte on-chip instruction cache.

The 68030 introduced an internal design change and a number of advanced features which make it a very powerful microprocessor. The internal design is split into two independent 32-bit processors to speed up processing. In addition, the 68030 adds intelligent data bus management. This allows the 68030 to determine whether the external device has an 8-, a 16-, or a 32-bit data bus and then to communicate with the device at its data bus width. An internal MMU allows the 68030 programmer to write software which addresses the full 4 Gbytes of memory space regardless of how much physical memory is actually present.

The 68040 adds an internal floating point arithmetic unit which substantially speeds up processing of complex arithmetic calculations. Additionally, internal design improvements allow the 68040 to carry out many of its instruction executions in a pipeline mode and therefore to speed up its operation.

The 68060 is the latest addition to the 68XXX family of advanced microprocessors. It uses a new superscalar execution technique which allows it to process more than one instruction per clock cycle. The floating point arithmetic unit has been improved to increase its speed substantially and the on-chip cache memory has been increased to 16 kbytes.

The 68XXX processors can work with a wide range of main memory. As you have just learned, the original 68000 can address 16 Mbytes to main memory, and the latest processors can address 4 Gbytes using their 32-bit address bus. The actual amount of memory attached to the processor is called the *physical memory*. If the processor does not have a full 16 Mbyte or 4 Gbyte of physical memory (which very few actually do!), then various virtual memory techniques are used to let the programmer write programs which act as though the processor has this memory. The virtual memory techniques cause information to be loaded from mass storage devices before the information is addressed by the processor.

Self-Test

Answer the following questions.

53. Internally, the 68XXX family of advanced microprocessors has a(n) ———————— architecture.

    a. 4-bit   *d.* 32-bit
    b. 8-bit   *e.* 64-bit
    c. 16-bit

54. The first microprocessor in the 68XXX family to support on-chip data cache memory was the
a. 68000　　　　d. 68030
b. 68010　　　　e. 68040
c. 68020　　　　f. 68060

55. The maximum physical memory space which can be addressed by the 68010 is
a. 16 bits　　　　d. 4 Gbytes
b. 32 bits　　　　e. All of the above
c. 16 Mbytes

56. An internal memory management unit (MMU) first appeared on the
a. 68000　　　　d. 68030
b. 68010　　　　e. 68040
c. 68020　　　　f. 68060

57. On-chip floating point arithmetic units first appeared on the
a. 68000　　　　d. 68030
b. 68010　　　　e. 68040
c. 68020　　　　f. 68060

58. If the processor you are using does not have enough physical memory for the program being used, you can use _____ to make the program believe that the processor has enough main memory for the program.
a. An on-chip floating point processor
b. Virtual memory capability
c. Cache memory
d. Any of the above

10-8　A PROGRAMMING MODEL FOR THE 68XXX FAMILY OF ADVANCED MICROPROCESSORS

The MC68000 microprocessor was introduced by Motorola and, like the 8088, has its own unique architecture; it is not a copy of a microcomputer architecture. The 68000 is different from the 8088 because it is a true 16-bit microprocessor; that is, it has a full 16-bit data bus. Additionally, the 68000 internal architecture uses 32-bit buses and registers.

The 68000 external address bus has 23 bits, A_1 to A_{23}. It can address 8,388,608 16-bit memory locations. Because each memory location has 2 bytes, the bus addresses 16,777,216 bytes. For this reason, the 68000 is said to have a 16-Mbyte memory address space. *Note:* Address bit A_0 selects the odd and even addresses, which are the two different bytes making up the 16-bit word. Therefore, address bit A_0 is used only internally

because all external data operations move 16-bit data. Likewise, the 68010 has a 23-bit address bus.

All new 68XXX processors, starting with the 68020, have a 32-bit external address bus and can therefore address 4 Gbytes of physical memory. Depending on the processor and the application, the memory may be organized as bytes, words, or long words. Wherever possible, the memory is organized as long words (32-bit words), because this means fewer fetches and faster operation.

As we learned at the beginning of this chapter, modern software requires an advanced microprocessor to allow privileged operation for certain parts of the software. In the 68XXX family, this is done by sectioning the processor into the *user* and *system* sections. The user section is where ordinary tasks run, and the system section is where the full power of the microprocessor is allowed. The 68XXX family programming model is broken down into the user and system sections, reflecting this architecture.

The user section of the programming model for all 68XXX processors is shown in Fig. 10-27 on page 270. There are two differences between the various members of the 68XXX family. First, the 68000 uses a 23-bit program counter (because it has only a 24-bit memory address bus). All other members of the family have a 32-bit program counter. Second, the 68040 adds seven 80-bit data registers and three 32-bit control registers for the on-chip floating point coprocessor.

The 68XXX family of advanced microprocessors has seventeen 32-bit registers, a 32-bit program counter (23 bits for the 68000), and a 16-bit status register. How can we say that the 68XXX processors have a memory-oriented architecture when they have so many registers? The answer is clear when we understand what the registers are for and how the registers are used.

The first eight registers (D_0 to D_7) are data registers. They work in the bit (1-bit), BCD (4-bit), byte (8-bit), word (16-bit), or long-word (32-bit) modes. Only the addressed low-order bits are changed during an operation on one of these data registers. For example, moving a byte of data into the D_4 register changes only bits 0 to 7. Bits 8 to 31 are not changed. However, moving a long data word into register D_4 changes bits 0 to 31.

The eight address registers (A_0 to A_7) either point to memory locations or their contents

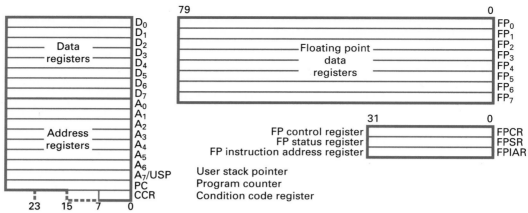

Fig. 10-27 The 68XXX user programming model. The left-hand side of this model is common for all 68XXX processors except that the program counter (PC) is only 23 bits wide in the 68000. The right-hand side of the model shows the seven 80-bit data registers and the three 32-bit control registers available to the programmer when a floating point unit is needed. *Adapted from Motorola literature.*

are used to help calculate the memory location address. The way they work depends on the kind of memory reference instruction used. Registers A_0 to A_6 can be used as stack pointers if extra stack pointers are needed. To do this, A_0 to A_6 are used with special instructions which automatically decrement or increment the register before or after it is used to address a memory location. Address register A_7 is generally used when an instruction calls for a stack pointer.

Any of the eight address registers can be used as a stack pointer or as a base register to compute a memory location address. The memory address registers can be used for either 16- or 32-bit address operations. However, any data written to the address registers always affect the entire register.

As we will see later, the 68XXX family of processors uses indexed instructions. Any of the seventeen 32-bit registers can be used as an index register in an indexed memory address operation.

The 68XXX processors have very few dedicated registers. A program counter is always needed, so this is a dedicated register. However, all other 68XXX registers are general-purpose. For example, all the 68XXX data registers can be used as accumulators, and any of the address registers can be stack pointers, index registers, or address base registers.

The 68XXX processor status register is shown in Fig. 10-28. The first 4 bits of this 16-bit register are the same as the 6802 flags. The fifth is the extend flag, which works with

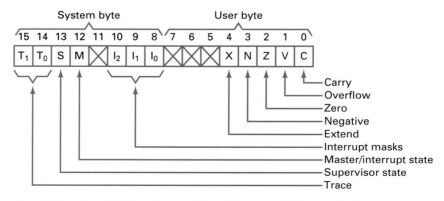

Fig. 10-28 The 68XXX status register. The lower 4 bits are identical to the status bits found in the 8-bit series of 6800 microprocessors. The upper byte contains bits new to the 68XXX processors and can be accessed only in the supervisor mode. *Adapted from Motorola literature.*

multiple precision arithmetic. The three interrupt mask bits (I2, I1, I0) let the programmer set one of eight interrupt priority levels. Like the X86 processors, the 68XXX processors have a trace mode. The 68XXX processors are in the trace mode when the trace bit (T) is set. *Note:* In the newer processors starting with the 68020, there are two trace bits (T_1 and T_0) which give additional control over the trace feature.

As we discussed earlier, the 68XXX family has two privilege states—user and supervisor. How do you know which of the two states the processor is in? The status register supervisor state flag (S) controls which state the processor is in. If S is cleared (logic 0) and you use the stack pointer, you will use the user's stack. If S is set (logic 1), and you use the stack pointer, you will use the supervisor's stack. Each state has its own stack pointer, so the programmer can easily switch between the user and supervisor states without tracking the state being used. This also provides protection, because a user programming error is less likely to contaminate the supervisor stack if the user program cannot address that area of memory.

When the processor is in the supervisor state, you have access to many more instructions and a number of system registers in the more advanced versions of the 68XXX family of processors. Figure 10-29 on page 272 shows the additional registers available with the different members of the 68XXX processors. As you can see, the number of registers available in the supervisor state grows with the newer processors. With the original 68000, there were only two registers in the supervisor extension to the programming model. These were a supervisor stack pointer and the upper byte of the flag register. As new memory management, cache memory, and arithmetic coprocessing features became part of the 68XXX architecture, the control registers for these features were added in the supervisor section of the programming model.

The 68XXX processors are always in one of three major processing states. These are the normal, exception, and halt processing states. When in the *normal processing state,* the 68XXX processors execute program instructions using either user privileges or supervisor privileges. When a 68XXX processor is interrupted or when certain interrupt-like events happen inside the processor, it goes into the *exception processing state.* It remains in the exception processing state until the

special condition is over. The conditions which can cause the 68XXX processor to enter the exception state are shown in Table 10-14 on page 273. In the *halt processing state,* the 68XXX processors do not execute any instructions; they wait for an interrupt.

Self-Test

Answer the following questions.

59. We say that the 68000 is a true 16-bit microprocessor because
    a. It uses a 23-bit bus which addresses 8 million 16-bit memory locations
    b. It uses 32-bit internal registers for addressing and data manipulation
    c. The data bus is 16-bits wide, giving the 68000 an external 16-bit data architecture
    d. All of the above

60. The 68XXX processor is able to expand easily to a 32-bit version because
    a. It uses a 23-bit bus which addresses 8 million 16-bit memory locations
    b. It uses 32-bit internal registers for addressing and data manipulation
    c. The data bus is 16-bits wide, giving the 68000 an external 16-bit data architecture
    d. All of the above

61. If you were to say that the 16-bit microprocessor architecture came from the 8-bit architecture, then you would say that the 68000 came from the
    a. Z80
    b. 8088
    c. 6802
    d. None of the above

62. We can say that the 68XXX family has a memory-oriented architecture because
    a. Its 17 registers can be divided into those which are used like accumulators and those which are used for memory addressing
    b. Memory-oriented architectures require two or more stack pointers, and two of the 68XXX registers are usually used as stack pointers
    c. You cannot address 1 Mbyte of memory with a 16-bit register, which means that the 16-bit memory addressing registers must be combined with another value to get the required 32-bit address word
    d. None of the above, because the X86 is really the advanced memory-oriented processor family

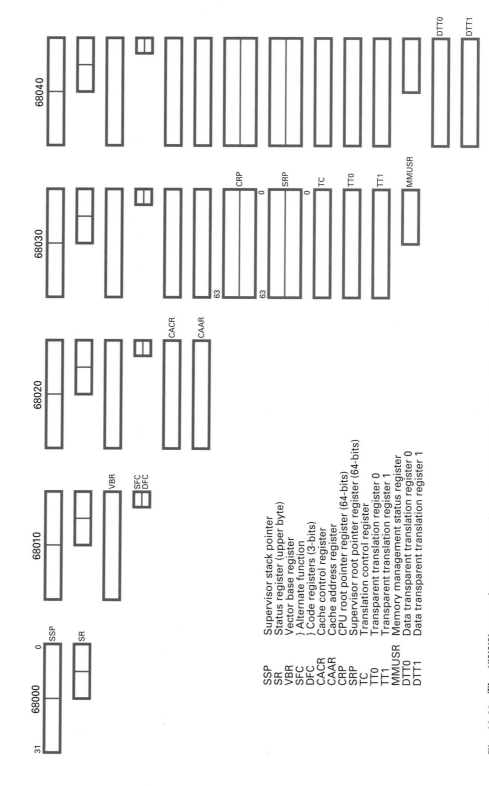

Fig. 10-29 The 68XXX supervisor programming model. This figure shows the growth of the 68XXX microprocessors as each new version is added to the family. *Note:* These registers can be from 3 bits wide to 64 bits wide, depending on the use and the processor.

SSP	Supervisor stack pointer
SR	Status register (upper byte)
VBR	Vector base register
SFC	} Alternate function
DFC	} Code registers (3-bits)
CACR	Cache control register
CAAR	Cache address register
CRP	CPU root pointer register (64-bits)
SRP	Supervisor root pointer register (64-bits)
TC	Translation control register
TT0	Transparent translation register 0
TT1	Transparent translation register 1
MMUSR	Memory management status register
DTT0	Data transparent translation register 0
DTT1	Data transparent translation register 1

Table 10-14 68000 Exception States

Type of Cause	Exception Type	Exception Caused by
External	Reset	$\overline{\text{RESET}}$ asserted
External	Interrupt	Valid interrupt asserted
External	Bus error	$\overline{\text{BERR}}$ asserted
External	Spurious interrupt	$\overline{\text{BERR}}$ asserted during interrupt acknowledge
Internal	Instruction	TRAP, TRAPV, CHK, DIVS, or DIVU instruction executed
Internal	Privilege violation	Privileged instruction executed in user state
Internal	Trace	In trace mode
Internal	Illegal address	Word or long-word address with odd address
Internal	Illegal instruction	Instruction does not exist
Internal	Unimplemented instruction	Instructions with 1010 to 1111 in OPCode

63. Match the following 68XXX word length descriptions with the word lengths
   a. Word   1. 32 bits
   b. Byte   2. 16 bits
   c. Long word   3. 8 bits
   d. Bit   4. 4 bits
   e. BCD   5. 1 bit
   f. Quad word   6. 64 bits

64. Each time a 68000 or 68010 processor addresses a memory location, it works with 16 bits of data. Because _____ we can say that the 68000 and 68010 processors address 16 million bytes of memory.
   a. It uses a 23-bit bus which addresses 8 million 16-bit memory locations, with each location having 2 bytes
   b. It uses a 32-bit address bus which addresses 4 billion 8-bit memory locations
   c. The data bus is 16 bits wide, giving the 68000 an external 16-bit data architecture
   d. All of the above

65. Because _____ we can say that the 68030 and 68040 processors address 4 Gbytes of memory.
   a. They use a 23-bit bus which addresses 8 million 16-bit memory locations, with each location having 2 bytes
   b. They use a 32-bit address bus which addresses 4 billion 8-bit memory locations
   c. The data bus is 16 bits wide, giving the 68000 an external 16-bit data architecture
   d. All of the above

66. Most of the 68XXX user registers are
   a. 32 bits wide
   b. Used for data processing
   c. General-purpose
   d. All of the above

67. When a 68XXX processor is in the supervisor state, it
   a. Is given the use of additional (privileged) instructions and registers
   b. Watches out for interrupts, but otherwise does nothing
   c. Enables the program counter in the 32-bit mode
   d. All of the above

68. When a 68XXX processor is servicing an interrupt, it is in the _____ processing state.
   a. Normal   c. Halt
   b. Exception   d. Reset

10-9 THE 68XXX ADDRESSING MODES

Like the X86 processors, the 68XXX processors have many different addressing modes which give the 68XXX instructions a great deal of power. In this section, we will look at the 68XXX addressing modes. Once you understand the addressing modes, you can study the 68XXX instruction set.

Like the X86 processors, the 68XXX processors allow the programmer to separate memory into areas for data and areas for instructions. This is done by using instructions which use different address registers for instruction and data fetches. The program counter is used for instruction fetches, and one of the address registers (A_0 to A_7) is used for data fetches.

Figure 10-30 on page 274 shows different ways the 68XXX processors can organize data in memory. The 68XXX processors can address bit, BCD, byte, word, long-word and

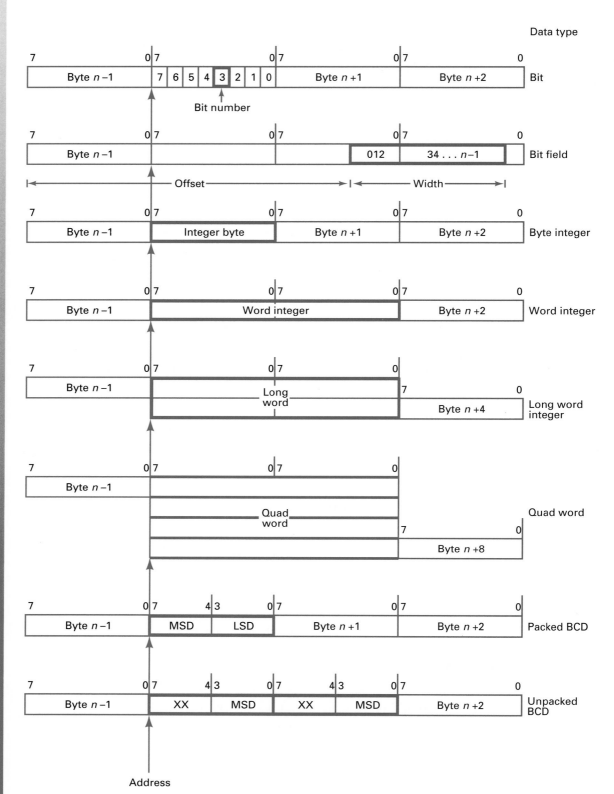

Fig. 10-30 **The organization of data in 68XXX memory locations. The left-hand vertical mark is the address point for each of these different-length data elements. The byte to the left of the address mark is at a lower memory address, and the byte to the right of the addressed element is at a higher memory address. Depending on the specific 68XXX processor used, physical memory supporting this structure can be 8, 16, or 32 bits wide.** *Adapted from Motorola literature.*

quad-word data in memory. For most 68XXX systems, the memory can be organized as bytes, 16-bit words, or 32-bit words (68020 and higher). Most system designs use the longest possible word as doing so speeds up memory access time.

There are six different kinds of addressing modes on the earlier processors and nine on the later versions (68020 and later). They are summarized in Table 10-15. Each mode can be used with many different kinds of instructions. However, the type of data (bit, BCD, byte, word, long word, or quad word) which the instruction operand uses depends on the type of data the operand can work with. All ad-

dressing modes do the same thing: They generate an effective address. The *effective address* is the actual address placed on the address bus to select a physical memory location. Figure 10-31 shows how the last six bits of an instruction are used to set the addressing mode and the register used.

The following paragraphs summarize the memory addressing modes shown in Table 10-15.

There are two *register direct addressing modes*. The *data register direct mode* (mode 000) lets you specify an instruction operand in one of the eight 32-bit data registers (D_0 to D_7). The *address register direct mode* (mode

Effective address

Register direct addressing mode

Data register direct mode

Address register direct mode

Table 10-15 68XXX Addressing Modes by Processor Type			
Addressing Modes	Mode Number	68000 and 68010 Syntax	68020, 68030, and 68040 Syntax
Register direct			
Data register direct	000	Dn	Dn
Address register direct	001	An	An
Register indirect			
Address register indirect	010	(An)	(An)
Address register indirect with postincrement	011	(An)+	(An)+
Address register indirect with predecrement	100	−(An)	−(An)
Address register indirect with displacement	101	(d_{16}, An)	(d_{16}, An)
Register indirect with index			
Address register indirect with index (8-bit displacement)	110	(d_8, An, Xn)	(d_8, An, Xn)
Address register indirect with index (base displacement)	110		(bd, An, Xn)
Memory indirect			
Memory indirect postindexed	110	N/A	([bd, An], Xn, od)
Memory indirect preindexed	110	N/A	([bd, An, Xn], od)
Program counter indirect with displacement	111	(d_{16}, PC)	(d_{16}, PC)
Program counter indirect with index			
PC indirect with index (8-bit displacement)	111	(d8, PC, Xn)	(d8, PC, Xn)
PC indirect with index (base displacement)	111	N/A	(bd, PC, Xn)
Program counter memory indirect			
PC memory indirect postindexed	111	N/A	([bd, PC], Xn, od)
PC memory indirect preindexed	111	N/A	([bd, PC, Xn], od)
Absolute			
Absolute short	111	(xxx), W	(xxx), W
Absolute long	111	(xxx), L	(xxx), L
Immediate	111	#(data)	#(data)

Notes:
Dn = Data Register D0—D7
An = Address Register A0—A7
d8, d16 = A 2's complement or sign-extended displacement; added as part of the effective address calculation
Xn = Address or data register used as an index register
bd = A 2's complement base displacement (16 or 32 bits)
od = Outer displacement
PC = Program counter
(data) = Immediate data (8, 16, or 32 bits)
( ) = Effective address
[ ] = Use as indirect access to long-word address
N/A = Not available on these processors

Register indirect
mode

Address register
indirect mode

Address register
indirect with
postincrement
mode

Address register
indirect with
predecrement
mode

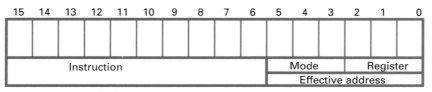

15	14	13	12	11	10	9	8	7	6	5	4	3	2	1	0

Instruction	Mode	Register
	Effective address	

Fig. 10-31 A 68XXX instruction word showing the part used for the instruction (bits 6 to 15), the part used to indicate the mode (bits 3, 4, and 5), and the part used to identify the register which will be used (bits 0, 1, and 2). *Adapted from Motorola literature.*

001) lets you specify an operand in one of the eight 32-bit address registers (A_0 to A_7). These two addressing modes are as simple as they seem. Their operation is shown in Fig. 10-32.

The *register indirect modes* (modes 010 to 110) specify an instruction operand in a specific memory location. That is, the address in the instruction points to the register, which, in turn, points to a memory location. There are four different register indirect modes. Each one of them uses one of the memory address registers (A_0 to A_7) to help point at the memory location you want to use.

You use the *address register indirect mode* (mode 010) when you want the address register to point to the memory location you want to work with. The instruction points to an address register (A_0 to A_7), and the register points to the memory location. In this case, the register contents are the effective address. Unlike the X86 processors, the 68XXX processors have 32-bit registers. This means that no other work

is needed to select a memory location. This operation is shown in Fig. 10-33.

The *address register indirect with postincrement mode* (mode 011) lets you use any address register as a stack pointer. After the memory address and data operations are finished, the value in the 32-bit address register is incremented so that it is pointing to the next stack position. The value is incremented by 1, 2, or 4, depending on the size of the data word specified in the instruction. This operation is shown in Fig. 10-34.

If there is a way to increment the address register, there must be a way to decrement the address register. This function is performed by the *address register indirect with predecrement mode* (mode 100). Before the operand is used, the value in the address register is decremented by 1, 2, or 4, again depending on the size of the data word specified in the instruction. By decrementing *before* the value in the address register is used, you can pop

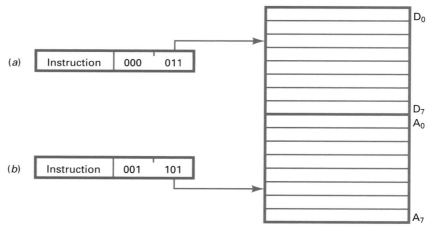

Fig. 10-32 Register direct addressing. The instruction operand is the contents of a register. (*a*) Data register direct. Here mode 000 indicates the contents of a data register (register D 011, in this figure). (*b*) Address register direct. Here mode 001 indicates the contents of an address register (register A 101, in this figure).

**Address register
indirect with
displacement
mode**

**Register indirect
with index mode**

**Memory indirect
mode**

**Address register
indirect with index
mode**

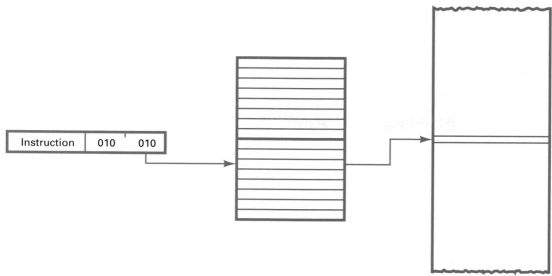

Fig. 10-33 **Register indirect addressing. The instruction operand is the memory location
pointed to by one of the seven address registers (A$_0$ to A$_7$).** *Note:* **The 68XXX processors
can point directly to any memory location within the 4-Gbyte address space because all
registers are 32 bits long.**

the next data off the stack. Again, the operation
is identical to that shown in Fig. 10-34, except
that the processor *decrements* the memory
address register contents by 1, 2, or 4 *before*
the addressing operation takes place.

As we have just seen, the memory address
registers can point directly at the memory
location; that is, the register contents are
the effective address. However, the register

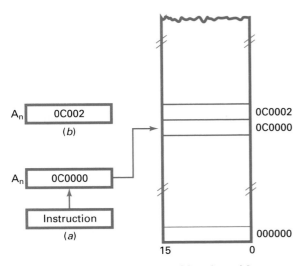

Fig. 10-34 **Register indirect addressing with
postdecrement. In (***a***) the instruction specifies the
contents of memory address register A$_n$. Therefore
memory location 0C0000 is selected. After the
memory location is addressed and the data is
fetched, the value in A$_n$ is incremented. (***b***) The
incremented value is 0C0002.**

contents can also be the base of a calculation
which points to the memory location. In this
case the effective address is calculated from
the register contents and some other data. A
number of modes use this approach.

The first is *address register indirect with
displacement mode* (mode 101). The value in
the address register is used as the base for the
address computation. A 16-bit number (the
second word in the instruction) is added to
the 32-bit base value using 2's complement
arithmetic. This gives an addressing range of
$-32,768$ to $+32,767$ bytes relative to the
current base values in the address register.
Figure 10-35 on page 278 shows how this
address is calculated.

The second is a group of addressing modes
(mode 110) called either the *register indirect
with index mode* or the *memory indirect mode*.
The first in this group is called the *address
register indirect with index mode* (mode 110). It
adds three different values together to compute
the effective address. The computation uses
values from

- The 32-bit address register (A$_0$ to A$_7$)
- The low-order 8 bits of the instruction sec-
  ond byte
- The contents of a 16-bit index register

Note: The index register is any one of the
68XXX's 17 different registers, all of which
can be used as index registers.

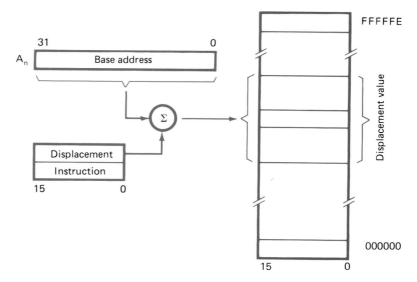

Fig. 10-35   **A computed memory address. The value in memory address register A_n is added to the 16-bit displacement in the second word of the instruction using 2's complement arithmetic. The address range is the value in A_n −32,768 to A_n + 32,767.**

The second in this group is the *address register indirect with index (base displacement) mode* (mode 110 as well); it is similar to the previous operation. Instead of adding the contents of a 16-bit index register to the address register value, the instruction uses the 32 bits of a base displacement value which come from the second word of the instruction. This mode is not on earlier processors.

Instructions which use these two addressing modes to give two offsets from the base register are very useful when addressing data in two-dimensional arrays. In this case the index can represent one dimension, and the offset the other. The result of using these addressing modes is shown in Fig. 10-36.

The third in this group, the *memory indirect* version of this mode (mode 110 as well), is found only on the newer processors. Both a postindexed and a preindexed version are available, so the programmer can put data into or get data from complex lists of data. The displacement is added to the value of an address register to point to a list in memory. The contents of this list are added to the value of an index register to find the effective address. The mode allows the creation of a three-dimensional list.

The *program counter with displacement mode* (mode 111) uses a two-word instruction. The effective address is the sum of the current program counter value and the 16-bit value in the second word of the instruction. This gives

the effective address a range of −32,768 to +32,767 bytes relative to the current base value in the program counter. Figure 10-37 on page 280 shows how this effective address is calculated.

The *program counter indirect with index mode* (mode 111) is also a two-word instruction but with two submodes. The first is the *program counter indirect with index (8-bit displacement)*. The current program counter value is added to the low-order 8 bits in the second word, and this value is added to the index register value. Figure 10-38 on page 280 shows how this mode calculates an effective address. In the new additions to this addressing mode group, the current program counter value is added to the second word in the two-word instruction, and this value is added to the index register value. This is called the *program counter indirect with index (base displacement)*. It is available only on 68XXX processors after the 68020.

The two *program counter memory indirect modes* are available only on the 68020 processors and later. They are identical to the memory indirect modes, except that they use the program counter rather than a memory address register.

There are two absolute addressing modes. The *absolute short address mode* uses a two-word instruction. The second 16-bit word is the effective address. The absolute short address modes specifies memory locations between

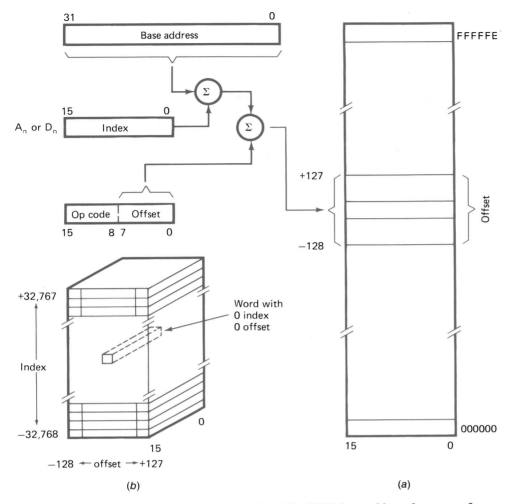

Fig. 10-36 (*a*) **Indirect addressing with an index. The 32-bit base address from one of the memory address registers (A$_n$) is combined (using 2's complement arithmetic) with a 16-bit value from any of the memory address registers, the data registers, or the program counter. This is called the *index value*. The indexed result is combined with the 8-bit offset from the instruction word second byte. (*b*) A model of a two-dimensional array of 16-bit data words. If both the index and the offset are zero, the word shown in the center of the array is selected.**

000000 and 007FFE and between FF8000 and FFFFFE. These are the lowest and highest 32 kbytes in the 68XXX address space. This instruction is used when the programmer needs short instructions for programming efficiency and long addressing is not needed.

The *absolute long address mode* is a three-word instruction. The last two 16-bit words make up the 32-bit effective address. The first word is the instruction, the second word is the high-order part of the address, and the third word is the low-order part of the address. The long- and short-address modes are shown in Fig. 10-39 on page 281.

The addressing modes also include the *immediate data mode*. As you might suspect,

the immediate data mode uses the next word or two words as data for the instruction. This is either a two- or three-word instruction, depending on the type of data referred to in the instruction. Bit, BCD, byte, and word data types use only a two-word instruction. However, long-word instructions need a three-word instruction.

Self-Test

Answer the following questions.

69. The 68XXX processors use separate registers to address data and instructions. The

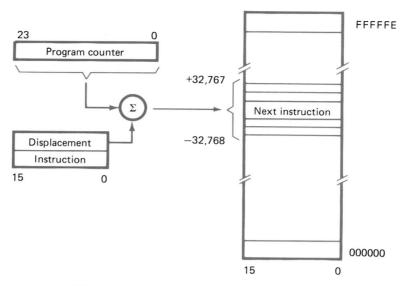

Fig. 10-37 Making a relative memory address. This memory addressing mode uses the program counter as the base address so that the programmer can select a memory location with a range of PC −32,768 to PC +32,767.

reason they use separate registers to address data and instructions is

a. To make flexible instructions

b. Because the instruction space is read-only, and the data space must be read-write

c. To let the programmer separate instructions and data

d. All of the above

70. The two-register direct addressing modes let you

a. Address either the data registers or the address registers

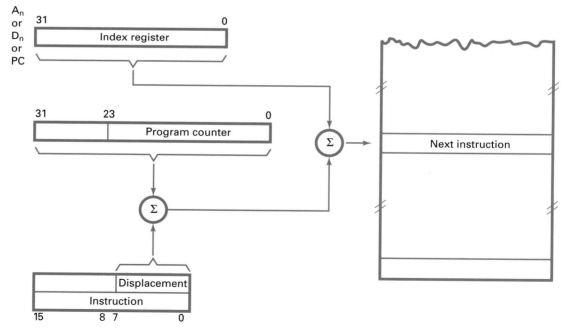

Fig. 10-38 Relative addressing using a displacement and an index. This mode allows the programmer to develop a two-dimensional array which is relative to the current instruction.

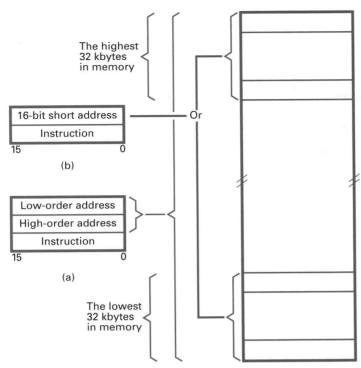

Fig. 10-39   The long and short addressing modes. (*a*) Long addressing uses the two 16-bit words following the instruction to address any memory location in the 68XXX memory space. (*b*) Short addressing uses a single 16-bit word following the instruction to address either the uppermost or the lowermost 32 kbytes of memory.

b. Address a memory location by pointing to it with an address register
c. Address a memory location by pointing to the memory location with a computed value
d. Decrement or increment the register value so that is can be used as a stack pointer

71. The register indirect addressing modes let you
a. Address either the data registers or the address registers
b. Address a memory location by pointing to it with an address register
c. Address a memory location by pointing to the memory location with a computed value
d. Decrement or increment the register value so that it can be used as a stack pointer

72. The displacement value used in 68XXX memory addressing calculations is taken from
a. The current program counter value

b. The current status register value
c. A register called an index register
d. The second word of the instruction

73. The address register indirect with postincrement and address register indirect with predecrement instructions let you
a. Address either the data registers or the address registers
b. Address a memory location by pointing to it with an address register
c. Address a memory location by pointing to the memory location with a computed value
d. Decrement or increment the register value so that it can be used as a stack pointer

74. The address memory indirect with index can be used to
a. Address data in an array
b. Store data for a 1024 × 768 point display
c. Address points in a large mutidimensional look-up table
d. All of the above

75. The special addressing mode called _____ lets you address any one of the 68XXX memory locations directly from the instruction
    a. Absolute short address
    b. Absolute long address
    c. Program counter with displacement
    d. Immediate data
76. When the address of the next instruction is used to compute the effective address, you are using the _____ memory address mode.
    a. Absolute short address
    b. Absolute long address
    c. Program counter with displacement
    d. Immediate data
77. The _____ memory address mode puts all 32 bits of data in the second two words of the instruction
    a. Absolute short address
    b. Absolute long address
    c. Program counter with displacement
    d. Immediate data

10-10 THE 68XXX INSTRUCTION SET

The 68XXX processors have many different instruction types. As each new member of the 68XXX family is added to the line, a few more instructions are also added to the 68XXX instruction set. However, the processors are always *upwardly compatible;* that is, software written for an earlier version will always run on a later version. *Note:* It is not true that software written for a new processor will necessarily run on an older processor. In fact, software written on the newer processor may use instructions which the older processor will not recognize, and therefore the processor may not be *downwardly compatible.*

In the previous section, we learned about the different addressing modes the 68XXX processors use to let these instructions work with a wide range of memory space and registers. In this section, we will look at the instructions, but you must keep the addressing modes in mind. It is the combination of the wide variety of instructions and addressing modes which makes the 68XXX processors so powerful.

A summary of the 68XXX instructions is given in Tables 10-16 through 10-29. These tables are organized by type of instruction. The first nine are the instructions available in the user mode. The remainder are those instructions only available in the supervisor mode. Within the user and supervisor subdivisions, the instructions are organized by their operational type. For each type, its table shows the instruction mnemonic, the instruction definition, and which of the 68XXX processors introduced the instruction to the 68XXX instruction set. *Note:* Once an instruction is introduced to the 68XXX instruction set, it is available to all newer processors. So, for example, the CMP2 instruction, which was introduced on the 68020 processor, is not available for the 68000 or the 68010 but is available on the 68020, 68030, 68040, and 68060 processors.

The following paragraphs provide general comments on different types of 68XXX instructions. As you read these paragraphs, review the instructions in the associated table. Be sure to read the notes which are given at the bottom of some of the tables so that you will understand how the symbols and abbreviations are used.

The data movement operations are shown in Table 10-16. The 68XXX addressing allows you to move data between any two locations which can contain data. There are both data move instructions and address move instructions. The data move instructions let you move byte, word, and long-word data. The address move instructions let you move only words and long words. Additionally, when you use an instruction to move address information, the 68XXX checks to be sure that the move does not create an illegal address. There are a few data movement instructions which are available only in the supervisor state. These are discussed under the SC (system control) instruction types.

The integer arithmetic operations are shown in Table 10-17. As with the X86 processor instructions, the 68XXX processor instructions include multiplication and division as well as addition and subtraction arithmetic operations. The extended add and subtract instructions let the programmer perform multiple precision arithmetic. They are the same as the add or subtract with carry instruction found in other microprocessors. However,

Table 10-16 The 68XXX Data Move Instructions

Mnemonic	Description	68000, 68010	68020	68030, 68040
EXG	Exchange registers	X	X	X
LEA	Load effective address	X	X	X
LINK	Link and allocate	X	X	X
MOVEA	Move address	X	X	X
MOVEP	Move peripheral	X	X	X
MOVEQ	Move quick	X	X	X
PEA	Push effective address	X	X	X
UNLK	Unlink	X	X	X

when you use an extended instruction, the zero flag (Z) is set only if the entire result is zero.

The multiply and divide instructions include the MULU and DIVU instructions, which work with unsigned numbers, and MULS and DIVS instructions, which use signed numbers. The multiply instructions use a word multiplier and multiplicand to generate a long-word product. The divide instructions use a long-word dividend and a word divisor to generate a word quotient and remainder. The DIVSL and DIVL instructions use a quad-word (64-bit) dividend and a long-word divisor to generate a long-word quotient and remainder.

You use the sign extend (EXT) instruction to perform arithmetic operations when working with two different length numbers. Executing the EXT instruction fills the unused part of the 32-bit long word with the sign bit from the

Table 10-17 The 68XXX Integer Arithmetic Instructions

Mnemonic	Description	68000, 68010	68020	68030, 68040
ADD	Add	X	X	X
ADDA	Add address	X	X	X
ADDQ	Add quick	X	X	X
ADDX	Add with extend	X	X	X
CLR	Clear	X	X	X
CMP	Compare	X	X	X
CMPA	Compare address	X	X	X
CMPI	Compare immediate	X	X	X
CMPM	Compare memory to memory	X	X	X
CMP2	Compare register against upper and lower bounds		X	X
DIVS	Signed divide			
DIVSL	Signed divide long		X	X
DIVU	Unsigned divide	X	X	X
DIVL	Unsigned divide long		X	X
EXT	Sign extend	X	X	X
EXTB	Sign extend bit		X	X
MULS	Signed multiply	X	X	X
MULU	Unsigned multiply	X	X	X
NEG	Negate	X	X	X
NEGX	Negate with extend	X	X	X
SUB	Subtract	X	X	X
SUBA	Subtract address	X	X	X
SUBI	Subtract immediate	X	X	X
SUBQ	Subtract quick	X	X	X
SUBX	Subtract with extend	X	X	X

short word. For example, Fig. 10-40 shows the subtraction of the BCD (4-bit) value 1010 from the byte (8-bit) value 010000000. The four most significant bits must be all logic 1s before the subtraction process works. Executing the EXT instruction converts 1010 into 11111010.

The compare (CMP) instructions perform the subtraction but do not save the result. The only register changed by a compare instruction is the status register. You can determine less than, equal to, and greater than relationships with this instruction. The CMP2 instruction compares the addressed value against two limits for bounds checking.

Two 68XXX instructions perform arithmetic operations, with one of the two operations set to zero. The negate (NEG and NEGX) instructions subtract the operand from zero. This generates a true 2's complement of the operand.

The standard four logical operations are AND, OR, EOR, and NOT and are shown in Table 10-18. The immediate versions of these instructions contain one of the two logical operands. The test (TST) instruction also uses zero as one operand and is also considered a 68XXX logical operation. The operand named in the TST instruction is compared to zero, and the status register bits are set. No other results are saved.

The shift and rotate instructions are shown in Table 10-19. All shift and rotate instructions can operate on a register or on memory. However, memory shifts and rotates must work on a word. Register shifts and rotates are much more flexible. All operand lengths are supported for registers, and you can specify how many shifts or rotates (1 to 64) are to take place when the instruction executes. This lets you perform a high-byte, low-byte swap, for example, using a rotate right (ROR) eight

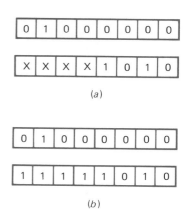

Fig. 10-40 (a) Two data words referenced by a subtraction instruction. The upper 4 bits of the nibble format word are unknown. (b) After the EXT instruction is executed, the upper 4 bits of the nibble are extended from the sign bit of the 4-bit word.

times instruction. With this instruction, you do not need some of the more limited swap instructions used by other microprocessors. The extended shift and rotate instructions let the programmer work with numbers used in multiple precision arithmetic operations.

The four bit manipulation (B) instructions (Table 10-20) let you test (BTST), set (BSET), clear (BCLR), or change (BCHG) any bit in any memory location or register. Each bit-oriented instruction must have not only an effective address but also a bit address. The bit address is given by the contents of an immediate word or by the contents of a specified register. Only the status register zero flag is changed by bit operations.

The 68020 introduced a series of bit field instructions. These are shown in Table 10-21. These instructions let the programmer work with a variable-length bit field; that is, the operand can be from 1 to 32 bits long. The insert and extract instructions let you put a bit into a field or get a bit from a field. Using

Table 10-18	The 68XXX Logical Instructions			
Mnemonic	Description	68000, 68010	68020	68030, 68040
AND	Logical AND	X	X	X
ANDI	Logical AND immediate	X	X	X
EOR	Logical exclusive-OR	X	X	X
NOT	Logical complement	X	X	X
OR	Logical inclusive-OR	X	X	X
TST	Test operand	X	X	X

Table 10-19   The 68XXX Shift and Rotate Instructions

Mnemonic	Description	68000, 68010	68020	68030, 68040
ASL, ASR	Arithmetic shift left and right	X	X	X
LSL, LSR	Logical shift left and right	X	X	X
ROL, ROR	Rotate left and right	X	X	X
ROXL, RORX	Rotate with extend left and right	X	X	X
SWAP	Swap register words	X	X	X

Table 10-20   The 68XXX Bit Manipulation Instructions

Mnemonic	Description	68000, 68010	68020	68030, 68040
BCHG	Test bit and change	X	X	X
BCLR	Test bit and clear	X	X	X
BSET	Test bit and set	X	X	X
BTST	Test bit	X	X	X

the 68XXX addressing capabilities, you can specify the operand length and the displacement from the operand address in bits.

Most 68XXX instructions can operate on byte, word, or long-word (operand size 8, 16, 32) data. The 68XXX processors can also operate on bit and BCD data. The BCD instructions are shown in Table 10-22. The BCD instructions let you perform arithmetic operations (add, subtract, and negate) on byte operands. Because these instructions work only on byte operands, you must use packed BCD data. The negate (NBCD) instruction generates either the 10's complement or the 9's complement of the operand, depending on the status register X flag. The 68020 introduced the PACK and UNPK instructions, which allow a programmer to convert ASCII or EBCDIC

Table 10-21   The 68XXX Bit Field Instructions

Mnemonic	Description	68000, 68010	68020	68030, 68040
BFCHG	Test bit field and change		X	X
BFCLR	Test bit field and clear		X	X
BFEXTS	Signed bit field extract		X	X
BFEXTU	Unsigned bit field extract		X	X
BFFFO	Bit field find first one		X	X
BFINS	Bit field insert		X	X
BFSET	Test bit field and set		X	X
BFTST	Test bit field		X	X

Table 10-22   The 68XXX BCD Instructions

Mnemonic	Description	68000, 68010	68020	68030, 68040
ABCD	Add decimal with extend			
NBCD	Negate decimal with extend	X	X	X
PACK	Pack BCD		X	X
SBCD	Subtract decimal with extend	X	X	X
UNPK	Unpack BCD		X	X

data strings into BCD data or to make BCD data into ASCII or EBCDIC data strings.

The 68XXX program control instructions let the programmer use subroutines and return from those subroutines. These are shown in Table 10-23. The conditional branch (B_{cc}) instructions let the programmer test for one of a number of different conditions and generate a new program counter value if the condition is true. The new program counter value is computed by adding a number to the current program counter value using 2's complement arithmetic. This gives the branch instructions a range of -128 to $+127$ bytes or $-32,768$ to $+32,767$ bytes relative to the current program counter value. Don't forget that the program counter is pointing to the next instruction, not to the one which is executing.

There are two special conditional instructions. The DB_{cc} tests the operand and branches if the condition is met. Each time the test is made, the instruction also decrements the low-order word in the register or the memory location if the condition is not met. This instruction lets you set up a counter which is tested and decremented in a single instruction instead of the usual two instructions needed by most microprocessors.

S_{cc} is used to set (FF) or clear (00) the operand byte if the tested condition is met or is not met. This instruction is used by many programmers to temporarily store a condition code when the condition is not going to be used for a while. Most programmers prefer to use true = FF and false = 00 instead of a single bit.

The BRA (branch always) and JMP (jump) instructions are the 68XXX unconditional program control instructions. As with the 6802, the branch instruction uses signed arithmetic to compute the offset. The 68XXX processors allow 8-, 16-, or 32-bit offsets depending on the processor word size. The jump instruction lets you jump to a new program location. The 68XXX can jump to any place within its memory space.

The 68XXX subroutine call instructions are BRS and JRS (branch to subroutine and jump to subroutine). The branch to subroutine instruction lets the programmer make an 8- or a 16-bit relative branch using the program counter as the base address. The jump to subroutine instruction can address any location in the 68XXX address space. The NOP instruction is used to consume processor time.

There are three 68XXX return instructions. The RTD instruction returns to the called subroutine, but the stack pointer is loaded with a new value, which is the current stack pointer value plus the displacement value from

Table 10-23 The 68XXX Program Control Instructions

Mnemonic	Description	68000, 68010	68020	68030, 68040
Bcc	Branch conditionally	X	X	X
BRA	Branch	X	X	X
BSR	Branch to subroutine	X	X	X
DBcc	Test condition, decrement, and branch	X	X	X
JMP	Jump	X	X	X
JSR	Jump to subroutine	X	X	X
NOP	No operation	X	X	X
RTD	Return and deallocate	X	X	X
RTR	Return and restore codes	X	X	X
RTS	Return from subroutine	X	X	X
S_{cc}	Set conditionally	X	X	X

cc = Condition code, which can be

CC — Carry clear	GE — Greater or equal
LS — Lower or same	PL — Plus
CS — Carry set	GT — Greater than
LT — Less than	T — Always true*
EQ — Equal	HI — Higher
MI — Minus	VC — Overflow clear
F — Never true*	LE — Less or equal
NE — Not equal	VS — Overflow set

* Not applicable to the B_{cc} or cpB_{cc} instructions.

the RTD instruction. This instruction is used to deallocate the stack. The RTR instruction returns the program to the point at which it was executing. It also loads the status register from the stack. If the first instruction in the subroutine saves the status register on the stack, this instruction restores the stack to the values it had when the call happened. RTS returns to the called subroutine, but the status register stays as it was during the subroutine.

The 68040 floating point processor provides a subset of the external floating point coprocessor functions. These are shown in Table 10-24. The floating point instructions are similar to the other arithmetic instructions we have used in the past. As we saw in the programming model, these instructions act on the seven 80-bit data register in the on-chip floating point arithmetic unit. These instructions are a subset of the instructions which are available if the external arithmetic coprocessor is used.

The following 68XXX instructions are available only when the processor is used in the supervisor mode. That is, they are privileged instructions and will not execute unless the S bit is set in the status register.

Table 10-25 on page 288 shows the data movement operations added for the 68040 processor. These introduce the ability to move blocks of data, multiple registers, and the special function registers.

The supervisory control instructions are made up of three different types of instructions. These are the privileged, trap-generating, and condition code register instructions. Again, these instructions execute only when the processor is in the supervisory state. The 68XXX processors are in the supervisory state when the status register S bit is set. The *privileged instructions* make major changes in the way the 68XXX processors operate. Therefore, these instructions work only when

Table 10-24	The 68XXX Floating Point Instructions			
Mnemonic	Description	68000, 68010	68020	68030, 68040
FABS	Floating point absolute value			X
FADD	Floating point add			X
FB$_{cc}$	Floating point branch			X
FCMP	Floating point compare			X
FDB$_{cc}$	Floating point decrement and branch			X
FDIV	Floating point divide			X
FMOVE	Move floating point register			X
FMOVEM	Move multiple floating point registers			X
FMUL	Floating point multiply			X
FNEG	Floating point negate			X
FRESTORE	Restore floating point internal state			X
FSAVE	Save floating point internal state			X
FS$_{cc}$	Floating point set according to condition			X
FSQRT	Floating point square root			X
FSUB	Floating point subtract			X
FTRAP$_{cc}$	Floating point trap on condition			X
FTST	Floating point test			X

cc = Condition code, which can be

CC — Carry clear GE — Greater or equal
LS — Lower or same PL — Plus
CS — Carry set GT — Greater than
LT — Less than T — Always true*
EQ — Equal HI — Higher
MI — Minus VC — Overflow clear
F — Never true* LE — Less or equal
NE — Not equal VS — Overflow set

* Not applicable to the B$_{cc}$ or cpB$_{cc}$ instructions.

Table 10-25   The 68XXX Special Data Movement Instructions

Mnemonic	Description	68000, 68010	68020	68030, 68040
MOVE16	16-byte block move			X
MOVE CCR	Move condition code register			X
MOVE SR	Move status register			X
MOVE USP	Move user stack pointer			X
MOVEM	Move multiple registers			X

the processors are in the supervisory state. If you write programs that can cause problems when they execute incorrect instructions, then start by running these programs in the user state. This way the program causes an interrupt if it tries to execute a privileged instruction. The instructions shown in Table 10-26 are privileged versions of common instructions.

The 68XXX *traps* are like interrupts. The trap happens when the trap mode is turned on and a special condition occurs. Traps cause the processor to move to the exception processing state. The program counter is then reloaded from the contents of a specified memory location. Like the interrupt, the trap memory locations are pointed to by vectors. Traps can be generated by certain internal error conditions or by executing the TRAP instruction. The TRAP instruction is like the 6802 software interrupt instruction. TRAP instructions are shown in Table 10-27.

The trap on overflow (TRAPV) generates a trap if the status register overflow bit (V) is set; otherwise, the next instruction in sequence executes. The CHK instruction checks a specified register to see whether its contents are below zero or above a specified number. Either condition generates a trap. This a useful instruction to make sure that data stay within given limits. The CHK2 instruction (68020

and higher) checks a specified register to see whether the content is above one number and below another number.

The *condition code instructions* let the programmer change the flags in the status register or save the status register contents.

A number of instructions have been added to the programmer control special functions in the new 68XXX processors. These are shown in Table 10-28.

The 68030 introduced the use of on-chip memory management. With an on-chip memory management unit, the 68030 processor allows the programmer access to virtual memory by changing logical addresses (a theoretical address anywhere in the processor's memory space) into physical addresses (memory locations which are physically in the system). This is done by using address translation tables which are stored in memory. To speed up operations, the most recently used translations are stored in an address translation cache (ATC) in the MMU. The PFLUSH and PTEST instructions are provided to allow the programmer control over the ATC contents.

The 68020 introduced the ability to support multiple coprocessors. The 68XXX processors which can support multiple coprocessors can use from one to eight coprocessors. The multiprocessor instructions allow the programmer

Table 10-26   The 68XXX Supervisory Control Instructions

Mnemonic	Description	68000, 68010	68020	68030, 68040
MOVEC	Move control register	X	X	X
MOVES	Move alternate address space	X	X	X
RESET	Reset external devices	X	X	X
RTE	Return from exception	X	X	X
STOP	Stop	X	X	X
MOVE	Move	X	X	X
ADDI	Add immediate	X	X	X
EORI	Logical exclusive-OR immediate	X	X	X
ORI	Logical inclusive-OR immediate	X	X	X

Table 10-27 The 68XXX Trap Instructions

Mnemonic	Description	68000, 68010	68020	68030, 68040
BKPT	Breakpoint		X	X
CHK	Check register against bounds	X	X	X
CHK2	Check register against upper and lower bounds		X	X
ILLEGAL	Take illegal instruction trap		X	X
TRAP	Trap	X	X	X
TRAPcc	Trap conditionally		X	X
TRAPV	Trap on overflow	X	X	X

Table 10-28 The 68XXX Special Functions Instructions

Mnemonic	Description	68000, 68010	68020	68030, 68040
PFLUSH (A, N)	Flush entry in the ATCs			X
PTEST	Test a logical address			X
CAS	Compare and swap operands		X	X
CAS2	Compare and swap dual operands		X	X
TAS	Test operand and set	X		X
CINV	Invalidate cache entries			X
CPUSH	Push then invalidate cache entries			X

to work with the coprocessors and in multiprocessor environments. The special instructions introduced in order to manage multiple processors are the CAS, CAS2, and TAS instructions.

The 68040 introduced on-chip cache memory and an on-chip floating point processor. The CASH instructions CINV and CPUSH are provided for on-chip cache memory support.

For most instructions, the programmer selects the register and addressing mode to be used. However, there are some cases when the instruction automatically selects a certain register. These are shown in Table 10-29. When these instructions are used, the addressing is automatically implied.

Table 10-29 shows three different stack abbreviations: the system stack pointer (SP), the user stack pointer (USP), and the supervisor stack pointer (SSP). Why do the 68XXX processors have these different stacks, and what is the difference between them?

Some 68XXX instructions use either the USP or the SSP, whichever is selected. When

Table 10-29 Implicit Instruction Addresses

Instruction	Implied Register
B$_{cc}$	PC
BRA	PC
BSR	PC, SP
CHK	SSP, SR
DB$_{cc}$	PC
DIVS	SSP, SR
DIVU	SSP, SR
JMP	PC
JSR	PC, SP
LINK	SP
MOVE CCR	SR
MOVE SR	SR
MOVE USP	USP
PEA	SP
RTE	PC, SP, SR
RTR	PC, SP, SR
RTS	PC, SP
TRAP	SSP, SR
TRAPV	SSP, SR
UNLK	SP

PC = program counter; SP = stack pointer (USP or SSP); SSP = system stack pointer; USP = user stack pointer; SR = status register this is the case, A_7 is called the *system stack pointer (SP)*. If the instruction uses only one of the two stack pointers, that pointer (either USP or SSP) is shown in the list.

As you can see, the 68XXX processors have a very powerful set of instructions made even more powerful by the addressing modes. Typically, this powerful instruction set is used to support high-level language processing, multitasking or multiprogramming operations, or operations involving a great deal of sophisticated logical or arithmetic processing, such as graphics or digital signal processing. Usually this powerful an instruction set is not needed to perform simple control applications, which are better done with the simpler, 8-bit microprocessors. However, as industry begins to use more and more sophisticated computations as part of process control systems, processors like the 68XXX family are being found in industrial control situations.

Self-Test

Answer the following questions.

78. If the 68XXX status register S flag is set and the PEA (push effective address onto the stack) instruction is executed, the processor
    a. Uses the user stack pointer to point to the stack
    b. Uses the supervisor stack pointer to point to the stack
    c. Either *a* or *b* is true
    d. Will not use the stack

79. The data movement instruction EXG exchanges the data in two registers. You would expect that this instruction would use only 32-bit operands instead of the usual 68XXX variable range of operand lengths because
    a. The 68000 can address only 16 Mbytes of memory
    b. 32 bits handle 4 bytes at a time
    c. The 68XXX processors use 32-bit registers
    d. All of the above

80. The 68XXX processors can perform
    a. Signed multiply and divide operations
    b. Unsigned multiply and divide operations
    c. Multiply and divide operations with 16-bit multipliers, multiplicands, divisors, quotients, and remainders and 32-bit products and dividends
    d. All of the above

81. The 68XXX sign extend (EXT) instruction lets the processor
    a. Check for nonzero numbers
    b. Extend the time required for a multiplication
    c. Generate a 2's complement of the operand
    d. Fill in the unused part of the addressed location with repeated sign bits

82. If register D5 contains 00F3D2CF and you execute a CLR D5 WORD instruction, you would expect the result to be
    a. 00000000   c. 00F30000
    b. 00F3D2CF   d. 00F3D200

83. The test (TST) instruction
    a. Subtracts the operand from zero and sets the status register bits but does not save the results of the subtraction
    b. Subtracts the operand from zero and sets the status register overflow bit but does not save the results of the subtraction
    c. Subtracts the operand from zero and sets the status register zero bit and saves the result of the subtraction in the operand
    d. Subtracts the operand from zero and sets the status register overflow bit and saves the result of the subtraction in the operand

84. The difference between the AND and the ANDI instructions is that
    a. ANDI is used for byte and word data and AND is used for long word
    b. ANDI has only one address operand; the second operand is contained in the second word or second and third words of the instruction
    c. ANDI is used only in the 6802
    d. ANDI returns the result to the effective address and AND returns the results to the indicated register

85. The 68XXX shift and rotate instructions
    a. Have versions which include the status register extend (X) bit
    b. Allow the programmer to specify the number of shifts up to 64
    c. Work on registers or memory locations
    d. All of the above

86. To complement a bit in a memory location, you would use the _____ instruction.
    a. BCHG   c. BCLR
    b. BTST   d. BSET

87. The _____ instruction places the complement of the indicated bit in the status register zero bit.

a. BCHG   c. BCLR
b. BTST   d. BSET

88. The 68XXX has instructions to _____ BCD numbers.
    a. Add and subtract
    b. Negate
    c. Multiply and divide
    d. a and b are true
    e. a, b, and c are true

89. The 68XXX branch instructions use _____ addressing.
    a. Relative
    b. Direct
    c. Register indirect
    d. All of the above

90. The program control instructions let the programmer
    a. Test for a number of conditions and generate a new program counter value if the condition is true
    b. Use subroutines and return from these subroutines
    c. Add a number to the program counter value using 2's complement arithmetic
    d. All of the above

91. The 68XXX processors have a special instruction which lets you set up loops. This _____ instruction combines the test and decrement operations in one instruction.
    a. DB_{cc} c. S_{cc}
    b. B_{cc} d. All of the above

92. The difference between the 68XXX branch and jump instructions is that branches
    a. Are used to start subroutines and jumps are not
    b. Use immediate addressing and jumps use relative addressing
    c. Cannot use a return instruction
    d. Use relative addressing and jumps can address the whole memory space

93. If you do not want a called subroutine to change the status register flags, you use a MOVE SP, -SP (move status register to the stack) when you enter the subroutine and you return from the subroutine with
    a. RET   c. RTS
    b. RTR   d. RTE

94. The privileged instructions can be executed if the
    a. Processor is not in the supervisor state
    b. Processor is in the supervisor state
    c. Processor is in the user state
    d. Status register (S) bit is cleared

95. The 68XXX trap instructions can be thought of as sophisticated

a. Move instructions
b. Software interrupts
c. Arithmetic instructions
d. Subroutine calls

10-11 THE 68XXX HARDWARE

Each of the 68XXX processors is sufficiently different to have different function and pinout diagrams. In this section, we will first look at the variety of packages used for the 68XXX processors, and then we will look at the functional diagrams.

The 68000 and the 68010 have common function and pinout diagrams (see Fig. 10-41 on page 292). As Fig. 10-41(b) shows, these processors use a 64-pin package. The 64 pins are used for the larger address and data buses and because the 68XXX processors do not use any multiplexed address and data lines.

The 68020 and newer 68XXX processors each have a unique function and pinout diagram. Because these processors support a full 32-bit address bus and a full 32-bit data bus as well as additional control signals, much larger packages with large pinout capability are required. Two packages are common. Figure 10-42 shows a typical *pin grid array package,* and Fig. 10-43 shows a typical *quad flat pack* (see page 293). The number of pins and the exact pinout depend on the processor (or peripheral part). For example, the 68020 is available in either a 128-pin pin grid array or a 132-pin quad flat pack. However, the 68040 requires a 179-pin pin grid array.

Both the 68000 and the 68010 are also available in quad flat packs and pin grid arrays. Table 10-30 on page 294 shows the packages used for the different members of the 68XXX family. The pin grid array is designed to allow the device to be plugged in to a special pin grid array socket, and therefore it can be removed if necessary. The quad flat pack is designed to be soldered directly to the printed circuit board and is not generally considered to be replaceable. Quad flat packs can be replaced with special tools, but they are certainly not field-serviceable.

Figure 10-44 on page 294 shows a generalized function diagram for the 68XXX family of microprocessors. The signal groups on the right side of the diagram are common to all members of the family. The number of lines in each group varies with the exact processor, but the functions are similar. The signal groups

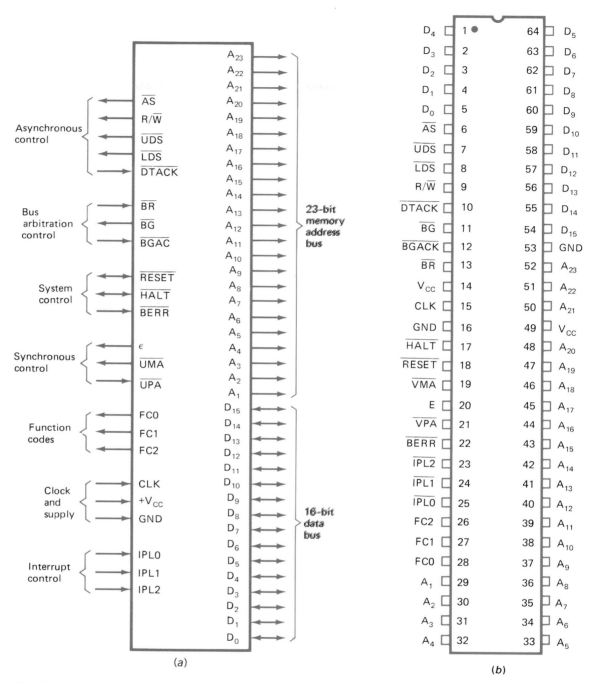

Fig. 10-41 (*a*) **The 68000 logical pinout groupings. Notice that the 68000 uses separate pins for its 23-pin memory address bus and its 16-pin data bus.** (*b*) **The 68000 in a 64-pin DIP.** *Adapted from Motorola literature.*

on the left side of the diagram are the ones which vary with the processor. Except for the M6800 peripheral control group, each of the other groups is added with the indicated generation of 68XXX processor and remains with newer versions of the processors.

The following paragraphs describe the functions performed by each of these groups. Refer to Figs. 10-41 and 10-44 as you study these signal descriptions.

The address bus is made up of unidirectional outputs which can be tristated. Normally,

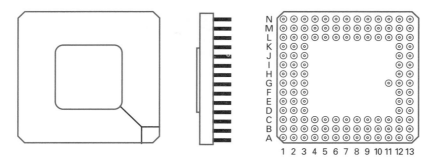

Fig. 10-42 **A 114-pin pin grid array package used for advanced mircoprocessors and peripheral devices.** *Note:* **Normally not all of the pins are dedicated to signals, because there are usually numerous V$_{cc}$ and ground pins to ensure that the current flow through any one pin is not too high and to distribute power to various parts of the integrated circuit.**

these address lines are used to select the desired memory location. During an interrupt operation, A_1 to A_3 shows the level of interrupt being serviced. Note that the 68XXX processors, like the 6802, use memory-mapped I/O.

The data bus is made up of lines D_1 to D_{15} (68000 and 68010) or D_0 to D_{31} (68020 and higher). This is a bidirectional bus which also can be tristated. During an interrupt, the exter-

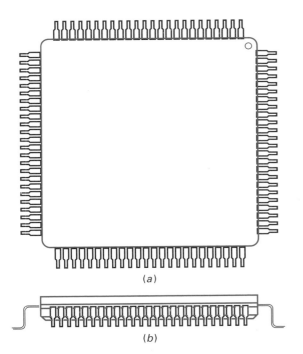

(a)

(b)

Fig. 10-43 **(a) A 132-pin quad flat pack.** *Note:* **The side profile (b) shows that the pins are formed to lie flat on printed circuit board pads. This is a surface-mount device, and therefore the pins do not go through the printed circuit board as they do with a DIP.**

nal interrupting device provides the interrupt vector on this bus.

The 68000, 68010, and 68020 have an address and data bus operation different from what is found on other microprocessors we have studied. The reason is that these processor buses operate synchronously; that is, each exchange with an external device waits until the external device tells the processor that it is done with the exchange. Other microprocessors have a synchronous bus; that is, the external devices must work synchronously with (in time with) the microprocessor bus timing. If a device is not fast enough to keep up with the microprocessor timing, it must ask the microprocessor to wait one clock period and try again.

The 68030 and newer processors offer three different types of bus operation: asynchronous, synchronous, and burst data transfers. The addition of two new types of bus operation allow the processor to transfer data faster when the external devices are fast enough to work in these modes. The *synchronous mode*, like the synchronous mode we have seen with other microprocessors, requires the external device to work at the processor speed rather than telling the processor when it is finished. The *burst data transfer* is a special type of bus operation which allows a continuous flow of data from memory into the on-chip cache memory until the cache is full.

The *asynchronous bus control lines* manage exchanges between external devices and the 68XXX address and data buses. This group has a line to indicate such conditions as the fact that the address lines hold a valid address, the start of the asynchronous bus cycle, the

Table 10-30   68XXX Processor Packages and Pin Count

Attribute	68000	68010	68020	68030	68040
Data lines	16	16	32	32	32
Address lines	23	23	32	32	32
Dual in-line package	64 pins	64 pins	N/A	N/A	N/A
Pin grid array	68 pins	68 pins	114 pins	128 pins	179 pins
Quad flat pack	68 pins	68 pins	132 pins	132 pins	—

size of the data transfer, and whether the processor is performing a read or write function.

The 68XXX processors support DMA (direct memory access) with the three bus arbitra-

tion control lines. The $\overline{\text{BR}}$ (bus request) line is used by an external device to tell the 68XXX processor that it wants to take control of the address and data buses. The 68XXX tells the external device that it can take control by

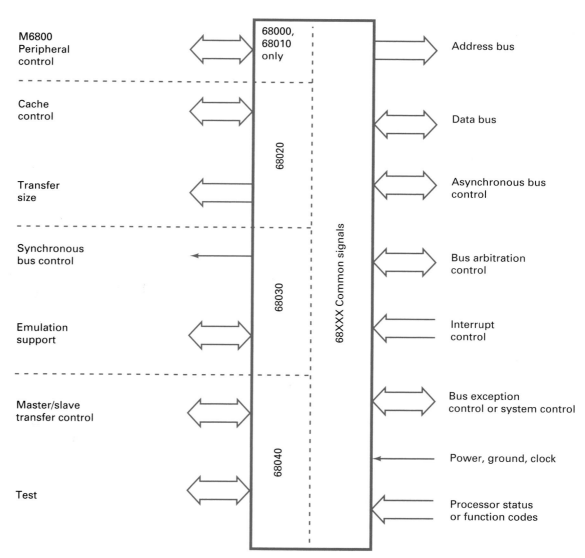

Fig. 10-44   A functional pinning breakout for the 68XXX processors. The signals on the right side are common to all members of the 68XXX family. However, the number of lines which make up each group may vary from processor to processor. The signals on the left side vary from processor to processor.

asserting $\overline{BG}$ (bus grant), and the external device can take control after it has asserted $\overline{BGACK}$ (bus grant acknowledge). The $\overline{BGACK}$ signal cannot be asserted until $\overline{AS}$ and $\overline{DTACK}$ are both not asserted; that is, until no device is using the address or data buses.

There are three interrupt inputs on a 68XXX processor. However, they are not three different inputs but a single 3-bit binary-coded interrupt. The binary codes indicate no interrupt (000 or level 0) or seven different levels of interrupt priority (001 to 111). This means that all external devices which interrupt a 68XXX processor must control all three interrupt priority level lines ($\overline{IPL0}$, $\overline{IPL1}$, or $\overline{IPL2}$). If all three lines are logic 0, there is no interrupt.

Bits 8, 9, and 10 of a 68XXX status register are called the *interrupt priority mask*. The programmer can set these bits to indicate the priority level of interrupt the 68XXX processor will recognize. Any interrupt with a priority equal to or less than the interrupt priority mask level is not recognized. Any interrupt with a priority greater than the interrupt priority mask level starts the 68XXX interrupt processing routine when the current instruction is completed.

When a 68XXX processor recognizes an interrupt, it starts an exception processing routine because an interrupt is one of the exception states. A copy of the current status register is saved on the system stack, and the processor is put in the privileged mode by setting the status register S bit. All exception mode processing must be done in the privileged state, so the full set of 68XXX instructions can be executed. Finally, the status register interrupt priority mask is set to the same priority level as the interrupt. This means that no other lower- or equal-priority device can interrupt the current interrupt service routine.

The 68XXX processors use interrupt vectors given by the external interrupting device. One of the first interrupt jobs is to ask the external interrupting device for its interrupt vector. This 8-bit number gives 256 different interrupt vectors. The 8-bit interrupt vector is multiplied by 4 and then is made into a physical address. A two-word (32-bit) number is fetched from this computed vector (the addressed memory location). This number is loaded into the program counter, and the interrupt service routine starts at this memory location.

The address calculated from the 8-bit vector is shown in Fig. 10-45. Table 10-31 on page 296 is a list of the vector assignments. As you can see, vectors 64 to 255 (memory locations 000100 to 0003FF) are reserved for the user-supplied starting address for the interrupt processing routines.

If the interrupt priority level is set to 7 (111), the interrupt recognition is a little different. All priority 7 interrupts are recognized even if the interrupt priority level is set to priority level 7. This gives the 68XXX processors a nonmaskable interrupt.

When the 68XXX processor returns from an interrupt, the RTE (return from exception) instruction is used. This returns the program counter, status register, and stack pointer to the values they had when the interrupt happened.

The 68XXX interrupt control system has many other features too detailed to discuss here. However, as you can see, the interrupt control system is designed to be used in a complex system where many interrupt vectors are needed and interrupt priorities are needed to make sure that high-speed external devices or important internal processor operations are serviced before lower-level interrupts are serviced. This is another example of how the advanced microprocessors are much more complex devices than the 8-bit microprocessors.

The 68XXX bus exception control or system control lines are used to reset or halt the processor and to tell the processor that a bus error happened. The reset input is a bidirectional input. As with other microproces-

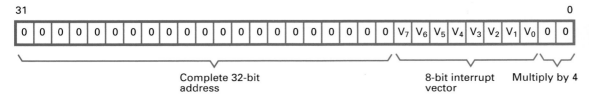

Fig. 10-45 Generating a complete address from the 8-bit interrupt vector supplied by an external device.

Table 10-31   Vector Assignments

Vector	Hex Address	Assignment
0	000	Reset initial SSP
—	004	Reset initial PC
2	008	Bus error
3	00C	Address error
4	010	Illegal instruction
5	014	Zero divide
6	018	CHK instruction
7	01C	TRAPV instruction
8	020	Privilege violation
9	024	Trace
10	028	Line 1010 emulator
11	02C	Line 1111 emulator
12	030	Reserved
13	034	Coprocessor protocol violation
14	038	Format error
15	03C	Uninitialized interrupt vector
16–23	04C–05F	Reserved by manufacturer
24	060	Spurious interrupt
25	064	Level 1 interrupt autovector
26	068	Level 2 interrupt autovector
27	06C	Level 3 interrupt autovector
28	070	Level 4 interrupt autovector
29	074	Level 5 interrupt autovector
30	078	Level 6 interrupt autovector
31	07C	Level 7 interrupt autovector
32–47	080–0BF	Trap instruction vectors
48	0C0	FPCP branch or set on unordered condition
49	0C4	FPCP inexact result
50	0C8	FPCP divide by zero
51	0CC	FPCP underflow
52	0D0	FPCP operand error
53	0D4	FPCP overflow
54	0D8	FPCP signaling NAN
55	0DC	Reserved
56	0E0	MMU configuration error
57	0E4 ⎫	
58	0E8 ⎬	Coprocessor
59–63	0C0–0FF	Reserved by manufacturer
64–225	100–3FF	User-assigned interrupt vectors

sors, asserting the $\overline{\text{RESET}}$ line causes the 68XXX processor to reset. Additionally, executing the reset instruction causes the 68XXX processor to assert the $\overline{\text{RESET}}$ line. This resets all the external devices connected to the 68XXX $\overline{\text{RESET}}$ line. However, the processor is not reset. This is implemented as two lines (a reset-in and a reset-out) in the 68040.

Likewise, the $\overline{\text{HALT}}$ line is a bidirectional line. Asserting $\overline{\text{HALT}}$ causes the processor to stop executing instructions at the completion of the current instruction. The processor remains halted until an interrupt is received.

The $\overline{\text{BERR}}$ (bus error) input is used to tell the processor that there is an external device error on the bus. A bus error is caused, for example, when an external device cannot respond to a bus request. This could be caused by addressing a device which is not there, for example.

The clock input needs to be driven with an externally generated square wave. The frequency of this square wave depends on the processor. Some models can use clock signals of 100 MHz or more; however, signals in the 10- to 50-MHz range are more common. The processors draw maximum power when they are operated with a maximum frequency clock. Typically 68XXX processors draw as little as 1 W to as much as 5 W from their 5 V dc power source.

The three processor status signals (FC0, FC1, and FC2) tell external devices what the state is (user or supervisor) and whether the processor is working with data or program information. This information is used by certain 68XXX support ICs to make sure that they are operating synchronously with the 68XXX processor.

The 68000 and 68010 have three 6802 peripheral control lines which let you add 6802 peripherals, such as ACIAs, PIAs, and PTMs, to 68000 and 68010 systems. *Note:* These 6802 peripheral control lines are not found on later members of the 68XXX family. Because the 6802 uses different bus control lines, they must be provided before the 68000 or 68010 can work with these devices. These processors provide enable (E), valid memory address ($\overline{\text{VMA}}$), and a special signal $\overline{\text{VPA}}$ (valid peripheral address). A signal on $\overline{\text{VPA}}$ tells the processor that the external device is a 6802 peripheral device and data signals must be synchronized with the enable timing signal.

Starting with the 68020, the 68XXX family of advanced microprocessors introduced on-chip cache memory. Depending on the size and sophistication of the on-chip cache memory, a number of external control lines are provided to control the cache and provide external information about its operation. In the simplest form, a cache disable line is available so that microprocessor emulator systems, used for microprocessor system development work, can defeat the cache memory, thus making testing easier.

The 68020 and later members of the 68XXX family also added a set of output lines which are used to indicate the number of bytes left in a data transfer. These signals are part of the processor's ability to dynamically adjust the bus size to fit the device connected to the bus.

The 68030 added synchronous data transfers to the 68XXX family. Prior to the 68030, all data transfers were asynchronous. The use of synchronous data transfers requires an added set of processor control lines which allow the processor and the peripheral device to synchronize their data transfer. Typically, a synchronous data transfer will be two-thirds faster than an asynchronous data transfer.

With increased processor complexity, the 68030 adds a number of specific control lines which allow the processor to work with a microprocessor emulator. Again, these signals are used in the development process.

A number of the 68XXX processors can be used in multiprocessor systems. To do this requires a great deal of control to make sure that data does not become mixed up between the processors. The 68040, with its large data and instruction on-chip cache memories, requires special control signals to make sure that the data does not become mixed.

The 68040 also adds a number of special control lines which allow an external device to aid in testing the processor. These are mostly used during manufacturing.

Self-Test

Answer the following questions.

96. The 68030 is not available in a DIP because
    a. The 68030 address bus has 32 bits
    b. The 68030 data bus is 32 bits wide
    c. There are too many control signals
    d. All of the above

97. Some 68XXX processors can transfer byte-wide data, word-wide data, or long-word-wide data by
    a. Sensing the data width from the peripheral
    b. Exchanging the upper and lower bytes
    c. Making multiple transfers if the data is more than byte-wide
    d. None of the above, because they transfer only word-wide and long-word data

98. All of the 68XXX processors use the _____ input for an external device to tell the processor that it has taken data from the data bus or the data it placed on the data bus is valid.
    a. $\overline{\text{DTACK}}$ c. $\overline{\text{BG}}$
    b. $\overline{\text{BR}}$ d. $\overline{\text{VMA}}$

99. The 68XXX $\overline{\text{RESET}}$ line is different from that in the X86 because
    a. It takes a logic 1 to reset the processor
    b. It takes a logic 0 to reset the processor
    c. The $\overline{\text{RESET}}$ line can be connected to external devices as well
    d. The $\overline{\text{RESET}}$ line is bidirectional and can output a reset signal as well as receive one

100. If the $\overline{\text{BGACK}}$ line is asserted, it means that
    a. The processor is acknowledging an interrupt
    b. The processor is communicating with a 6800 peripheral
    c. An external device requested use of the bus, received a bus grant signal from the processor, and no one else was using the buses
    d. the processor just executed a reset instruction, which causes it to reset external peripheral devices

101. If both the 68XXX halt and the bus error inputs are asserted, the processor completes the current bus cycle and then remains halted until the $\overline{\text{BERR}}$ (bus error) and $\overline{\text{HALT}}$ lines are no longer asserted. When the $\overline{\text{HALT}}$ line is no longer asserted, the bus cycle is rerun. This lets the processor work with slow external devices. This is very much like the Z80 _____ input.
    a. $\overline{\text{HALT}}$ *c.* $\overline{\text{WAIT}}$
    b. $\overline{\text{MREQ}}$ *d.* $\overline{\text{BUSACK}}$

102. The 68000 has special control signals which allow it to work with
    a. 6802 peripherals
    b. TTL signals
    c. A multiplexed address data bus
    d. Separate program and data memory space

103. The three 68XXX interrupt inputs (IPL0, IPL1, and IPL2) are used to
    a. Select the reset, maskable, and nonmaskable interrupts
    b. Give the processor an interrupt priority level
    c. Provide instruction processing levels
    d. All of the above

104. If the status register interrupt priority bits are set to 101, a 68XXX processor will
    a. Not respond at all; this is the interrupt defeat setting
    b. Respond to interrupts at priority level 4 and below
    c. Respond to interrupts at priority level 5 and up
    d. Not respond to interrupts at priority level 5 and below

105. When a 68XXX processor responds to an interrupt, the processor
    a. Is in the exception state
    b. Gets its vector from the interrupting device
    c. Is put into the privileged status
    d. All of the above

SUMMARY

1. The first 16-bit microprocessors were not outgrowths of 8-bit microprocessors but, rather, entirely new architectures.
2. The 32-bit microprocessors are outgrowths or extensions of the 16-bit microprocessors.
3. A typical use for an advanced microprocessor is as the CPU for a general-purpose computer.
4. The power and speed of an advanced microprocessor are required to achieve reasonable performance in a graphic display system.
5. A common use for advanced microprocessors is in support of systems which use a multiuser or multitasking operating system.
6. Operating system support usually means that the processor has a supervisory, protected, or privileged mode, which possesses the processor's most powerful capabilities.
7. Multiuser and multitask systems require protection systems, usually as a part of the processor, to keep one user or task from interfering with another user or task.
8. Often large software systems require more memory than is physically present on the processor. The advanced microprocessors can support these systems by using virtual memory techniques which allow automatic switching of software from mass storage to make the system look as if it is of maximum memory size.
9. The first Intel advanced microprocessors were introduced in the late 1970s. They were the 8088 and the 8086.
10. Internally, both the 8088 and the 8086 are 16-bit microprocessors. Externally, the 8088 has an 8-bit bus and the 8086 has a 16-bit bus. Both can address 1 Mbyte of memory.
11. The first IBM PC was introduced in 1981 and was based on the 8088.

12. The 286 was the basis for the IBM PC/ AT. The 286 supports up to 16 Mbytes of memory, real and protected modes, and virtual memory management.

13. The 386 is the first Intel 32-bit microprocessor. It supports up to 4 Gbytes of memory, high speeds, and advanced virtual memory.

14. There are a number of 386 versions. The 386DX is the base model. The 386SX uses a 16-bit data bus. The 386SL is a lower-power version with power management. The 386DX2 and 386DX4 are clock-doubling and clock-tripling models.

15. The 486 integrates a 386, 8-kbyte cache memory, and a math coprocessor into one IC. The 486DX is the base model. The 486SX does not have a math coprocessor. The DX2 and DX4 are clock-doubling and clock-tripling models.

16. The Pentium followed the 486. Although it is a 32-bit microprocessor, it has a 64-bit data bus. It has an 8-kbyte code cache and an 8-kbyte data cache.

17. The X86 base programming model has eight general-purpose 16- or 32-bit registers, four or six 16-bit segment registers, a flags register, and an instruction pointer.

18. All X86 processors have some level of code prefetching or code caching.

19. X86 instructions can address immediate data, registers, or memory.

20. An X86 instruction can operate on one or two operands. The result is stored in one of the operands.

21. When addressing memory, the X86 processors can use either direct or indirect addressing modes.

22. The direct addressing process adds the logical address (which in this case is the displacement from the instruction) to the segment base address from a segment register to produce the physical address.

23. The segment base address is created by taking the selected segment register contents and shifting them 4 bits to the left (multiplying by 10H or 16_{10}).

24. Indirect addressing creates the logical address through a series of computations using data from index and pointer registers which may be added to the instruction displacement.

25. The X86 memory addressing technique creates 64-kbyte memory segments in the 16-bit processors and 1-Mbyte or 4-Mbyte segments in the 32-bit processors.

26. Normally, a specific segment register is used with a specific type of operation. For example, the stack segment is built using the stack segment register.

27. Starting with the 286, the X86 processors have real and protected modes.

28. Real-mode memory addressing is limited to 1 Mbyte.

29. Protected-mode addressing uses the processor's full memory addressing capability.

30. In the protected mode, the segment register points to a segment descriptor table in memory. Data in the segment descriptor table includes the segment base address, the segment length, and segment rights.

31. Segment rights include segment privilege level, segment granularity, and segment type.

32. The X86 processors have four kinds of data movement instructions:
    • General-purpose instructions perform moves, push, pop, and exchange
    • Accumulator-specific instructions move data into or out of the A register
    • Address-object instructions load specific registers from memory
    • Flag instructions work with the flag register

33. The X86 arithmetic instructions include addition, subtraction, multiplication, and division. Multiply and divide can work with signed or unsigned numbers. There are provisions for working with packed and unpacked BCD numbers.

34. The logical instructions include NOT, AND, OR, and exclusive-OR. The TEST instruction is an AND without result except the status bit. The logical instructions also include the shift and rotate instructions.

35. The string manipulation instructions let you move blocks of data, search the data for specific data, or load blocks of memory. The repeat section of the string instructions lets you continue an instruction until the loop termination condition occurs.

36. The bit manipulation instructions let you set, clear, or test any bit in any word.

37. The control transfer instructions include many different conditional jump instructions which test status register bits and jump relative to a new location if the test is met. They also include the simple unconditional jump and the subroutine call instruction and its return.

38. The X86 instruction set includes a group of instructions specifically designed to help high-level languages run on the processor. These instructions provide special limit testing, stack operations, and byte ordering.

39. The processor control instructions let you work with bits in the flag register and perform external synchronization control.

40. There are significant hardware differences between the early (8088 and 8086) and newer (286 and later) X86 processors. The early processors use a multiplexed address-data bus and are packaged in 40-pin DIPs. The newer processors use large packages and do not multiplex signals.

41. The 8088 and 8086 can be operated in either minimum or maximum mode. This mode determines how much external hardware will be needed to support the processor.

42. The 286 and newer processors are packaged in high pin-count pin grid array, flat packs, or PLCCs. These four-sided packages have from 60 to nearly 300 pins.

43. The X86 processors require an external interrupt vector for all defined interrupts. There are 256 different interrupt vectors, with 32 reserved for internally generated interrupts and 224 available for user-defined externally generated interrupts.

44. X86 system implementations which use the PC architecture are moving toward few VLSI chip configurations.

45. Late in 1976, Motorola introduced the first advanced microprocessor in IC form with an architecture developed specifically as a microprocessor. It was called the Motorola MC68000.

46. The 68XXX is the CPU used in Apple's Macintosh series of microcomputers, in a number of CAD/CAE (computer aided-design/computer-aided engineering) workstations, and in many advanced industrial control products.

47. The 68000 and 68010 have an internal 32-bit architecture, a 16-bit data bus, and a 24-bit memory address bus. They can address only 16 Mbytes of physical memory. Special instructions added to the 68010 give it virtual memory addressing capability.

48. The 68020 introduced the 32-bit data bus, used a coprocessor interface which supports up to eight coprocessors, and added a 256-byte on-chip instruction cache.

49. The 68030 internal design consists of two independent 32-bit processors for fast processing. Intelligent data bus management allows 8-, 16, or 32-bit external data bus operation. An internal MMU allows the 68030 to address 4 Gbytes of memory space in real or virtual memory mode.

50. The 68040 adds an internal floating point arithmetic unit. Internal design improvements allow the 68040 to carry out many of its instruction executions in a pipeline mode and therefore speed up its operation.

51. The 68060 uses a new superscalar execution technique which allows it to process more than one instruction per clock cycle. The floating point arithmetic unit has substantially increased speed, and the on-chip cache memory has been increased to 16 kbytes.

52. The 68XXX processor programming models are divided into the user and system sections. The user section is where ordinary tasks run, and the system section is where the full power of the microprocessor is allowed.

53. The 68XXX family of advanced microprocessors have seventeen 32-bit registers, a 32-bit program counter (23 bits for the 68000), and a 16-bit status register.

54. The first eight registers (D_0 to D_7) are data registers. They work in the bit (1-bit), BCD (4-bit), byte (8-bit), word (16-bit) or long-word (32-bit) modes.

55. The eight address registers (A_0 to A_7) either point to memory locations or their contents are used to help calculate the memory location address.

56. The instruction pointer is a dedicated register.

57. The status register includes the flag which switches the 68XXX processor from the user to supervisor state.

58. Each state has its own stack pointer so that the programmer can easily switch between the user and supervisor states without tracking the state being used.

59. When the processor is in the supervisor state, you have access to many more instructions and a number of system registers in the more advanced versions of the 68XXX family.

60. The 68XXX processors are always in one of three major processing states: normal, exception, and halt. The normal processing state uses program instructions

with user or supervisor privileges. An interrupt or certain interrupt-like events put the processor into the exception-processing state. In the halt processing state, the 68XXX processors do not execute any instructions. They wait for an interrupt.

61. The 68XXX processors allow separate memory areas for data and instructions. This is done by using instructions which use different address registers for instruction and data fetches. The program counter is used for instruction fetches, and one of the address registers (A_0 to A_7) is used for data fetches.

62. There are two register direct addressing modes. The data register direct mode specifies an instruction operand in one of the eight 32-bit data registers (D_0 to D_7). The address register direct mode specifies an operand in one of the eight 32-bit address registers (A_0 to A_7).

63. The register indirect modes specify an instruction operand in a specific memory location. There are four different register indirect modes. Each one of these four modes uses one of the memory address registers (A_0 to A_7) to point at the addressed memory location.

64. The register contents can also be the base of a calculation which points to the memory location. The effective address is calculated from the register contents and some other data. Instructions which use these addressing modes to give two offsets from the base register are very useful when addressing data in two-dimensional arrays.

65. The data movement operations move data between any two locations which can contain data. There are both data move instructions and address move instructions. The data move instructions move byte, word, and long-word data. The address move instructions move only words and long words.

66. The integer arithmetic operations include multiplication and division as well as addition and subtraction. The extended add and subtract instructions perform multiple precision arithmetic. With the extended instruction, the zero flag (Z) is set only if the entire result is zero.

67. The four logical operations are AND, OR, EOR, and NOT. The test instruction (TST) uses zero as one operand. The operand named in the TST instruction is compared to zero, and the status register bits are set. No other results are saved.

68. The shift and rotate instructions can operate on a register or on memory. Memory shifts and rotates must work on a word. Register shifts and rotates are much more flexible.

69. The bit manipulation instructions test (BTST), set (BSET), clear (BCLR), or change (BCHG) any bit in any memory location or register. Each bit-oriented instruction must have not only an effective address but also a bit address.

70. The 68XXX program control instructions allow the programmer to use subroutines and return from those subroutines. The conditional branch (B_{cc}) instructions let the programmer test for one of a number of different conditions, and generate a new program counter value if the condition is true.

71. The 68040 floating point processor provides a subset of the external floating point coprocessor functions. These instructions act on the seven 80-bit data registers in the on-chip floating point arithmetic unit.

72. The supervisory control instructions are the privileged, trap generating, and condition code register instructions. These instructions execute only when the processor is in the supervisory state. A program causes an interrupt if it tries to execute a privileged instruction.

73. The 68000 and the 68010 use a 64-pin DIP package. The 64 pins are used for the larger address and data buses. The 68XXX processors do not use multiplexed address and data lines.

74. The 68XXX address bus is made up of unidirectional outputs which can be tristated. Normally, these address lines are used to select the desired memory location. During an interrupt operation, A_1 to A_3 show the level of interrupt being serviced. The 68XXX processors, like the 6802, use memory-mapped I/O.

75. The data bus is made up of lines D_1 to D_{15} (68000 and 68010) or D_0 to D_{31} (68020 and higher). This is a bidirectional bus which can be tristated. During an interrupt, the external interrupting device provides the interrupt vector on this bus.

76. The 68000, 68010, and 68020 address and data buses operate synchronously. Each exchange with an external device waits

until the external device tells the processor that it is done with the exchange.

77. The 68030 and newer processors offer three different types of bus operation: asynchronous, synchronous, and burst data transfers. The two new bus operations allow the processor to transfer data faster when the external devices are fast enough to work in these modes. The synchronous mode requires the external device to work at the processor speed rather than telling the processor when it is finished. The burst data transfer allows a continuous flow of data from memory into the on-chip cache memory until the cache is full.

78. There are three interrupt inputs on a 68XXX processor which are single 3-bit binary-coded interrupts. The binary codes indicate no interrupt (000 or level 0) or seven different levels of interrupt priority (001 to 111). All external devices which interrupt a 68XXX processor must control all three interrupt priority-level lines. If all three lines are logic 0, there is no interrupt.

79. Bits 8, 9, and 10 of a 68XXX status register are the interrupt priority mask, and they indicate the priority level of interrupt the 68XXX processor will recognize. Any interrupt with a priority equal to or less than the interrupt priority mask level is not recognized. Any interrupt with a priority greater than the interrupt priority mask level starts the 68XXX interrupt processing routine when the current instruction is completed.

80. The 68XXX processors use interrupt vectors given by the external interrupting device. An 8-bit number gives 256 different interrupt vectors. The 8-bit interrupt vector is multiplied by 4 and then is made into a physical address. A two-word (32-bit) number is fetched from this computed vector (the addressed memory location). This number is loaded into the program counter, and the interrupt service routine starts at this memory location.

CHAPTER REVIEW QUESTIONS

Answer the following questions.

10-1. The first advanced microprocessors were
  a. Introduced in the late 1970s
  b. First introduced as 16-bit architectures and then expanded to 32-bit architectures
  c. New 16-bit architectures developed specifically for implementation as 16-bit microprocessors
  d. All of the above

10-2. The major difference between the Intel 8088 and the 8086 is that the 8086 has
  a. All 64-bit internal registers
  b. A 32-bit external data bus
  c. A 16-bit external data bus
  d. Lower power consumption because it was designed later

10-3. The 486DX2 is different from the 8088 because it has
  a. All 64-bit internal registers
  b. A 32-bit external data bus
  c. A 16-bit external data bus
  d. Lower power consumption because it was designed later

10-4. Often an application which uses an advanced microprocessor is configured as a
  a. Single-chip microprocessor controller
  b. General-purpose computer-microcontroller
  c. Limited-function multichip controller
  d. Any of the above

10-5. One common use for advanced microprocessors is graphic applications because
    a. The slower speeds which come from systems based on 8-bit microprocessors can cause operator frustration
    b. Graphic applications often treat the presentation as a large array, and these large arrays require a great deal of computing power to handle them
    c. These displays are often in color, which adds another dimension to the array that must be quickly processed
    d. All of the above

10-6. When a great deal of _____, the X86 processors are often supported with a numeric coprocessor.
    a. Memory is to be addressed
    b. Fast I/O is to dealt with
    c. Data movement between memory locations is required
    d. Complex mathematical computations are required

10-7. A 32-bit microprocessor is inherently faster than a 16-bit microprocessor because
    a. It has a faster clock and therefore a faster instruction cycle
    b. It uses words which are twice as long and therefore can do twice as much with them in the same amount of time
    c. It has a much richer instruction set and therefore can perform more complex actions with each instruction
    d. All of the above

10-8. One of the reasons the X86 processors are so fast is that they have separate ALUs for ordinary computations and
    a. Fast I/O operations
    b. Work with graphic applications
    c. Calculate memory address values
    d. Each of the above as needed

10-9. The Pentium processor has
    a. An 8-kbyte code cache   *c.* A 64-bit data bus
    b. An 8-kbyte data cache   *d.* All of the above

10-10. The X86 memory address values are based on the current segment register value. The value in this register must be _____ before it can be used to address a memory location.
    a. Shifted four places to the left
    b. Combined with the logical address from the instruction
    c. Used as the address for the segment descriptor table if the processor is in the protected mode
    d. All of the above

10-11. Which of the following is not addressed by a separate segment register?
    a. Data
    b. The stack
    c. Code
    d. All have separate segment registers

10-12. The logical address for code addresses is found in the _____ register.
    a. Stack pointer   *d.* Destination index
    b. Instruction pointer   *e.* Base pointer
    c. Source index

10-13. The logical address for the place you temporarily store data on a first-in–last-out basis is found in the _____ register.
    a. Stack pointer   *d.* Destination index
    b. Instruction pointer   *e.* Base pointer
    c. Source index

10-14. The logical address for an instruction which moves data into or out of memory is found in the _____ register.
  a. Stack pointer
  b. Instruction pointer
  c. Source index
  d. Destination index
  e. Base pointer

10-15. If you do not change the segment register value, you are limited to _____ of program space.
  a. 1 Mbyte
  b. 64 kbytes
  c. 4 Gbytes
  d. Each of the above is correct depending on the X86 processor being used and the conditions under which it is being used

10-16. The X86 A, B, C, and D registers
  a. Can be 8-, 16-, or 32-bit (in the 386 and above) registers
  b. Can all be the operand in an I/O instruction
  c. Must be used as a part of a memory reference instruction
  d. Can be operated in stack pointer mode

10-17. The stack and base pointer registers
  a. Cannot be operated in the byte mode
  b. Are used to point to the current operating location in a stack
  c. Autodecrement or autoincrement (depending on the instruction) when they are used to address memory
  d. All of the above

10-18. The X86 was expanded to have two extra data segment registers starting with the
  a. 8088
  b. 286
  c. 386
  d. 486
  e. Pentium

10-19. The segment descriptor table stores information about a segment to be used in the protected mode including the
  a. Size of the segment
  b. Segment base address
  c. Segment priority level
  d. All of the above

10-20. If the segment descriptor table has the _____ bit set, the processor knows that it must initiate a special routine to load the desired segment into main memory from mass storage before it allows the processor to address the target segment.
  a. Granularity
  b. Segment not present
  c. Segment type
  d. All these bits must be set

10-21. The displacement value in an X86 instruction
  a. Is used as immediate data
  b. Is not controlled by the programmer
  c. Is available only for very special instructions
  d. Becomes part of the physical address once it is added to the shifted segment value

10-22. If you want to move data into an X86 processor from an external device, you can use the _____ instruction.
  a. IN
  b. MOVE
  c. XLAT
  d. Any of the above

10-23. _____ is an instruction specifically designed to let you perform table-based data translations.
  a. PUSH
  b. POP
  c. XLAT
  d. LEA

10-24. The X86 arithmetic instructions work on
  a. Signed and unsigned numbers
  b. Packed BCD data
  c. ASCII data
  d. All of the above

10-25. The purpose of the convert byte to word (CBW) instruction is to
  a. Allow a proper calculation with data from a shorter word
  b. Allow calculations with data such as externally provided 8-bit numbers
  c. Fill the unused bits with a replication of the number's sign bit
  d. All of the above

10-26. The double precision shift instruction was introduced on the _____ processor.
  a. 8088/8086   d. 486
  b. 286   e. Pentium
  c. 386

10-27. The TEST instruction is the same as the AND instruction except that
  a. The result is stored in the second operand
  b. The result is stored in the first operand
  c. No result is stored except in the flag bit
  d. Each of the above is correct depending on the X86 processor being used and the conditions under which it is being used

10-28. The string manipulation instructions let you perform
  a. The movement of blocks of data between areas of memory
  b. A data compare between the value stored in a register and the values in a block of data
  c. Repeated instructions until a given condition exists
  d. All of the above

10-29. You are inputting a string of data from an external device and you need to watch for an end-of-file (EOF) character and terminate the load operation when EOF is detected. To do this, you would probably use the _____ instruction.
  a. MOVS   d. REP
  b. CMPS   e. REPE
  c. SCAS   f. REPNE

10-30. The X86 conditional transfers cause the instruction pointer to be set to a new value
  a. If the test condition is true
  b. By a jump to the new address
  c. Through the calculation which is relative to the current instruction pointer value
  d. All of the above

10-31. If you wish to unconditionally and permanently transfer program execution to a distant memory location, you would probably use
  a. The call procedure instruction
  b. The jump short instruction
  c. A jump long instruction
  d. Any of the above—use depends on the amount of memory available

10-32. If you want an X86 processor to "go to sleep" until an external event occurs on one of the interrupt lines, you execute the _____ instruction.
  a. HLT   d. LOCK
  b. WAIT   e. NOP
  c. ESC

10-33. If you want to pass an instruction to an external device such as a numeric coprocessor, and you do not want the X86 processor to execute the instructions as they are fetched, you execute the _____ instruction to prepare the processor.
  a. HLT   d. LOCK
  b. WAIT   e. NOP
  c. ESC

10-34. The introduction of the protected mode in the X86 processors included new instructions to
    a. Load and store the segment descriptor register
    b. Load the machine status word into a general register
    c. Load the register which deals with the TSS
    d. All of the above

10-35. One of the major hardware differences between the 8088 and the 286 and later processors in the X86 family is that the
    a. 286 and later models do not use a multiplexed address-data bus
    b. 8088 and 8086 are available in a 40-pin DIP, whereas the other models require a PGA, flat pack, or PLCC
    c. 8088 and 8086 have a minimum and maximum mode
    d. All of the above

10-36. An interrupt signal applied to the INTR input will be acted on only if _____ and the signal stays on INTR until it is acknowledged on INTA.
    a. The interrupt vector is placed on the data bus by the interrupting device
    b. IF is set
    c. The processor is not halted
    d. All of the above conditions are in effect

10-37. In addition to new versions of the X86 family, there are derivations of the existing versions. These derivations are often processors combined with memory, I/O, and other functions. This is very understandable when we compare this trend with the
    a. Development of 8-bit microcontrollers from 8-bit microprocessors
    b. Use of VLSI building blocks to reduce the chip count for a PC
    c. Evolution of the 486 from the 386
    d. All of the above

10-38. The 68XXX architecture is an evolution of the
    a. PDP-11 minicomputer
    b. 8088
    c. 6802
    d. None of the above; it is unique

10-39. The 68000 has a 16-bit external data bus which makes it a
    a. Microprocessor
    b. Slower processor than the 68010
    c. True 16-bit microprocessor
    d. All of the above

10-40. Internally, all members of the 68XXX family have a(n) _____ architecture.
    a. 4-bit
    b. 8-bit
    c. 16-bit
    d. 32-bit
    e. 64-bit

10-41. The _____ introduced the 32-bit data bus to the 68XXX family.
    a. 68000
    b. 68010
    c. 68020
    d. 68030
    e. 68040
    f. 68060

10-42. Briefly explain why the 68XXX archiecture is really much simpler than the X86 architecture when the 68XXX has seventeen 32-bit registers in the user model.

10-43. List the different kinds (lengths) of data the 68XXX processor data registers work with. What is each different length called?

10-44. Briefly explain how two of the five different memory address register modes of 68XXX generate an effective address for a memory location.

10-45. In a 68XXX processor, register A_7 is
    a. One of the eight address registers of the 68XXX

b. The user stack pointer

c. Not used if the status register supervisor state bit is set

d. All of the above

10-46. Briefly explain what different things happen in the normal, exception, and halt states for a 68XXX processor.

10-47. If you need an additional stack, you need an additional stack pointer. When you are using a 68XXX processor, you can create the additional stack pointer by using one of the address registers in the _____ addressing mode.

a. Register direct

b. Register indirect

c. Register indirect with either predecrement or postincrement

d. Register indirect with either displacement or index register with offset

10-48. With the 68XXX processor, the best register addressing mode for putting data into an array is the _____ mode.

a. Register direct

b. Register indirect

c. Register indirect with either predecrement or postincrement

d. Register indirect with either displacement or index register with offset

10-49. To modify data in a 68XXX address register, you would use the _____ mode.

a. Data register direct

b. Address register direct

c. Data register indirect

d. Address register indirect

10-50. If an instruction is to address any memory location within the 4-Gbyte address space of the 68030, the address portion of the instruction needs to be _____ words long.

a. One   c. Three

b. Two   d. Four

10-51. The 68XXX processors have two different kinds of relative addressing. One lets you add a 16-bit signed displacement of the _____, and the other uses an 8-bit offset plus an index register.

a. Program counter   c. Address register

b. Data register   d. Stack pointer

10-52. Instructions in the 68XXX instruction set which cannot be executed when the supervisor state bit is not set are called _____ instructions.

a. Privileged   c. Data transfer

b. Arithmetic   d. Logical

10-53. Like the 6802, the 68XXX processors do not use any _____ instructions.

a. Jump   c. Branch

b. Return   d. Call

10-54. All 68XXX processors have synchronous bus operation. Briefly explain what this means.

10-55. Some advanced versions of the 68XXX processors have modes of bus operation other than synchronous. Briefly explain what these modes are and why they are used.

10-56. What do the three different 68XXX interrupt lines do?

10-57. What does it mean when all of the 68XXX interrupt lines are at a logic 1?

10-58. Where do the 68XXX processors get their interrupt vectors?

10-59. The 68XXX processors use the 8-bit interrupt vector to

a. Load and store the segment descriptor register

b. Load the machine status word into a general register

c. Load the register which deals with the TSS

d. All of the above

CRITICAL THINKING QUESTIONS

10-1. Both the Intel X86 family and the Motorola 68XXX family came along at about the same time in history. What do you think were the driving factors which pushed the introduction of advanced microprocessors at this time?

10-2. In this chapter we have reviewed many of the things which can be done better, faster, and with greater precision by using advanced microprocessors. However, there are disadvantages to using an advanced microprocessor. What do you think some of these disadvantages are?

10-3. IBM chose the Intel family of advanced microprocessors for its line of PCs. Apple chose the Motorola family of advanced microprocessors for its Macintosh line of PCs. What major differences between these two processors do you think shows up between the IBM and Apple products? Why? Is there any real difference? If not, explain why the two companies made the choice they did.

10-4. The X86 processors started as 16-bit microprocessors and evolved into 32-bit microprocessors. How has this affected the growth from the early models to the most recent models?

10-5. Why do you think the 386 processor changed the size of a segment from 64 kbytes to a minimum of 1 Mbyte?

10-6. The 68XXX family of advanced microprocessors uses memory-mapped I/O. What are the advantages and disadvantages of memory mapped I/O?

10-7. The 68XXX processors are in either the user or the supervisor state. What are the corresponding states in an X86 processor? Compare the states.

Answers to Self-Tests

1. c	29. e	57. f	80. d
2. c	30. a	58. b	81. d
3. b	31. a	59. c	82. c
4. d	32. d	60. b	83. a
5. a	33. a	61. c	84. b
6. f	34. d	62. a	85. d
7. h	35. c	63. **a.** 2	86. a
8. a	36. b	**b.** 3	87. b
9. c	37. d	**c.** 1	88. d
10. c	38. d	**d.** 5	89. a
11. e	39. d	**e.** 4	90. d
12. b	40. a	**f.** 6	91. a
13. d	41. b	64. a	92. d
14. d	42. b	65. b	93. b
15. d	43. c	66. d	94. b
16. c	44. c	67. a	95. b
17. a	45. d	68. b	96. d
18. a	46. d	69. c	97. a
19. c	47. b	70. a	98. a
20. h	48. a	71. b	99. d
21. d	49. b	72. d	100. c
22. b	50. a	73. d	101. c
23. c	51. a	74. d	102. a
24. b	52. d	75. b	103. b
25. d	53. d	76. c	104. d
26. b	54. d	77. d	105. d
27. d	55. c	78. b	
28. a	56. d	79. c	

CHAPTER 11

Memory

■

CHAPTER OBJECTIVES

This chapter will help you to:

1. *Identify* read-write and read-only memory characteristics and systems.
2. *Compare* static and dynamic memories.
3. *Predict* the results of using various memory systems.
4. *Differentiate* between EPROMs and EAROMs.
5. *Name* and *explain* microprocessor memory extension techniques.

To make a microprocessor do something, we must have a place to store the program which tells it what to do, and we must have a place to store the data it works on and the answers it makes. All of this takes place in the microprocessor's main memory. Today a microprocessor's main memory is made up of many memory ICs. These ICs make up both read-write and read-only random access memory (RAM). Sometimes the microprocessor memory space is not big enough to store all the information needed. In this case, memory expansion techniques are used to get sufficient memory.

■

11-1 RANDOM-ACCESS READ-WRITE MEMORIES

The work we have done so far has shown us that the microprocessor must be able to read the instructions for a given program and store the data generated by that program. We have found that we must be able to read data from memory and write data to memory.

In today's microprocessor, we often find that there is only one addressable memory space. We call this the microprocessor's *main memory*. At least some of the microprocessor's main memory must be read-write memory. We must be able to put data into the memory locations or take data from them.

Almost all of the main memory in general-purpose microcomputer systems is read-write memory. Program instructions are stored in memory by using a memory-write process. Later, when executing the program, the microprocessor reads the instructions from memory. Data, of course, is also written into or read from memory locations.

The term *random-access memory* simply means a memory system in which any memory location can be accessed as easily as any other memory location. Most of today's microprocessor memory systems are random access. In fact, *RAM* (*random-access memory*) is so common in microcomputer systems that we often unthinkingly call all of the microprocessor's main memory RAM. It is important to remember that the term *RAM* does not tell you whether the memory is read-write or read-only. However, today the term RAM is often used to imply read-write RAM. Often when referring to an integrated circuit (IC) or device, we simply say RAM when we mean RAM IC (random-access memory integrated circuit).

If a memory system is not random access, then it is sequential access or some mixture of both. To access a memory location in a *sequential-access memory*, you must access every memory location between the present memory location you are addressing and the memory location that you want to read from or write to. Sequential-access memory is no longer used for microprocessor main memory.

For example, information recorded on a reel of magnetic tape is stored in sequential-access form. Sequential access is mostly used for mass storage of data. Sequential access is very satisfactory for storing large quantities of data that we do not need to get to very quickly.

Later, in Chap. 12, when we study mass storage systems, we will find that some mass storage devices, such as the floppy disk, use a combination of random access and sequential access. That is, these memory systems allow us to randomly access large blocks of the memory system and to use sequential access within these blocks. Although this system does not give us access as fast as does a completely random-access memory, it is much faster access than a completely sequential-access memory.

When we are working with memory systems, we often speak of the *memory access time* and the *memory cycle time*. Both of these are measures of the memory system's performance. The two measures go hand in hand—that is, the faster the access time, the faster the memory cycle time.

One kind of memory access time tells us how long a memory system takes to place information on the data bus after the desired memory location is addressed. This is called the memory's *read access time* or, simply, *read time*. *Write access time* measures how long it takes the memory system to write data into the addressed memory location.

Memory access time depends on the way that the memory system is built and on the speed of the memory system's circuits. For example, an integrated-circuit memory system might have a 50-nanosecond (50-ns) access time. That is, the time between addressing any memory location and the placing of data at the memory device's output is 50 ns (0.05 microseconds, μs).

On the other hand, the access time for magnetic tape can be quite long. In fact, access time depends on where the address memory location is found on the tape. If the addressed memory location is very near the beginning of the tape, the access time is *relatively* short. If the addressed memory location is toward the end of the tape, the access time is long. Usually we speak of the average access time for sequential access devices. For example, a tape that requires 40 s to run from one end to the other has an average access time of 20 s.

We also speak of the *memory system's cycle time*. The memory system's cycle time is the shortest possible time between two operations which access the memory. The memory system's cycle time depends on other microcomputer system timing factors. These other timing factors may keep the memory from being used for a set period after each memory access.

We often use two other terms when speaking of microcomputer memory systems. We speak of either *volatile* or *nonvolatile* memory. Simply, a nonvolatile memory retains its data when power is removed from the microcomputer system. *Volatile memory loses its data when power is removed from the microcomputer system.*

Since the microprocessor can do nothing without instructions, it is obvious that it must have some nonvolatile memory. That is, it must have a place to start once power is turned on. However, the amount of nonvolatile main memory needed does not have to be great. All that is necessary is to store a short program in main memory if all the other program instructions are available from a mass storage subsystem. This short program makes the microcomputer load the rest of the instructions into main memory from the nonvolatile mass storage device. As we have learned, this short program is called a *bootstrap* or *boot program*. As we shall see later, this boot program is often stored in some form of ROM.

You can also see that all programming instructions must be stored in some kind of nonvolatile memory. Otherwise, you would have to recreate the program each time you run it. Writing a program on paper is one kind of nonvolatile memory storage. This type of storage obviously has very slow memory cycle time.

Magnetic tape and disk are also nonvolatile memories. Core memory is also nonvolatile memory. Semiconductor read-write memory is volatile unless it has a battery backup.

Many *semiconductor main memory* systems now have battery backup. This gives them temporary nonvolatility. These memory systems can last through power failures as long as their batteries last. In a large computer, the batteries may only last for 10 or 15 minutes or the batteries may be able to sustain the system through several days' power failure. Generally speaking, battery backup does not give permanent nonvolatility. This is because the power drawn by the semiconductor memories is still great enough to require an unreasonably large battery in order to maintain memory for long periods.

Core memory was once the most popular computer main memory. In a core memory, each data bit is stored in a small, doughnut-shaped permanent magnet. The magnet is magnetized in one or two directions. The direction of the magnetism depends on the direction of the currents flowing through the memory wires. Core memory systems are not often used today, for they are very large, expensive, and use a lot of current. The heat this generates is difficult to handle. Today, computers are usually built with semiconductor memory systems rather than core memory systems.

Two major semiconductor technologies are used to build integrated-circuit memory devices, which are used to build memory systems for today's computers. These are the same two technologies used to make other digital integrated circuits—the bipolar and the MOS (metal-oxide semiconductor) technologies.

Bipolar memories are seldom used with microprocessor systems. The advantage of bipolar memories is their very fast access time. They do have a number of disadvantages when compared with MOS memories. They draw a great deal of power, and there are fewer memory bits for the same size silicon chip. The bipolar semiconductor fabrication process is much more complicated than the MOS process. This makes bipolar memory much more expensive than MOS memory. For these reasons, bipolar memory is used only for applications which can afford its great cost.

MOS memory is by far the most common microcomputer memory. There are two different ways to construct MOS memory integrated circuits. MOS memory circuits are either static or dynamic. Often, you will see the abbreviations SRAM and DRAM used for static RAM and dynamic RAM.

Static memory systems are simpler to build, especially for small memory systems. They are much easier to service. *Dynamic memory* systems use lower-cost integrated circuits but require more support circuits. Also, dynamic memories must be refreshed regularly. They are usually used in larger memory systems.

Both static and dynamic semiconductor technologies are steadily improving. This means that larger memories are being made on the same size silicon wafer. As a result, memory is becoming less expensive. As memory becomes less expensive, microcomputer systems become more *memory-intensive*. That is, the systems contain much more memory to do the same job. Of course, the systems begin to do the job in a different way when more memory is available.

For example, when we study programming in Chap. 14, we will learn that there are basically two different kinds of programs. There are *assembly language programs* and *high-level language programs*. An assembly language program makes very compact code, but it takes a great deal of programming time. High-level languages take much less development time but take much more memory space than assembly language programs.

Knowing that the cost of semiconductor memory is rapidly dropping, you can see why many manufacturers will choose to design microprocessor-based systems using high-level languages. When the cost of extra memory becomes less than the cost of extra engineering time to create the assembly language programs, manufacturers choose to use the extra memory.

Self-Test

Answer the following questions.

1. Main memory in a modern microcomputer system is usually
   a. Random access   *c.* Semiconductor
   b. Read-write   *d.* All of the above
2. The performance of a memory system is often measured by the time needed to place data on the microcomputer's data bus after the memory location has been addressed. This performance is measured by specifying the microcomputer system's
   a. Cycle time
   b. Sequential access
   c. Access time
   d. Random access
3. You would expect that _____ is a random-access device.
   a. Magnetic tape
   b. Read-only memory (ROM)
   c. A floppy disk
   d. All of the above
4. Electrically alterable read-only memories (EAROMs) are often used in microprocessor-based intelligent terminals. They are used to store data needed by the terminals at a given place over a period of time and through many power on/off cycles. You would expect EAROMs to be chosen because of their
   a. Nonvolatility
   b. Low power consumption

Core memory

Bipolar memory

MOS memory

Static memory

Dynamic memory

Assembly language programs

Memory-intensive systems

High-level language programs

c. Volatility

d. Ultrasmall size

5. A typical dedicated microprocessor system only has about 8 to 16 kbytes of main memory. Frequently, static MOS memory is used for such an application. In today's personal computer, memory systems of 8 Mbytes are not uncommon. Typically, you would expect to find _____ memories used for a system this large.

a. Static   c. Bipolar

b. Dynamic   d. Read-only

6. Typically, sequential-access memory devices are used in microcomputer systems to provide

a. True random access

b. Fast access time

c. Mass storage

d. All of the above

7. Nonvolatility for a conventional MOS RAM can be provided by

a. Battery backup

b. Core memory

c. Low power dissipation

d. Bipolar memory parts

8. Microprocessor systems are becoming memory-intensive because

a. Memory devices are becoming low cost

b. Programmers are using more high-level language programming

c. Microprocessor systems are being asked to do bigger and bigger jobs

d. All of the above

11-2 STATIC AND DYNAMIC MEMORIES

Most static memories are built using MOS technology, but some are built using bipolar technology. The MOS technologies used to construct memory parts usually come in two forms: NMOS and CMOS. The NMOS parts are much simpler to construct than the CMOS parts. However, the CMOS parts draw considerably less power and therefore are excellent candidates for memory systems which must use battery backup to make them nonvolatile.

In this section, we will take a brief look at how both types of NMOS static RAM are built and how bipolar static memory devices are built. We will also look at some common integrated-circuit memory device organizations.

Figure 11-1 shows a simplified schematic of a *bipolar memory cell*. A memory cell stores

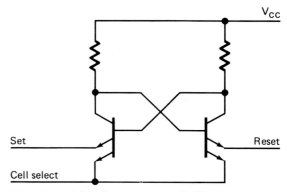

Fig. 11-1   A bipolar memory cell. The two multiple-emitter transistors are cross-coupled to make a simple flip-flop. The cell is set or reset by asserting the appropriate emitter. Asserting the CELL SELECT line makes a valid low-impedance output.

1 bit of information. This cell is implemented using TTL (transistor-transistor logic) multiple-emitter technology. As you can tell by looking at this schematic, the memory cell is nothing more than a simple *flip-flop*. This flip-flop is either set or reset to store the data. Once it is set, it stays set until it is reset or until it loses power.

Large numbers of these memory cells are organized on a row-and-column basis. Figure 11-2 shows the *row-and-column organization* of a 16,384-bit memory chip. This basic block diagram of the memory part is the same for both bipolar and MOS static and dynamic memories.

Each chip (integrated circuit) has 14 address lines, A_0 through A_{13}. These address lines are connected to the row-and-column address decoders. The first seven address lines, A_0 to A_6, are connected to the column decoder. The column decoder decodes a 7-bit address to indicate one of 128 columns. The second seven address lines, A_7 to A_{13}, are connected to the row decoder. The row decoder decodes a 7-bit address to indicate one of 128 rows. Where the decoded row-and-column outputs cross, they select the desired individual memory cell. Simple arithmetic shows that there are 128×128, or 16,384, crossings. Therefore, this memory system can address any of the memory chip's 16,384 different cells. Once a cell is selected, the system may either write data to the cell or read data from the cell.

Figure 11-3 shows a simplified schematic of an NMOS static memory cell. Looking at this schematic, you can see that this cell is also a simple flip-flop. Like most MOS devices, this

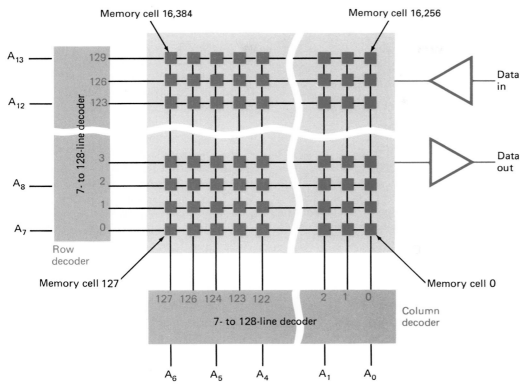

Memory cell 16,384   Memory cell 16,256

A_{13} — 129

A_{12} — 123

7- to 128-line decoder

Data in

A_8 — 2

A_7 — 0

Data out

Row decoder

Memory cell 127   Memory cell 0

127 126 124 123 122   2 1 0

7- to 128-line decoder

Column decoder

A_6 A_5 A_4 A_1 A_0

Fig. 11-2   The row-and-column addressing for a 16,384-cell memory device. Each cell has inputs that enable the cell when the row and column lines are both active. The data in and data out lines are connected to each cell.

one is implemented by using fixed-bias MOS transistors for the drain loads, instead of resistors. Additional MOS transistors are used to couple the data into and out of the selected cell.

Like the bipolar memory cell, the MOS static memory cell holds binary information because of the flip-flop or cross-coupled gate design. The cell retains the set or reset state as long as power is applied.

By comparison, the MOS dynamic memory cell is simpler. This is shown in Fig. 11-4 on the next page. In this schematic, you can see that the dynamic cell does the same job with only half as many transistors. Obviously, a dynamic memory integrated circuit can have twice as many memory bits as a static memory circuit the same size.

The MOS dynamic memory cell is not built like a flip-flop. The capacitance shown in the

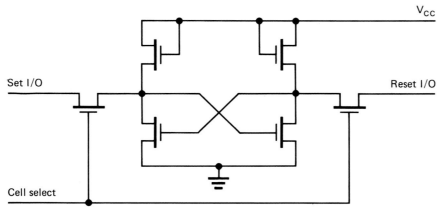

Fig. 11-3   A static memory cell built with MOS transistors. Again note the cross-coupled design, which is a flip-flop.

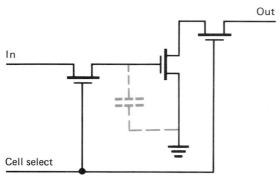

Fig. 11-4 A dynamic memory cell built with MOS transistors. The capacitor is the MOS transistor's input capacitance. A charge stored on this capacitor makes the cell hold a logic 1. No charge is a logic 0.

schematic is the MOS transistor's input capacity. This very small capacitance is used as the dynamic MOS memory cell's storage. The capacitance stores binary data by holding its charge, a logic 1 or a logic 0, for a few milliseconds. After this time, the data must be rewritten into the cell.

Rewriting the data in a dynamic MOS memory cell is called *refreshing* the data. The dynamic memory cell is refreshed each time that it is accessed. However, there is no guarantee that every bit will be refreshed by the microprocessor's normal memory activity. For example, the microprocessor might spend more than a few milliseconds in a simple timing loop that uses only a few memory locations. During this time, all the other memory words are not accessed. For this reason, memory systems that use MOS dynamic RAM have what is called *refresh logic*. This logic automatically accesses each column every few tenths of a millisecond. Dynamic memories are built so that simply accessing the column refreshes all the bits in the column. The refresh logic must be coordinated with other microprocessor activities so that it does not interfere with them. For example, if the microprocessor tries to access the memory while memory is being refreshed, the refresh logic must give the microprocessor priority. However, the refresh logic must be intelligent enough to know that a particular section of memory has not been refreshed within the required time and must perform the refresh before the data is lost.

Figure 11-5 shows two different pinouts, or ways of packaging, MOS integrated static memories for use in microprocessor systems. Figure 11-5(a) is a 2-kbyte SRAM. This IC also

stores 16,384 bits of data. However, the data is organized as 2048 eight-bit data words. This so-called bytewide organization is chosen for its obvious ease of use with 8-bit microcomputer systems. As you can imagine, only one device is needed to build a 2-kbyte microprocessor memory system.

Each of these devices has 11 address lines. Eleven address lines allows the selection of one of the 2048 bytes ($2^{11} = 2048$). The chip select line allows the use of a 12-bit address to selectively address one particular device. In such a system, the 12th bit is connected directly to the chip select pin. When the particular device's chip select pin is active, the other devices' chip select pins must not be active.

In addition to the address lines and the chip select line, the device has eight data lines. These data lines either receive data from the bus or send data to the bus, depending on the commands received at other control inputs. These other control inputs are the read-write input and the output enable. The read-write input tells the memory device to take the data from the eight inputs and write it to memory or to read data from the memory. If data is being read from the memory, then the signal must also be placed on the output enable pin to tell the device that it is to supply the data to the memory output pins.

Figure 11-5(b) shows a common 65,536-bit integrated-circuit SRAM. You can see this device also has address, data, chip select, output enable, and read-write lines. This IC is simply four times larger than the one in Fig. 11-5(a). The IC is usually called an 8K × 8 (pronounced 8K by 8) SRAM.

The 2K by 8 and 8K by 8 SRAM ICs are available in both NMOS and CMOS technologies. As you learned earlier, the CMOS device is more expensive but typically draws less power. For example, data may be retained for a number of years in an 8K by 8 CMOS SRAM device with a single 1.5-V lithium battery.

The bytewide static RAM is also designed to be "pin-for-pin" compatible with bytewide ROMs. As we shall learn, the bytewide ROM does not use a read-write pin, but otherwise the devices are essentially identical. This means that a microprocessor-based system may have memory sockets which can be filled with either bytewide static RAMs or bytewide ROMs.

Additionally, many static and dynamic RAMs are built to be interchangeable with different-sized parts. That is, a printed circuit

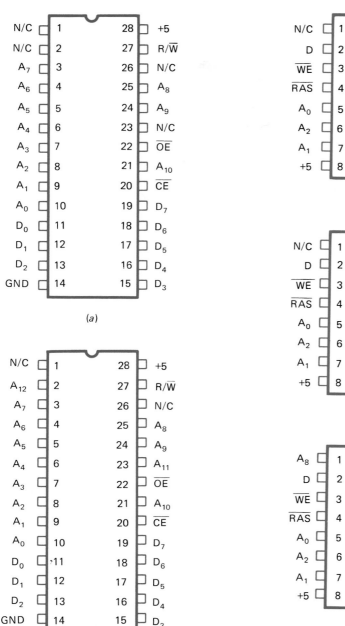

(a)

(b)

Fig. 11-5 Two popular organizations for static memory devices (ICs). These are called the bytewide devices. (a) is the 2K × 8 static device, and (b) is the 8K × 8 static device. Each time one of these devices is addressed, a byte of data present on the D_0 to D_7 pins is written into the eight addressed cells, or 8 bits of data contained in the eight addressed cells is output on pins D_0 to D_7.

(a)

(b)

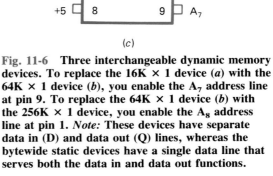

(c)

Fig. 11-6 Three interchangeable dynamic memory devices. To replace the 16K × 1 device (a) with the 64K × 1 device (b), you enable the A_7 address line at pin 9. To replace the 64K × 1 device (b) with the 256K × 1 device, you enable the A_8 address line at pin 1. *Note:* These devices have separate data in (D) and data out (Q) lines, whereas the bytewide static devices have a single data line that serves both the data in and data out functions.

board can be laid out so that the memory integrated-circuit positions can take either 2- or 8-kbytewide static RAMs. Such a layout permits configuring the microcomputer system memory to the customer's requirements. Often you may be asked to service a board that

uses a selection of different integrated-circuit memories. You may also be asked to change the memory capacity of the microcomputer memory system.

Figure 11-6 shows the pinouts of three DRAMs. These are 16K, 64K, and 256K devices. As you can see, these chips are packaged in *16-pin dual in-line packages (DIPs)*. In order to minimize the number of pins, a special memory addressing technique is used. To explain this technique, we use the 16K example.

Figure 11-7 shows a diagram of the decoding system used. At first glance, this block diagram looks exactly like that of the memory shown in Fig. 11-2. The 16K memory uses a 7- to 128-line column decoder. It also uses a 7- to 128-line row decoder. Each decoder has a 7-bit latch. To address a memory bit in this device, use two cycles. First, the chip's seven address pins are connected to the memory system's lower seven address lines (lines A_0 through A_6). A signal then strobes the memory chip's $\overline{\text{ROW-ADDRESS STROBE}}$ input. Strobing $\overline{\text{ROW-ADDRESS STROBE}}$ stores the seven address bits in the row-decoder latch.

Next, the microcomputer's upper seven address bits A_7 through A_{13} are connected to the chip's seven address inputs. This time, $\overline{\text{COLUMN-ADDRESS STROBE}}$ is strobed. These seven address bits are stored in the

column-decoder latch. All 14 address bits are now stored in the latches. This decodes one of the 16,386 memory cells. This technique is called *multiplexed addressing*.

Each of the DRAMs shown in Fig. 11-6 has a common pinout. Each chip has data input (D) on pin 2 and data output (Q) on pin 14. Remember, these devices store only the single bit of information at each memory address location. This means that eight devices are needed to store 1 byte of data. You can see that these chips use separate data input and data output lines.

Each chip has a $\overline{\text{WRITE ENABLE}}$. You assert this to write data into the addressed memory location. If this is not asserted, the chip is in the read mode.

There is a $\overline{\text{WRITE ENABLE STROBE}}$ (RAS) and a $\overline{\text{ROW-ADDRESS STROBE}}$ (CAS) on each chip. All three chips use multiplexed addressing.

The 16K DRAM has seven address lines, A_0 through A_6. The 64K DRAM has eight address lines, A_0 through A_7. The 256K DRAM has nine address lines, A_0 through A_8. You can see that using each of these address lines twice

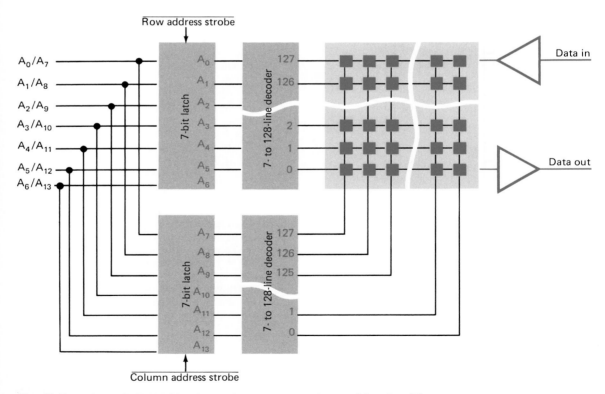

Fig. 11-7 A dynamic RAM IC using multiplexed row-column addressing. The $\overline{\text{ROW-ADDRESS STROBE}}$ line is asserted to store address bits A_0 to A_6 in the row-decoder latch. The $\overline{\text{COLUMN-ADDRESS STROBE}}$ line is asserted to store address bits A_7 to A_{13} in the column-decoder latch.

gives these DRAMs the address ranges that they need. These ranges are

$$2^7 \times 2^7 = 16,384 \quad (16K)$$

$$2^8 \times 2^8 = 65,536 \quad (64K)$$

$$2^9 \times 2^9 = 262,144 \quad (256K)$$

Only the address lines A_0 through A_6 are common. The 16K by 1 DRAM has no connections on pins 1 and 9. Pin 9 becomes the A_7 line on the 64K by 1 DRAM, and pin 1 becomes the A_8 line on the 256K by 1 DRAM. These devices do not provide a chip select function. If this is required, you must use external logic.

All three devices also have two pins allocated for power. Pin 16 serves as the ground or return pin, and pin 8 is the V_{CC} or +5-V pin. You will note that the power supply pinout is the opposite to the conventional upper right-hand, lower left-hand pinouts you have become used to with conventional logic devices.

Today, both 1-megabyte and 4-megabyte devices are used in advanced microprocessor-based systems. Although the 16-, 64-, and 256-kbyte devices were able to follow a common pinout, this was not possible with the larger parts.

Figure 11-8 shows the pinouts for 1-Mbit and 4-Mbit devices. As you can see, these devices are housed in larger packages in order to accommodate the additional memory addressing pins.

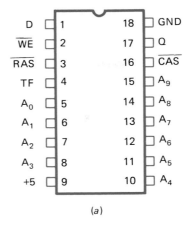

(a)

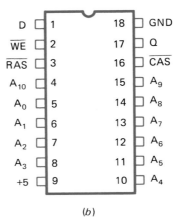

(b)

Fig. 11-8 Two interchangeable dynamic RAM ICs. (a) One pinout for the 1M × 1. (b) One pinout for the 4M × 1 device.

Self-Test

Answer the following questions.

9. An SRAM stores its information in a
   a. Capacitor   c. Bipolar circuit
   b. Flip-flop   d. MOS circuit
10. A DRAM needs special external circuits to perform
   a. Row decoding   c. Refreshing
   b. Column decoding   d. Cell addressing
11. A 64-kbit memory device can be organized as
   a. 64K × 1   c. 16K × 4
   b. 8K × 8   d. All of the above
12. The chip select lines let two memory devices operate with common _____ lines.
   a. Power   c. Address
   b. Ground   d. Inverter
13. The larger (1 and 4 Mbyte) DRAMs use
   a. − 5 V
   b. + 12 V

   c. Bipolar construction
   d. Multiplexed addressing _____
14. Asserting a memory chip's $\overline{\text{WRITE}}$ line will cause data to be
   a. Stored in the addressed cells
   b. Addressed
   c. Placed on the data bus
   d. Latched
15. A semiconductor memory device with multiplexed addressing needs two strobe signals. What do they do? Why are these chips built with this extra complication?
16. Explain how a static memory device stores data and how a dynamic memory device stores data.
17. Why does a DRAM need refresh logic?
18. In what situation would you choose a more expensive CMOS static memory device rather than a lower-cost NMOS static memory device?

11-3 TWO MEMORY SYSTEMS

In the last section, we looked at MOS static and dynamic memory ICs. How are these actually used to build a microcomputer's memory system? What additional circuits are needed on the memory card? In this section, we will look at two different memory systems. First, we will look at a simple static memory system that might be used with a typical 8-bit microprocessor. Second, we will look at a block diagram of a memory card which might be used with a personal computer. This card uses three sets of 256-kbyte DRAMs to provide a 768-kbyte memory.

Figure 11-9 is a simplified schematic of a small microcomputer memory system. In this schematic, we use two 8K by 8 static memory chips to build a 16-kbyte memory system. As shown, this system uses 8 kbytes of SRAM. However, it could be configured to use all RAM, all ROM, or any combination of RAM and ROM.

Keep in mind that this is a simplified schematic. Depending on the particular microcomputer's bus structure, the exact memory chips used, and the discrete TTL devices that are available, a few more integrated circuits may be used to build a real memory system.

The simplified bus connections are shown on the left-hand side of the schematic. Starting from the top, you can see that the first 16 connections are the system memory address lines A_0 through A_{15}. The next eight lines are the system bidirectional data bus lines D_0 through D_7.

To be functional, the memory system needs four more bus lines. The $\overline{\text{READ}}$ (NOT READ) line is asserted to make the memory system place data on the microcomputer's data bus. The next line, $\overline{\text{WRITE}}$ (NOT WRITE), is asserted when the microcomputer writes data to a memory location. The terms $\overline{\text{READ}}$ and $\overline{\text{WRITE}}$ indicate that these lines are asserted by a low voltage level. This is done to make the system's electrical design easier. The last two lines supply power and ground to the integrated circuits on the memory card.

As you can see, each address line is buffered. This is usually done by using small- or medium-scale TTL integrated circuits designed especially for buffering. The *buffers* isolate the microcomputer's bus from the integrated-circuit memory's address lines. In some small systems, these buffers may not be used. Typically, single-board systems where the microprocessor and its memory circuits are on the same printed circuit board do not use address or data buffering.

The buffered memory address lines A_0 through A_{12} are connected directly to the integrated-circuit memories.

Address line A_{13} (remember that A_{13} is the 14th address line) selects either the first chip or the second chip. Memory address lines A_0 through A_{12} can address any location in an 8192-byte range. Address line A_{13} selects either the first 8192 bytes or the second 8192 bytes. The A_{13} memory address signal is passed through an OR gate and is then directly connected to the first memory chip. The A_{13} memory signal is connected to the second memory chip through an inverter and an OR gate. Therefore, when one chip is selected, the other is not selected. The OR gate lets you deselect both chips regardless of the A_{13} signal.

Address lines A_{14} and A_{15} are connected to a 2-bit comparator. The other 2 bits come from a 2-bit switch. The memory address lines A_{14} and A_{15} are compared to the switch setting by the magnitude comparator. This tells you which 16-kbyte block of memory the microcomputer is addressing. For example, if address lines A_{14} and A_{15} are both logic 0, and switches are also both logic 0 (closed), the magnitude comparator outputs a logic 0 to the OR gate. If address lines A_{14} and A_{15} are both logic 1, then the magnitude comparator outputs a logic 1.

In the above example, we use the all-logic-0 switch setting. We have decided that this memory system is to respond to addresses in the 0- to 16-kbyte range. If, for example, the switches are set to 1 and 0, this memory system responds to addresses in the 16- to 32-kbyte range. Figure 11-10 on page 320 shows a memory map of the four 16-kbyte blocks which this memory system may occupy.

The logic 0 from the magnitude comparator's output is used to enable the OR gates in series with address line A_{13}. A logic 1 to the OR gates will guarantee a logic 1 at their outputs. Because a logic 0 is needed to enable the chips, a logic 1 disables them. No $\overline{\text{CHIP SELECT}}$ signal is permitted at all if the address is not in the range of 0 to 16K.

We now see how the address selection process works. Now we must look at how to get data into and out of the memory. Data lines D_0 through D_7 are connected to the integrated-circuit bus transceivers. These TTL integrated circuits perform two functions.

First, the bus transceivers buffer the microcomputer's data bus lines from the compo-

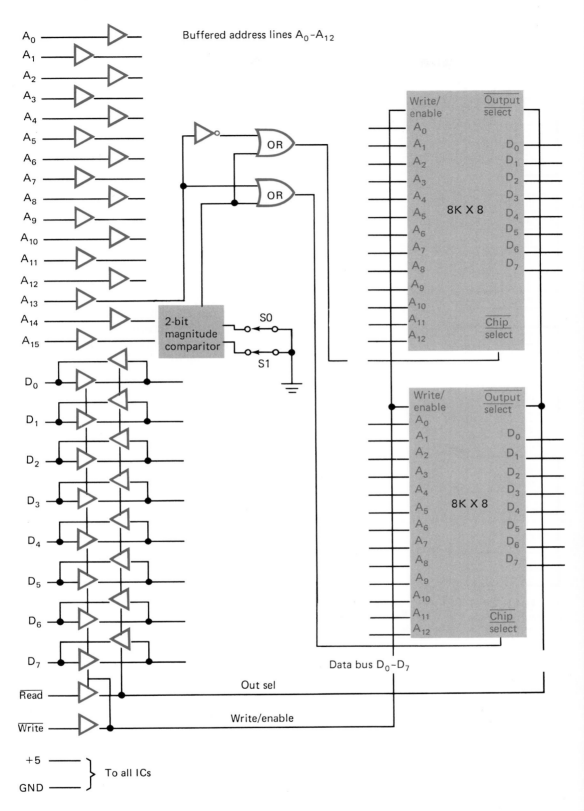

Fig. 11-9 A 16-kbyte memory system using two 8K × 8 memory devices. The
WRITE ENABLE line is asserted when you want to write a byte into memory. The OUT-
PUT SELECT line is asserted when you want the memory devices to place data on the
data bus; otherwise the memory device's D_0 to D_7 (data in, data out lines) are in a high-
impedance state.

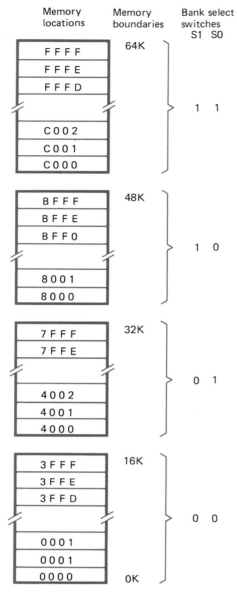

Fig. 11-10   The 16K memory boundaries for the memory system in Fig. 11-9. Each 16-kbyte increment is selected by setting bank select switches S0 and S1.

the memory chips. When the output buffer control signal is asserted, data from the memory chips is placed on the microcomputer's data bus.

The data buffer control signal and the memory chip's READ/$\overline{\text{WRITE}}$ and $\overline{\text{OUTPUT SELECT}}$ signals are all controlled by the $\overline{\text{READ}}$ and $\overline{\text{WRITE}}$ bus signals.

Let us look at an example of a read operation. An address is placed on memory address lines A_0 through A_{15}. The memory addressing logic selects a memory location in one of the two memory chips. The memory $\overline{\text{READ}}$ line is asserted. The buffered memory $\overline{\text{READ}}$ signal asserts the chip's $\overline{\text{OUTPUT SELECT}}$ lines. The selected chip places the data at the addressed memory location on the memory card's internal data bus. The $\overline{\text{READ}}$ signal also connects the memory card's internal data bus to the microcomputer's data bus lines D_0 through D_7. The information from the address location is moved from the memory card's data bus to the microcomputer's data bus and into the microprocessor.

As you can see, the sequence of operations is as follows:

1. The microprocessor puts valid information on the address lines.
2. The microprocessor's control logic generates a $\overline{\text{READ}}$ signal.
3. The addressed memory location places its data on the chip outputs.
4. The chip outputs a buffered byte which is connected to the microcomputer's data bus.
5. The microprocessor inputs the data through the data bus port and transfers the data to the location designated.

Now let us take a look at a memory-write operation. The sequence is the same; only the names are changed.

1. The microprocessor puts valid information on the memory address lines.
2. The microprocessor's control logic generates a $\overline{\text{WRITE}}$ signal.
3. The microprocessor outputs data from the data bus port.
4. The microcomputer's data bus is buffered and connected to the memory chip data inputs.
5. The data is written into the selected memory locations.

The schematic shown in Fig. 11-9 uses either two 8K × 8 SRAMs or a ROM-RAM mix. To

nents on the memory card. Second, these transceivers can connect the card's signals to the microcomputer's data bus. This lets the memory system place data on or receive data from the microcomputer's data bus. As you can see, the input buffers and the output buffers each have a common control signal. Normally, this buffering operation is performed with a single integrated circuit called an *octal* (eight lines) *data bus transceiver*. When the input buffer control signal is asserted, data on the microcomputer's data bus is connected to build this memory, using DRAMs, you would need extra circuits. These circuits are needed to refresh the dynamic memory chips about every 300 μs. The use of dynamic memory increases the card's *overhead*. Overhead refers to the extra integrated circuits that do not store data but are necessary to make the memory work. In Fig. 11-9, the bus-buffering and the address-decoding integrated circuits are all overhead. They perform needed functions, but they do not store data.

Figure 11-11 on the next page shows another form of memory system. This is the memory system used with a typical personal computer (PC). When you study this figure, you will immediately notice some major differences between this memory system and the memory system for the 8-bit computer.

First, this system uses 20 memory address lines (A_0 through A_{19}). The microprocessor which this memory card works with is capable of addressing 1 Mbyte of memory. Second, there are a number of extra blocks of logic on this diagram. These blocks of logic handle such functions as row-and-column selection, refresh generation, and parity checking. Third, the memory system, which is made up of $256K \times 1$ chips, uses 27, rather than the expected 24, chips.

Let us look at each one of these differences so that we can understand its function. As noted above, this memory card is designed to operate in a PC system. The original PC used the Intel 8088 microprocessor. Although internally it is a 16-bit microprocessor, its external data communication takes place over an 8-bit data bus. However, it does use a 20-bit memory bus so that it may address 1 Mbyte of memory. Although most of the personal computers being produced today use full 16- or 32-bit microprocessors, with even greater addressing capability and 16- and 32-bit data buses, the memory system shown here is still in use.

As is typical with most large PC systems, this memory system uses DRAMs. Therefore, additional logic circuits are required to address the dynamic RAM chips and to ensure that they are refreshed. The addressing logic splits the microcomputer's address bus into row-and-column select signals at the appropriate time. The row-and-column select timing is generated from logic which monitors the memory-write and memory-read signals from the microcomputer's bus. Additionally, these circuits automatically generate column select signals for the DRAMs when the memory is not being ac-

cessed by the microcomputer. This automatic generation of column select signals ensures that the DRAMs are refreshed on a regular basis.

As the memory devices become larger and larger, the internal components (the MOS transistors) become smaller and smaller. This means that smaller and smaller capacitors are used to store the charge, which indicates that a logic 1 or logic 0 is being stored. All this means that the memory devices become more and more susceptible to losing their data because of something which happens outside the chip. For example, the occasional wandering neutron which exists from solar radiation can easily penetrate the computer housing and the integrated-circuit device itself. If it should strike the capacitor, it will discharge it. In order to compensate for the potential of random data loss, the personal computer often uses a 9-bit memory to store an 8-bit word. The ninth bit is a parity bit. The parity bit checks the incoming or outgoing data word to see if an odd number of logic 1s are present. If they are not, it generates one. This is placed in the ninth memory location. If, when reading data from the memory, an odd number of data bits is not found at the addressed memory location, including the corresponding parity bit location, the parity logic generates an error signal. Additional logic is required as the use of memory parity checking is under the programmer's control and may be turned on or off at will.

Needless to say, the inclusion of parity also explains why each bank of data memory is made up of nine chips rather than the expected eight. Note that in 16-bit memory systems, it is common to include 2 parity bits. That is, parity is checked on a byte basis rather than on a word basis.

As you would expect, the PC memory card contains the usual address and data buffering circuits. These circuits also decouple the memory card from the microcomputer's bus, as they do in an 8-bit system. Because it is not uncommon to have relatively large printed circuit boards in personal computers, it is often customary to include other circuits on the memory card. Often, the memory card includes one or more serial ports, a parallel port, a clock, or other such I/O circuits. Although these circuits share the use of the bus interface logic, they do not otherwise share any common logic with a memory system.

Depending on the size of the memory system, the system card may or may not have switches,

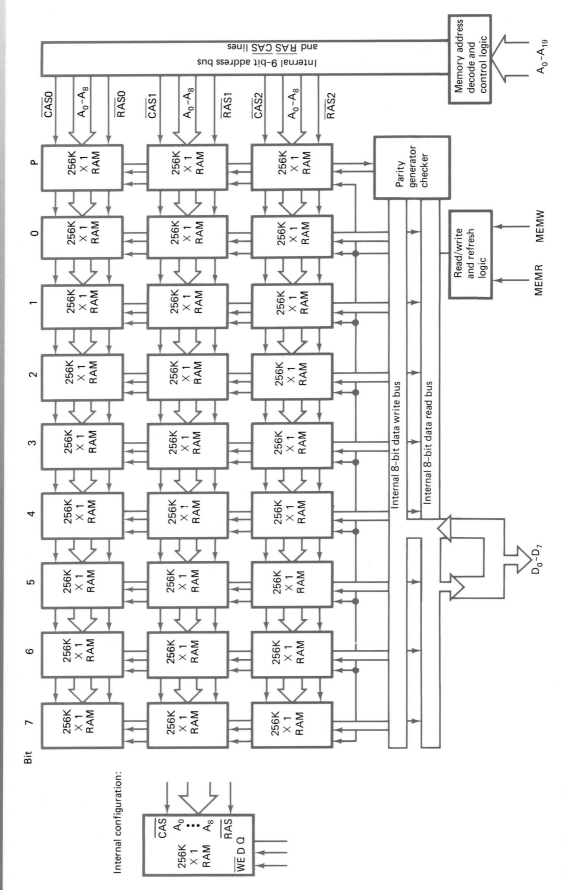

Fig. 11-11 A simplified schematic for a memory card using 256K × 1 dynamic RAM ICs. This card uses nine ICs per byte of data stored so that parity error checking can also be included.

or user-changeable jumpers to place the memory within the overall memory address range.

Self-Test

Answer the following questions.

19. What circuits and devices would change if the memory system of Fig. 11-9 used eight 16K × 1 SRAMs?
20. If the devices in question 19 are replaced with 64K × 1 dynamic memory ICs, what does the increased memory design do to the memory card's overhead? Why?
21. Using diagrams like Fig. 11-10, show how the memory bank select would divide up a memory system like the one in Fig. 11-9 if it was built with 2K × 8 ICs.
22. Why is it not desirable to use DRAMs on small memory systems?
23. Although the system memory card in Fig. 11-11 uses three banks of 256K memory parts for a total of 768 kbytes of memory, the system is only able to address a maximum of 620 kbytes of memory. How many memory locations are unused? Is there a less wasteful way to build this memory using a combination of 256K × 1 and 64K × 1 parts?

11-4 ROMS, EPROMS, AND EAROMS

The ROM (read-only memory) is a very important part of any microprocessor system. As its name tells you, the ROM is a memory device which, once set with a given bit pattern, cannot be changed. Often microprocessor systems use ROM because the systems always execute the same program. The program instructions are stored in the ROM. This storage overcomes the semiconductor RAM's volatility problem. *Every microprocessor system must have some ROM, because every system must have at least enough built-in program to load its RAM with a program from a mass storage device such as magnetic tape or disk.*

It is important to understand that although we give the name RAM to static and dynamic read-write memory devices, that does not mean that the ROMs that we are using are not also random-access devices. In fact, most ROMs are addressed identically to the way we address static and dynamic RAMs, and therefore their addressing techniques are also random-access.

There are four different types of ROM. Each of the four types is used for a different application. This section looks at these four types of devices.

First, we will look at the simple ROM. When we speak of a ROM, we mean a device with a bit pattern permanently fixed by the semiconductor manufacturer. Often, this is called a *mask-programmed* ROM. The bit pattern is fixed by the masking part of the integrated-circuit manufacturing process. The bit pattern is fixed according to the customer's specifications. ROMs are only used in high-volume applications, because custom mask designs are expensive. Of course, once a part is masked, it may never be changed.

Second, we will look at the EPROM. The *EPROM* (*erasable programmable read-only memory*) can be programmed, erased, and reprogrammed by the user. Although these devices are somewhat more expensive than ROMs, EPROMs do let the user change the bit pattern as needed. Also, they allow manufacturers to change product software at the last minute before production. Often, EPROMs are supplied with a microprocessor system whose functions will be changed as the user, by experience, decides exactly what the system should do.

Third, we will look at the *EAROM* (*electrically alterable read-only memory*). The EAROM can be programmed and altered electrically. Unlike the EPROM, the EAROM does not need an outside device in order to be erased.

Figure 11-12 on page 324 shows a very simple ROM. This ROM could be built using only diodes and a TTL decoder. Since this ROM contains four 8-bit words, it is a 32-bit ROM. As you can see, this ROM has diodes at each bit that is to be a logic 0.

The two-line to four-line decoder places a logic 0 on the selected row. Each output line goes to logic 0 if a diode connects the output data column to the selected row.

Most of today's ROMs use MOS technology. ROMs are available today which are constructed from both the NMOS and CMOS processes. Again, the difference is cost versus power. A few bipolar ROMs are built for use where very high speed operation is needed. In NMOS ROMs, an MOS transistor takes the place of the diode. A very simple four-word ROM using MOS devices is shown in Fig. 11-13 on page 325. Since each word has 4 bits, this is a 16-bit ROM.

As you can see, each data column is connected to four of the ROM's MOS transistors. An MOS transistor on each column is used as

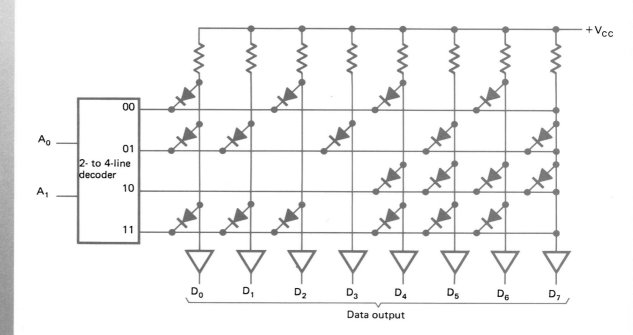

Fig. 11-12 A simple four-word (4-byte) ROM. This diode ROM has a total of 32 bits. Each bit that has a diode at the matrix crosspoint generates a logic 0 at that data output line when the word is addressed. Each bit that does not have a diode at its crosspoint generates a logic 1 at that data output line when the word is addressed.

ROM contents

Word	Binary								Hexadecimal
	D_0	D_1	D_2	D_3	D_4	D_5	D_6	D_7	
00	0	1	0	1	0	1	0	1	55
01	0	0	1	0	1	0	1	0	2A
10	1	1	1	1	0	0	0	0	F0
11	0	0	0	1	0	0	0	1	11

a pull-up. Some of the MOS transistor's gates are connected to the row select lines. When the row is selected, it is pulled high. If the row is connected to an MOS transistor's gate, the MOS transistor is turned on. This pulls the row to logic 0 (ground). This causes a logic 0 at the ROM's output. If the gate is not connected to the row select line, then the transistor is not turned on. This causes a logic 1 at the ROM's output.

ROMs are made by mask programming. As you know, the active parts of an integrated circuit are interconnected by depositing very thin layers of metal. These thin layers of metal, like the layers of metal on a printed circuit board, are the integrated circuit's internal wiring. The final layer of metallization for a mask-programed ROM is set by the customer. It is this thin metallized layer that connects the gates of some transistors to the row select lines. The transistors are not connected to

those row select lines on which a logic 1 is wanted at the output.

Mask-programmed ROMs are made by a high-volume process. As a rule, they are not used unless the manufacturer intends to produce many thousands of the same ROM. The ROMs used in microprocessor-based toys, TV games, home computers, and other such high-volume consumer products use mask-programmed ROMs.

The EPROM is built like the mask-programmed ROM. That is, it uses MOS transistors in an array like the one shown in Fig. 11-13. However, there are a few differences between the EPROM and the ROM. First, all transistors in the EPROM are connected to the select rows. Second, the transistors are designed so that a high voltage can be applied to the transistor's gate. When a high voltage is applied, the transistor is changed so that it is always at a high-impedance state. Third, this

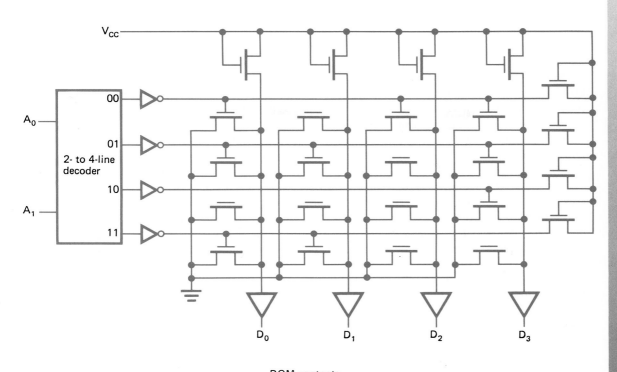

ROM contents

Word	Binary				Hexadecimal
	D_0	D_1	D_2	D_3	
00	0	1	0	0	4
10	0	0	1	0	2
01	1	1	1	0	E
11	0	0	1	1	3

Fig. 11-13 **A simple four-word (half-byte) ROM. This ROM shows how MOS transistors are used to make a mask-programmable ROM. The integrated-circuit process connects the MOS transistor's gate to the ROW SELECT line if the output bit is to be a logic 0.**

characteristic can be reversed by shining an intense ultraviolet light (UV) on the transistor.

The EPROM can be programmed in the field by using a device which provides the high-voltage pulses as well as the address and data. Once the EPROM is programmed, it remains programmed until a high-intensity ultraviolet light is shown through the EPROM's quartz window. Ultraviolet light erases the EPROM's stored bit pattern. Each bit is left at a logic 1. Figure 11-14 shows the EPROM's physical construction.

A third type of ROM is used in some situations. It is called electrically alterable ROM (EAROM). The EAROM is constructed very much like the EPROM. However, the EAROM has one unique characteristic. The contents of the EAROM can be altered by applying special electrical signals for a certain time. Usually the right signals must be applied for a few

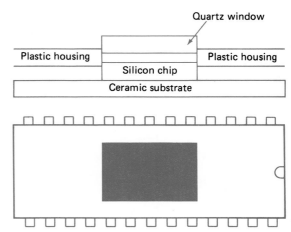

Fig. 11-14 **The internal structure of a UV EPROM. The quartz window is located directly above the silicon integrated-circuit chip. When ultraviolet light shines on the quartz window, the light passes through the window and erases the data stored in the ROM.**

milliseconds, so the writing process is quite slow when compared with the read process. The EAROM is not as permanent as the EPROM. However, it can be changed by the customer while it is installed in the circuit, and it is nonvolatile.

EAROMs are finding uses in such applications as channel selectors on TV sets and in storage of temporary setup information in video terminals.

To summarize, we have looked at three different ROMs. The mask-programmed ROM is permanent. It is also the least expensive ROM and offers the most bits. The EPROM is both field programmable and reusable. It is a little less dense than a ROM. The EAROM is electrically alterable. It is the most expensive and the least dense ROM.

Figure 11-15 shows pinouts of three popular EPROMs. The 2764 is a 64-kbit (8K $\times$ 8) UV-erasable PROM. The 27128 is a 128-kbit (16K $\times$ 8) UV-erasable PROM. The 27256 is a 256-kbit (32K $\times$ 8) UV-erasable PROM. As you can see, all three of these chips are organized in the bytewide mode. Therefore, they are pin-compatible with equivalent SRAM ICs. Because all three devices have a common pinout, one device can be removed and replaced by a model with a higher bit density if more ROM is needed.

The address lines are shown as A_0 through A_{12}, A_{13}, or A_{14}, and the data is output on lines D_0 through D_7. All three devices use a single +5-V supply at pin 28 and ground at pin 14. A $\overline{\text{CHIP ENABLE}}$ ($\overline{\text{CE}}$) pin allows the user to select the desired device. The $\overline{\text{OUTPUT ENABLE}}$ ($\overline{\text{OE}}$) places data on output pins D_0 through D_7 when $\overline{\text{ENABLE}}$ is asserted. To program the device, we apply address information to the address pins, data to D_0 through D_7, and a 12-V pulse of the specified width to pin 1 (V_{PP}). Additionally, the 2764 and 27128 require that pin 27 ($\overline{\text{PGM}}$) be asserted as well as having $\overline{\text{CHIP ENABLE}}$ asserted.

In the erase mode, the EPROM is exposed through its quartz window to a special UV lamp for 15 to 20 min. This is the same UV energy that it would receive from being exposed for three years to fluorescent lighting or to one week of direct sunlight. You will see covered quartz windows in some EPROMs to prevent accidental erasure.

With three different kinds of ROM and significant variations in density, there are many different ROMs available. There are some mask-programmed ROMs that can be directly inserted into the same circuit used by an EPROM. This is very convenient, because it lets the manufacturer develop the device and place it on the market, using EPROMs. Pro-

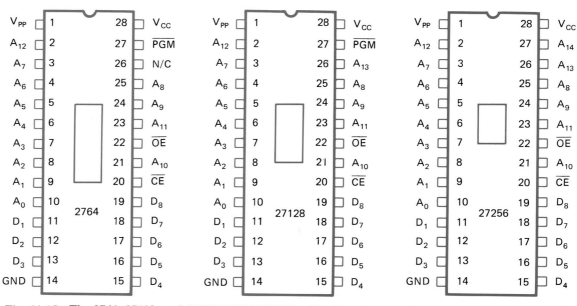

Fig. 11-15 The 2764, 27128, and 27256 MOS EPROMs. The V_{PP} (programming output) is not used during normal operation. It is reserved for use in a special EPROM programming circuit. Note that the three EPROMs are upward-compatible.

gramming problems can be corrected by reprogramming the EPROMs. Later, when the software design is perfected, the EPROMs can be replaced by the less expensive mask-programmed ROMs.

There are also some preprogrammed mask-programmed ROMs. These ROMs are available with certain standard tables. For example, you can find ROMs for use with CRT-based terminals. These ROMs can generate all the alphanumeric characters by using a dot matrix pattern.

Self-Test

Answer the following questions.

24. The highest-cost read-only memory in the following group is
    a. Mask-programmed ROM
    b. Erasable programmable ROM
    c. Electrically alterable ROM
    d. Static MOS RAMs
25. Although you can order ROMs, EPROMs, and EAROMs with your custom bit pattern already in them, the _____ is the only one that uses an integrated-circuit manufacturing process to set the bit pattern.
    a. Mask-programmed ROM
    b. Erasable programmable ROM
    c. Electrically alterable ROM
    d. Static CMOS RAM
26. The 27512 is the 512-kbit version of the 2700 series family of UV EPROMs. You would expect that it has _____ address lines.

    a. 12   *d.* 15
    b. 13   *e.* 16
    c. 14
27. Why do you think the EPROM is the most popular read-only memory for development work?
28. Explain why a microcomputer must have some read-only memory.
29. Explain why it is very unlikely that a microcomputer will have EPROMs for all of its main memory.

11-5 DIRECT MEMORY ACCESS

So far, all our discussions of input/output data transfers have been about programmed data transfers. That is, the data is moved from the input port to the accumulator and then to a memory location. Program data transfers out of memory require the transfer of data from memory into the accumulator, and then from the accumulator into the output register.

Unfortunately, the program data transfer is a slow process. The fact that the data transfer process is slow usually causes a problem only when we have to transfer large amounts of data. For example, suppose you wish to transfer 16 kbytes of data from a magnetic tape to main memory. The magnetic tape drive stores 800 bytes/in. of tape. The drive moves the tape at 125 in./s. Sixteen kbytes of data are contained on 20.5 in. of tape. This means that all 16,384 bytes of data can be transferred in 0.16384 s. In other words, the data being transferred at a rate of 100,000 bytes/s.

Figure 11-16 shows a short program that we might use to transfer data from a magnetic tape

Symbolic address	Op code		Operand	Comment
DATAIN	LDA	D	44H	;Load D with terminal counter
	LRP	B	0400H	;Point BC to memory file
INAGN	IN		01H	;Input data to accumulator
	STI	A		;Store accumulator in file
	IRP	B		;Point BC to next memory
	MOV	A,B		;Put B into A
	CMP	D		;16K bytes yet?
	JNZ		INAGN	;No, input more data
	RET			;Yes, all Done. Exit

Fig. 11-16 A short subroutine to load 16,384 bytes of data into a memory file. This program is given in a form for use with an assembler. To assemble this program, you would need to use an ORG instruction and to specify a starting address as the ORG's operand. The "H" after the operand tells the assembler that the number is in hexadecimal format.

by using program data transfers. Once the program is written, we can count the microprocessor's cycle and see how long the program will take to do the job. We will start the 16K data block at memory location 0400. The 16K data transfer will be complete at memory location 4400.

We will start the program data transfer subroutine at memory location DATAIN. The subroutine is assigned symbolic addresses, so we can combine it with any program we wish. When we assemble the subroutine, the symbolic address DATAIN will be translated into an absolute address. The first two program instructions load the D register with 44 and point BC register pair to the bottom of the memory file (0400). The D register will be compared with the B register at the end of each store. When the D register is incremented to 44, we will exit this routine. That is, the 16,384th byte will have been loaded in memory location 4400.

We begin the routine at the symbolic address INAGN. An input instruction transfers the data from I/O port 01 to the accumulator. A load indirect instruction stores the accumulator's contents in memory location 0400. We now use and IRP B instruction to increment the BC register pair.

The instruction LD A,B moves register B's contents into the accumulator. With register B's contents in the accumulator, we can now compare register D with the accumulator. We can do this by using a CMP D instruction.

What we are doing is checking to see if the last increment instruction pointed the BC register pair to memory location 4400. If it did, the job is complete. If it did not, we must go

back. We use the JNZ instruction to check the status register's zero bit. If the status register's zero bit is not set to logic 1, the job is not complete. We jump back to INAGN. When the job is complete, the status register's zero bit will be set to logic 1. Then the jump instruction will not execute on the 16,384th operation.

Figure 11-17 shows this program's timing. Remember, most of the program must execute 16,384 times to fill up the file. If we assume that our microprocessor has a 1-μs instruction cycle, then we need 294,915 μs to do the job. That's almost 0.3 s. A job that should have been done in 0.16 s took almost twice as long because we used a program data transfer.

Even if we used a microprocessor with a 0.5-μs instruction cycle, the processor would barely be able to keep up with the data coming from the tape.

The purpose of this example is to show the need for DMA. *DMA stands for direct memory access.* It is one way that we can accomplish high-speed transfers into memory. For example, we know that a good microprocessor memory has an 80-ns access time. With such a short access time, the microprocessor would be able to make 12 million or more transfers per second. However, we have seen an example where using a program data transfer sets an upper limit of about 25,000 transfers per second.

In most cases, DMA is accomplished by using a separate DMA controller. The microprocessor must be disabled during the DMA process. To start the DMA process, the microprocessor loads an external register in the DMA controller with the data file's starting ad-

Instruction		Number of times used	μs at 1μs per CPU cycle	Total μs
LDA	D	1	3	3
LRP	B	1	5	5
IN		16,384	5	81,920
STI	A	16,384	2	32,768
IRP	B	16,384	2	32,768
MOV	A,B	16,384	2	32,768
CMP	D	16,384	2	32,768
JNZ		16,384	5	81,920
				294,920

Fig. 11-17 Timing the programmed data transfer listed in Fig. 11-16. The JNZ instruction does not jump on the last operation.

dress. The microprocessor also loads the terminal count register with the total number of bytes to be transferred. The microprocessor disables the address and data buses and gives memory system control to the DMA controller. The DMA controller places sequential addresses on the microprocessor's memory bus and issues read-write pulses. As each byte is transferred, the terminal count register is decremented. When this register is decremented to 0, it tells the external device the data transfer is complete.

Because this is a limited application, a special-purpose hardware controller can do it very quickly. Today, DMA transfers take place with speeds close to the memory cycle time. Once the DMA controller has finished transferring data into or out of memory, the DMA controller gives control back to the microprocessor. Of course, the microprocessor cannot accomplish any other function during the time that a DMA transfer is taking place in memory. This is caused by two conditions. First, the microprocessor's memory is being used for a data transfer. It is not available to supply program instructions or receive the results of computations. Second, the typical DMA process requires that the microprocessor place its memory address bus and data bus in a high-impedance condition. This high-impedance condition allows the DMA controller and the memory system to control the bus but prevents the microprocessor from providing any bus control.

Self-Test

Answer the following questions.

30. Assume that the DMA controller and the memory system let you make 1 million transfers per second. If we transfer data from an 800-bpi 75-in./s tape with a DMA controller, which device will limit the transfer's speed: the tape or the DMA system?

31. What limits the system's speed if the tape transfer in question 30 is a 1600-bpi 125-in./s drive?

32. The programmed data transfer uses only one memory transfer per byte of data. Why is it so much slower than a DMA transfer?

33. One DMA controller chip will control four different DMA channels. This controller has eight 16-bit registers. What are they for?

11-6 PAGING AND OTHER MEMORY EXTENSION TECHNIQUES

When most microprocessors were first designed, they were given a memory address range which the designers of the time felt was larger than the size of any memory system that would be needed. As the cost of memory devices has come down, microprocessor system designers have been designing microprocessor systems with larger and larger memories. Today, it is not uncommon to have a microprocessor system which has significantly more memory than the microprocessor can address. How does the microprocessor address more memory than its address space will allow?

To do this, the microprocessor uses *paging techniques*. A paging system for a Z80-based system is shown in Fig. 11-18 on page 330. Here, the lower 32 kbytes of memory are always directly addressed by the microprocessor. However, up to four different 32-kbyte blocks of memory may be used in the upper 32-kbyte address space. That is to say, this system has four 32-kbyte pages of *bank-switched* memory. The banks of memory are switched in as needed. Obviously, a location within one of the banks of memory cannot be accessed as quickly as one in the fixed bank of memory.

When using a bank-switched system such as the one shown in Fig. 11-18, the programmer must first determine that the bank of memory containing the desired data is switched into the microprocessor's address space. Typically, this means that the programmer must access external hardware connected to the I/O bus. This hardware manages the memory's bank-switching circuits. In sophisticated bank-switched systems, this external hardware is known as the *memory management unit* (MMU).

Paging, bank-switching, and memory management are all different names for a process which has been used for some time in microcomputers. Special hardware to perform the MMU function is now becoming available for certain microprocessors. In many cases, however, the microprocessor hardware system designer must create the bank-switched logic from discrete components.

Figure 11-19 on page 330 shows a memory map for a typical personal computer with a 1-Mbyte upper addressing limit. As you can see from examining this memory map, the first 640 kbytes of memory space are allocated to

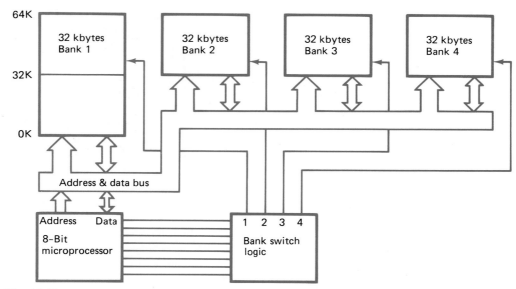

Fig. 11-18 **A bank-switched memory system. This system allows the programmer to store program segments or data in one of four different 32-kbyte blocks of memory. The main program and the bank-switching information are located in bank 0 (0–32K). The bank-switch logic is accessed as an I/O port.**

the user. The memory space lying between 640 kbytes and 1 Mbyte is occupied by the video RAM, the boot ROM, and other ROMs which support dedicated controllers.

In order to allow the user to have more memory than the 640K limit allows, an *expanded memory specification* (EMS) has been developed for the PC.

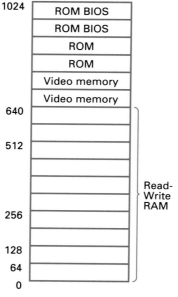

Fig. 11-19 **A memory map for a conventional PC. This shows the 640 kbytes of working RAM and how the remaining address space (640K–1024K) is used.**

A memory map for the EMS standard is shown in Fig. 11-20. As you can see, this standard allocates four 16K blocks of memory address space between 832K and 896K. These are referred to as four 16-kbyte blocks rather than one 64-kbyte block, since they are used as independent 16-kbyte blocks. The extended memory standard allows the user to install a special memory board in the computer which can be accessed through special I/O instructions. These I/O instructions map any of the EMS board's 16-kbyte pages into one of the four 16-kbyte segments of the PC's address space between 832K and 896K. As Fig. 11-20 shows, the user can "map" any 16-kbyte page from the EMS memory into any of the four different 16-kbyte spaces.

This system effectively allows the PC user to work with programs which require many megabytes of data storage and yet still work within a limited address range of the original 8088 microprocessor.

Self-Test

Answer the following questions.

34. One way to increase the size of memory available to a personal computer is to use
a. Program data transfers
b. An interrupt-driven controller
c. An extended memory system
d. DMA

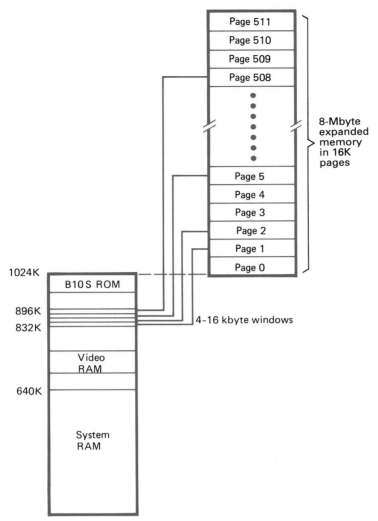

Fig. 11-20 **A memory map for a PC which uses the expanded memory specification (EMS) to access more memory locations than the 640K shown in Fig. 11-19. There are two ways to access the extra memory: using a single 16-kbyte block or using four 16-kbyte blocks.**

35. Why is it necessary to keep some part of the microprocessor's memory in a non-switched bank?
36. A memory management unit is
    a. Hardware which lets you use 256K chips
    b. Hardware which allows you to bank-select memory
    c. Hardware which controls the DMA process
    d. All of the above

SUMMARY

1. Some of the microprocessor's main memory must be read-write memory.
2. RAM stands for random-access memory. Random access means that you can ad-dress one single memory location just as quickly as you can address any other.
3. Memory access time tells how long a memory system takes to place information

on the data bus after the desired location is addressed.

4. Memory cycle time is a measure of the shortest time between two successive memory access operations.

5. Volatile memory loses its data when there is a power failure. Nonvolatile memory does not need power to keep the data which is stored in it.

6. Microprocessor bootstrap or boot programs must be stored in nonvolatile main memory. They are used to load other executable programs from mass storage. If there is no mass storage on a microprocessor-based system, then all programs must be stored in nonvolatile main memory.

7. Semiconductor read-write RAM is volatile memory. Core memory systems are nonvolatile read-write RAM.

8. Semiconductor memories are built by using both the bipolar and the MOS processes. MOS memory parts are the most common for microprocessor systems.

9. Static memories are constructed by using a flip-flop to store each bit of data.

10. Dynamic memories are constructed by using the input capacitance of an MOS transistor to store each bit of data. Dynamic memories have greater density and lower cost than static memories. Dynamic memories will hold a charge only for a few milliseconds and then they must be refreshed.

11. Semiconductor read-write RAMs use row-and-column decoders to select the addressed bit from the array of data storage cells.

12. Memory ICs come in bitwide and bytewide forms.

13. Many memory ICs are designed so that the lower-bit-density ICs can be replaced with higher-bit-density ICs. For example, a 2K × 8 SRAM IC can be replaced by an 8K × 8 SRAM IC. Many bytewide RAMs can be replaced by bytewide ROMs, and vice versa.

14. High-density DRAMs use multiplexed addressing to reduce the package pin count. Multiplexed addressing means that the IC loads the first half of the address data and then the second half of the address data. It does this when the row-address strobe and column-address strobe control lines are strobed.

15. A memory system has overhead. This refers to the devices that do not actually store data but which are needed to make the system work.

16. A ROM is a read-only memory. It is used to provide a microprocessor system with nonvolatile main memory. There are three different kinds of ROMs in common use: mask-programmed ROMs, erasable programmable ROMs (EPROMs), and electrically alterable ROMs (EAROMs).

17. Mask-programmed ROMs are MOS devices which have a preprogrammed bit pattern that is loaded into the ROM when the IC is made. They are used in high-volume applications where low cost is very important and the information stored in the ROM will not change. They are the highest-density ROMs and therefore give the most data or program storage capacity.

18. The EPROM can be programmed, erased, and reprogrammed in the field. EPROMs are programmed by applying data and address information to the EPROM along with a programming signal. They are erased by exposure to a strong ultraviolet light. The EPROM is the most popular ROM for microprocessor systems.

19. The EAROM can be reprogrammed by applying address and new data to the EAROM along with a programming signal. The programming process is slow, but the EAROM can be read as fast as any of the ROMs. The EAROM is not as permanent or as dense as the mask-programmed ROM or the EPROM.

20. Some ROMs are pin interchangeable with EPROMs, and some series of ROM parts allow pin interchangeability with higher-density versions.

21. Direct memory access (DMA) is a high-speed process which allows the user to transfer data directly with the microprocessor's main memory locations without using a programmed data transfer. DMA is often used to read data from or load data onto such devices as disk and tape drives.

22. The DMA controller takes control of the microprocessor's data and address buses during the data transfer. The microprocessor is inactive during a DMA transfer.

23. A microprocessor can address more main memory locations than its memory addressing capacity allows by switching banks of memory devices into and out of a part of its main memory address space. This technique is called paging, bank switching, or memory management.

24. Some microprocessor systems use a separate controller to perform memory management and help the programmer keep track of information about memory. This logic is called the memory management unit (MMU).

25. The PC uses a standardized form of bank switching in order to allow the programmer access to more than the 640-kbyte limit of main memory allowed by the standard PC memory map. This special form of memory address extension is called the expanded memory specification (EMS).

CHAPTER REVIEW QUESTIONS

Answer the following questions.

11-1. Which term does not apply to semiconductor read-write RAM?
    a. Dynamic   *c.* ROM
    b. Static   *d.* Volatile

11-2. When a microprocessor system has mass storage, nonvolatile memory is usually used to store the
    a. Bootstrap program   *c.* Instructions
    b. Main program(s)   *d.* All of the above

11-3. Semiconductor read-write memory is
    a. Volatile   *c.* Dynamic
    b. Static   *d.* Nonvolatile

11-4. Memory access time is the time between
    a. Refresh cycles
    b. Addressing a memory location and the time the data is placed on the data bus
    c. Successive memory access operations
    d. DMA transfers

11-5. A microcomputer system must have some
    a. ROM   *c.* I/O ports
    b. Read-write memory *d.* All of the above

11-6. Semiconductor memory devices that use the _____ process are the most common.
    a. Dynamic   *c.* MOS
    b. Bipolar   *d.* Static

11-7. Static RAMs store data in a
    a. ROM   *c.* Flip-flop
    b. SRAM   *d.* Capacitor

11-8. A dynamic memory must be
    a. Refreshed   *c.* Sequential-access
    b. Random-access   *d.* All of the above

11-9. A 64K × 1 memory IC and an 8K × 8 memory IC are both _____-bit devices.
    a. 16K   *c.* 64K
    b. 32K   *d.* 256K

11-10. A dynamic memory stores its data in a
    a. ROM   *c.* Flip-flop
    b. DRAM   *d.* Capacitor

11-11. Compared with a memory system built with static memory ICs, a memory system built with dynamic memory ICs of the same density will have _____ overhead devices.
    a. No   *c.* Less
    b. More   *d.* The same

11-12. The higher-density dynamic memory ICs use _____ addressing.
  a. Multiplexed   c. Binary
  b. Row-column   d. All of the above

11-13. A ROM is a _____ memory device.
  a. Volatile   c. Dynamic
  b. Nonvolatile   d. All of the above

11-14. In a microprocessor-based system, an EPROM is used as a _____ memory device
  a. Mask-programmed   c. Random-access read-write
  b. Bipolar   d. Random-access read-only

11-15. A mask-programmed ROM is a _____ device.
  a. Field programmable
  b. Random-access read-write
  c. Permanently programmed (by the semiconductor manufacturer)
  d. Volatile

11-16. The initials EAROM stand for
  a. Easily altered read-only memory
  b. Equally available read-only memory
  c. Electrically alterable read-only memory
  d. Wrong; it is electrically alterable random-access memory (EARAM)

11-17. Direct memory access is used to
  a. Transfer data to memory via the accumulator
  b. Transfer data out of memory via the accumulator
  c. Accomplish a high-speed data transfer between system memory and another external device
  d. a and b are correct

11-18. Briefly explain the purpose of the $\overline{\text{CHIP ENABLE}}$ control line on a memory device. Is the $\overline{\text{CHIP ENABLE}}$ control line limited to either read-write memory or ROM? Explain your answer.

11-19. How can you get nonvolatility with semiconductor read-write memory devices? Why is this only limited protection? What is the best semiconductor process to use for this purpose?

11-20. Briefly explain the difference between RAM and sequential-access memory.

11-21. A DMA controller has two registers that are programmed by the microprocessor before a DMA transfer starts. What do these two registers do? Why do you suppose they are 16-bit registers on a DMA controller which is designed to work with an 8-bit microprocessor?

11-22. Why does the microprocessor go into a hold state during a DMA transfer?

11-23. Briefly explain how an 8-bit microprocessor with a 64-kbyte memory space can address 128K memory locations. What is the name given to the hardware which manages this memory addressing process?

CRITICAL THINKING QUESTIONS

11-1. The main memory system is a very critical part of the overall microprocessor system. In what ways does the microprocessor's main memory system allow or limit what the microprocessor system can be asked to do?

11-2. Most personal computers use semiconductor memory which is *not* backed up by battery and is therefore volatile. How can a PC work without having nonvolatile memory?

11-3. Some memory ICs are constructed using dynamic techniques, but they have built-in refresh circuits. These are sometimes called pseudo-static memory integrated circuits. Why?

11-4. What would have happened to memory ICs if the multiplexed addressing technique had not been developed?

11-5. Often you see a label taped to the top of an EPROM. The information typed on this label tells you the version number of the information programmed into the EPROM. You will also often find this label pasted over the EPROM's quartz window. Why?

11-6. What were some of the driving forces behind the development of memory addressing techniques which allow a microprocessor to address memory systems which are larger than its basic address range says it can address?

Answers to Self-Tests

1. *d*
2. *c*
3. *b*
4. *a*
5. *b*
6. *c*
7. *a*
8. *d*
9. *b*
10. *c*
11. *d*
12. *c*
13. *d*
14. *a*

15. One of the two strobe signals loads the row decoder, and the other loads the column decoder. This extra complication reduces the number of address pins needed on the memory IC and therefore lets the manufacturer use a lower-cost package.

16. The static memory devices (ICs) store data in flip-flop circuits. The dynamic memory devices (ICs) store data as a charge or lack of charge on the MOS transistor's input capacitance.

17. If dynamic memory devices are not refreshed regularly (every millisecond or so), the charge on the capacitor is lost and therefore the data is lost.

18. You would choose a CMOS memory device when you create a nonvolatile memory system with

battery backup because CMOS memory devices draw very little current from the battery.

19. The chip select logic (the two OR gates and the inverter) would reduce to a single OR gate. This is because the 16K × 1 devices have 12 address lines rather than the 11 address lines found on the 8K × 8 devices. The number of memory devices would increase from two to eight.

20. The dynamic memory devices require refresh logic; therefore, this increases the amount of overhead logic used.

21. See Fig. 11-21 on page 336.

22. Because the dynamic memory devices require the additional overhead of refresh logic, which may not be economically justified for a system with a small amount of main memory.

23. A total of 768K − 640K = 148K memory locations are unused when three banks of 256K memory devices are used to create the memory system for a conventional PC. The memory system could be built with two banks of 256K devices and two banks of 64K parts. This would provide exactly 640K memory locations

(256K + 256K + 64K + 64K = 640K); however, this requires eight more ICs (nine more in a system with parity), so it may not be more efficient.

24. *c*

25. *a*

26. *e*

27. Because the EPROM can be erased and reprogrammed when errors are found.

28. The microprocessor cannot do anything without instructions. The microprocessor must have at least enough instructions in ROM so that it can read the needed program instructions from an external mass storage device.

29. The microprocessor must have some memory locations in which it can store variable data. These must be read-write memory locations.

30. DMA = 1,000,000 transfers per second Tape = 800 bpi × 75 in/s = 60,000 transfers per second. Therefore, the tape is the limiting factor.

31. Tape = 1600 bpi × 125 in.s = 200,000 transfers per second. The tape is still the limiting factor.

32. Because the microprocessor must perform an I/O read and cycle count for each data

transfer. Also, the transfer data rate is limited by the microprocessor's instruction cycle time, not just by the memory cycle time.

33. Four of the registers are used to hold the starting address for the data transfer, and the other four registers are used to hold the terminal counts for the data transfers.

34. *c*

35. The microprocessor must have some nonswitched memory that holds the switching program. If all of the memory was switched, the microprocessor would not know how to return to the original configuration.

36. *b*

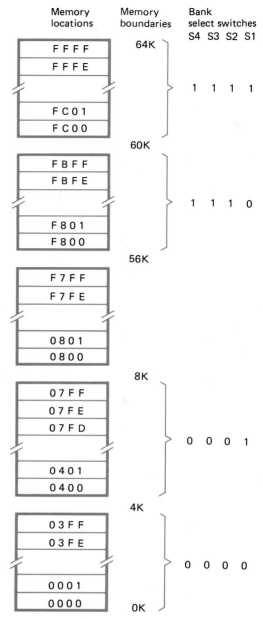

Fig. 11-21   **See self-test question 21.**

CHAPTER 12

Mass Storage

∎

CHAPTER OBJECTIVES

This chapter will help you to:

1. *Name* and *explain* the most commonly used mass storage devices.
2. *Understand* how data is stored on magnetic devices.
3. *Recognize* the fundamental components used to build mass storage devices.
4. *Understand* installation and basic maintenance of mass storage devices.
5. *Explain* the advantages and limitations of optical disk drives.

———————

As we learned in the previous chapter, to make a microprocessor do something, we must have a place to store the program which tells it what to do and we must have a place to store the data it works on and the answers it makes. For any one activity (program) this takes place in the microprocessor's main memory. But what if we want the microprocessor to do different things at different times, or what if the programs are so big they cannot fit in main memory? To handle this situation, we use mass storage devices. For most microprocessor-based systems (including PCs) the most common forms of mass storage devices are the floppy disk and the Winchester (hard) drive.

∎

12-1 AN INTRODUCTION TO MASS STORAGE

Your first question may be "Why have mass storage devices at all?" This is a reasonable question because in Chap. 11 we learned that main memory is becoming very low cost. We also learned that a modern microprocessor may have a memory address space with many megabytes or even gigabytes of storage.

Not all microprocessor-based systems use *mass storage*. For example, the simple microprocessor-based controller which uses a 16-kbyte EPROM for program storage and a 2-kbyte SRAM for data and working storage certainly does not require mass storage subsystems. On the other hand, we all know that every microprocessor-based personal computer uses mass storage subsystems. Additionally, any study of industrial microprocessor-based systems will show a large number of the more complex systems have mass storage sub-systems. Why are these subsystems required? What functions do they accomplish?

There are three different reasons for mass storage subsystems to be added to microprocessor-based products. First, there is a need to have long-term nonvolatile storage. Programs take a great deal of time and effort to write. We cannot afford to lose them. They must be stored in a nonvolatile medium.

Second, the programs must be stored in an economical manner. That is, they must be stored for the absolute lowest cost per bit. Although both ROM and battery-backed RAM are becoming very low cost, neither one is the most economical way to provide mass storage.

Third is the requirement for versatility. Many microprocessor-based systems are designed to perform multiple functions. When they change function, they are loaded with a different program. Some form of very easily updated mass storage is necessary to accomplish this function. Once again, neither ROM nor battery-backed RAM gives the needed flexibility. This is especially true because future updates may be unknown at the time of production.

One common characteristic in each of the above three reasons is the use of machine-readable, nonvolatile media. The *medium* is where you write the programs and data. The requirement that the medium be *machine-readable* means that some very low cost methods of nonvolatile mass storage cannot be used.

For example, neither data dumped to a printer nor a hand-written program listing can be read by the microcomputer. Therefore they are not useful forms of mass storage. In this chapter we will spend most of our time looking at one particular type of mass storage medium—*magnetic media.*

Any mass storage subsystem has three major parts: the medium, the drive, and the controller. We write the information on and read the information from the medium. The *drive* moves the medium so that it can be read, and performs the read-write operation. The *controller* connects (interfaces) the drive to the microcomputer system. In the following sections, we will look at each of these different parts which make up mass storage subsystems.

Many different forms of magnetic media are used in the computer industry. Early computers used magnetic drums. Later they used magnetic tape, and then they added magnetic disk-drives to replace the drums. All of these devices use magnetic storage principles to capture the data or program information.

The computer industry has not always used magnetic media for its mass storage. In the past, very low cost mass storage often used paper media. Holes punched in the paper made the data machine-readable. Paper media came in two forms: paper tape and paper cards. Both paper tape and computer cards are now only used in special applications. In most general applications, paper media have been replaced by a very low cost magnetic medium called *floppy disks.*

The floppy disk was popularized as a mass storage medium by the rapid growth of the microcomputer. As microcomputers grew from low-volume hobbyist devices to high-volume consumer and commercial products, the requirement for low-cost mass storage became great. As is often the case, this demand created a new industry. This industry took the principles of magnetic storage and the princi-

ples of the disk drive and created a very low cost medium and drive called the *floppy disk* and the *floppy-disk drive.*

The floppy-disk drive became popular in the late 1970s. At that time a drive cost over $1000. Today, the consumer price of a single floppy-disk drive, with greatly enhanced storage capabilities, is under $50.

Obviously, this great a price decrease creates a tremendous increase in product volume. What does this mean to you? If you are installing, servicing, or designing microcomputer-based systems, it is very likely that these systems will use one or more floppy-disk drives.

Typically, a floppy disk is limited to less than a few Mbytes of storage capacity. Although a few Mbytes sounds like a lot of storage, it is very easy to find microcomputer applications which need more than a few Mbytes of online mass storage. The term *online mass storage* refers to the mass storage which can be accessed by the computer at any time without help from a human. A floppy-disk drive loaded with a floppy disk and connected to the microcomputer is online storage. The data on the floppy disk installed in the drive is the online data.

Because of the growing requirement for even greater amounts of mass storage, additional research is continuously being undertaken. This research led to the development of a mass storage device called the *Winchester-disk drive.* The Winchester-disk drive is often called a *hard drive* (the terms are synonymous). The Winchester-disk drive allows the user to store much greater quantities of online data than does the floppy-disk drive. The Winchester-disk drive typically does not offer removable media as the floppy-disk drive does. However, it does provide a very low cost method to store hundreds of megabytes or thousands of megabytes (gigabytes) of data or program information. Typically, the Winchester-disk drive is 4 to 20 times as expensive as a floppy-disk drive.

As you might expect, the Winchester-drive does not always have enough storage capacity. There is continued research to create mass storage devices which store ever greater amounts of information. One such device, introduced in the 1980s, is the *optical-disk drive,* which grew out of the audio compact disk (CD). This storage uses optical rather than magnetic techniques and provides much greater storage density. However, there are special problems in producing an optical device which can be read from and written into.

The development of very high density mass storage devices with nonremovable media forced another development—the *streaming tape*. A streaming-tape system is used as a high-density backup mass storage system.

Why do we need backup storage systems? For example, the person operating the microcomputer system must do a great deal of work to generate 120 Mbytes of data or program information. Therefore, a Winchester-disk drive with 120 Mbytes of information stored on it becomes very valuable. A sudden loss of that information through accidental erasure or other damage can be a major financial loss. As you know, the medium in a Winchester system is nonremovable. To protect valuable information, we make a copy onto a removable medium. This duplicate is called the *backup copy*. The backup copy can be stored away from the microcomputer.

Slow-speed high-density storage systems are used to back up such devices as the nonremovable Winchester disk. Typically, magnetic media are chosen for this operation. The backup medium is usually a tape in a cartridge or a floppy disk. Although very slow and strictly sequential-access, tape backup allows the user to protect against accidental loss of data on the mass storage device.

Self-Test

Answer the following questions.

1. Microcomputer systems require mass storage because
   a. Microcomputer systems need long-term nonvolatile storage
   b. Programs and data must be stored economically
   c. If a microprocessor-based system is to be used for many purposes, it must be easily reloaded with new programs and data
   d. All of the the above
2. One requirement for most mass storage systems is that they use
   a. Floppy-disk drives
   b. Machine-readable media
   c. Some form of backup
   d. All of the above
3. Information printed on paper is a form of low-cost mass storage; however it lacks
   a. Nonvolatility

   b. Data capacity
   c. A reasonable life expectancy
   d. Machine readability
4. List the three parts which make up any mass storage subsystem that is attached to a microprocessor-based system.
5. Briefly explain the function performed by each part of the mass storage subsystem in your answer to the previous question.
6. What is meant by online storage?
7. What are two major differences between a Winchester-based mass storage subsystem and a floppy-disk-based mass storage subsystem?
8. What is the major reason for needing separate mass storage subsystems devoted to backup?

12-2 BASIC MAGNETIC STORAGE TECHNIQUES

Before we can explore specific magnetic mass storage devices such as floppy, Winchester, or tape systems, we must understand how magnetic media store digital information. In this section, we will review the fundamental principles of reading and writing data magnetically.

As you know from other science courses, any ferrous (iron) material may be magnetized. Simply, this means that materials with a high iron content will be attracted to a magnet. If the magnet is left on the material long enough, the material also becomes magnetized.

How does magnetization occur? Figure 12-1 shows a block of magnetic material which has not been magnetized. When this magnetic material is brought near a magnetic field, it becomes magnetized. Our studies in physical science tell us that the magnetic dipoles within the material line up with the external magnet. This is shown in Fig. 12-2 on the next page.

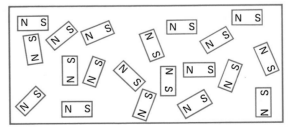

Fig. 12-1 **A microscopic view of magnetic material such as that used for magnetic recording. Note that the millions of magnetic dipoles within the ferromagnetic surface material are in a random pattern.**

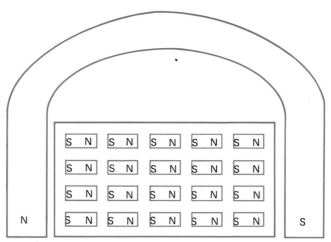

Fig. 12-2 An external magnet aligns the magnetic dipoles within the ferromagnetic material. When the external magnet is taken away, the magnetic dipoles stay aligned—the material is magnetized.

As you can see from this diagram, the magnetic dipoles act like thousands of tiny little magnets. The south poles of the magnetic dipoles are attracted to the north pole of the external magnet. Likewise, the north poles of the magnetic dipoles are attracted to the south pole of the external magnet.

When the external magnetizing force is removed, the magnetic dipoles stay lined up. We say then that the material is magnetized. It now acts just like a permanent magnet and has a magnetic field of its own, as shown in Fig. 12-3.

This new magnet, like any permanent magnet, has both north and south poles. The direction of the north and south poles depends on the direction of the magnetizing force. It is this principle which allows us to write digital information to a magnetic tape or disk. *We write a logic 1 as magnetization in one direction and a logic 0 as magnetization in the opposite direction.*

How is a magnetic medium constructed? It is made up of a very thin layer of magnetic material deposited on thin plastic. A cross section of magnetic tape is shown in Fig. 12-4.

To write information to the media, we use a specially constructed electromagnet called a *write head*. The north-south poles on this magnet are separated by a slit only a few thousandths of an inch wide. This slit is called the *head gap*. To read stored data, the medium is passed over a *read head*. This generates tiny electrical signals in the head windings. Figure 12-5 shows a section of magnetic tape being written to by a write head. It also shows the information being picked up by a separate read head.

We are all familiar with the audio cassette recorder. In a cassette recorder, the magnetic tape moves past a record-playback head. Dur-

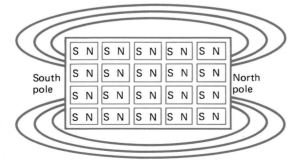

Fig. 12-3 The magnetic material after the external magnetizing force is removed. Note that the magnetic dipoles remain aligned as if the magnet were still in place. This material is now a permanent magnet.

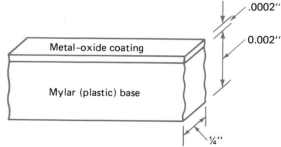

Fig. 12-4 A section of magnetic medium (quarter-inch tape). The magnetic material (metal oxide about 0.0002 in. thick) is a coating on the 0.002-in.-thick plastic (Mylar) base.

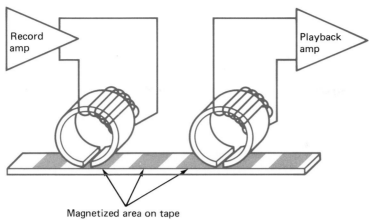

Fig. 12-5 **The read and write heads in action. Electrical signals connected to the write head generate a small magnetic field in the area of the head gap. When a magnetic tape is passed by this gap (left-hand side), the tape is recorded. When a previously recorded tape passes by the read head gap (right-hand side), a very low level electrical signal is generated in the read head winding. This signal is amplified and processed to become a series of TTL-compatible logic 1s and logic 0s.**

ing the recording process, we put information on the tape, and during the playback process we take information from the tape.

The world of digital magnetic mass storage is very much the same. The magnetic medium moves across a read-write head. Either logic 1s or logic 0s are written onto the medium. Later, the magnetic medium moves across the head again. This time the magnetized spots on the medium generate small electrical signals in the head. These signals tell us that there are logic 1s and logic 0s on the medium. Different direction signals are generated for logic 1s and logic 0s.

The basics of magnetic storage are really quite simple. However, many complications must be overcome to make high-density, high-reliability magnetic mass storage. For example, the read-write head gap must be very small to read or write high-density (number of bits per inch) data. As the head gap narrows, the amount of magnetic field available to magnetize the magnetic medium becomes smaller. That is, there is less magnetic energy available to do the work. This means that the head must be in very close contact with the magnetic medium. Even then, only very low levels of magnetization are actually written onto the magnetic medium.

When the recorded medium passes a read head, the changing magnetic field caused by the moving medium generates very small elec-

trical signals in the head winding. These small electrical signals are amplified, shaped, and processed to generate high-level (TTL-compatible) logic signals. These logic signals are 1s or 0s, depending on the data originally stored on the medium.

During the read-write process, many hundreds of millions of bits of information are taken off and put on magnetic media by using the techniques we just discussed. There is a relatively high probability that some bits will not be written or read back properly. This causes errors in the data or program information.

Errors may occur in the read-write process for many different reasons. Errors can be permanent or temporary. For example, if the magnetic medium is defective, no information is written in the first place. If there is a small speck of dust on the magnetic medium and it forces the medium away from the head, the bit is either not written or not read.

The information on magnetic media can be easily destroyed if exposed to a magnet. As you learned earlier, the magnetic fields which actually write the information are quite weak. Therefore, any other magnetic signals brought into close contact with the medium erase or corrupt the data. Additionally, the signals being generated in the read head are also very weak signals. This means that they can be easily interfered with by other, outside electrical signals.

All this means that it is relatively easy to have errors in the data taken from a magnetic medium. Because this is true, most of these systems are designed with a great deal of error checking. Earlier, we learned that memory circuits and some communications circuits use an error-checking technique called *parity checking*. Typically, parity checking adds one error detection bit per byte of data. If the parity does not check, an error is discovered. However, this is not good enough error checking for data written on magnetic media.

Reading and writing on magnetic media require more sophisticated error-checking techniques. Some systems even use *error detection and correction* (*EDAC*) techniques. One common technique used in magnetic storage is the *cyclical redundancy check* (CRC).

CRC-16 is one common version of the cyclical redundancy check. It treats the data as the coefficients of a binary polynomial. This binary polynomial is divided by a known 16th-order binary polynomial ($x^{16} + x^{12} + x^5 + x$). The remainder is saved. The data and the remainder are then stored on the magnetic medium. When the data is recovered, it is once again divided by the known 16th-order polynomial and the remainders are compared. If the remainders are equal, the data was properly read from the magnetic medium. If the remainders do not match, an error occurred and the data must be reread.

CRC-16 is not the only error-checking technique used to make sure the data taken from the magnetic medium is error free. It is, however, a very common one. It is discussed here to show the level of sophistication used to ensure that data are safely captured on the magnetic media.

The way that data are written on the magnetic media is called the *format*. For example, one format only writes changes in data. In other words, no individual logic 1 nor individual logic 0 is written separately. The system only writes *changes* from a logic 1 to a logic 0 or from a logic 0 to a logic 1.

Each format has its own advantages and disadvantages. For example, some formats write more data on the same length of magnetic medium than others. That is, different formats can have different *densities*. One very common format is *double density*. As is true everywhere else, one does not get something for nothing. The higher-density formats typically have greater error rates, and the read-write electronics cost more money.

Self-Test

Answer the following questions.

9. When a material is exposed to a magnetic field and its magnetic dipoles line up with the magnet's field and remain lined up after the external magnetic field is taken away, we say that the material is
   a. Magnetized   c. Nonmagnetic
   b. Conductive   d. All of the above

10. When we store digital information on magnetic media, the difference between a logic 1 and a logic 0 is that
    a. Logic 1s use plastic and logic 0s use Mylar
    b. Logic 1s are magnetized and logic 0s are not
    c. Logic 1s are strongly magnetized areas and logic 0s are weakly magnetized areas, or vice versa
    d. Logic 1s are magnetized with their poles in one direction, and logic 0s are magnetized with their poles in the opposite direction

11. Tapes and disks used for mass storage are made using a thin coating of _____ material on a nonmagnetic surface.
    a. Nonmagnetic   c. Nonferrous
    b. Ferrous   d. Plastic

12. We use a special electromagnet to write the logic 1s and logic 0s on magnetic media. This is called the
    a. Head   c. Playback
    b. Recorder   d. Head gap

13. The very fine slit between the read-write electromagnet's north and south poles is called the
    a. Head   c. Playback
    b. Recorder   d. Head gap

14. When media with stored digital information moves across this electromagnet, changes in the direction of the magnetized spots generate
    a. Excess surface wear
    b. Tiny electrical signals in the head windings
    c. Additional magnetization in the media
    d. Alignment of the magnetic dipoles

15. Because the head gap is very small, the magnetic field used to write digital information on the magnetic medium is quite small. This means that
    a. The medium must be kept very close to the head to ensure that the information is stored on the medium

b. It is very easy for external electric and magnetic fields to cause errors in the read-write process

c. A small speck of dust can push the medium away from the head and cause a read or write error

d. All of the above

16. Because it is quite easy to have read and write errors, magnetic mass storage systems use
   a. Sophisticated error-checking techniques
   b. Separate read and write heads
   c. Dust-free environments
   d. 2 bytes of data

17. The way data are written on the magnetic medium is called
   a. The cyclical redundancy check
   b. Parity error checking
   c. The format
   d. A system of logic 1s and logic 0s

12-3 TAPES AND DISKS

Two styles of magnetic mass storage media have become quite popular during the history of digital computing. They are tape and disk. Each has its own unique advantages as well as its own unique disadvantages. We will briefly look at each of them in order to understand the major reasons for choosing one style over the other.

Magnetic tape was the earliest form of removable-medium high-density magnetic mass storage. Before magnetic tape was used to store digital information, magnetic tape was being used for analog information in the form of audio signals. Actually, audio magnetic tape was preceded by audio on magnetic wire during the late 1930s.

You must use *sequential access* to get information stored on magnetic tape. For example, the data you want is at the tape midpoint. To get it you must pass by all the information between the point you are currently at and the tape midpoint. This is shown in Figure 12-6(*a*). If you are at one end of the tape and the information that you want is at the other end of the tape, you must travel the complete length of the tape to get to the desired data, as shown in Fig. 12-6(*b*).

Most magnetic tape systems have fast forward and reverse modes to speed up long searches. However, they are only two to five times faster than the read-write speed. Additionally, you cannot read or write data when you use fast forward or reverse. Therefore to find data which is a long way away, you first fast forward to the approximate place where the target data is located. You then slow the tape to the read speed and read data to be sure that you are at the target data.

For example, suppose your 2400-ft tape is positioned at the starting point and the data you want is at the 1200-ft point. You fast forward at 180 in./s to the 1185-ft point and slow to read

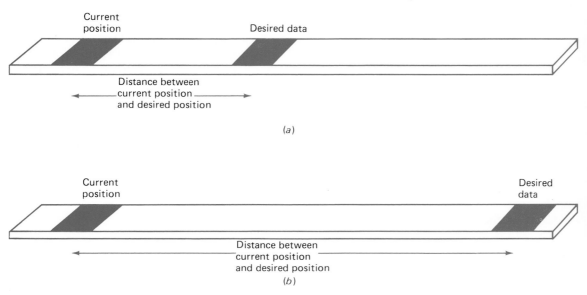

(a)

(b)

Fig. 12-6 Sequential access of data on a magnetic tape. (*a*) To get to the data in the middle of the tape, you must pass all the data between the middle and the current position. (*b*) To get data at the end, you must pass all of the data between the current position and the end.

speed (60 in./s) for the last 15 ft. The time for this is

$$\frac{1185 \text{ ft} \times 12 \text{ in./ft}}{180 \text{ in./s}} + \frac{15 \text{ ft} \times 12 \text{ in./ft}}{60 \text{ in./s}} = 82 \text{ s}$$

If the data you need is at the end of the tape, the time is

$$\frac{2385 \text{ ft} \times 12 \text{ in./ft}}{180 \text{ in./s}} + \frac{15 \text{ ft} \times 12 \text{ in./ft}}{60 \text{ in./s}} = 162 \text{ s}$$

As you can see, getting to the end of the tape takes almost twice as long as getting to the middle of the tape, even when you use fast forward. Both are long times to wait for data.

Even with the disadvantages of sequential access, magnetic tape is still popular. Tape is extremely inexpensive, many different tapes can be used on one tape drive, and they can be written at a relatively high density, which allows many bits to be written on an inch of tape.

There are many applications which require faster access to online data in a mass storage device than can be provided with a magnetic tape system. The *magnetic disk* was invented to solve this problem.

How does the magnetic disk mass-storage subsystem provide high-density storage but much more rapid access time? The medium is in a different physical style. If you think of the magnetic tape in very simple terms, you think of one long string of logic 1s and logic 0s. To access any particular data, we must pass all the data between our present position and the place on the tape where the desired data lies.

Disks look like 45-rpm records, except that 45-rpm records use a single spiral groove. Here the data is written as a series of separate, closed rings, each slightly smaller than the next. These are called *concentric rings*. Each of these rings is called a *track*. Typically, each track is broken into a number of sections called *sectors*. This is shown in Fig. 12-7.

In operation, the disk spins, and the head is moved into position over the track of data you want. Once the head is positioned over a particular track, the time to access any data on that track is the same as the time needed to access the data on the same length of magnetic tape. If the data is at the end of the track and the head is positioned at the beginning of the track, we must let the disk make nearly one complete revolution to bring the data under the head.

If, however, the data is on track 15, and you are currently positioned on track 32, it is only necessary to step the head from track 32 to track 15 before beginning this process. That is,

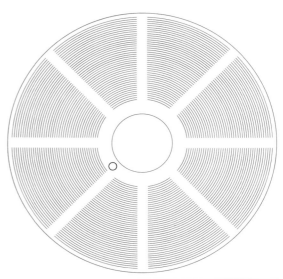

Fig. 12-7 The physical format of a magnetic disk. Note that each of the data tracks is a concentric ring. These rings are called a track. Each track is broken into a number of different sections. The small hole near the center opening is the sector ID hole which marks the beginning of sector 0.

it is not necessary to pass over all the data on tracks 32,31,...,17,16 in order to reach the data on track 15.

Once the head is positioned over the correct track, the *average access time* is the time required for the disk to turn half of a rotation. This is because on the average, the data you want is found within half of the distance that separates the two ends of the track.

For example, you are using a disk drive which turns at 300 rpm and has a track-to-track stepping time of 6 ms and requires 15 ms to settle on the selected track. There are 40 tracks. The time to reach data in the middle of track 19 when you are at track 00 is

$$20 \text{ tracks} \times 6 \text{ ms/track} = 120 \text{ ms}$$

$$120 \text{ ms} + 15 \text{ ms settling time} = 135 \text{ ms}$$

$$\frac{300 \text{ rpm}}{60 \text{ s/min}} = 5 \text{ rps} = 200 \text{ ms/rev}$$

Therefore

$$\tfrac{1}{2} \text{ revolution} = 100 \text{ ms}$$

Thus

$$135 \text{ ms} + 100 \text{ ms} = 235 \text{ ms}$$

The time to reach data on the last track is

$$40 \text{ tracks} \times 6 \text{ ms/track} = 240 \text{ ms}$$

$$240 \text{ ms} + 15 \text{ ms settling time} = 255 \text{ ms}$$

$$\frac{300 \text{ rpm}}{60 \text{ s/min}} = 5 \text{ rps} = 200 \text{ ms/rev}$$

Therefore

$$\frac{1}{2} \text{ revolution} = 100 \text{ ms}$$

Thus

$$255 \text{ ms} + 100 \text{ ms} = 355 \text{ ms}$$

As you can see, the time to reach data on a floppy disk is relatively short compared with a tape system.

The magnetic disk does not have complete random access. However, it is much faster than an all-sequential-access system. Once again, one does not get something for nothing. Typically, the amount of data stored on disk is much less than can be stored on magnetic tape. Also, the physical format of a disk makes both the head and the medium move. This means that alignment becomes a much more critical issue in the product.

Both disk and tape are popular storage devices. Each serves a very worthwhile purpose and both have costs within easy reach of microcomputer owners. In the following sections, we will take a detailed look at the drive support electronics and a number of different disk and tape systems.

Self-Test

Answer the following questions.

18. Two common styles of magnetic media are
    a. Logic 1s and logic 0s
    b. Tapes and disks
    c. Drums replaced by disks
    d. Tapes and drums
19. The main disadvantage of tape systems is that they
    a. Use sequential access and are therefore quite fast
    b. Use sequential access and are therefore slow
    c. Have low-cost removable media
    d. All of the above
20. The main advantage of tape is that it
    a. Uses inexpensive tape cartridges
    b. Can contain a lot of data
    c. Has low-cost removable media
    d. All of the above
21. The main advantage of disk versus tape is that it
    a. Is generally more error-free
    b. Has much faster access to the data
    c. Does not hold as much data and therefore is not as valuable
    d. Cannot be written over

22. The data on a disk are written in a series of
    a. Spirals
    b. Concentric tracks
    c. Grooves
    d. Concentric tracks broken into sectors
23. Why is the average access time for a tape system one half of the time that is needed to run the tape from one end to the other?
24. The access time for a disk is made up of two separate access times. What are they?

12-4 MASS STORAGE SUPPORT ELECTRONICS

A mass storage system is normally supported by two different electronic packages. They are called the *drive electronics* and the *controller*. These two packages work closely together.

A typical controller can only interface one type of drive to the microcomputer. A different controller may be needed for tape, floppy, and Winchester drives. These two electronic packages are usually located in two different places. The drive electronics are usually packaged with the drive, and the controller is usually installed in the microprocessor-based system. Frequently, you will find that one controller can work with two or more drives of the same type.

A *drive package* has motors to rotate the disk and position the head, solenoids to load the head (position it against the disk), sensors to indicate disk position, and the read-write head. Each of these electromechanical components needs or generates different levels of analog signals. The drive electronics package converts the incoming data and command signals from the controller into analog signals at the right levels to run the different parts of the drive.

For example, TTL command signals are sent to the drive. At the drive, these TTL command signals are converted into high-level analog signals. The high-level analog signals run the drive's motors, which operate on 12 V. Likewise, the TTL data signals must be converted into write signals before they can drive the write head.

The drive must send a number of signals to the controller. For example, the microcomputer needs to know that the disk is loaded and ready. You cannot write to or read from a disk which is not there. This disk-ready condition is detected by an LED-disk phototransistor pair. The output of the phototransistor is a TTL-level signal which is buffered and sent to the controller.

The *read signal path* uses a high-gain wideband amplifier to strengthen the very small signals coming from the head. This amplifier is usually placed very close to the head. The amplified data signals are converted from their analog form to digital signals at TTL levels. Once this is done, the signals can be sent to the controller. At the controller, the data is error-checked and deformatted.

The controller interfaces (connects) the mass storage subsystem to the microcomputer system. In microcomputer systems such as the personal computer, controllers are bus-compatible. They are provided in card form, so they can plug into the microcomputer's bus. The drive is connected to the controller with a multiconductor cable. The multiconductor cable is a special-purpose bus for that particular type of drive.

The controller gives the microprocessor-based product control over the drive and information about the drive. The microcomputer uses command words sent to an I/O port to tell the drive what action it wants done. It does this by putting the command words in the controller's registers. The controller provides information to the microcomputer about current drive conditions. It does this with words left in the controller's status registers.

For example, a tape drive can be told to move the tape forward, backward, fast forward, or fast reverse. This is done by sending command words to the controller. Additionally, the programmer can, by examining a status register within the controller, determine the current tape direction and speed and the current footage count.

The controller also provides an interface between the microcomputer and the data signals being sent to and received from the drive. Typically, data is sent to and from the drive one bit at a time.

The controller may buffer large blocks of data before finally passing them to the microcomputer. Likewise, the controller may receive large blocks of data from the microcomputer before actually sending it to the drive to be written to the medium. The controller may communicate with the microprocessor-based system using DMA between the memory on the controller and the microprocessor-based system's main memory.

Self-Test

Answer the following questions.

25. The two electronic packages used to support a mass storage system on a microprocessor-based system are
    a. The microprocessor and the controller
    b. The drive electronics and the microprocessor
    c. The controller and the drive electronics
    d. All of the above
26. Briefly explain why we say that the function of the drive electronics is to interface the analog and electromechanical devices to TTL controller signals.
27. List three control signals that a controller sends to the drive electronics and briefly tell what they do.
28. List three status signals that a controller receives from the drive electronics, and briefly tell what information they give the controller.
29. What function does the controller perform?
30. Why can one controller manage more than one drive?

12-5 THE FLOPPY DISK

As we learned earlier, the floppy disk became popular as a mass storage subsystem for microcomputers. This is because the floppy-disk drive and medium are inexpensive. The rapid growth of the low-cost microprocessor-based personal computer allowed the disk manufacturers to make a tradeoff. They traded the much lower cost floppy-disk design for the very high density storage of hard-disk systems. In other words, the microcomputer can use the cheaper mass storage, even though it stores less data than a hard disk.

Why is the floppy disk cheaper than the hard disk? In order to answer this question, we must briefly look at how a hard disk is constructed. Hard-disk construction starts with a large nonmagnetic metal platter. It is like a large (18-to 24-in.) 33 rpm record without the grooves. The platter surface is carefully machined until it is flat to within a few millionths of an inch. The platter surface is coated with a magnetic material a few millionths of an inch thick.

The platter is built into a box. This box has a motor to spin the platter and another motor to move the heads over the selected track. It also has the read-write heads, read-write electronics, and motor control electronics. The completed box is called a *hard-disk drive*. It can be interfaced to a computer with a controller. The main elements of a hard-disk drive are shown in Fig. 12-8(*a*).

In operation, the magnetic read-write head comes very close to the magnetic surface. Typically, the head "flies" a few millionths of an inch above the surface of the magnetic media, as shown in Fig. 12-8(*b*). The head does not touch the magnetic surface. Both the surface and the head wear too fast if the head touches the continuously rotating surface. Unlike the magnetic tape, the platter is always moving even if you are not reading or writing data to it. If the platter does not rotate continuously, the disk access time must include the time needed to bring the platter up to its working speed of many thousands of revolutions per minute.

A hard-disk drive is quite expensive. It is expensive to create the precision platter. It is expensive to maintain the distance from the platter to the head at a microscopic fraction of an inch. It is also expensive to keep the platter in a clean and vibration-free environment. All of this expensive precision is needed to keep the head from becoming too far away from the surface of the disk or from crashing into the surface of the disk and being destroyed.

A hard-disk drive has rapid access time and can store relatively large quantities of data. When the disk spins at 6000 rpm, all the data on one track passes under the head in 10 ms. This gives an average access time of 5 ms. A typical hard-disk drive system can store many

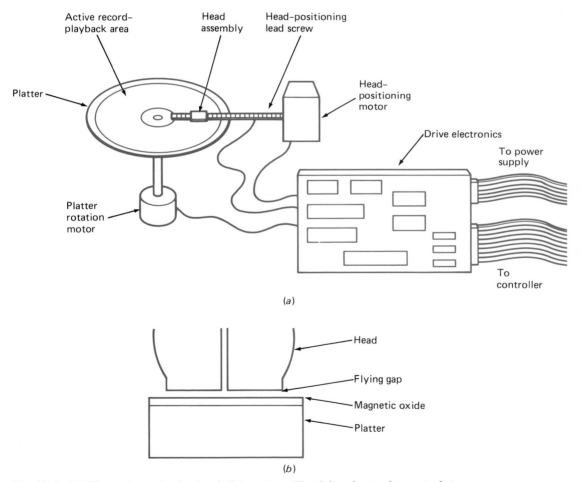

(a)

(b)

Fig. 12-8 (*a*) The main parts of a hard-disk system. The drive electronics controls two motors, one to rotate the disk and one to move the head to the correct track. The read-write head is also connected to the drive electronics, where read-write amplifiers interface the head with TTL-level signals. (*b*) The read-write head flies above the disk platter. The gap is a few millionths of an inch.

megabytes or even gigabytes of data and costs many thousands of dollars.

The floppy-disk drive borrowed technology from both magnetic tape systems and the hard-disk drive. The medium comes from magnetic-tape technology and the basic drive construction comes from hard-disk technology.

The floppy-disk medium does not start with a machined aluminum disk. It uses a large sheet of Mylar with a magnetic oxide coating, just like tape. This is cut into a doughnut shape and slipped into a paper or plastic envelope with openings. This makes a floppy disk or floppy diskette as shown in Fig. 12-9.

Because the magnetic medium is flexible (or floppy), the media can be pressed directly against the head without causing the damage that occurs when a hard disk contacts the read-write head. In order to keep head and media wear under control, the disk speed is much slower. Where hard disks have speeds of thousands of revolutions per minute, the floppy disk turns at 300 rpm. Additionally, the floppy disk does not have the density used with hard-disk drives. This means it holds much less data than its hard-disk equivalent. However, all this tradeoff allows the floppy-disk drive to be made very inexpensively.

The floppy-disk drive uses medium which is easily removable and very cheap. The typical floppy disk is a dollar or less, whereas removable hard disks start at a few hundred dollars. Even though you cannot put great amounts of data on a floppy disk, you can easily afford to have many different floppy disks to use with one floppy-disk drive.

There are three sizes of floppy disks in general use. The original floppy disk is an 8-in. disk. That is, the envelope in which the magnetic media is packaged is 8 in. square. This format was very popular in the late 1970s and early 1980s. It decreased in popularity when the personal computer was introduced with the 5.25-in. floppy-disk drives. The 3.5-in. floppy-disk drives were introduced in the mid-1980s. Although the 3.5-in. floppy disk is housed in a plastic case rather than a paper envelope, its operating principles are the same. Today the 3.5-in. floppy disk is the most popular form.

Floppy disks use one of three different formats when writing data to or reading data from the disk. These are called the *single-density, double-density,* and *very high density* formats. The single-density format uses frequency modulation (FM) to write the data, and the double-density format uses modified frequency modulation (MFM) to write the data. Figure 12-10 shows data being written and read using FM and MFM.

Technology introduced in the early 1980s let floppy-disk drives read and write data on both sides of the floppy disk. This doubles the disk's data capacity. *Double-sided floppy disks* require two read-write heads rather than one. However, the cost of a head is low compared with the cost of a second disk drive.

Figure 12-11 is a cross-sectional view of the floppy-disk drive with a floppy disk mounted

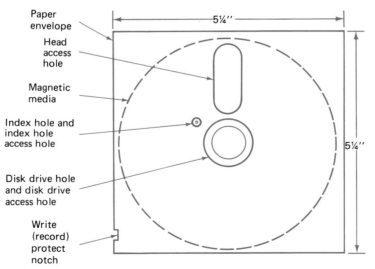

Fig. 12-9 **The floppy disk. This drawing shows openings for the drive ring, the head, and the index (sector identification) hole. When the read-protect notch is covered, the system will not write on the disk.**

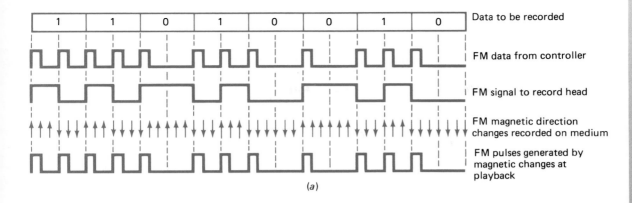

Data to be recorded

FM data from controller

FM signal to record head

FM magnetic direction
changes recorded on medium

FM pulses generated by
magnetic changes at
playback

(a)

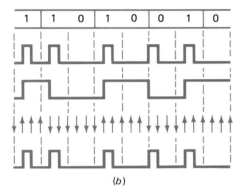

(b)

Fig. 12-10 Reading and writing data on magnetic medium using the FM (frequency modulation) and MFM (modified frequency modulation) formats. (*a*) FM creates a single-density recording. (*b*) MFM creates a double-density format. The changes in magnetic field on the medium create pulses in the read head.

**Sector
identification hole**

**Hard-sectored
floppy disks**

**Soft-sectored
floppy disks**

(installed) and clamped. The heads are shown in the loaded position, pressed against the floppy disk.

One question you may ask is "How do you know when you are at the start of a track?" In other words, how do you know when sector 0 is starting? The floppy-disk drive uses a *sector identification hole*. This is shown in Fig. 12-9. When the sector identification hole passes by a detector (Fig. 12-12 on page 351), the floppy disk is starting to place the first sector under the head.

Some floppy disks use many sector identification holes. Each hole marks the beginning of a new sector. A double hole marks the beginning of the first sector. These are called *hard-sectored floppy disks*. Floppy disks which use a single hole are called *soft-sectored floppy disks*.

As you learned before, floppy disks can be 8 in., 5.25 in., and 3.5 in. Each size can have different density formats, sectoring, and side usage. This results in many different types of floppy disk with many different capacities.

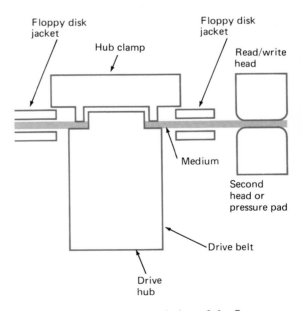

Fig. 12-11 A cross-sectional view of the floppy disk clamped in the floppy-disk drive. This view also shows two read-write heads pressed (loaded) against the floppy disk.

Users purchase floppy disks based on the size and formats. A number of common ones are shown in Table 12-1. The data density for a floppy disk is given by the number of bytes which the floppy disk can hold. Floppy-disk capacity extends from 90 kbytes to over 1 Mbyte. As new read-write electronics and media technologies become available, floppy disks are being created in higher and higher densities.

The blank floppy disk, or completely erased floppy disk, is called an *unformatted disk*. Most systems require that you *format* the floppy disk before the floppy disk can be put into general use. *Formatting* writes each sector of each track with standard null data at the correct density for the selected format.

Additionally, formatting may put information onto the disk in certain critical tracks and sectors. For example, formatting an MS-DOS disk for a personal computer builds a blank file directory. In use the directory is used to enter the names and starting locations of program and data files. The formatting process also builds a *file allocation table (FAT)*, which is used to point to the rest of the sectors used by a file and puts a boot-up program on the disk.

Once the formatting information is written onto the disk, the disk controller can then find the individual sectors it needs to use to write and read data and programs.

A block diagram of a typical floppy-disk drive is shown in Fig. 12-12. It requires two different sets of electrical connections. One connection is a special-purpose bus which runs between the floppy-disk drives and the floppy-disk controller. This is normally a 34-conductor ribbon cable with alternate strands connected to ground to make a shield.

The ungrounded conductors transfer data and control signals. Usually, these are TTL levels which only allow a short ribbon cable from the controller to the floppy-disk drive. Therefore, the drive must be located within 3 to 7 ft of the controller.

Four wires are *drive select signals*. These signals allow four floppy-disk drives to be connected to one controller. Although only one floppy-disk drive may communicate with the controller at any time, it is more economical to have a single controller manage a number of different drives.

The rest of the signals in the floppy-disk controller bus are two data lines, six command lines, and four status lines. Data to and from the floppy-disk drive is sent 1 bit at a time. There is one serial line for data from the floppy-disk drive and one serial line for data to the floppy-disk drive.

The six command lines control the drive read-write status, turn the motor on or off,

Table 12-1	Floppy-Disk Formats						
Disk Size (in.)	Sides	Recording Density	Number of Sectors per Track	Number of Tracks per Side	Sector ID Type	Bytes per Sector	Data Capacity Bytes*
8	Single	Single	10	77	Hard	256	197K
8	Single	Single	13	77	Soft	256	256K
8	Single	Double	26	77	Soft	256	512K
8	Double	Double	26	77	Soft	256	1024K
8	Single	Double	8	77	Soft	1024	612K
8	Double	Double	8	77	Soft	1024	1.26M
5.25	Single	Single	10	40	Hard	256	102K
5.25	Single	Single	8	40	Soft	256	160K
5.25	Double	Single	8	40	Soft	256	320K
5.25	Single	Double	8	40	Soft	512	320K
5.25	Double	Double	8	40	Soft	512	640K
5.25	Single	Double	9	40	Soft	512	180K
5.25	Double	Double	9	40	Soft	512	360K
5.25	Double	Double	15	80	Soft	512	1.2M
5.25	Double	Double	9	80	Soft	512	720K
3.5	Double	Double	9	80	Soft	512	720K
3.5	Double	Double	9	160	Soft	512	1.44K

*The disk capacity is calculated by multiplying
Sides × sectors/track × tracks/side × bytes/sector

The capacity is rounded to a commonly used nearby whole number to describe the disk size. The actual data capacity is also limited by the disk formatting process.

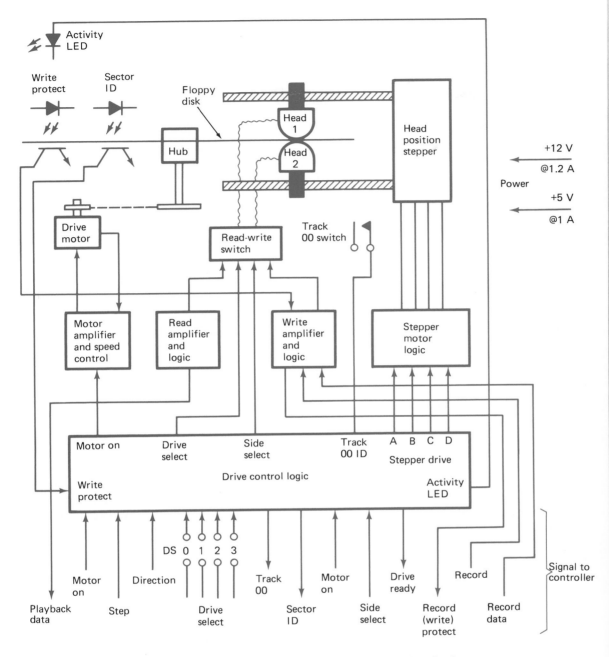

Fig. 12-12 A block diagram of a typical floppy-disk drive. This diagram breaks the drive electronics into smaller sections which perform the different functions. Note the sector ID and write-protect LED photodetector pairs. The correct drive select (DS) pair must be connected to address the desired drive.

move (step) the head, tell the head which direction (in or out) to move, allow the drive to generate record signals, and select the side being used. The status lines tell the floppy-disk controller the status of the selected floppy-disk drive. The drive status signals tell the controller when the sector index hole passes the detector, that the drive is ready, that the head is at track 00, and that a *write-protected floppy disk* is in the drive.

Another set of connections supplies power to the floppy-disk drive. Power for a floppy-disk drive includes +5 V at approximately 1 A for the drive electronics and a +12-V source at 1.2 A. The +12-V supply is needed to operate the motors. The drive motor spins the floppy-disk at 300 rpm. Another motor steps the head from the outside track (track 00) to the innermost track. Frequently this job is done with a stepper motor, but some drives

Voice-coil mechanism

Hard error

Soft error

use a *voice-coil mechanism* for rapid head positioning.

Usually floppy-disk drives may be mounted in any position. Operationally you must make sure that they are not subjected to mechanical shock, especially when operating, and are kept at reasonable room temperatures. Excessive shock and temperatures may cause the drive to operate improperly.

Although the floppy-disk is a relatively rugged device, you can damage it. The floppy-disk has a plastic base. Therefore, excessive temperatures may warp the base material. Also, excessive temperatures may cause damage to the jacket, warping the disk and making it difficult, if not impossible, to read.

The floppy disk has access holes so that the read-write heads can come in direct contact with the medium. It is possible for a user to scratch the magnetic surface in these exposed areas. If the surface is scratched, it is likely that a number of bits of information on each track will be destroyed. To prevent this, you should always keep the floppy disk stored in its protective jacket or the head access door closed on 3.5-in. diskettes. Obviously, the floppy disk must be kept away from sources of strong magnetic fields, since these can erase the disk. Note that tools such as screwdrivers or power tools that use permanent-magnet motors are a very common source of high-intensity magnetic fields.

When servicing a floppy-disk-based system, you may find people who have mistreated floppy disks regularly and do not seem to have had problems. But you should encourage them to use proper floppy-disk-handling procedures, because mistreating them will, sooner or later, result in the loss of valuable data.

When discussing disk errors, we often speak of hard and soft errors. What do these terms mean? A *hard error* is one which occurs over and over again. For example, suppose a floppy-disk is scratched through its head-access holes. If this happens, and the scratch destroys some data on the disk, this is a hard error. That is, the information is lost forever. No matter how many times the disk is read, the information will never reappear.

On the other hand, if a speck of dust falls on the floppy-disk surface, when the disk is being read the speck of dust may come between the head and the disk surface. During this one revolution, the head is pushed away from the magnetic surface. This time the information is lost. However, the data may be properly read

the next time the head passes over this section of the track if the dust is moved. This is a *soft error*.

Self-Test

Answer the following questions.

31. The floppy-disk is a tradeoff of price versus performance. In the floppy-disk, the performance feature of the hard disk given up for a much lower cost is
   a. Data capacity   c. Data rates
   b. Access time   d. All of the above
32. Why do you suppose we say there is a ''head crash'' if the head on a hard-disk system comes into contact with the spinning disk?
33. Why does the disk spin all of the time instead of starting and stopping as you read each block of data?
34. Using the floppy disk instead of a hard disk means that the
   a. Head can be pressed directly against the surface instead of flying a few microinches above it
   b. Disk speed will not cause any wear, so hard disk speeds can be maintained
   c. Media has a high data capacity
   d. All of the above
35. List the two sizes of floppy disks commonly used for microcomputer systems.
36. Two common formats are FM and MFM. What do these abbreviations stand for? What is the basic difference between these two formats?
37. How does the floppy-disk drive let the controller know when it is starting to read the first sector?
38. How does the floppy-disk drive let the controller know when the head is positioned at track 00?
39. Briefly explain the difference between a hard-sectored disk and a soft-sectored disk.
40. Before a new disk can be used to store data or programs, it must be
   a. Formatted   c. Magnetized
   b. Demagnetized   d. Sectored
41. A floppy-disk drive has four drive select inputs. These allow
   a. The controller to write MFM data to the disk
   b. The controller to tell when a disk is loaded in the drive
   c. One controller to manage up to four drives simultaneously

d. One controller to select which of four drives it will manage

42. A floppy-disk drive requires a signal to step the head and another associated signal to
    a. Turn on the drive motor
    b. Tell the head to step forward or backward
    c. Allow the drive to generate WRITE signals
    d. Select the side of the disk that is to be used

43. Floppy-disk damage can be caused by
    a. Overheating
    b. Scratches through the head access hole
    c. Local external magnetic fields
    d. All of the above

44. A floppy disk is unable to properly read the data for two revolutions of the disk, but reads the data successfully on the third pass. This is an example of
    a. Formatting
    b. Hard errors
    c. Soft errors
    d. Permanent disk damage

12-6 THE WINCHESTER-DISK DRIVE

As we learned earlier, the floppy-disk drive provides low-cost data storage at the expense of speed and data capacity. As microcomputer systems began to develop and find their way into more commercial applications, these speed and data capacity limitations became a problem. The requirement for additional speed and additional data capacity led to the development of the Winchester-disk drive.

The Winchester-disk drive is a hard-disk system. That is, the basic construction of the disk is very much like the original hard disks. A nonmagnetic metallic platter is machined to a very precise level. A coating of magnetic material is deposited on this surface to provide a place to write the data. During operation, the read-write head is positioned a few millionths of an inch above the surface of the disk.

Because the head is not in contact with the surface, the disk can be operated at higher speeds, giving faster access time. Also, because this form of disk construction is more precise, the Winchester-disk drive can support many more tracks per inch and higher data densities. This means the Winchester-disk drive can hold more data in the same space.

It is also quite common to use more than one platter and both surfaces of each platter in a Winchester-disk drive. A four-platter drive is shown in Fig. 12-13 on page 354. You can see that the Winchester actually has four disks or *platters*. Each platter has two surfaces. Therefore, this Winchester provides eight times the capacity of a single-sided platter.

Each platter has the same set of tracks. The stack of tracks that all have the same track number is called a *cylinder*. In this case, 8 bits of data may be written into each position on a cylinder.

The Winchester-disk drive provides much faster access time because the rotational speed is 10 or more times greater than that of a floppy disk. It also holds much more data. Common Winchester drives are available with data densities of 80, 120, 160, 320, 640, and 1200 Mbytes.

Although the Winchester drive is more expensive than the floppy-disk drive, many users are willing to pay the extra money for the improved data density and access time. The Winchester, however, is still much less expensive than a hard disk used in large computer systems. What makes the Winchester less expensive than a conventional hard-disk drive?

There are a number of differences between the Winchester-disk drive and the commercial hard-disk system. These make the Winchester cost less than a conventional hard-disk drive.

First, the Winchester-disk drive is physically smaller. This means it holds less data but is cheaper to build because the precision mechanical parts are not as large. Typical Winchester-disk drive systems are packaged in similar-sized housings as 8-, 5.25-, and 3.5-in. floppy-disk drives.

Second, the Winchester spins at lower speeds. This means it has slower access times and is cheaper to build.

Third, the Winchester uses a sealed head–platter mechanism. The head and platters are sealed in an evacuated environment. They are built into the disk drive, which means that the Winchester drive has nonremovable media. This means that the media, unlike that of a floppy-disk drive, cannot be taken out of the drive and new media inserted. If you want to save the data externally to the Winchester-disk drive, it must be copied onto a floppy-disk or magnetic-tape system. If the disk becomes full, some data must be removed in order to make additional room.

The Winchester-disk drive system requires a special disk controller. The disk controller

Platter

Cylinder

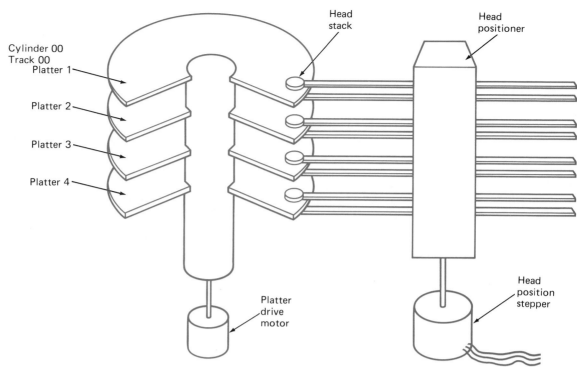

Cylinder 00
Track 00
Platter 1

Platter 2

Platter 3

Platter 4

Head stack

Head positioner

Platter drive motor

Head position stepper

Fig. 12-13   Inside the Winchester-disk drive. You can see that the drive uses four plat-ters and eight heads. This means that 8 bits of information can be recorded on each point on a track position. The track position extended through all four platters is called a cyl-inder.

which operates a floppy-disk drive cannot op-erate a Winchester-disk drive. Typically, Winchester controllers are more expensive than floppy-disk controllers. This is because they must handle higher data rates and have more complex control electronics.

Like the floppy-disk drive, the Winchester requires a motor to spin the platter and some way to move the head or heads from cylinder to cylinder. Head movement may use a step-per motor or, for very fast access drives, a voice-coil system.

The Winchester drive construction makes it much more easily damaged by mechanical shock and vibration than a floppy-disk drive. Care must be taken to make sure that no me-chanical shock occurs to the system, especially when the unit is powered-down and the system is being moved.

Winchester-disk drive systems have a spe-cial cylinder called a *parking cylinder*. In this position the head assembly is parked at a cyl-inder which provides protection against me-chanical shock and vibration during shipping. Usually, the microprocessor-based system which uses this Winchester-disk drive system has special software called the *ship* or *park rou-*

tine. The routine moves the heads to the ship-ping cylinder and then shuts the drive and processor off. This protects the heads until they are accessed again. Newer Winchester-disk drive systems automatically park on power down.

The 3.5-in. package has become one of the most popular for both floppy-disk drives and Winchester-disk drives. This package, which is somewhat wider than 3.5 in. so that it may accept a 3.5-in. floppy disk, is fairly standard in its size and mounting. The original 3.5-in. floppy disks were introduced in a 4-in.-high package. This has become known as the *full-height drive*. Both half-height and one-third-height drives are now common in the 5.25-in. configuration. The 3.5-in. configuration is found in the half-height format.

It is not uncommon for a field service person to be requested to remove a single, full-height 5.25-in. floppy-disk drive and replace it with one half-height 5.25-in. floppy-disk drive and one half-height Winchester-disk drive, for ex-ample. This means that the old floppy-disk controller may be kept and a new Winchester-disk controller must be installed. Alterna-tively, the old floppy-disk drive controller may be removed and a new combination floppy-disk drive and Winchester-disk drive controller may be installed.

Usually, installation of a controller requires the individual perform some controller "configuration." Configuring a controller usually consists of moving jumpers or setting switches on the controller. The configuration process tells the controller what type of drive it is connected to, the controller's memory or I/O address within the microprocessor's address space, and provides other similar information. This is information which cannot be determined until the microprocessor-based system, the system software, the controller, and the selected drive are brought together as a complete package.

When a new Winchester-disk drive and its controller are installed, there is additional work needed to prepare the Winchester for use. Like the floppy disk, the surface of the Winchester disk must be formatted.

Self-Test

Answer the following questions.

45. The Winchester-disk drive came into being because
    a. Nonremovable media are easier to use
    b. The floppy disk became too expensive
    c. They give the same data capacities and data speeds as hard disks
    d. The floppy's data capacity is too low, data speed is too slow, and hard disks are too expensive
46. The basic characteristic of a Winchester-disk drive which forces you to use a backup system for data safety and data overflow is
    a. Low capacity of the floppy disk
    b. High capacity of a tape system
    c. The Winchester's nonremovable media
    d. That when the Winchester becomes full, you must erase some data to make more room
47. The major difference between a Winchester-disk system and a hard-disk system is
    a. That the media are housed in a non-removable evacuated area
    b. The Winchester's rotational speed of 3600 rpm compared with the floppy's speed of 300 rpm
    c. The use of multiple platters to increase data storage capacity

    d. That the Winchester head flies a few microinches above the surface of the platter
48. The group of tracks that are all the same distance from the edge of the platters is called
    a. The head-positioning stepper
    b. A cylinder
    c. Track 00
    d. A sector
49. The purpose of the Winchester's park cylinder is to
    a. Provide a place to store the heads during the time between read accesses
    b. Provide a place to store the heads whenever dc power is turned off
    c. Provide a place to store the heads when the Winchester disk drive is physically moved
    d. All of the above
50. When you install a new Winchester-disk drive and its controller in a microprocessor-based system, you can expect to
    a. Be able to partition the drive
    b. Configure the controller to match the drive and the operating system's I/O address and memory space
    c. Format the Winchester
    d. All of the above
51. In addition to the work in the question above, on the Winchester-disk drive and its controller, you can also expect to
    a. Remove one of the system's floppy-disk drives
    b. Install a new floppy-disk controller
    c. Make changes to the system operating software to tell it that the Winchester is there and to tell it how you have configured it
    d. Lengthen all of the drive-related control cables

12-7 MAGNETIC-TAPE STORAGE

As we learned earlier in this chapter, magnetic-tape storage is really just another physical format for magnetic media. It uses the same read-write principles as a disk drive. The advantage of using magnetic tape is that you get very large quantities of storage which use a very low cost removable medium. It is quite easy to find magnetic-tape data storage subsystems which store 30 to 240 Mbytes of information on relatively small cassettes or cartridges.

One of the major uses of magnetic tape subsystems in microcomputers is providing a backup for Winchester-disk drives. As we learned in the section on Winchesters, the Winchester-disk drive uses nonremovable media. If the user wants to lower the risk of losing valuable data stored on a Winchester-disk drive, the data must be backed up. This means the data must be copied onto another mass storage device.

One can make a backup by creating a second file on the same disk. The second file is identical to the first file. This protects the user against loss of data due to a sector failure, but not against loss of data due to a failure of the disk drive or controller or major media failure.

Microprocessor-based systems and microcomputer systems which use Winchester-disk drive systems usually use one of two different kinds of backup. One way to back up a Winchester is to write the data onto floppy disks. This takes many disks because an 80-Mbyte Winchester holds almost 60 times as much data as a 1.4-Mbyte, 3.5-in. floppy disk. However, this works if backup is needed only for a few selected files.

The second technique uses a high-capacity magnetic-tape subsystem to store backup data. As you learned earlier, it is quite possible for a magnetic tape to have as much storage capacity as a 120- or 240-Mbyte drive.

The use of *digital magnetic-tape systems* for storing computer data is almost as old as computers themselves. In fact, magnetic-tape drives are older than disk drives. However, the type of magnetic-tape drive commonly used on large computer systems is quite expensive. Its cost comes from two factors.

First, these drives hold large quantities of data and write the data at a fast rate. This means that the tape drive must handle a great deal of tape and must move it rapidly. Second, most of the magnetic-tape systems used for computer work are incremental. This means that the tape can start and stop for each block of data that is written. Typically, a block of data is 128, 256, or 512 bytes long. The ability to stop and start on each block of data adds a great deal of expense. The tape must be handled with expensive mechanisms so that it can be brought up to speed quickly and stopped quickly without damage.

The need for incremental reading and writing goes away when the only use for the tapes is making a backup. When you make a backup, the process can start at track 00, sector 0 and continuously write data to the tape until all of the data on the disk drive is copied to the magnetic tape. Under these conditions, we can let the tape move continuously. This means we can use a much lower cost tape transport called a *streaming-tape drive.*

As we noted earlier, there is a great deal of similarity between magnetic-tape systems and magnetic-disk systems. Both write information on magnetic surfaces by magnetizing a small portion of the surface in one direction or the other. As you know, the major difference between tape and disk is the physical format. With a tape, the data is written in one continuous track; the data on a disk drive is written in many short tracks.

The more complex tape systems used in the computer industry have multiple heads. That is, they write more than one track at a time. Typically they write either seven or nine tracks at a time, with nine tracks being the more common. These systems use a half-inch-wide magnetic tape and a reel-to-reel tape drive.

One reason that magnetic tape systems can store so much data is the use of multitrack heads. For example, a magnetic tape mass storage subsystem which writes data at a density of 1600 bpi is actually writing the data on the tape at a density of 1600 bytes/in. It uses nine-track heads. This means that a 2400-ft reel of tape can hold over 46 million bytes of data.

1600 bytes/in. × 12 in./ft × 2400 ft/reel = 46,080,000 bytes/reel

A 9600-ft reel of tape has a capacity of

1600 bytes/in × 12 in./ft × 9600 ft/reel = 184,320,000 bytes/reel

Physical formats for streaming-tape drives use a cassette or cartridge to hold the quarter-inch- or eighth-inch-wide tape. The tape system may use multiple tracks or a single track. One way to do multiple-track taping is to remove the tape and put it back in so that additional data is written on the opposite side. Because the streaming-tape systems cannot work as incremental-tape systems, it is not unusual to require two passes to create and verify the backup material.

An alternative form of tape system is occasionally used for very low cost systems. This uses a conventional audio cassette. The information on the tape is not digital but consists of two different tones—one for logic 1s and the other for logic 0s. Although these systems have very low capacity and limited data speed (usually 120 characters per second), they do serve a purpose in some very low cost systems. Occasionally, you will find such systems in low-cost industrial applications.

Self-Test

Answer the following questions.

52. One of the major uses of magnetic-tape mass storage is
    a. Writing incremental data
    b. Providing fast-access data storage
    c. Backing up nonremovable media
    d. Replacing floppy-disk drives
53. When a magnetic-tape system is only used to back up a disk drive, there is no need for _____ and therefore, one can use a streaming-tape drive.
    a. Writing incremental data
    b. Providing fast-access data storage
    c. Backing up nonremovable media
    d. Replacing floppy-disk drives
54. Data can be stored by using an audio recorder. The data rates are very slow because the data is stored as
    a. A series of tones, one for logic 1s and another for logic 0s
    b. A group of short tracks
    c. Magnetic reversals for logic 1s and logic 0s
    d. Different amplitude signals for logic 1s and logic 0s

12-8 OPTICAL STORAGE DEVICES

Engineers working with computer systems are always searching for improved mass storage capabilities. The objectives are always faster access, more data storage, and lower cost. During the mid-1980s, a form of mass storage using optical techniques began to emerge. These *optical-disk drives* came as an outgrowth of the *compact-disk* (*CD*) technology developed for audio reproduction. The compact disk stores audio information, whereas the optical storage techniques were being developed for digital systems.

One of the greatest features of optical storage systems is their tremendous capacities. A 12-in. double-sided disk can store as much as 2 Gbytes of information, and a 5-in. CD-size optical disk can store more than 600 Mbytes of data. There is no reason to believe that optical storage capacities will not increase as future technology in this area is developed.

What can one do with this much storage? One example, introduced by a personal computer manufacturer, is the storage of a complete encyclopedia on an optical disk. With the encyclopedia on disk, the microcomputer software can search the disk for key words, references, definitions, and other research points.

The optical disk stores information as small pits in the disk's highly polished surface. The pits are detected by shining a laser at the disk and detecting the light which is bounced back. One difficulty with optical storage is writing data onto the disk. To be truly useful, a mass storage system needs to have both read and write capability. Writing data onto an optical disk is a much more difficult process than reading data from an optical disk.

Three different types of optical-disk drive products are in use:

Optical ROM (OROM)
Write once–read many (WORM)
Write many–read always (WMRA)

The lowest-cost optical storage devices are the OROMs. They can be mass-produced very much like compact disks are produced for the audio market.

The WORM optical storage devices are very good for such applications as archiving data. Once the data is written on the disk, it is saved forever. Usually, writing is much slower than reading.

The WMRA optical storage device is the closest to the other mass storage devices we have studied in this chapter. The WMRA allows the user to write data onto the optical disk, erase the data, and later write new data on the disk. As optical disk technology improves, the time required to write to a WMRA optical storage device will decrease, and the number of times that the device can be written to will increase.

Self-Test

Answer the following questions.

The optical mass storage devices are an outgrowth of _____ technology.
a. Floppy-disk
b. Winchester
c. CD
d. Magnetic-tape

Optical-disk drives

Compact disk (CD)

Optical ROM (OROM)

Write once–read many (WORM)

Write many–read always (WMRA)

56. The greatest advantage of optical mass storage devices is
   a. Very high speed operation
   b. The ability to store an entire encyclopedia
   c. The fact that they are read-only devices
   d. Their ability to store very large amounts of information at a very low cost
57. Currently, optical mass storage devices are available in the _____ format.
   a. Optical ROM (OROM)
   b. Write once–read many (WORM)
   c. Write many–read always (WMRA)
   d. All of the above
58. One of the uses for the WORM format is
   a. Storing encyclopedias
   b. Creating archival storage
   c. High-volume online working mass storage
   d. All of the above

SUMMARY

1. There are three reasons why we add mass storage to microprocessor-based systems:
   a. To give long-term nonvolatile storage for programs and data
   b. To provide a more economical storage than ROM or battery-backed RAM
   c. To give universal systems versatility by letting them load the programs that they need when they need them
2. To be useful, a mass storage system must keep the information in machine-readable form.
3. Any machine-readable mass storage system has three parts:
   a. The medium, where the information is written
   b. The drive, which handles the medium
   c. A controller, to interface the drive to the microprocessor
4. Paper with holes punched in it to represent the data has been a popular form of mass storage in the past.
5. The floppy disk is the most popular mass storage system for the personal computer and is a very low cost system.
6. We say that a mass storage system provides online data when the data can be accessed by the microprocessor without help from a human.
7. The Winchester-disk drive has greater storage capacity and faster access than the floppy disk, but does not have removable media. Winchester-disk drives are also called hard drives in personal computer systems.
8. A backup tape system is used to make a copy of data and programs on inexpensive removable medium so that the copy can be stored in a safe place.
9. Streaming-tape mass storage systems are very popular backup devices.
10. Magnetization occurs in an iron-based ma-terial when it is exposed to the magnetic field of an external magnet for enough time.
11. When we write digital information on magnetic media, logic 1s are written as magnetized spots with their magnetic fields going in one direction, and logic 0s are written as magnetized spots with their magnetic fields going in the opposite direction.
12. Digital magnetic data storage uses a special electromagnet to read and write the data. This is called a read-write head.
13. Because the magnetizing signals are very weak, the head must be very close to the medium before data can be written to the medium or read from the medium.
14. The weak magnetic fields mean that it is easy to have reading or writing errors. Magnetic mass storage systems use sophisticated error detection processes to detect these errors.
15. The format is the way that data is written onto the magnetic medium. Two formats, called single-density and double-density, are commonly used.
16. Tape and disk are the two popular physical formats for magnetic mass storage media.
17. Tape is a sequential-access medium. This means we must pass by all the information which lies between where we are and where we want to go. Its access time is limited by how long it takes to pass over all the data in between.
18. The advantages of tape are its high data storage capacity, very low cost, and re-movability.
19. A disk system writes data in a series of concentric rings called tracks. The tracks are broken into a number of sectors.
20. Disk access time is made up of the time needed to step from track to track and the time to turn the disk. The disk access pro-

cess gives the disk a much quicker access to data than a magnetic tape has.

21. The average time to access data on any one track is half the time for the disk to spin one complete revolution.

22. Online mass storage systems require two electronic packages: the drive electronics and the controller.

23. The drive has the electromechanical components such as motors, heads, detectors, and switches. The drive electronics interfaces these components to the TTL special bus which communicates with the controller.

24. The controller interfaces one or more drives to the microprocessor-based system. It provides data error checking and formatting, converts command words into individual signals, and converts individual status signals into status words for the microprocessor.

25. Typically, one controller can manage a number of drives of the same type.

26. A hard-disk system flies the read-write head a few millionths of an inch over a precisely machined platter coated with magnetic metal oxide.

27. If the hard-disk head contacts the surface of the platter, it "crashes" and is destroyed.

28. The disk's platters rotate continuously so that the disk access time does not include the time needed to bring the platters up to operating speed.

29. The floppy disk uses tape media technology and hard-disk media-handling technology.

30. Because the floppy disk is a coated plastic medium, it is pressed directly against the read-write head.

31. The floppy-disk drive is inexpensive and uses inexpensive removable medium. It has a much lower data capacity and longer access times than a hard disk.

32. Two sizes of floppy disk are in common use: 5.25 and 3.5 in.

33. Most floppy-disk drives use a double-sided, double-density format to get the maximum data density. Data densities run from a low of 100 kbytes to nearly 1.5 Mbytes.

34. As the floppy disk spins, a sector identification hole is used to tell the controller when a sector is starting.

35. Floppy-disk systems which use a sector identification hole to mark the beginning of each sector are called hard-sector disks. Floppy-disk systems which use a sector

identification hole to mark the beginning of the first sector and count on timing to identify the remaining sectors in the track are called soft-sectored systems.

36. Typically, a single-density floppy-disk drive uses the FM (frequency modulation) format, and a double-density system uses the MFM (modified frequency modulation) format.

37. Before a new disk can be used, it must be formatted. Formatting writes blank information into every sector of every track. In some systems, formatting places the operating system and/or a boot-up program on the disk and sets up the file directory and file location tables so that the operating system software can create data and program files on the disk.

38. A floppy-disk drive is connected to the controller with a multiconductor special bus. This bus carries the data to and from the drive as well as control and status information. It also provides the drive addressing signals so that a single drive is selected when more than one drive is connected to the controller.

39. Typically, floppy-disk drives use +5 V at 1 A to run the drive electronics and +12 V at 1.2 A to run the drive's motors.

40. Both floppy-disk drives and floppy disks are rugged devices, but they can be damaged if not treated carefully. The floppy disk should be kept away from heat and magnetic fields and must not be scratched.

41. There are two different kinds of errors found with magnetic storage devices. A soft read or write error can be corrected by trying again. A hard error cannot be corrected by retrying.

42. The Winchester-disk drive was developed to give the microcomputer user a faster-access, higher-capacity online mass storage system with lower cost than the conventional hard-disk drive.

43. The read-write head on a Winchester-disk drive flies a few millionths of an inch above the platter. The heads and platters are packaged in a sealed evacuated environment. This makes the Winchester's media nonremovable.

44. The Winchester-disk drive may have one or more platters, and each platter may have tracks on both sides. All of the tracks which are the same distance from the edge of the platter are called a cylinder.

45. The Winchester's platters rotate 10 times faster than the floppy disk. This gives the

Winchester a much faster access time than the floppy disk has. The precision disk allows it a much higher data density than the floppy disk has.

46. The Winchester is lower in cost than the hard-disk drive because it is smaller, spins at a lower speed, and uses nonremovable media.

47. Because of the precision characteristics of the Winchester-disk drive, it can be easily damaged by shock and vibration. When the drive is moved, the heads are moved to a special parking cylinder where they are protected.

48. Both floppy-disk drives and Winchester-disk drives are packaged in standard 5.25-in. and 3.5-in. packages. These packages come in full-height, half-height, and one-third height sizes.

49. Magnetic-tape mass storage uses the same principles as the disk storage systems; its medium just uses a different physical format.

50. A major use for tape in microcomputer systems is as a backup device for Winchester-disk drives.

51. The streaming-tape drive is a very low cost implementation of a tape drive because this drive does not need the expensive mechanisms which an incremental-tape drive does.

52. The more complex tape systems used on computer systems use multiple-track heads and incremental storage. Typically, these tape systems have nine-track heads, which give them great density.

53. A very low cost tape system uses the audio tape recorder. Instead of writing logic 1s and logic 0s directly, tones are used to represent the logic levels.

54. Optical-disk storage has the advantage of extremely large data storage capacity. The digital information is written as pits in the surface of the disk, which are detected by a beam of light from a laser.

55. Optical-disk storage techniques have been derived from techniques developed for the audio compact-disk players.

56. Optical disks have been developed in three different forms: optical ROM (OROM), write once–read many (WORM), and write many–read always (WMRA).

57. The OROM is the lowest in cost. It is used to supply information which does not need to be changed.

58. The WORM is useful for archival uses where there is only a need to write on the disk once.

59. The WMRA is intended to provide an optical mass storage system which can be used in the same way as any read-write magnetic disk.

CHAPTER REVIEW QUESTIONS

Answer the following questions.

12-1. Which of the following is not a reason that we add mass storage to microprocessor-based equipment?
 a. To give long-term nonvolatile storage for programs and data
 b. To provide more economical storage than ROM or battery-backed RAM
 c. To give universal system versatility by letting the system load any program it needs
 d. All of the above

12-2. A useful mass storage system keeps the data
 a. In ROM
 b. In machine-readable form
 c. Only in battery-backed RAM due to power failures
 d. On floppy disks or half-inch tape

12-3. Which of the following is not one of the basic three parts of a mass storage system?
 a. The medium
 b. The controller
 c. The interface cable
 d. The drive

12-4. Paper-tape mass storage is not often used for today's microprocessor-based systems because
   a. Paper tape is bulky and hard to store
   b. The system is very slow compared with magnetic media
   c. The medium is easily damaged and wears quickly
   d. All of the above

12-5. One of the most popular mass storage systems for the personal computer is the
   a. Optical-disk drive   *c.* Floppy-disk drive
   b. Streaming-tape drive   *d.* Paper-tape punch

12-6. A mass storage system is _____ when the data on it can be accessed by the microprocessor-based system without having an operator work with the medium.
   a. Very high density   *c.* On line
   b. Optically coupled   *d.* Human-interfaced

12-7. A major advantage of the Winchester mass storage system is that it _____ than the floppy-disk system.
   a. Has more storage and is faster   *c.* Is significantly less expensive
   b. Has more storage but is slower   *d.* Is freer from errors

12-8. The major purpose of making a backup of a Winchester or any other mass storage device is to
   a. Make sure that a copy is kept in a safe place in case the original is destroyed
   b. Free up space on a Winchester when it becomes too full
   c. Allow more frequently used data to be on line
   d. All of the above

12-9. One very popular device for backing up a Winchester disk-drive system is the
   a. OROM   *c.* Battery-backed RAM
   b. Streaming tape   *d.* All of the above

12-10. Exposing iron-based material to a magnetic field for a period of time causes it to become
   a. A mass storage device   *c.* Prone to errors
   b. Demagnetized   *d.* Magnetized

12-11. The read-write head used for digital magnetic data storage works with digital information recorded on the magnetic medium as
   a. Two levels of magnetic intensity
   b. Two different directions of magnetization
   c. One very intense magnetized spot and one weak one
   d. All of the above

12-12. The read-write head must be very close to the medium during the read operation because
   a. You must avoid head wear
   b. Antimagnetic dirt could cause errors if it comes between the head and the medium
   c. The signals from other tracks mix up the head which is far away from the track it is listening to
   d. The recorded signals are very weak, and the head must be close to the medium to hear them

12-13. Magnetic mass storage systems have sophisticated error detection and correction systems
   a. So they can use low-cost media
   b. Because the record and playback signals are weak and therefore can have errors
   c. To be sure they transfer good data directly into the microprocessor memory system
   d. All of the above

12-14. When you hear the terms *single-density* and *double-density,* someone is speaking of
   a. The weight of the floppy disk
   b. The kind of tape used for streaming backup
   c. Two different formats for writing data onto magnetic medium
   d. The different kinds of ROMs used for booting a system

12-15. Magnetic tape is a _____ system, and floppy disks are more of a _____ system. (Choose two of the four answers.)
   a. Reliable
   b. Random-access
   c. High-density
   d. Sequential-access

12-16. The two major advantages of magnetic tape are _____ and _____ . (Choose two of the four answers.)
   a. Its high data storage capacity
   b. Its very low cost per bit of stored information
   c. That it is a removable medium
   d. All of the above

12-17. Each track on a disk is broken into
   a. 128 bytes
   b. 256 bytes
   c. Sectors
   d. A format

12-18. The average amount of time required to access data on a disk drive depends on
   a. One half the time needed to make a complete revolution
   b. The time needed to step from track to track
   c. a and b are true
   d. The time needed to go from one end of a track to the other end

12-19. Typically, the drive includes an electronics package which has
   a. Error-checking logic
   b. Status information for the microprocessor
   c. Motor drive circuits and the read-write switch and amplifier
   d. All of the above

12-20. The mass storage controller is usually found
   a. Mounted in the microprocessor-based system, where it can manage a number of drives
   b. At the end of a long ribbon cable attached to the microprocessor bus
   c. Mounted directly to the drive to avoid long cables
   d. Any of the above may be correct, depending on the density of the recording used for the system

12-21. The reason that a hard-disk system flies the head a few microinches above the medium is
   a. To avoid the very strong magnetic signals from the aluminum disk
   b. To avoid the excessive wear that would occur if the head was in contact with a disk turning that fast
   c. Because head crashes are just too expensive
   d. Both a and b are correct

12-22. The floppy disk borrowed its medium from _____ and its medium-handling technology from _____ . (Choose two of the four following answers.)
   a. Hard-disk drives
   b. Magnetic drums
   c. Microprocessor designs
   d. Magnetic tape

12-23. The floppy disk's plastic-coated medium allows
   a. Very high recording densities
   b. The head to be pressed against the medium without causing a crash
   c. A very short life
   d. The same number of bits per track as a Winchester-disk drive

12-24. Compared to the Winchester-disk system, the floppy-disk system
    a. Is much less expensive   *c.* Holds much less data
    b. Is 10 times or more slower   *d.* All of the above

12-25. Currently, the popular size of floppy disks for personal computers is
    _____ in.
    a. 8   *c.* 3.5
    b. 5.25   *d.* All of the above

12-26. With a double-sided, double-density 5.25-in. floppy disk, you can expect
to hold _____ of data.
    a. 720 kbytes   *c.* 320 kbytes
    b. 1.2 Mbytes   *d.* 256 kbytes

12-27. A soft-sectored floppy disk
    a. Holds less data than a hard-sectored disk
    b. Only uses a single hole to mark sector 0; all other sectors on the track
        are identified by rotational timing
    c. Uses one hole for each sector on the track and two holes close together to identify sector 0.
    d. Is not a commonly used form of floppy disk today.

12-28. The MFM (modified frequency modulation) format is used
    a. To read or write data on double-density media
    b. To read or write data on single-density media
    c. To overcome high-noise environments
    d. With hard-sectored drives

12-29. You format a disk to
    a. Break in the new surface
    b. Create double-density media from single-density media
    c. Ensure a full erasure of improper data
    d. Put initial clock and data on the disk so the system can use it

12-30. The multiconductor cable connecting the floppy-disk drive and its controller carries
    a. Drive addressing information
    b. Read-write data
    c. Status and control information
    d. All of the above

12-31. Although a floppy disk is a relatively rugged device, it can be damaged
    a. If it is scratched
    b. If it is exposed to excessive heat
    c. By bringing it near a magnetic field
    d. All of the above

12-32. The difference between a soft error and a hard error is that
    a. You can recover from a soft error
    b. You can recover from a hard error
    c. You can recover from both hard and soft errors, but hard error recovery takes three or four more tries
    d. Hard errors mean damage to the read-write head

12-33. One of the reasons that the Winchester-disk drive is much cheaper than
a conventional hard-disk drive is that it
    a. Spins at a lower speed
    b. Uses sealed nonremovable media
    c. Generally has a much lower data capacity
    d. All of the above

12-34. All of the tracks on a Winchester which are the same distance from the
edge of a platter are said to be
    a. A uniform distance   *c.* Error-free
    b. In one cylinder   *d.* All of the above

12-35. The Winchester-disk drive uses a parking cylinder
 - a. To protect the heads from shipping damage
 - b. When it must keep track of bad sectors
 - c. On dates
 - d. To give an error-free cylinder for data storage

12-36. The streaming-tape drive
 - a. Has high data capacity and is therefore good as a backup device
 - b. Is low cost because it avoids the expense of start-stop mechanisms and electronics
 - c. Is a sequential-access device, which means that the average time to access any data is half the time to get from one end of the tape to the other
 - d. All of the above

12-37. The major advantage of optical-disk storage
 - a. Is its very rapid access time
 - b. Is its extremely high data capacity
 - c. Is its ability to go through the read-write process many times
 - d. Will not be known for some time to come because this is still an experimental device

12-38. Currently, the optical ROM (OROM) is the lowest-cost optical disk. This is a good storage device for such applications as
 - a. An online encyclopedia
 - b. One-time archival recording
 - c. Slow mass storage to back up fast Winchesters
 - d. Times when you only need to write on the disk once

CRITICAL THINKING QUESTIONS

12-1. Early personal computers often used simple audio cassette recorders to record a string of tones which represented bits of information. Why do you think these systems were popular in the early days of personal computing, and why have they been replaced by the floppy disk?

12-2. Often, programs written for today's microprocessor-based systems are so large that the entire program will not fit into main memory. In this case, sections of the program, called *overlays,* are stored on disk and called (or swapped) into main memory when they are needed. Why do overlays have to be placed in on line storage? Why is a Winchester preferred to a floppy disk for overlay storage?

12-3. Obviously, faster speeds and higher-density magnetic recording techniques improve all kinds of magnetic storage systems. What do you think might begin to limit the speed and density which can be achieved from a magnetic storage system?

12-4. Typically, drive electronics are located close to the medium and the heads, and control electronics are located close to the microprocessor-based system's bus. Why do you think these locations are chosen?

12-5. Although there can be a wide variety of floppy disk formats, only a few are used. What is the advantage of having "common" floppy disk formats?

12-6. Most sophisticated microprocessor-based systems today have both a removable mass storage system and a nonremovable mass storage system (usually a floppy disk and a hard disk). Why are both needed? Why shouldn't the system be constructed with just the hard disk, since the floppy disk is slower and holds much less data?

Answers to Self-Tests

1. *d* 2. *b*   3. *d*
4. The three parts which make up any mass storage subsystem are the medium, drive, and controller.
5. The medium holds the information, the drive handles the medium and moves it so that all of the information can be accessed, and the controller manages the flow of data and the operation of the drive.
6. We say that storage is on line if the microprocessor-based system can access the data in the mass storage system without the help of a human.
7. Some of the major differences between the Winchester and floppy-disk mass storage systems are
   The Winchester is much more expensive
   The Winchester holds a great deal more data
   The floppy disk is much slower
   The floppy disk uses removable media
   The floppy disk uses low-cost media
8. A separate mass storage system devoted to backup is needed when the major online mass storage system does not have removable media. The backup provides two functions in this case. First, it provides a copy of valuable data in case the data on the online system is lost. Second, it allows an overloaded online system to move infrequently used data to offline storage.

9. *a*	14. *b*	19. *b*
10. *d*	15. *d*	20. *d*
11. *b*	16. *a*	21. *b*
12. *a*	17. *c*	22. *d*
13. *d*	18. *b*	

23. The average access time of a tape is half the time needed to run the tape from one end to the other because the data is scattered uniformly across the length of the tape. This means that some data is found at one end and some at the other, and the average data request is in the middle.
24. The access time for a disk is made up of the track-to-track stepping time and half the disk's rotational time.
25. *c*
26. We say that the function of the drive electronics is to interface the analog and electromechanical devices to the TTL controller signals because the drive electronics connects the read-write head, the drive motor, the head positioner, the disk sector detectors, and so on, to the controller TTL signals.
27. Controller commands going to the drive include
   The drive address to select the desired drive
   The motor start signal to turn the drive on
   Head step information to position the head
   Step direction to move the head in or out with a step
   Medium side selection to select one of two sides
   Write signals to tell the head to write data on the medium
   Head load signals which position the head against the medium
28. Status signals sent from the drive to the controller include
   An indication that the medium is loaded in the drive
   Sector 0 identification
   Track 00 identification
   Drive ready
   Medium write-protected
29. The controller manages one or more drives and interfaces them with the microcomputer.
30. A controller can manage more than one drive because it has addressing capability. Using this capability, the controller can address the drive it wishes to talk to and then address another drive with different information.
31. *d*
32. When the head on a hard-disk drive comes in contact with the spinning disk, the head is destroyed because there is no flexibility in the medium. This is called a head crash.
33. The disk spins all the time so that the rotational access time does not include any time for the disk to get up to speed.
34. *a*
35. The two common sizes for floppy disks are 5.25 and 3.5 in.
36. FM stands for frequency modulation, and MFM stands for modified frequency modulation. MFM records twice as much data in the same space as FM does.
37. The floppy-disk drive lets the controller know it is starting to read the first sector (sector 0) by sending a pulse on the sector ID line. The pulse is generated by detecting a hole in the floppy disk.
38. The floppy-disk drive has a switch which is closed when the head is pulled back to the track 00 position.
39. A hard-sectored disk has a hole to identify the beginning of each sector. A soft-sectored disk identifies the beginning of sector 0 with a hole, and then all other sectors are identified by time.

40. *a*	47. *a*	54. *a*
41. *d*	48. *b*	55. *c*
42. *b*	49. *c*	56. *d*
43. *d*	50. *d*	57. *d*
44. *c*	51. *c*	58. *b*
45. *d*	52. *c*	
46. *c*	53. *a*	

CHAPTER 13

Microprocessor I/O

∎

CHAPTER OBJECTIVES

This chapter will help you to:

1. *Explain* the use of the ASCII code and data communication protocols for data transmission.
2. *Compare* and *identify* parallel and serial communications.
3. *Define* and *explain* modem and modem standards.
4. *Name* and *describe* common and special input/output (I/O) devices.
5. *Understand* digital-to-analog and analog-to-digital conversion.

───────────

If a microprocessor-based system is to be of value, we must get data into the system and take results from the system. We use I/O devices and communications circuits to move data into and out of a microprocessor-based system. There are many different kinds of data and many different ways to transmit the data. Short-distance applications use parallel data circuits, and long-distance communications rely on serial data circuits. Special devices called *modems* are used to send data over telephone circuits, and other devices are used to interface with the analog world.

∎

13-1 AN INTRODUCTION TO DATA COMMUNICATIONS

We will spend a large part of this chapter learning about data communications—the different ways that microprocessor-based systems talk to each other and to other devices. Each way has its advantages and disadvantages. We will look at these for different types of data communications.

Two common standards tell what the data bytes mean. These are ASCII and EBCDIC. ASCII is widely used with mini- and microcomputer. *EBCDIC (Extended Binary-Coded Decimal Interchange Code)* is almost exclusively used to communicate with IBM mainframe computers and minicomputers. We will not study EBCDIC in this book.

A standard code used to represent alphanumeric characters in data transmission is the *American Standard Code for Information Interchange*. Its acronym (ASCII) is pronounced ASK-key. A standard ASCII character has 8 data bits. Seven bits are used for the 128 different characters, as

$$2^7 = 128$$

Figure 13-1 is a table of the 7-bit ASCII characters. Unfortunately, some of the control codes (00 to 1F) are known by more than one name. However, the codes for the alphanumeric characters, mathematical operators, and punctuation symbols (20 to 7F) are standardized. Each character, its 7-bit binary code, and hexadecimal and binary numbers are given in this table. The numbers in this table presume the eighth data bit is stuck at logic 0. Where a character is given in mnemonic form, the mnemonic is shown in the figure in black letters.

Almost all data transmissions send 8-bit words. The eighth bit in an ASCII data word has several uses. First, it may be transmitted as a permanent logic 0 or as a permanent logic 1. This is called a *stuck bit*. Second, it may be used as an error code. Third, it may be used to indicate an alternative meaning for the other 7 bits.

Most Significant Bits or Digit

Binary →	0000	0001	0010	0011	0100	0101	0110	0111
Hex →	0	1	2	3	4	5	6	7
0000 / 0	NULL NOTHING	DATA LINK ESCAPE CONTROL P	SPACE	0	@	P	`	p
0001 / 1	START OF HEADING CONTROL A	DEVICE CONT. 1 CONTROL Q*	!	1	A	Q	a	q
0010 / 2	START OF TEXT CONTROL B	DEVICE CONT. 2 CONTROL R	"	2	B	R	b	r
0011 / 3	END OF TEXT CONTROL C	DEVICE CONT. 3 CONTROL S†	#	3	C	S	c	s
0100 / 4	END OF TRANSMISSION CONTROL D	DEVICE CONT. 4 CONTROL T	$	4	D	T	d	t
0101 / 5	ENQUIRY CONTROL E	NOT ACKNOWLEDGED CONTROL U	%	5	E	U	e	u
0110 / 6	ACKNOWLEDGE CONTROL F	SYNCHRONIZE CONTROL V	&	6	F	V	f	v
0111 / 7	BELL CONTROL G	END OF TRANS. BLOCK CONTROL W	'	7	G	W	g	w
1000 / 8	BACK SPACE CONTROL H	CANCEL CONTROL X	(	8	H	X	h	x
1001 / 9	HORIZ TAB CONTROL I	END OF MEDIUM CONTROL Y	)	9	I	Y	i	y
1010 / A	LINE FEED CONTROL J	SUBSTITUTE CONTROL Z	*	:	J	Z	j	z
1011 / B	VERT TAB CONTROL K	ESCAPE	+	;	K	[	k	{
1100 / C	FORM FEED CONTROL L	FILE SEPARATOR	,	<	L	\	l	|
1101 / D	CARRIAGE RET CONTROL M	GROUP SEPARATOR	-	=	M	]	m	}
1110 / E	SHIFT OUT CONTROL N	RECORD SEPARATOR	.	>	N	↑	n	~
1111 / F	SHIFT IN CONTROL O	UNIT SEPARATOR	/	?	O	←	o	DELETE

Least Significant Bits or Digit

*also called XON
†also called XOFF

Fig. 13-1   A table of ASCII codes. The binary and hexadecimal codes shown here presume that the eighth bit is a logic 0. Many of the control signals are often given by a mnemonic. The mnemonics are shown as black letters in the full description. For example, the start of text mnemonic is STX, and the device control 1 mnemonic is DC1.

From page 366:

Extended Binary-Coded Decimal Interchange Code (EBCDIC)

American Standard Code for Information Interchange (ASCII)

Stuck bit

There are many different rules for controlling the organization and flow of data. These are called the *data communication protocols*.

There are different levels for each data communications protocol. For example, the ASCII character set is really a data communications protocol. It tells you how the data within a byte is organized and what it means. ASCII and EBCDIC are character protocols.

There must be a protocol for the hardware level in data transmission. That is, both ends of a data communications link must use the same electrical signals for the data lines if they are to communicate with each other. The transmission of data is often controlled by additional handshaking lines. These must also match at both ends of the transmission circuit. For example, the receiving microprocessor has a handshake line called CLEAR TO SEND. It uses this line to control the flow of data so that it does not become overloaded. The sending microprocessor watches this line and does not send data until it is told it can.

At the next level, there must be a protocol for the sequence of characters which are exchanged. One of the most common commands you wish to send is a command telling the sending microprocessor when to start the flow of data and when to stop the flow of data. If you do not have hardware handshaking to control the data flow, this must be done with control codes added to the data stream.

ASCII specifies two characters for this purpose. They are the DC1 (11H) and DC3 (13H). They are also known as control Q and control S, or XON and XOFF. These signals must be generated by the microprocessor system receiving the data and must be recognized by the microprocessor system sending the data.

This is a simple data communications protocol. Often, much more complex data communications protocols are used. If you must add control characters to the data, you increase the communications *overhead*. That is, some of the transmission time is taken up with control information rather than data. If you just use XON and XOFF, the overhead is low. If you use one of the much more complex data communication protocols, the percentage of nondata time (overhead) becomes larger.

Data communications can be either synchronous or asynchronous. In *synchronous data transmissions*, once the transmission starts, the time from one data byte to the next is known. *Asynchronous data transmission* means that the time between data bytes is random and can-

not be predicted by the sending or receiving system. Special software in the microprocessor must be used to synchronize these random (asynchronous) data words with the microprocessor operation. Most microprocessor-based systems and personal computers use asynchronous data communications.

Some microcomputer-to-mainframe communications use synchronous data transmission, transmitting a frame of characters at one time. The frame starts with some synchronization characters followed by a header. The characters in the header identify the frame. This is followed by a block of data. The block of data is then followed by some closing characters and an error checking code. Figure 13-2 shows a synchronous data frame.

Different synchronous data transmission protocols are used. One of the common protocols used by most IBM mainframes is called *synchronous data-link control* (SDLC). An earlier synchronous protocol is called *Bisync*, for *binary synchronous transmissions*. Typically, SDLC and bisync transmissions are used only when the communications involve microcomputers or mainframe computers.

High-level data-link control (*HDLC*) and X.25 are standards established by the International Standards Organization (ISO). These standards are data-independent (transparent) synchronous sytems. Because they are data-independent, they send data of any length. HDLC and X.25 are mostly used for *LAN* (*local area network*) and satellite communications systems.

Self-Test

Answer the following questions.

1. We use a data communications protocol so that the system on the receiving end knows
   a. How to control the flow of data
   b. The organization of the data
   c. The meaning of the data
   d. All of the above
2. ASCII has 128 characters. Each is coded as a _____ -bit binary word.
   a. 6   c. 8
   b. 7   d. 128
3. Give the hexadecimal ASCII codes for the following characters, assuming the eighth bit is a logic 0. Also give the hexadecimal ASCII codes when an eighth (MSB) even-parity bit is added.

Bisynch frame

S Y N	S Y N	S O H	Heading	D L E	S T X	Transparent data	D L E	I T B	BCC	D L E	S T X	Transparent data	D L E	E T B	BCC	P A D

Synchronous control characters

Character	Description
SYN	Synchronous idle
PAD	Start of frame pad
PAD	End of frame pad
DLE	Data line escape
ENQ	Enquiry
SOH	Start of heading
STX	Start of text
ITB	End of intermediate block
ETB	End of transmission (block)
ETX	End of text
BCC	Block check characters

Fig. 13-2 **A synchronous data communications protocol frame.** *Note:* **The characters in this frame are EBCDIC not ASCII.**

a. 6
b. A
c. ↑
d. L
e. p
f. 4
g. *
h. a
i. 0
j. Bell
k. $
l. b
m. Control D
n. Space

4. Briefly explain the difference between an asynchronous data transmission and a synchronous data transmission.

5. The X.25 packet data protocol is frequently used for LANs. This is probably because the X.25 protocol offers
   a. Very sophisticated data transmissions
   b. A standard 25-pin D-connector
   c. A good 8-bit parallel data path
   d. Compatibility with older industrial equipment

13-2 PARALLEL I/O

Many kinds of interfaces connect data to the microprocessor. The simplest is a parallel interface. All other kinds of I/O ports start with a parallel I/O and then convert the parallel data to some other form.

Figure 13-3 on page 370 is an 8-bit *parallel I/O card* block diagram. This card permits inputting one 8-bit data word. Note many circuits are the same as those for the memory card in Chap. 11—specifically the address bus buffers, data bus buffers, and the address decoding logic.

Address lines A_0–A_7 are buffered and drive the I/O address decoder which decodes three

addresses: DATA IN, DATA OUT, and STATUS. For example, DATA IN might be addressed as I/O port 00, DATA OUT might be addressed as I/O port 01, and STATUS might be addressed as I/O port 02.

I/O READ and I/O WRITE are like the memory READ and WRITE lines but they are used during the I/O operations. You can see I/O READ and I/O WRITE are enabled by address decoder outputs. This shows I/O WRITE is enabled only when the DATA OUT address is decoded. I/O READ is enabled when either the DATA IN or STATUS addresses are decoded.

The microcomputer data bus signals are bidirectionally buffered. This block diagram shows two 8-bit buffers, one for data into the card and one for data out of the card. These buffers are enabled by the I/O READ and I/O WRITE lines when these lines are enabled by the card's address decoder.

This parallel I/O card has a 2-bit status register. The status register allows the microprocessor to test the I/O port status. Additionally, four control lines from the status register to an external device help control data transfers. These four handshaking lines are: DATA OUT AVAILABLE, OUTPUT DATA RECEIVED, INPUT DATA AVAILABLE, and INPUT DATA RECEIVED.

When the microprocessor sends data to the I/O port, it addresses DATA OUT and asserts the bus signal I/O WRITE. This transfers data from the microprocessor bus to the 8-bit data latch and sets the output status bit. Setting this

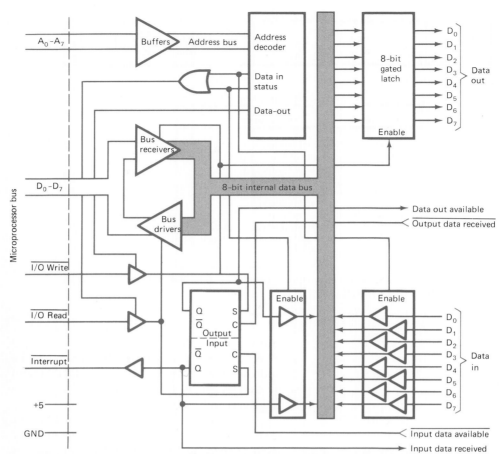

Fig. 13-3 A parallel I/O card transfers 8 bits of data to or from the microcomputer data bus and an external device. Output data is latched so that it remains stable after the transfer. The input data is the data on the input lines at the exact moment the transfer executes.

bit puts a logic 1 on DATA OUTPUT AVAILABLE, telling an external device data is available. When the external device takes the data, it asserts OUTPUT DATA RECEIVED, which clears the output status bit. Later, the microprocessor can read the I/O port status register and, if bit 0 is cleared, it knows the external device took the data.

If an external device wants to send data to the microprocessor, it asserts INPUT DATA AVAILABLE. This clears the input status bit which sends an interrupt to the microprocessor. The microprocessor polls the status registers in each I/O port until it finds one with bit 7 cleared. To input data from the interrupting I/O port, the microprocessor addresses I/O port DATA IN and asserts I/O READ. This enables the data input buffers, placing the input data on the I/O port's internal bus. It also enables the bus drivers to transfer this data to the microprocessor data bus. Asserting I/O READ also sets the input status bit, which

places a logic 1 on INPUT DATA RECEIVED. An external device can check INPUT DATA RECEIVED. If INPUT DATA RECEIVED is a logic 1, the microprocessor has taken the data. If it is a logic 0, the external device must keep the data on the input port because the microprocessor has not read it.

Any time the microprocessor wishes to know the I/O port status, it addresses STATUS and asserts I/O READ. This enables the input and output status bit drivers, placing the output status bit data on D_0 and the input status bit data on D_7 of the internal data bus. It also transfers this data to the microprocessor bus.

Some I/O cards are built using discrete TTL SSI and MSI devices. Other cards use LSI microprocessor support circuits specially designed for use as I/O devices. Frequently, a number of 8-bit parallel I/O ports are included on a single-chip microprocessor or on I/O support circuits. These support circuits make it possible to build low-chip-count microprocessor-based systems. I/O support circuits for specific microprocessors are discussed in the chapters on the specific microprocessor.

The parallel I/O port is the basic microcomputer I/O port. If the data is to be converted into another form, it is done on the I/O side of the parallel port. That is, the microprocessor communicates with a parallel I/O port. The parallel I/O port then communicates with external devices which convert the data into high-power signals for long parallel transmissions, serial data, data formatted for a disk, analog signals, and so on.

How is data from the parallel I/O port connected to external devices? One way uses a multiconductor cable with eight lines for incoming data, eight lines for outgoing data, and a few handshaking lines. This simple parallel I/O data transfer is used on many systems for short distances. The I/O connection is limited to the length of cable the I/O circuits can drive. For simple TTL drivers, this is 1 or 2 m.

Even when high-power drivers are used, cable lengths are limited to about 50 m. The cable capacitance limits the cable length. The capacitance increases as cable length increases. When the capacitance is more than the integrated-circuit drivers can overcome, reliable signals are not transmitted to the receiving end.

A standard interface for transferring parallel data between electronic test instruments and microcomputer-based systems is known as the *IEEE-488* or GPIB (general-purpose interface bus). This bus concept was developed by Hewlett-Packard as the HP-IB (Hewlett-Packard interface bus) and standardized by the Institute of Electrical and Electronics Engineers (IEEE) as the IEEE-488. The IEEE-488 standard is widely used in the United States, Canada, and Europe. Several microprocessor manufacturers have LSI support circuits to interface their microprocessors to the IEEE-488 bus.

Devices attached to the IEEE-488 bus are listeners, listener/talkers, or controllers. Each device in an IEEE-488 system is assigned an address. Figure 13-4 shows a simple example of a microcomputer-based IEEE-488 system. In this system, the microcomputer is in control. The microcomputer tells the voltmeter (a talker/listener) what ranges and functions to select. The microcomputer also tells the voltmeter when to put data on the bus. Both the printer (a listener) and the microcomputer (the controller) receive data. The printer simply prints data. The microcomputer puts additional data on the bus telling the printer to add calculated parameters to the raw data.

The IEEE-488 bus can transfer data at rates of 1 Mbyte/s or more at a distance of 50 m. The bus speed is limited by the speed of the slowest device connected to bus. The IEEE-488 data communications protocol is quite complex.

Two other forms of parallel interfaces are common with personal computer systems. These are the Centronics interface and the *small computer systems interface (SCSI)*.

The *Centronics interface* is named after the printer manufacturer who originally used the interface. Today this is often called the PC's parallel port. The signals in a Centronics interface are shown in Fig. 13-5 on page 372. The Centronics interface may be limited to outputting data from the microcomputer. However, most personal computers allow bidirectional data transfers. Typically, a Centronics interface is used to transfer information from a microcomputer to a printer. In addition to eight data lines, there are handshaking lines which control the transfer of data from the

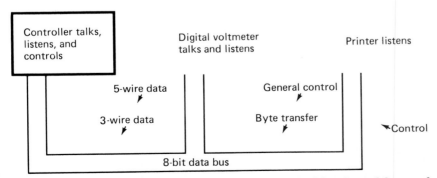

Fig. 13-4 **The IEEE-488 parallel data bus. This standard bus is used for parallel data transfers between instruments and intelligent controllers such as microcomputers.**

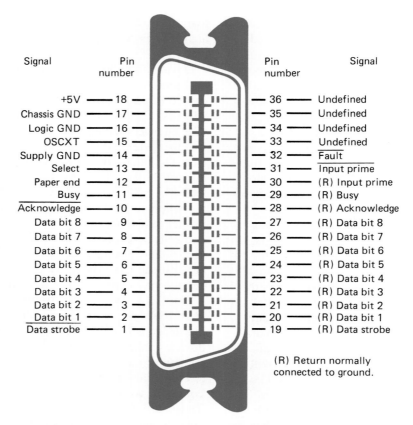

Signal	Pin number		Pin number	Signal
+5V	18		36	Undefined
Chassis GND	17		35	Undefined
Logic GND	16		34	Undefined
OSCXT	15		33	Undefined
Supply GND	14		32	$\overline{\text{Fault}}$
Select	13		31	$\overline{\text{Input prime}}$
Paper end	12		30	(R) Input prime
Busy	11		29	(R) Busy
$\overline{\text{Acknowledge}}$	10		28	(R) Acknowledge
Data bit 8	9		27	(R) Data bit 8
Data bit 7	8		26	(R) Data bit 7
Data bit 6	7		25	(R) Data bit 6
Data bit 5	6		24	(R) Data bit 5
Data bit 4	5		23	(R) Data bit 4
Data bit 3	4		22	(R) Data bit 3
Data bit 2	3		21	(R) Data bit 2
$\overline{\text{Data bit 1}}$	2		20	(R) Data bit 1
$\overline{\text{Data strobe}}$	1		19	(R) Data strobe

(R) Return normally connected to ground.

Fig. 13-5 **The Centronics parallel interface using a 36-pin connector. Although mostly a printer interface, it can connect any 8-bit parallel information-and-control (handshaking) signals from the microcomputer to a peripheral.**

microcomputer to the printer. Usually, the Centronics data port is driven by TTL, which limits the cable length to 10 feet or less.

The SCSI interface provides a lower-cost parallel data transfer than does IEEE-488. SCSI is mostly used to connect microcomputers to mass storage peripherals such as Winchester-disk drives, optical drives (CD ROMs), and streaming-tape drives. As shown in Fig. 13-6, SCSI provides addressing, eight bidirectional data lines, and a few handshaking lines.

Self-Test

Answer the following questions.

6. In Fig. 13-3, $\overline{\text{I/O READ}}$ is used to
   a. Read data to the I/O port outputs
   b. Read data from the I/O port inputs and into the microprocessor
   c. Read a memory-mapped I/O port
   d. Load the I/O port status register

7. Parallel I/O port handshaking lines are used to
   a. Transfer data into the port
   b. Transfer data out of the port
   c. Help control the transfer
   d. Enable the status register

8. The $\overline{\text{INPUT DATA AVAILABLE}}$ line in Fig. 13-3 is asserted when
   a. An external device can input data
   b. The microprocessor has taken the data
   c. An external device has taken data from the output
   d. An external device has data for the microprocessor

9. A parallel I/O port uses an address decoder to
   a. Select the addressed port
   b. Select the memory location for transfer
   c. Memory-map the I/O port
   d. Buffer the address lines

10. The DATA OUT AVAILABLE line is a logic 1 when
    a. An external device can input data
    b. The microprocessor has taken the data

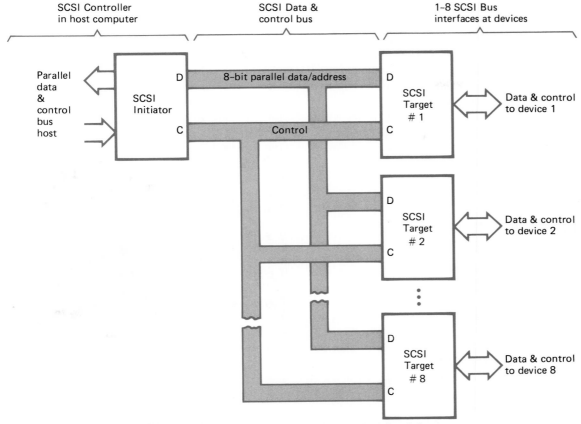

SCSI Controller
in host computer

SCSI Data &
control bus

1–8 SCSI Bus
interfaces at devices

Fig. 13-6 **The SCSI interface. This parallel I/O system allows bidirectional data transfers with addressed external devices under the control of different handshaking signals.**

  c. An external device has taken data from the output
  d. An external device has data for the microprocessor

11. How can you use a memory-mapped I/O on a microprocessor that has separate I/O and processing?

12. Does the parallel I/O port described in this section use a programmed data transfer or DMA? Explain.

13. Below is a list of some typical instruments that have IEEE-488 interfaces. Mark the instruments as talkers, talkers/listeners, or controllers.
  a. Signal generator
  b. Power meter
  c. Digital LCR bridge
  d. Programmable attenuator
  e. Intelligent terminal
  f. Printer
  g. DC power supply
  h. Digital clock
  i. Thermometer
  j. Electronic counter

13-3 SERIAL COMMUNICATIONS

The following sections look at the world of serial communications. Many microcomputer peripherals are connected to microcomputers with serial data transmission systems, and most microcomputer-to-microcomputer communications use *serial communications*. Why are many devices connected by serial data lines when the microcomputer's microprocessor uses parallel data transfers?

The reason is the expense and inconvenience of long parallel communications lines. As we learned earlier, the working distance of a common TTL parallel I/O is 1 or 2 m. Cable capacitance limits unbuffered high-speed data transfers to this length. Longer-distance transmissions need special long-distance driving (buffer) circuits. Because a parallel data transfer needs one wire for each bit in the data word, bytewide parallel data transfers are eight times as expensive as serial data transfers. A long-distance parallel data transfer needs eight times as many drivers, receivers, and wires as a long-distance serial data transfer.

When we study *serial data transmission,* there are three parts we must learn. First, we must learn how parallel data are converted to serial data and how serial data are converted back into parallel data. Second, we must learn what kinds of electrical transmission circuits and signals are used to send data over long distances. Third, we must learn how we know what data are being sent and how we control the data transmission. These rules are the serial data transmission protocols.

13-4 THE SERIAL INTERFACE AND THE UART

Because the data in a microprocessor is in parallel form, a serial I/O data transfer must start and finish with parallel data. However, parallel-to-serial conversion is easily done. The parallel data is loaded into a shift register. The shift register is then clocked. The data comes out of the shift register LSB 1 bit at a time (1 bit for each clock cycle). The first bit of a serial transmission is the data LSB (least significant bit). The second bit is the next LSB, the third is the third LSB, and so on. The last data bit transmitted is the MSB. The parallel-to-serial conversion process is shown in Fig. 13-7.

Receiving serial data and converting it to parallel data is the reverse operation. The serial data is shifted into a shift register. After all bits are clocked into the shift register, the data is taken out of the shift register in parallel and put into the microprocessor system.

The device that performs this parallel-to-serial and serial-to-parallel conversion is called a *universal asynchronous receiver-transmitter* (UART). The UART is a large-scale integrated circuit. In addition to performing the serial-to-parallel and parallel-to-serial conversions, the UART also has control and monitoring functions.

Transmitting an 8-bit data word with UARTs actually requires sending 10 bits. The first bit tells the receiving UART that the data word is coming. The last bit tells the receiving UART that the data word is finished. Figure 13-8 shows an 8-bit data word with a start bit and a stop bit added. The start bit is always a logic 0, and the stop bit is always a logic 1. The UART adds a start bit and a stop bit to the transmitted data.

The UART data transmission speed is called its *baud rate.* The baud rate, or signaling rate, tells how many bits are transmitted per second. For example, a 1200-baud transmission takes place at a rate of 120 ten-bit characters (1 start bit, 8 data bits, and 1 stop bit) per second. Figure 13-9 shows a table of common UART signaling rates. It is important to understand that the baud rate is a measure of a single word's signaling rate. Later you will see why baud rate cannot be measured for more than one word in asynchronous transmissions.

The 110-baud signaling rate uses a different format from the other rates. The 110-baud rate has 1 start bit and 2 stop bits. This makes the data word 11 bits long. Figure 13-10 shows an 11-bit word used for a 110-baud system. This signal was popular with most mechanical teleprinters but is rapidly becoming obsolete.

When the eighth bit is used as error detection code, it is called a *parity bit.* Most UARTs can generate or detect odd or even parity. Even parity sets the parity bit to logic 1 or logic 0, so the 8 bits of data have an even number of logic 1s. Odd parity sets the parity bit to logic 0 or logic 1, so an odd number of logic 1s are in the word.

Odd or even parity is used to detect transmission errors. When a data word is received, the UART tests it for odd or even parity. If the UART detects the wrong parity, the UART status register parity bit is set to show the error. The receiving microprocessor software can then ask for a retransmission. *Parity catches 50 per-*

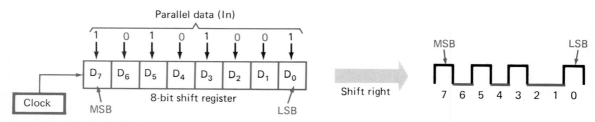

Fig. 13-7 **Transmitting parallel data over a serial line. The LSB is sent first, and then the next LSB is sent. The process continues until the MSB is sent.**

Stop	D_7	D_6	D_5	D_4	D_3	D_2	D_1	D_0	Start

Fig. 13-8 An 8-bit data word with 1 start bit and 1 stop bit. The start bit will always be a logic 0, and the stop bit will always be a logic 1.

Stop	Stop	D_7	D_6	D_5	D_4	D_3	D_2	D_1	D_0	Start

Fig. 13-10 The 11-bit serial data word. Most commonly this is only used for 110-baud transmissions. The 11-bit code is very common for electromechanical teleprinters. Some early electromechanical teleprinters required 1.5 stop bits.

cent of the errors. It catches only odd errors. If an even number of bits are in error, the parity does not change and the error is not detected.

Figure 13-11 shows how a noisy serial data transmission can cause a parity error. Bit 4 is changed by noise from a logic 1 to a logic 0, and therefore incorrect data is received.

The UART also detects *framing errors.* Framing errors happen when the UART misses the start bit. This means that the UART receives the transmission with the start and stop bits in the wrong place.

Figure 13-12 on page 376 shows a received word with a framing error. You can see that the received data word is not aligned with the UART's shift register. Noise on the signal line is a common cause of framing errors.

Figure 13-13 on page 376 is a block diagram of a simple UART. The diagram shows that the UART can be divided into four major parts: the transmit section, the receive section, the status section, and the control logic section.

The transmit section has two parts: the *transmitted data output buffer* and the transmit register. The *transmit register* shifts out the start bit, the data bits D_0 to D_7, and finally the stop bit or bits on the serial data (out) line. Data in 8-bit form is loaded into the transmit data output buffer by the trailing edge of the signal asserting the DATA INPUT STROBE control line. The leading edge of this signal causes the

serial data transmission to start; that is, to start the shifting process. Figure 13-14 on page 377 shows this action. The data is shifted out of the shift register onto the serial output line.

The receiver is almost a mirror image of the transmitter. Data enters through serial data (in). The receive register shifts 10 or 11 times after it detects a start bit. Once the receiver has stopped shifting, the 8 data bits are transferred to the received data output buffer. Asserting the DATA OUTPUT ENABLE STROBE causes the transfer.

Both the transmit shift register and the receive shift register are driven by a clock signal that runs 16 or 64 times the selected baud rate. This clock rate ensures that the shifting clock is never more than $\frac{1}{32}$ or $\frac{1}{128}$ of a bit width out of synchronization with the input signal. This degree of synchronization is maintained even though the baud-rate clock and the incoming data are completely asynchronous.

Looking at Fig. 13-13, you can see the UART status register outputs. The status word can be loaded into the received data word out-

Baud rate	Bytes per second
110	10
150	15
300	30
600	60
1,200	120
2,400	240
4,800	480
9,600	960
19,200	1,920
38,400	3,840

Fig. 13-9 The common serial data transmission signaling rates (baud rates). Other rates are used for special situations.

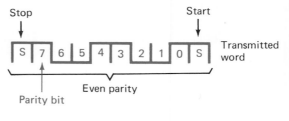

Fig. 13-11 A serial transmission with a single bit error detected by the parity bit. The noise during transmission changes bit 4 from a logic 1 to a logic 0, causing incorrect data to be received.

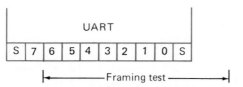

Fig. 13-12 A framing error. The incoming data does not line up with the UART start and stop positions. Usually a framing error is caused by noise.

put buffer. This loading is done by asserting the STATUS OUTPUT STROBE control line.

The UART status register provides the following information:

1. *Overrun (OR):* A logic 1 in this bit indicates the current data word overran the previous data word. That is, the previous data word never transferred into the received data output buffer.

2. *Framing error (FE):* A logic 1 in this bit indicates the UART did not find a stop bit. The UART probably missed the real start bit and made a false start on one of the data bits.

3. *Parity error (PE):* A logic 1 in this bit means the UART detected an odd- or even-parity error. The kind of parity to be detected is set by the UART control logic before the data is received.

4. *Transmit buffer empty (TBE):* A logic 1 in this bit indicates the next data word can be loaded into the transmit data output buffer. In other words, the previous word has been transmitted.

5. *Data available (DA):* A logic 1 in this status bit indicates a new data word in the receiver. The word can be sent to the microprocessor by asserting the data output strobe.

After the UART status word is in the received data output buffer, you can transfer it to the microprocessor accumulator. There, you can use compare instructions and other tests to see if the proper status bits are set.

For example (using the status word shown in Fig. 13-15), if a data word is received with no errors, the status word is

$$0000\ 0010 = 02H$$

You can test for this status word by using a compare 02H. The compare instruction sets the

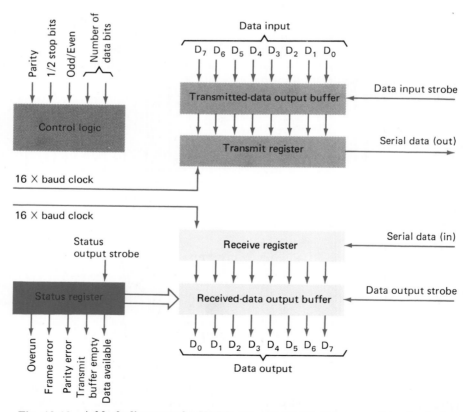

Fig. 13-13 A block diagram of a UART. These basic blocks are used in stand-alone UARTs and in special UARTs connected directly to a microprocessor.

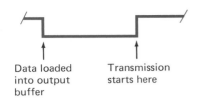

Data loaded into output buffer   Transmission starts here

Fig. 13-14   The DATA INPUT STROBE signal. This signal causes one operation on the trailing edge and one on the leading edge.

zero bit of the microprocessor status register to a logic 1 if the only status bit set is the data available bit. If any other bits are set, the test fails. Failure of this test tells the microprocessor that there is a problem with the data in the UART. Additional software can be used to tell just what the problem is or to request a retransmission of the data word.

The UART control register lets you set the UART operating mode (again, see Fig. 13-13). The two data control bits let you select 5, 6, 7, or 8 data bits per received data word. As you know, 7 data bits are common for ASCII transmissions. Five data bits are used for early teleprinter transmissions, and 6 data bits are used for special compressed transmissions. Eight data bits are used when the parity bit is not transmitted, but the user wishes to transmit a full byte of information with each serial word.

The *odd/even control bit* lets you select either odd or even parity. This control bit is used only when the no parity bit turns the parity function on. The parity bit is placed in the transmitted data word's last data bit. This is the eighth data bit when you are transmitting a 7-bit ASCII character. Setting the parity control also tells the UART what parity check to make on the incoming data.

The 10/11 stop bit lets you select a transmission format with 1 or 2 stop bits. Remember, 110-baud signaling uses 2 stop bits. Frequently, there is an additional control bit giving the option to select 1.5 stop bits. This is used with the old 5-bit transmission standard, which used a 1.5-bit stop signal.

UARTs are often used with a special baud-rate generator IC. This IC generates a times 16 (16 ×) or times 64 (64 ×) signal for all the standard baud rates. The basic signal comes from a crystal oscillator, so the baud rate is very stable and accurate. Frequently, the input signal for the baud-rate generator is generated from the microprocessor clock signal by using some form of a divider. A programmable baud-rate generator allows the UART to operate over a wide range of baud rates under program control.

The UART operates asynchronously with the microprocessor. This means that the data rate for the time that the data is transmitted has no relationship to the microprocessor timing. Each operates with its own timing signals. If the microprocessor and the UART use the same clock, this is done to save on clock hardware, not to synchronize their operation.

Frequently you will hear the term *asynchronous data transmission* applied to the UART. This has a little different meaning. It means that the time from one data word to the next is unknown. As an example, think of data being generated by a person at a terminal keyboard. Each asynchronous character is generated as a key is depressed. The spacing between one character and the next is completely unpredictable. However, once the transmission of a single ASCII character starts, the baud rate remains the same as long as that character is being transmitted. That is, the data within an ASCII character is synchronous from bit to bit, but the data from one character to the next is asynchronous because its relative timing is unknown.

Figure 13-16 (page 378) shows the serial transmission of an ASCII character from a UART. The start bit is always a logic 0 and always follows a random time at logic 1. The data word D_0 to D_7 is a combination of logic 1s and logic 0s, depending on the data being sent. The parity bit may be a logic 1 or a logic 0, and the stop bit is always a logic 1. The output stays at a logic 1 until the next start bit is sent.

Figure 13-17 (page 378) shows the serial transmission of four characters and two important characteristics of this form of asynchronous transmission. First, the time between characters is random. A start bit can follow immediately after the previous character's stop bit, or an interval of any length may pass before the next start bit. Second, the interval between characters is always at the stop level.

0	0	Overrun	Frame error	Parity error	Transmit buffer empty	Data available	0

Fig. 13-15   The UART's status word.

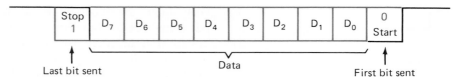

Fig. 13-16 The serial transmission of an ASCII character. Start is always a logic 0 (called a Space), and stop is always a logic 1 (called a Mark). When not in use, the line is marking.

How does the microprocessor receive data which is coming from a UART sending asynchronous characters? How does the microprocessor tell when the next character will be received when the timing to the next character is completely random? The answer is that the microprocessor must check the UART status register on a regular basis. If the UART status register data available bit is set, the microprocessor then reads the UART data. If the UART status register data available bit is not set, then the microprocessor must keep checking the status register until the data available bit is set.

The flowchart in Fig. 13-18 shows this operation. As you can see, the microprocessor can be in this *wait loop* indefinitely. The following example follows Fig. 13-18.

Suppose you are waiting for a 1200-baud transmission. At 10 bits/word, the maximum data transfer rate is 120 characters/s. If the microprocessor instructions (IN, ANI, JNZ) have 1-μs CPU cycles, they execute in a total of 13 μs. (Both the IN and the JNZ instruction in this example take five CPU cycles, and the ANI instruction takes three CPU cycles.) With the 13-μs loop time and 8.3 ms for sending a character, the microprocessor loops 640 times waiting for the next time that the data available status bit is set.

Of course, the program loops more than 640 times if the second ASCII character does not immediately follow the first character. The next character probably will not follow immediately if the data is coming from a terminal, since few people can type at 120 characters/s. That rate is equivalent to typing at 1440 words/min!

The wait problem is reduced by using the UART data available output to interrupt the microprocessor. As we have learned, the interrupt signal causes the microprocessor to execute a special interrupt-handling routine. In this interrupt-handling routine, started by the UART, an in instruction loads the current value of the UART status word into the accumulator. This gives the microprocessor the data necessary to complete the input service routine.

Self-Test

Answer the following questions.

14. The UART performs
    a. A serial-to-parallel conversion
    b. A parallel-to-serial conversion
    c. Control and monitoring functions
    d. All of the above
15. A signaling rate of _____ baud is not a common speed.
    a. 300   c. 256
    b. 150   d. 1200
16. A signal of _____ baud has 2 stop bits.
    a. 110
    b. 150
    c. 300
    d. 600
17. The _____ is not a part of a UART.
    a. Control and logic function
    b. Transmit shift register
    c. Address decoder
    d. Receive shift register

Fig. 13-17 The serial transmission of four asynchronous characters. The first three characters are randomly spaced. The fourth character starts as soon as transmission of the third is complete. The line is marking between characters.

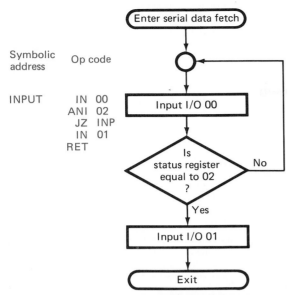

```
Symbolic    Op code
address

INPUT       IN  00
            ANI 02
            JZ  INP
            IN  01
            RET
```

Fig. 13-18 A routine for handling a serial I/O port. Once the contents of the UART's status register are in the accumulator, an AND IMMEDIATE 02 masks all but the data available bit and clears the microprocessor status register zero bit if data is available.

18. Write a brief description for each of the following terms. If the term is an acronym, include its expanded name.

   a. UART   f. Framing error
   b. Baud   g. Start bit
   c. ASCII   h. Parity bit
   d. 16 × clock   i. Stop bit
   e. Overrun

13-5 SERIAL COMMUNICATION LINES

In the last section we looked at the UART. We saw that the UART is used to convert parallel data to serial data and serial data to parallel data. In this section, we will look at the lines used to transmit serial data. We will see how they are used and what standards apply to them.

Serial ports are usually connected to other serial ports with one of three standard serial line types. By far the most common serial line is one which meets the *EIA RS-232 standard*. Simply, it is known as RS-232. RS-232 lines are used on most terminals, printers, modems, and other devices used within 50 feet or less.

RS-232 calls for logic 1 to be a +3-V or higher signal. The logic 0 signal is to be −3 V or lower. To generate these signals, an RS-232 transmit IC usually needs to operate from a ±12-V supply. RS-232 uses a single wire for transmitted data and a single wire for received data. Each signal is referenced to signal ground. The standard also specifies a number of handshaking signals used to control the transmit and receive data signals.

The RS-232 signals and the commonly used 25-pin and 9-pin D-connectors are shown in the diagram in Fig. 13-19 (*a* and *b*) on page 380. Although RS-232 cabling often uses these connectors, they are not part of the standard. Because the 25-pin D-connector requires a great deal of panel space, the 9-pin configuration is used on many personal computers.

The RS-232 standard allows a maximum transmission distance of 50 ft at a data rate of 9600 baud. Longer transmissions may be used at slower data rates, and the transmission distance is shortened for data-transmission rates above 9600 baud. You will often find this 50-ft limit exceeded as today's RS-232 transmitters exceed the drive specifications.

The *EIA RS-422 standard* specifies a different set of electrical signals for data transmission. The RS-422 standard uses *differential signaling*. A pair of wires carries the transmitted data, and another pair of wires carries the receive data. Typically, each is a twisted pair. This minimizes electromagnetic interference from signals on these wires. Twisting the wires also helps the RS-422 signals to reject interference from outside sources.

RS-422 specifies that a logic 1 is a positive difference between these two wires. That is, wire A is at a higher voltage than wire B. Unlike RS-232, which uses both positive and negative signaling voltages, the RS-422 signals are between 0 and +5 V. This means RS-422 does not require supply voltages different from those already used in the system for the microprocessor and logic circuits.

At the receiving end, data signals are the difference in voltage between the two wires in the twisted pair. The absolute voltage of the twisted pair with respect to ground does not matter unless the absolute voltage is more than the receiver can handle. This means that RS-422 allows the transmission line to pick up a great deal of "common-mode" interference and still remain operational. *Common-mode interference* changes the voltage of both wires by some amount. However, the common-mode voltage does not change the voltage difference between them. Therefore, data (which is being sent as the voltage difference between the wires) still gets through.

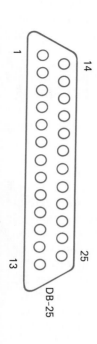

25 Pin	EIA–RS232 Circuit	CCITT-V.24 Circuit	RS232 Description	Signal type & direction
1*	AA	101	Protective ground	Ground
7*	AB	102	Signal ground/common return	Ground/common
2*	BA	103	Transmitted data	Data to DCE
3*	BB	104	Received data	Data from DCE
4*	CA	105	Request to send	Control to DCE
5*	CB	106	Clear to send	Control from DCE
6*	CC	107	Data set ready	Control from DCE
20*	CD	108,2	Data terminal ready	Control to DCE
22	CE	125	Ring indicator	Control from DCE
8*	CF	109	Received line signal detector	Control from DCE
21	CG	110	Signal quality detector	Control from DCE
23	CH	111	Data signal rate selector (DTE)	Control to DCE
23	CI	112	Data signal rate selector (DCE)	Control from DCE
24	DA	113	Transmitter signal element timing (DTE)	Timing to DCE
15	DB	114	Transmitter signal element timing (DCE)	Timing from DCE
17	DD	115	Receiver signal element timing (DCE)	Timing from DCE
14	SBA	118	Secondary transmitted data	
16	SBB	119	Secondary received data	
19	SCA	120	Secondary request to send	Control to DCE
13	SCB	121	Secondary clear to send	Control from DCE
12	SCF	122	Secondary received line signal detector	Control from DCE
11 18 25			Undefined	

(a)

9 Pin	EIA RS-232 Circuit	CCITT V.24 Circuit	RS-232 Description	Signal type & direction
5	AB	102	Signal ground/common return	Ground/common
2	BB	104	Received data	Data from DCE
3	BA	103	Transmitted data	Data to DCE
1	CF	109	Received line signal detector	Control from DCE
4	CD	108, 2	Data terminal ready	Control to DCE
6	CC	107	Data set ready	Control from DCE
7	CA	105	Request to send	Control to DCE
8	CB	106	Clear to send	Control from DCE
9	CE	125	Ring indicator	Control from DCE

(b)

Fig. 13-19 (*a*) The 25-pin DB-25 connector and pinout commonly used with RS-232 lines. All of the 25 different lines are shown in this drawing. However, one very common RS-232 cable only uses the nine wires marked with an asterisk (*). (*b*) The nine-pin DB-9 connector and pinout used with newer equipment where there is a shortage of panel space.

Unlike RS-232, RS-422 only specifies a transmit pair and a receive pair. No control lines are specified and there is no connector convention commonly used. A typical RS-422 differential circuit is shown in Fig. 13-20.

With no hardware handshaking, all of the data communications protocol is done by adding control words to the data stream. You cannot have control information transmitted in parallel with the data. This is because there are no separate wires for control information as with RS-232. This lack of separate hardware handshaking means RS-422 transmissions have higher overhead than RS-232 transmissions.

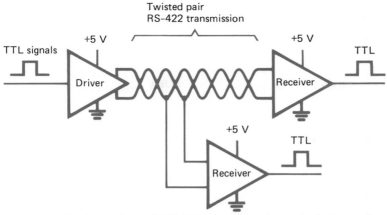

Fig. 13-20 A commonly used RS-422 system showing a single transmitter and multiple receivers.

RS-422 differential signaling transmissions allow faster signaling rates and much greater distances than the single-ended RS-232 system. RS-422 permits transmissions at 38.4K baud over a distance of 4000 ft. At 38.4K baud, RS-232 is limited to a distance of 12.5 ft.

Because of its differential signaling, which gives it a high common-mode noise rejection and long signaling distances, RS-422 is popular in industrial applications. RS-422 is not as popular as RS-232 for local transmissions. Here interference is not great, and RS-422 does not provide the control lines often used by printers, modems, terminals, and other serial devices. However, RS-232 is used by Apple computers to talk to their peripherals.

A third form of serial transmission is called the *20-mA current loop*. This is an early form of serial signaling not used extensively today. It is still popular in some factory systems because of its great noise immunity and simplicity. A typical 20-mA current loop is shown in Fig. 13-21. The 20-mA current loop is different from the RS-232 or RS-422 systems because it uses a constant-current source instead of a voltage source.

As long as the loop is closed, a constant 20-mA current flows. Signaling makes (logic 1) and breaks (logic 0) the loop. The data gets through even with large resistance changes in the transmission line. It also lets the line reject changes caused by outside interfering signals.

The 20-mA loop can send data over greater distances than RS-232 can with even greater noise immunity than RS-422. Although the 20-mA loop can transmit much further than RS-232 can, the loop has disadvantages. It has no control signals and is much less standardized than RS-232. The 20-mA loop has no conventionally used connector, and a high-voltage source is needed to keep a constant 20-mA over a wide range of line and load resistances. Typically, 20-mA loops can be as long as 1000 ft and are used for signaling speeds up to 300 baud.

Because of these disadvantages, the 20-mA loop is usually only used in manufacturing plants where control lines must be run 1000 ft or so near very noisy power and control lines. Additionally, the 20-mA loop is used when a microprocessor-based system is interfaced to

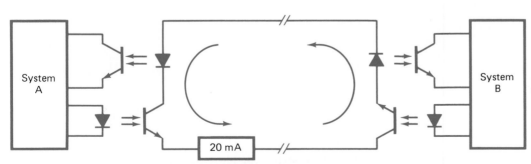

Fig. 13-21 A wiring diagram for a typical 20-mA current-loop system. Note that this system only uses two wires. This means that only one device can transmit at a time.

an older piece of equipment with a 20-mA current-loop interface.

When a high degree of isolation is required between the external device and the microprocessor-based system, *optoisolators* provide the isolation. The optoisolators protect the UART, the microprocessor, and other system logic from high voltages and ground-loop signals that might accidentally become connected to the transmission lines. This is not uncommon in industrial environments. Figure 13-22 shows a typical EIA RS-232 transmitter and receiver constructed with optoisolators. Optoisolation is also used with parallel signals or any other form of communication with the microprocessor.

Other high-speed serial transmission line standards are used with microprocessor systems, especially when communicating with mainframes or over local area networks (LANs). Data transmissions using synchronous protocols often use coaxial transmission lines or fiber-optic lines. Because of the closely controlled conditions available in these transmission lines, these systems use data rates as high as 100 Mbit/s. These standards provide high-speed serial data transmissions when high-volume data is needed.

Self-Test

Answer the following questions.

19. Which of the following is not one of the major serial transmission standards?
    a. EIA RS-232
    b. EIA RS-422
    c. IEEE-488
    d. HDLC

20. IEEE-488, RS-232, and RS-422 all have the common characteristic of using
    a. Multiconductor cables for transmission lines.
    b. Standardized connectors.
    c. Differential signaling.
    d. Very long (> 1000-ft) cables.

21. The RS-232 standard specifies that the signaling voltage will swing a minimum of _____ for a logic 1 to a logic 0.
    a. ±3 V   c. A 5-V differential
    b. ±20 mA   d. ±12 V

22. An RS-422 serial transmission line is characterized by
    a. Differential signaling
    b. Two twisted-pair cables, one for transmit and one for receive
    c. Long-distance signaling at relatively high speeds compared to RS-232 signaling
    d. All of the above

23. RS-422 is very useful in situations which
    a. Require high noise immunity
    b. Use a standardized 25-pin D-connector
    c. Require a parallel connection to a printer
    d. Require additional handshaking lines in addition to the transmit and receive pairs

24. Briefly describe the electrical characteristics of a serial communications line using the RS-232 standard and one using the RS-422 standard.

25. Why is it likely that an RS-422 transmission will have a higher overhead than will an RS-232 transmission? Could an RS-232

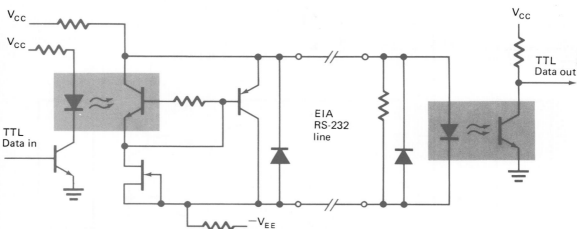

Fig. 13-22 An RS-232 line with optoisolators to keep ground-loop currents and high voltages from damaging the microelectronics. The reverse-biased diodes serve to suppress high-voltage transient signals.

transmission have the same overhead as an RS-422 transmission?

26. Rank the following data transmission methods from 1 (the one with the shortest usable distance) to 5 (the one with the longest usable distance).
   a. 20-mA current loop
   b. EIA RS-232
   c. EIA RS-422
   d. TTL parallel I/O
   e. Microcomputer internal bus

27. We can say that the UART timing is asynchronous with the microcomputer. We can also say that the transmitted characters are often asynchronous with respect to each other. What does this mean? Why are there two different uses of the term *asynchronous?*

28. Which of the following communication standards is not usually used for communications from a microcomputer to a simple printer?
   a. SDLC   c. X.25
   b. HDLC   d. All of the above

13-6 MODEMS

As we learned in the section on serial transmissions, each standard has a cable length limitation. For some standards, this is relatively short and for others it is fairly long, but in any case there is a definite limit.

What do computer users do when they need to communicate over greater distances? They use a device called a *modem.* A pair of modems lets the user run a serial line as far as needed.

The term *modem* combines the words *modulator* and *demodulator.* Simply, the modem modulator circuit converts binary data into tones. At the other end of the transmission line the demodulator circuit converts the tones back into binary data. By this technique, a pair of modems can send digital information over long distances. The modem allows personal computers to communicate with mainframes and each other, and it allows a terminal to connect to a mainframe or a personal computer.

Modems permit any two devices to communicate over a single pair of dedicated wires or the switched telephone network using a serial data communications protocol. The switched network comprises all the telephones any individual can reach by dialing a number. The switched telephone network is not only connected by wires, but it also uses microwave and satellite transmissions. This means it cannot be used to send binary data without first converting the data to tones, since these are ac-coupled circuits.

The simple modem converts digital information into two tones. One common standard converts the RS-232 signals into a high-frequency tone (1270 Hz) for a logic 1 and a low-frequency tone (1070 Hz) for a logic 0. These tones are then transmitted over a telephone line or any long serial line. Again the demodulator portion of the modem converts received tones from the telephone line back to RS-232 signals.

Figure 13-23 shows a microprocessor connected to a parallel port which, in turn, is connected to a UART connected to a modem. The modem on the first system is connected to a modem on a second system by a telephone or dedicated line. This way two microprocessor-based systems can communicate even though they are separated by a long distance.

The modem simply converts a logic-level signal to a *frequency-shift-keyed (FSK) signal.* The FSK signal can pass through ac coupling devices, such as transformers and capacitors often used in long telephone circuits. Modems can also transmit data by radio, as shown in Fig. 13-24. Data transmission by radio and modem is often used for industrial data gathering systems and for the wireless devices which are coming into use for personal computers. The system shown in Fig. 13-24 can also

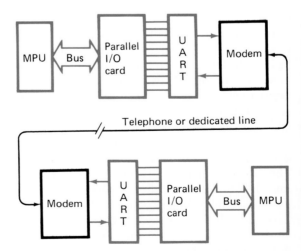

Fig. 13-23 **Using modems to communicate over long distances. The modems send and receive digital signals by using different tones for logic 1s and logic 0s. Modems can be used over the switched telephone network or over dedicated lines.**

Bell 103 standard

Originate tones

Answer tones

Full-duplex transmission

Half-duplex mode

Connect signal

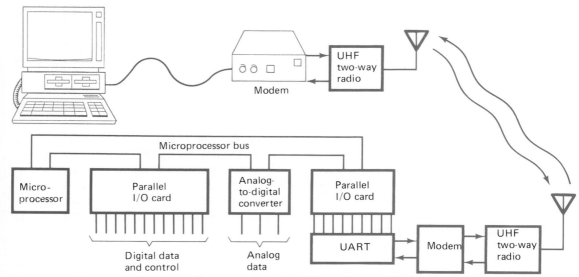

Fig. 13-24 Using microprocessors, parallel ports, UARTs, analog-to-digital converters, modems, and UHF radios to gather and analyze remote data. This system can also provide control signals from the parallel I/O card as well as gathering digital data via the parallel I/O card. The system is controlled by and data presented on a microcomputer.

provide control signals from the parallel I/O port.

There are a number of different standards for modem transmission. Obviously, some form of standard is needed so that one modem can talk to another modem. These standards were developed by the telephone company, and they have become known as the Bell standards. Two common modem standards used for microcomputer systems are the Bell 103 standard and the Bell 212 standard.

The *Bell 103 standard* uses the FSK transmission standard discussed earlier. That is, a Mark signal of 1270 Hz is the high-frequency tone and a Space tone of 1070 Hz is the low-frequency tone. These tones are called *originate tones* because they are sent by the modem originating the data transmission. Another set of tones (Mark = 2225 Hz and Space = 2025 Hz) is called the *answer tones*. Using these two tone sets, one telephone line lets data be sent two ways at the same time. The two tone sets are shown in the chart in Fig. 13-25. This is called *full-duplex transmission*. Sending binary data (logic 1s and logic 0s) over a single wire is called the *half-duplex mode* because data can only be sent in one direction at a time.

Although this form of FSK transmission is capable of operating from 0 to 600 baud, the Bell 103 standard is typically used at 110 or 300 baud. Early personal and microcomputer

systems used 300 baud. This gives the user an effective data rate of 30 characters/s.

There are no control lines which go with the modem signals on the telephone line. How does control information get passed? The answer is the control information must be sent as special characters inserted in the data. Remember, the modem is a relatively simple device. It only passes the data which you give it. Therefore, control characters must be generated by the applications software driving the serial port connected to the modem.

Modems do generate special hardware handshaking signals. For example, a modem connected to a conventional telephone line can listen for and indicate the presence of a ring signal. Additionally, most modems output a control signal, indicating they heard the tone from another modem. This signal is called the *connect signal*. Some modems output a signal

Logic level	Originate	Answer
Logic 1	1270 Hz	2225 Hz
Logic 0	1070 Hz	2025 Hz

Fig. 13-25 The tones used to implement the Bell 103 standard. Note that there are two sets of tones. At any one time there will be just one of the two tones from each set. Which tone depends on whether a logic 0 or logic 1 is being sent.

which indicates that they are turned on and ready to accept data.

You should note that these are local control signals, not control characters put into the data stream. The control signals are handshaking lines which are part of the RS-232 lines. These handshaking signals are communicated to the microcomputer by the RS-232 control lines. However, any data control signals which must be transmitted to the other modem via the telephone line must be encoded into control characters and inserted into the data stream being transmitted to the other modem. This increases the communications overhead.

Today, users find the 30-character/s rate of the Bell 103 too slow. The faster Bell standard becoming popular in the microcomputing world is *Bell 212*. This standard specifies a 1200-baud or 2400-baud signaling rate.

These signaling rates are too fast to use the simple FSK modulation used by the Bell 103 modems. The Bell 212 standard specifies a new modulation technique called *quadrature phase shift keying* (QPSK). The Bell 212 standard specifies two tones—one for transmitting (originate) and the other for receiving (answer). Data is transmitted by changing the phase of the transmitted tone in 90° steps. The phase changes only happen once every two bit times. Therefore, each phase change must transmit 2 bits of information as shown in Fig. 13-26.

Many users find the 2400-baud data rate too slow for some applications. For them, 9600-baud (960-character/s) operation is becoming popular. Here even more sophisticated modulation schemes are being used. This data rate allows only a few cycles of the transmitted tone to be sent during the time for each bit group.

This means 9600-baud transmissions need "clean" telephone lines. That is, the transmission strength must remain constant over a few cycles, and the lines must be low noise. Signal strength changes or noise can cause the modem to "see" phase shifts and make up data which are not there.

Today, most telephone lines (including the long-distance lines) are relatively noise-free. However, the signal strength (amplitude) of the signals passed through these lines may vary too much. It is common for sophisticated modems to include autoequalization. Autoequalization circuits let the modem correct for ongoing changes in the signal strength of the telephone line. With autoequalization, a modem may be able communicate over long-distance lines which would keep a nonautoequalized modem from communicating.

Self-Test

Answer the following questions.

29. The reason modems are used is
    a. Because many communications circuits are too long for any of the data communications electrical standards to work over
    b. Telephone circuits may be ac-coupled somewhere in the circuit
    c. That you can use the switched telephone network to reach anyone else who has a modem connected to the telephone system
    d. All of the above
30. The term *modem* is a contraction of _____ and _____ (Choose two of the four answers.)
    a. Mobile   c. Modulator
    b. Distance   d. Demodulator
31. Modems which allow both ends of the communications circuit to send data to each other at the same time use
    a. 1270-Hz tones
    b. Frequency- or phase-modulated tones
    c. Half-duplex signals
    d. Full-duplex signals

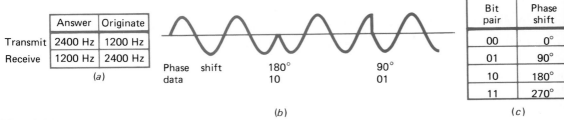

	Answer	Originate
Transmit	2400 Hz	1200 Hz
Receive	1200 Hz	2400 Hz

(a)

Phase shift data   180°   90°
                    10   01

(b)

Bit pair	Phase shift
00	0°
01	90°
10	180°
11	270°

(c)

Fig. 13-26 (a) The two tones used to implement the Bell 212 standard. One is used to send data in one direction, and the other sends data in the other direction. (b) The phase of the tone signal is changed to send the logic 1s and logic 0s. (c) Phase shifts to send bit pairs.

32. A modem coverts RS-232 signals into _____, which are converted back into RS-232 signals at the receiving end of the circuit.
    a. 1270-Hz tones
    b. Frequency- or phase-modulated tones
    c. Half-duplex signals
    d. Full-duplex signals

33. Modems which use the Bell 103 standard usually transmit data at a maximum rate of _____ baud.
    a. 110   c. 1200
    b. 300   d. 2400

34. Modems which use the Bell 212 standard usually transmit data at a maximum rate of _____ baud.
    a. 110
    b. 300
    c. 1200
    d. 4600

35. You would expect that the reason the data communications protocol for use with a modem circuit might have higher overhead than a local nine-wire RS-232 communications circuit is that
    a. The nine-wire RS-232 circuit will have higher overhead
    b. The modem-to-modem connection has fewer handshaking lines
    c. The modem-to-modem connection has no handshaking lines; the only connection is the data line
    d. All of the above

36. Amplitude (signal strength) or noise problems on a data communications circuit may cause
    a. A higher overhead because some of the data may have to be resent due to data transmission errors
    b. Data transmission errors
    c. The need for autoequalization circuits to cut down on signal strength variations on the data signals
    d. All of the above

13-7 INPUT/OUTPUT DEVICES

Many kinds of devices move data into and out of the microprocessor. The choice of the input/output device depends on the source of the data the microprocessor is to work on. For example, if the data is to come from magnetic media, then either a tape drive or a disk drive should be used as the I/O device. On the other hand, if people are to be entering and retrieving data, then a keyboard and a display are probably the best choices for an I/O device. In this section, we will look at a number of input/output devices built to communicate with people.

To start a discussion of I/O devices, we will separate them into two major groups. First, there are the devices that communicate with people. Second, there are the devices that communicate with machines. As you can imagine, the devices that communicate with people have some special characteristics. They must use standard alphanumeric characters plus a few special mathematical symbols and punctuation marks. In general, communications with people must be slow. Input comes from a keyboard. Output is on a display.

On the other hand, communications between the microprocessor and a machine can be in many different codes. Machines usually communicate with the microprocessor at high speeds. In most cases, machines input data to the microprocessor in the same form in which they accept data.

The most common device that enables people to input data to the microprocessor is the *keyboard*. Keyboards used with microprocessor systems are usually one of three different kinds. The first is the simple numeric keypad. Most electronic calculators and simple microprocessor-based control products use numeric keypads.

Second is the full alphanumeric keyboard. Microcomputers and terminals have full alphanumeric keyboards. Keyboards of this kind transmit all the alphanumeric characters (both uppercase and lowercase), 40 to 60 special characters, and another dozen special functions. Typically, these keyboards use 90 to 100 keys. The special characters include the mathematical operators, punctuation marks, editing keys, and a few control codes.

The third kind of keyboard is the dedicated, or special-purpose, keyboard. This keyboard is custom-designed for the application at hand. These keyboards are used on special-purpose microprocessor-based systems. For example, the keyboard on a microprocessor-based environmental control system may not have any alphanumeric input. It may simply have keys marked AIR CONDITIONING, HEATING, BLOWERS, and PUMPS.

Most keyboards work in the same basic way. Their operation is like the operation shown by the simplified schematic in Fig. 13-27. Here we see the familiar row-and-column matrix. The keyboard columns are scanned. First a signal is placed on column 1, then the signal moves to column 2, then to column 3, and then to column 4. Then the entire scan operation repeats itself.

Scanning may be done by the microprocessor itself, by an integrated-circuit shift register, or, as shown in Fig. 13-27, by a binary counter and decoder. The rows are connected to the columns by the keyboard switches. In this example, there are 16 switches. This keyboard is called a *4 × 4* (four-by-four) *matrix*.

The column scan and row output lines are connected to an encoder circuit. This has two functions. First, the contact closure is scanned a number of times. This is done to make sure that there is a valid contact closure. Multiple scanning avoids false contact closures. This function is called contact *debouncing*. For example, after 10 scans find the same contact closed, the chances are great that the particular column is connected to the particular row. That is, the key where the row and column cross is depressed.

Once we are sure there is a valid contact closure, we can begin the keyboard encoding process. The encoding circuit now takes over. The encoding circuit is a special ROM. It takes the row and column information and generates the needed parallel or serial output signals.

The keyboard logic for microprocessor-based systems is usually made in one of two ways. First, some keyboards are made to be scanned and encoded by a microprocessor. Very simple keyboards may be scanned by the microcomputer system's own microprocessor. If so, the keyboard matrix is connected to one of the microprocessor I/O ports. A subroutine does the scanning, the debouncing, and the encoding.

On very complicated keyboards, a second microprocessor may be used to do the scanning, debouncing, and encoding. This is the way keyboards for personal computers are built. A simple single-chip microprocessor such as the Intel 8048 or 8049 performs the keyboard management functions. It receives power through the coiled cable coming from the computer and generates serial signals which tell the computer what key is depressed.

The second way of making keyboard logic is to use custom LSI. Since keyboard commu-

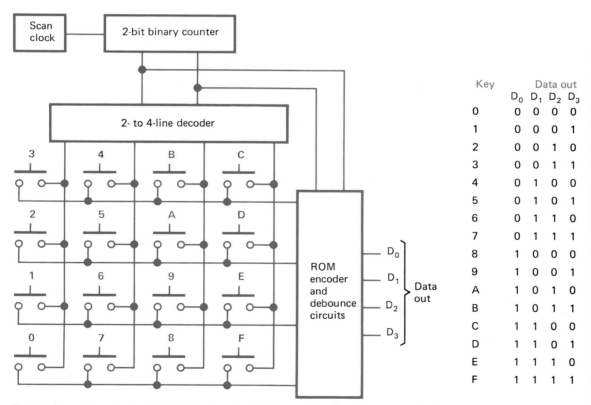

Fig. 13-27   A hexadecimal 4 × 4 keyboard and its encoder. The binary counter is driven by a 400-Hz clock so that the keyboard is scanned once every 10 ms. The debounce logic waits until it detects three contact closures one after the other before loading the data out.

Key	D_0	D_1	D_2	D_3
0	0	0	0	0
1	0	0	0	1
2	0	0	1	0
3	0	0	1	1
4	0	1	0	0
5	0	1	0	1
6	0	1	1	0
7	0	1	1	1
8	1	0	0	0
9	1	0	0	1
A	1	0	1	0
B	1	0	1	1
C	1	1	0	0
D	1	1	0	1
E	1	1	1	0
F	1	1	1	1

nications are so common, a number of special chips for making keyboards have been developed. These custom chips are usually used with intermediate keyboards. That is, they are used with keyboards which are more complex than a simple adding machine keyboard and contain fewer keys than a full alphanumeric keyboard for a personal computer.

Where there is a keyboard, there is usually a *display*. The display presents information using alphanumeric and special characters. Like the keyboard, the display can be one of several different kinds.

First, are the simple numeric-only displays. Seven-segment displays such as those shown in Fig. 13-28(*a*) are often used on calculators and other products which only need a limited display. The seven-segment display can give limited alphabetic ("alpha") information. Some of the alphabetic characters which can be created with a seven-segment display are shown in Fig. 13-28(*b*), which shows the 16 characters used to display hexadecimal information. Using all seven segments in different combinations, the numeric-only display can spell a few words. Some examples are shown in Fig. 13-28(*c*).

The second kind of display uses the 5 × 7 or 5 × 9 *dot matrix*. The 5 × 7 dot matrix can form uppercase alphanumeric characters. The 5 × 9 dot matrix can form both uppercase and lowercase alphanumeric characters. Figure 13-29 shows samples of some of these charac-

ters. Figure 13-29(*a*) shows the 5 × 7 dot matrix and how it forms the upper case characters A, B, C, 7, and %. Figure 13-29(*b*) shows the 5 × 9 dot matrix and how it forms the characters A, a, (, ↑ , and *.

The 5 × 7 and 5 × 9 dot-matrix displays are available in single-character or multiple-character liquid-crystal display (LCD), light-emitting diode (LED), vacuum fluorescent, and neon displays. One of the most common uses of dot-matrix displays is to produce the characters on a cathode-ray tube (CRT). The familiar computer terminal is a good example. The video display on a personal computer also uses a dot matrix to generate the characters in the character mode.

The third kind of display is the *custom display*. Custom displays do not usually display alphanumeric characters. Instead, they show special labels which suit the system functions. Returning to the environmental control system example, you might see panels which light up with labels such as AIR CONDITIONING ON, MOTOR OVERHEAT, WATER HIGH, WATER LOW, or PUMP ON.

These displays may be created in different ways. One common technique is to use a custom LCD. Because the LCD uses a photographic process, it is very easy to have any "segment" of the display be any shape desired. An LCD segment can be a simple straight line in a seven-segment group, or it can be a complete word such as OVERLOAD or

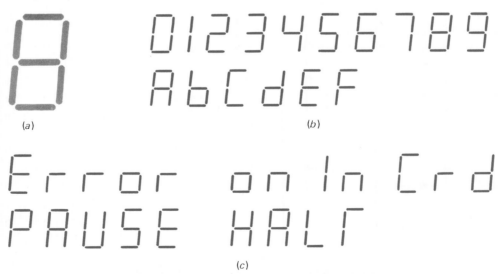

(*a*)   (*b*)

(*c*)

Fig. 13-28 **The seven-segment display.** (*a*) **The basic seven segments.** (*b*) **The hexadecimal characters 0 through F.** (*c*) **A few words that can be spelled using a seven-segment display.**

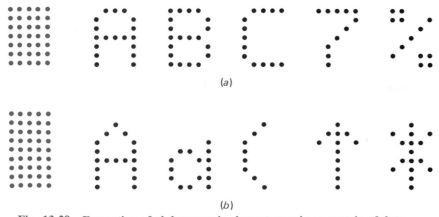

(a)

(b)

Fig. 13-29 Formation of alphanumeric characters using a matrix of dots. (*a*) The 5 × 7 dot matrix, which only can display uppercase characters. (*b*) The popular 5 × 9 dot matrix, which displays the full ASCII character set.

Soft copy

Hard copy

Dumb terminal

Smart terminal

Printer

ERROR. Custom LCDs are very popular for special-purpose microprocessor-based products. A typical example is shown in Fig. 13-30.

Alphanumeric displays are called *soft copy* or *hard copy*. The terminal is a soft-copy display. The terminal has an alphanumeric keyboard and a display built around a CRT. Data is serially transmitted to and from the terminal. The display can show 24 to 48 rows each having 80 to 132 columns of characters. Note, occasionally the terminal is called a CRT, which as you have learned, is really just the name of one of its many components and therefore not a proper name for a terminal.

When a keyboard and a display are not built into a microprocessor-based system, and full alphanumeric input and output are needed, this function is usually performed by adding a terminal to the system. The terminal connects to the microprocessor system via a serial port. There are two different kinds of terminals in common use today. They are called the *dumb terminal* and the *smart terminal*.

The dumb terminal simply paints characters on the CRT as they are received. If the send-ing device fails to issue a carriage return and a linefeed character after the 80th character is sent, the dumb terminal may well fail to display the rest of the characters that are sent. Once the 24th line has been typed on the screen, the dumb terminal usually scrolls the first line of data off of the screen and creates a new 24th line containing the 25th line of data. Dumb terminals usually have simple keyboards which are limited to generating upper- and lowercase alphanumeric characters, punctuation, and a few special symbols.

As the use of microprocessors has become more widespread, the smart terminals have appeared. The smart terminal intelligence comes from one or more microprocessors performing special functions in the terminal. The smart terminal can do some data manipulating and data formatting on its own.

The dumb terminal simply transmits each character as it is typed on the keyboard, and displays each character as it is received. Obviously, the dumb terminal has no intelligence. However, the dumb terminal's logic may be implemented by a microprocessor. What the dumb terminal lacks is smart features.

The third common kind of alphanumeric display is the *printer*. The printer produces hard copy. Hard copy simply means output printed on paper. One of the very common ways to transfer an image to paper uses an inked ribbon. Other printing techniques use ink jets or photocopier (laser) techniques. Many printers use a 7 × 9 dot matrix. The very high speed printers and the printers which produce very high quality characters use a print head with preformed characters like a typewriter.

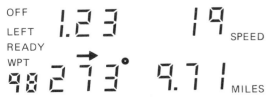

Fig. 13-30 A custom LCD. Note that some of the segments are straight lines used to make up seven-segment characters. Other segments spell out full words. (*Courtesy of Heath Company.*)

Self-Test

Answer the following questions.

37. A terminal displays its output on a CRT. Its local input is from
    a. A 10-digit keypad
    b. An alphanumeric keyboard
    c. A 5 × 7 dot matrix
    d. A 4 × 4 key matrix
38. A special-purpose keyboard has
    a. 10 keys
    b. 16 keys
    c. 57 keys
    d. Any number of keys
39. A terminal uses a 5 × 9 dot matrix to form its character set. You would expect that it could display
    a. The seven-segment numeric characters plus A, B, C, D, E, and F
    b. All the uppercase alphanumeric characters
    c. All the uppercase and lowercase alphanumeric characters.
    d. 132 columns by 48 rows
40. Why do you think keyboards use the row-column scanning circuit instead of using a ROM with one input for each key?
41. How would you use seven-segment displays to display these words?
    a. Open   **e.** And
    b. Run   **f.** Or
    c. Up   **g.** Pulse
    d. Off
42. (*a*) Why can you read the words in Fig. 13-28, even though some of the characters may not be their normal shapes? (*b*) What would happen if some of these characters were displayed alone?
43. Briefly explain the difference between a hard-copy and a soft-copy display. Give an example of each.

13-8 DIGITAL-TO-ANALOG AND ANALOG-TO-DIGITAL INTERFACES

As you have learned in your study of electronics, most real-world signals are analog signals. That is, their amplitude (strength) is between nothing and some high level. You may think of analog signals as the opposite of digital signals, which can only be one of two levels. They are either High or Low.

Typically, transducers generate analog signals. A *transducer* is a device which converts a physical phenomenon into a voltage or current. For example, a thermocouple is a temperature transducer. It converts heat energy into voltage. The voltage which is generated is proportional to the temperature of the thermocouple. Figure 13-31 shows a curve which illustrates how the thermocouple works.

Scientists and engineers use many different transducers to convert physical parameters into electrical signals so that they can measure the physical parameters. For example, there are transducers to measure temperature, position, pressure, flow rate, oxygen content, wind speed, speed, and acceleration, to name a few.

How are these analog transducer outputs converted into digital information for use by a microprocessor-based system? We use an *analog-to-digital (A/D) converter.*

Just as there are transducers which convert physical properties into electrical signals, there are transducers which convert electrical signals into physical phenomena. For example, a motor converts electrical energy into rotary mechanical motion. A linear motor converts electrical signals into linear mechanical motion, and a resistor converts electrical energy into heat. Again, scientists and engineers use transducers to convert electrical signals into physical changes to drive experiments or industrial systems. If these experiments or industrial systems are to be driven from microcomputer or

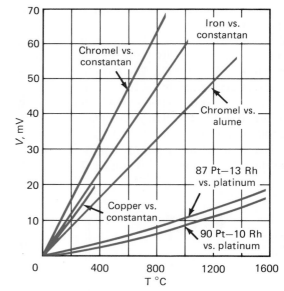

Fig. 13-31 The output voltage of a thermocouple versus temperature. The output voltage is proportional to temperature, but the relationship is nonlinear. It requires some intelligence to convert the voltage into linear temperature units.

other microprocessor-based systems, digital words in the microprocessor must be converted into analog signals to drive the transducers.

How are these digital signals converted into analog signals? The answer is a *digital-to-analog (D/A) converter*.

In this section, we will look at the different ways that A/D and D/A converters are built. We will also look at the ways they are connected to microprocessor-based systems.

Figure 13-32 shows a simple D/A converter. Although this D/A converter only converts 3 bits of digital information into eight different levels, it does illustrate how a D/A converter works. An 8-, 10-, 12-, or even 16-bit D/A converter uses the same principles; there are simply more nodes in the ladder.

As you can see from the diagram, the D/A converter is made up of four basic parts: a voltage reference, a binary resistive ladder, analog switches, and an output amplifier. Simply, the analog switches select the amount of current sent to the operational-amplifier (op-amp) summing junction. The operational amplifier converts this binary weighted current into a binary weighted voltage.

The amount of binary weighted current injected into the summing junction depends on which switches are closed. If no switches are closed, no current flows into the summing junction, and no output voltage is required to generate an equal and opposite current to keep the summing junction balanced.

If the LSB switch is closed, the 1000-Ω resistor is connected from the negative 1-V reference to the op-amp summing function. A current of negative 1 mA flows through the resistor. This is equalized at the summing junction by a 1-mA current coming from the feedback resistor. The output of the op amp must rise to 1 V to generate this 1-mA current.

If the MSB is closed, a 250-Ω resistor is connected to the summing junction. This causes a summing current of negative 4 mA. The output voltage of the op amp must rise to 4 V. If both the LSB and the MSB switches are closed, this generates a summing current of negative 5 mA and the output must rise to 5 V.

This system can generate eight different voltages from 0 to 7 V, depending on which switches are closed. If the D/A analog converter has eight switches and resistors, 256 different voltages can be generated.

Usually, a D/A converter uses the R-2R network shown in Fig. 13-33 on page 392 instead of a series of binary-related resistors. An R-2R network generates the binary currents but does not need precision resistors which vary 256:1. Resistors over this wide a range become very difficult to build.

The accuracy of the D/A conversion depends on the resistor accuracy, the switch quality, the voltage reference source stability and accuracy, and the operational-amplifier quality. Today, extremely high quality D/A converters are built using integrated-circuit technology which combines all of these components in a single IC. One of the important specifications for a D/A converter is how fast

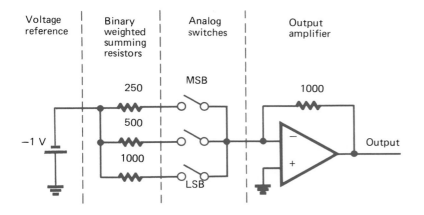

Switch operation			Output
MSB		LSB	Volts
0	0	0	0
0	0	1	1
0	1	0	2
0	1	1	3
1	0	0	4
1	0	1	5
1	1	0	6
1	1	1	7

0 = open
1 = closed

Fig. 13-32   A simple digital-to-analog converter using a switched set of binary weighted resisters to generate the binary weighted analog output. The disadvantage of this converter is the wide range of resistances needed for a high-resolution digital-to-analog converter.

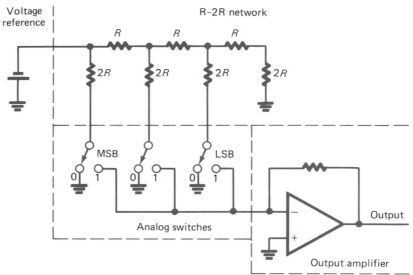

Fig. 13-33 An R-2R digital-to-analog converter. The advantage of this network is that the maximum variation between resistor values is 2:1. Its disadvantage is that it requires single-pole–double-throw switches.

it can respond to changes in the binary signals at its input. Today, integrated-circuit D/A converters can respond to changes in the input binary words in well under 50 ns.

A simple *A/D converter* uses a D/A converter, two voltage comparators, and some control logic, as shown in Fig. 13-34. In this circuit the D/A converter creates a voltage equal to the unknown analog signal. When the two voltages are equal, the output of the OR gate is logic 0. This means the binary number driving the D/A converter generates an analog voltage at the output of the D/A converter equal to the unknown analog voltage. If the

output of the D/A converter is either greater than or less than the unknown voltage, the High and Low comparators tell the A/D converter control logic to correct the 8-bit binary number.

There are two types of A/D converters which use this basic circuit. They are called the *continous balance A/D converter* and the *successive approximation A/D converter*. The only difference between these two types is the algorithm they use to balance the unknown voltage with a known voltage.

The continuous balance A/D converter never stops the conversion process. The binary num-

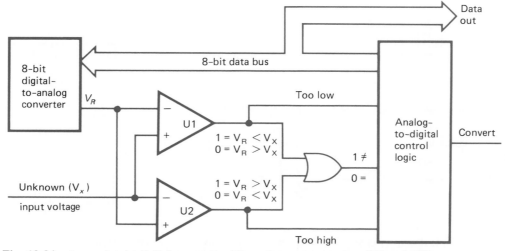

Fig. 13-34 An analog-to-digital converter. The voltage comparators (U1 and U2) are used to tell when the digital-to-analog converter output is as close as it can be to the unknown input voltage.

ber driving the D/A converter is continuously changing to keep the converter balanced. The microprocessor may read the unknown voltage at any time by sampling the digital word driving the D/A converter. The logic to accomplish this may be a special LSI circuit. It is usually not efficient to have the microprocessor running the balance operation.

The successive approximation A/D converter performs a new balancing operation each time the microprocessor requests a measurement. The algorithm is shown in the following steps.

1. The MSB is turned on. If the comparators show that the voltage generated is greater than the unknown voltage, the MSB is turned off. If the comparators show that the voltage generated is less than the unknown voltage, the MSB is left on. This sets the value for the MSB.
2. The second MSB is turned on. If the comparators show that the voltage generated is greater than the unknown voltage, the second MSB is turned off. If the comparators show that the voltage generated is less than the unknown voltage, the second MSB is left on. This process sets the value for the second MSB.

3. The third MSB is turned on. If the comparators show that the voltage generated is greater than the unknown voltage, then the third MSB is turned off. If the comparators show that the voltage generated is less than the unknown voltage, then the third MSB is left on. This process sets the value for the third MSB.

This process continues until each bit is tested and then set or cleared. The process ends once the LSB is tested.

Although the successive approximation A/D converter does not operate quite as quickly as the continuous balance A/D converter, it is a simple algorithm easily run by a microprocessor. It does not require a great deal of external logic and is simpler to implement.

The fastest form of A/D converter is called the *flash converter*. A simple 2-bit flash converter is shown in Fig. 13-35.

As you can see from this diagram, the flash converter simply compares the unknown voltage to the voltages from a binary divider. There is one comparator for each output from the binary resistive divider. A 2-bit flash converter uses three comparators, a 3-bit flash converter uses seven comparators, and an 8-bit flash converter uses 255 comparators.

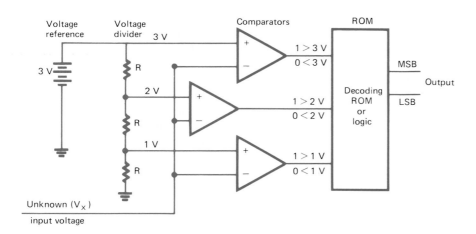

Input	Output	
voltage	LSB	MSB
0 V	0	0
1 V	1	0
2 V	0	1
3 V	1	1

Fig. 13-35 A 2-bit flash converter. This is the fastest analog-to-digital converter design. Its main disadvantage is cost because there are 2^{n-1} voltage comparators and resistors in the reference string for an *n*-bit measurement.

Obviously, an 8-bit flash converter is a very difficult product to build. It takes a large integrated circuit using both analog and digital technology. Flash converters can operate at tens or even hundreds of megasamples per second. They are very fast and equally expensive. Typically, they are used for high-speed instrumentation and other fast operations such as the digitization of video signals which require 15 to 20 megasamples/s.

Both the continuous balance converter and the successive approximation converter take time to operate. This is because the D/A converters can only step from one value to another as fast as the microprocessor generates output instructions and the D/A converter settles. This means an 8-bit successive approximation A/D converter operating at a 1-μs step time requires 8 μs to make a conversion.

During this 8 μs the unknown analog voltage must remain absolutely still. If it changes, the entire measurement can be invalid. For example, suppose the unknown voltage was exactly one LSB above the MSB. And, further suppose that the unknown voltage changes to one LSB below the MSB during the process of measurement. When the measurement starts, the MSB tested first, and it is turned on because the unknown voltage is greater than the MSB voltage. At the end of the measurement, balance never happens because one MSB is now greater than the unknown voltage.

To prevent this from happening, a *sample-and-hold circuit* captures the unknown voltage just before the measurement starts. A sample-and-hold circuit stores the unknown voltage on a capacitor. Using high-speed analog switches, the capacitor is briefly coupled to the unknown voltage. A circuit which performs this function is shown in Fig. 13-36. Once the capacitor is charged to the unknown voltage, the switch is opened. The voltage on the capacitor stays at

this value until the switch closes again. Usually, high-speed field-effect transistor (FET) analog switches are used in these applications.

In fast A/D converter work, the switch may close for only a few hundred nanoseconds, just enough time to sample the unknown voltage. Obviously, the voltage stored on the capacitor is the unknown voltage at the input when the switch closes. Therefore, the shorter the sampling window, the more accurately the sample-and-hold represents the unknown voltage.

Self-Test

Answer the following questions.

44. We must be able to convert analog signals into digital signals and digital signals into analog signals because
    a. Temperature measurements require the use of a thermocouple
    b. Most of the real world is analog, and if we wish to monitor it and control it with a microprocessor, we need to work with analog signals
    c. An analog signal gives us a much higher degree of resolution than a digital signal
    d. All of the above reasons are contributing factors

45. A transducer
    a. Converts an electrical signal into a physical phenomenon
    b. Converts a physical phenomenon into an electrical signal
    c. Is used to control or measure physical phenomena
    d. All of the above

46. An 8-bit D/A converter generates _____ different analog levels, depending on the binary word driving it.
    a. 256   *c.* 128
    b. 8   *d.* 3

47. The basic principle behind a D/A converter is to
    a. Balance an unknown voltage against a known voltage generated from a digital source which always hunts for the closest match
    b. Generate a binary weighted voltage or current by switching a binary resistor network with a digital source
    c. Find the balance between an unknown voltage and a known voltage by coming closer and closer to the unknown voltage in binary steps

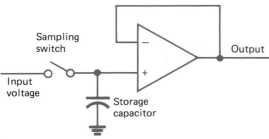

Fig. 13-36 A sample-and-hold circuit. A necessary part of any analog-to-digital converter to keep the unknown voltage steady during the measurement process.

d. Store an unknown voltage so that it will not vary during the measurement process

48. The basic principle behind a continuous balance A/D converter is to
   a. Balance an unknown voltage against a known voltage generated from a digital source which always hunts for the closest match
   b. Generate a binary weighted voltage or current by switching a binary resistor network with a digital source
   c. Find the balance between an unknown voltage and a known voltage by coming closer and closer to the unknown voltage in binary steps
   d. Store an unknown voltage so that it will not vary during the measurement process

49. The basic principle behind a successive approximation A/D converter is to
   a. Balance an unknown voltage against a known voltage generated from a digital source which always hunts for the closest match
   b. Generate a binary weighted voltage or current by switching a binary resistor network with a digital source
   c. Find the balance between an unknown voltage and a known voltage by coming closer and closer to the unknown voltage in binary steps
   d. Store an unknown voltage so that it will not vary during the measurement process

50. The basic principle behind a sample and hold circuit is to
   a. Balance an unknown voltage against a known voltage generated from a digital source which always hunts for the closest match
   b. Generate a binary weighted voltage or current by switching a binary resistor network with a digital source
   c. Find the balance between an unknown voltage and a known voltage by coming closer and closer to the unknown voltage in binary steps
   d. Store an unknown voltage so that it will not vary during the measurement process

51. Briefly explain why a flash A/D converter is so much faster than either the continuous balance or successive approximation A/D converters.

13-9 SPECIAL I/O DEVICES

In this section, we will look at some special I/O devices used with microprocessor-based systems and especially with personal computer systems. Most of these devices make it easier for human beings to communicate with computers. For example, some of the special I/O devices which we will look at generate sounds and even human-sounding speech. These devices can be used to communicate with humans in a situation where reading is difficult, or they can even be used with the visually handicapped. Another area of special I/O devices is called *speech recognition*. Speech recognition circuits allow the human being to input data to the computer using conventional speech rather than a keyboard.

As you know, the most complicated forms of computer interface are those which work with human beings. Most machine interfaces are quite simple when compared to human interfaces. This is because each human being is somewhat different from other human beings. Therefore, each human's method of communication is just a little bit different. This little bit of difference makes it very difficult for the computer to know that each one is trying to say the same thing.

Sound generation is one of the simplest forms of special I/O. The simplest form of sound generation consists of one or more oscillators which may be turned on and off by the computer. Many simple systems use one or two frequencies to alert an operator to a particular condition. Even if a single frequency is used, that frequency may be modulated by turning the signal on and off or by varying its intensity. This form of simple modulation can convey reasonable amounts of information. For example, a series of short beeps can mean one thing, a series of loud long beeps can mean another, and a continuous tone can mean yet another.

The simple tone can be enhanced by creating it with a *voltage-controlled oscillator (VCO)*. The voltage-controlled oscillator changes frequency with an applied voltage. If the voltage-controlled oscillator is driven by a D/A converter, the computer can cause the oscillator to operate over a wide range of frequencies. When this is combined with a programmable attenuator, one can generate a wide range frequencies with a wide range of amplitudes. If a number of these oscillators-attenuators are combined into a system, it is possible to create music and other complicated sounds.

Speech recognition

Voltage-controlled oscillator (VCO)

In order to create complex sounds, the microprocessor must be able to control the frequency of the signal, the purity of the signal (that is, the number and strength of harmonics in the tone), and the shape of the envelope. The envelope is how fast the tone turns on, how long it stays on, and how fast it turns off. These characteristics are referred to as *voicing*. If the microprocessor can fully control the voicing characteristics of a tone generator, extremely complex sounds can be created.

To create sounds like the human voice, special electronic circuits called *voice synthesizers* are used. Researchers have found that the elements of speech can be created by generating a series of sounds called *phonemes* (prounounced fo-nems). Phonemes are the sounds we generate for all of the different combinations of letters. For example, there is a phoneme for the sound created by the *ay* in the word *day*. There is another phoneme for the sound created by the *a* in *dad*.

The electronic circuit which creates these phonemes starts by generating a random noise. The envelope of this random noise is selected from a special table within the speech synthesizer. By carefully controlling the envelope, we can create a wide range of phonemes. These phonemes can be combined to create sounds very much like human speech. Figure 13-37 shows a list of commonly used phonemes, including the phoneme's symbol, the duration of the phoneme, and an example of a word which uses the phoneme.

There are a wide variety of phoneme generators available today. Those which are relatively simple can be understood by a human

Phonemes (organized alphabetically)

Phoneme symbol	Duration (ms)	Example word	Phoneme symbol	Duration (ms)	Example word
A	185	day	K	80	trick
A1	103	made	L	103	land
A2	71	made	M	103	mat
AE	185	dad	N	80	sun
AE1	103	after	NG	121	thing
AH	250	mop	O	185	cold
AH1	146	father	O1	121	aboard
AH2	71	honest	O2	80	for
AW	250	call	OO	185	book
AW1	146	lawful	OO1	103	looking
AW2	90	salty	P	103	past
AY	65	day	R	90	red
B	71	bag	S	90	pass
CH	71	chop	SH	121	shop
D	55	fade	T	71	tap
DT	47	butter	TH	71	thin
E	185	meet	THV	80	the
E1	121	be	U	185	move
EH	185	get	U1	90	you
EH1	121	heavy	UH	185	cup
EH2	71	enlist	UH1	103	uncle
EH3	59	jacket	UH2	71	about
ER	146	bird	UH3	47	mission
F	103	fast	V	71	van
G	71	get	W	80	win
H	71	hello	Y	103	any
I	185	pin	Y1	80	yard
I1	121	inhibit	Z	71	zoo
I2	80	inhibit	ZH	90	azure
I3	55	inhibit			
IU	59	you			
J	47	judge			

Fig. 13-37 **The speech phonemes. These are the different sounds which make up human speech.** *(Courtesy of Heath Company)*

being but have a very computerlike sound. Those which sound fairly human are either quite complex, or need generate only a few phonemes. These generators create only a limited list of words such as the numerals 0 through 9. There is still a great deal of development which must be done in order to produce low-cost, high-quality speech from microprocessor-based systems. However, it is becoming much more common among specialized interfaces and will be used more and more frequently.

The most complex form of microprocessor I/O is speech recognition. As we have just learned, human speech is complex, consisting of many phonemes in many different combinations. As you can imagine, the same phonemes generated by different human beings are quite different from one another. If a speech recognition system is to work properly, the circuits must recognize all of the different ways any one phoneme can sound.

Typically, speech recognition systems must be taught by each operator. The operator repeats a standard phrase or sentence for the speech recognition equipment a number of times. In the process, the speech recognition equipment learns how this particular individual pronounces certain phonemes on that particular day. The process must take into account the speaker's current physical condition. For example, a person with a head cold speaks quite differently than without a cold. This severely affects how the speech recognition system works.

Speech recognition techniques usually attempt to recognize the different phonemes. Alternatively, speech may be filtered into a series of different frequency bands within the audio spectrum. The amount of energy within each of these energy bands and the time at which the energy occurs can be used to identify a particular word.

Speech recognition systems operate very slowly and become even slower when their vocabularies become large. It is, for example, almost impossible with today's technology to expect a microprocessor-based system to recognize the more than 40,000 words used by a technical person describing a complex subject such as microprocessor systems, medical situations, or legal contracts.

There is another group of special I/O devices designed to translate hand movements into microprocessor inputs. Often these I/O devices are found on video games. Two of these, the *joystick* and the *track ball*, are very common.

They are used to move a cursor on a CRT or to control the position of a character.

Both joystick and track-ball movements are converted to mechanical *x* and *y* axes movements. These mechanical movements in the *x* and *y* axes are then converted into digital signals for the microprocessor by a number of different devices. For example, a joystick may control two variable resistors. One changes with motions in the *x* axis and the other with motions in the *y* axis. These variable resistors are connected to a voltage source, so their output voltages are proportional to the *x* and *y* motions. Two A/D converters are used to convert these analog signals into position information for the microprocessor.

Another very common form of converter changes rotary motion into a series of pulses. For example, Fig. 13-38 on page 398 shows a disk with a series of slots in its outer edge. As the disk is turned, the slots permit light from an LED to pass through to a photodetector on the other side. If the disk is turned at a continuous rate, the output of the photodetector is a square wave. The faster the disk is turned, the higher the frequency of the square wave. Therefore, the frequency of the square wave tells us how fast the disk is turning.

How do we know which direction the disk is turning? By adding a second sensor, and positioning the two sensors so that the two square waves have a 90° phase relationship (Fig. 13-38(b)), we can determine direction. If the leading edge of square wave A (1) happens before the leading edge of square wave B (2), the disk is turning clockwise. However, if (2) happens before (1), the disk is turning counterclockwise.

Two of these rotary converters may be used, one in the *x* axis and one in the *y* axis, to indicate the position of a track ball or joystick. Additionally, a rotary converter may be used to indicate the position of a dial, such as the tuning knob on a digitally tuned radio.

Another common form of position encoder for personal computers is the *mouse*. The mouse is simply a track ball which moves with the operator. Usually, the mouse has a serial output. Different characters are generated to indicate movement in the *x* and *y* directions. Additionally, the mouse usually has two or three buttons which the operator can depress to indicate action is to take place. The advantage of a mouse is that it is not built into a surface like a joystick or track ball is to be. All that is required to operate a mouse is a flat surface.

Joystick

Track ball

Mouse

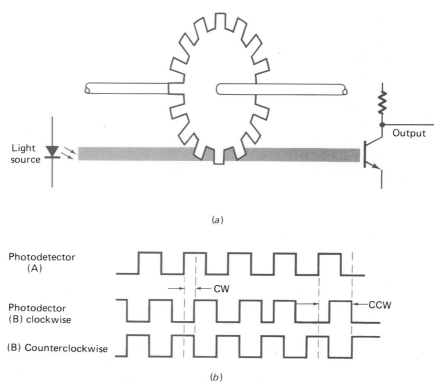

Fig. 13-38 A slotted-disk rotary-motion-to-digital converter. (*a*) The disk, LED (light source), and detector. (*b*) The waveforms from two detectors. The detectors are physically moved until the output waveforms from a rotating disk are 90° out of phase. The microprocessor can then determine both direction and rate.

Positioning devices such as the joystick, track ball, and mouse find their greatest use in systems where the operator receives immediate feedback showing the position of the device. That is, the best application provides relative motion indication rather than absolute position indication. Although there are a number of similar devices, you will find their operation is similar to one of the devices discussed.

Self-Test

Answer the following questions.

52. Speech synthesizers, speech recognition circuits, track balls, and the mouse are all examples of
    a. Special I/O devices which make it easier for people to interface with computers
    b. Devices which use special I/O ports
    c. I/O technology which is like a keyboard
    d. All of the above

53. By varying the _____ and the _____ of a simple tone generator, you can give an operator a wide range of information with sounds. (Choose two of the four answers.)
    a. Frequency   *c*. Voicing
    b. Phonemes   *d*. Duration

54. The sounds we make as we pronounce words are called different
    a. Frequencies
    b. Phonemes
    c. Voicings
    d. All of the above

55. Briefly explain why a speech recognition system needs to be trained regularly by each operator.

56. What is the common purpose of the joystick, track ball, and mouse?

57. A microprocessor-based shortwave receiver uses a knob which will turn through 360° of rotation as many times as you want to tune the radio to the desired frequency. How do you think the shaft rotation is converted into digital information for the radio's control microprocessor?

1. The two commonly used standards to define what bytes mean in data communications are ASCII and EBCDIC.

2. ASCII is the American standard code for information interchange. Seven of the eight bits are used for 128 different characters. The last bit can be fixed, a parity bit to permit some error checking, or it may indicate a different meaning for the first 7 bits.

3. EBCDIC is extended binary-coded decimal interchange code. It is mostly used to communicate with IBM mainframe computers.

4. The way we control and organize communications data is called the communications protocol. Different data communications protocols are used in microprocessor systems. Some are simple, and some are very complicated.

5. Different communications protocols are used to define the hardware or signaling level, the character level, and the control level.

6. Data communications are either synchronous or asynchronous. The data words in a synchronous transmission have a known time between them. The data words in an asynchronous transmission have a random time between them.

7. Most microcomputer-related data communications protocols are based on asynchronous transmissions. Communications with mainframe computers often use synchronous communications.

8. The parallel I/O port is the most basic I/O port. All other types of I/O ports build on a parallel I/O port.

9. The parallel I/O port uses many of the same parts as a memory system, such as address bus buffers, data bus buffers, and address decoding logic.

10. The parallel I/O port's status register stores information the microprocessor reads to find the condition of the port.

11. The parallel port uses handshaking lines to control the flow of data into and out of the I/O port. These lines are used to tell an external device that the data at the I/O port output is valid or, for an external device, to tell the I/O port it has data to transfer into the I/O port.

12. Most parallel I/O ports use simple TTL drivers. These can drive 1 or 2 m of cable connecting the I/O signals to an external device.

13. The IEEE-488 standard defines a special parallel I/O bus used to transfer data with instruments. Each device on the IEEE-488 bus has an address, and each device is a talker, listener, or a controller.

14. The Centronics parallel interface is a standard parallel I/O port used on personal computers. Often this I/O port is used to send data to a printer.

15. The SCSI (small computer systems interface) is a parallel I/O system which allows a microcomputer system to address and transfer data with a number of peripheral devices. Usually, the SCSI interface is used to connect a microcomputer to mass storage devices.

16. Serial communications are used because they are a low-cost way to send data over long distances.

17. To send data over a serial transmission line, we must convert the parallel data in the microprocessor to serial data. This is done with a device called a UART.

18. The UART (universal asynchronous receiver-transmitter) is built around a shift register which performs the parallel-to-serial and serial-to-parallel conversions.

19. A UART adds start and stop bits to 8-bit data words. When receiving data, the UART uses the start bit to tell that another data word is coming. The stop bit tells the UART that the word is done.

20. The data signaling rate in bits per second is called the baud rate. The baud rate only tells you the signaling rate within a single character for asynchronous transmissions.

21. If you are sending 7-bit ASCII data, the UART can add a parity bit for error checking. A UART may be set to generate and check for either even or odd parity.

22. A parity check can only catch odd (50 percent) errors. Any even number of errors will not be caught by the parity check.

23. The UART status register is used to control and monitor the UART. Status bits are used to indicate data overrun, framing error, parity error, transmit buffer empty, and data available. Control bits let you program the data word length, odd or even parity, parity on or off, and 1 or 2 stop bits.

24. UARTs are driven with a clock signal which is 16 or 64 times greater than the baud rate. This makes sure that although the UART timing is asynchronous to the incoming data, it will never be more than 1/16 of a bit width out of synchronization with the incoming data.

25. The time between data words in an asynchronous transmission is random. The microprocessor uses a combination of interrupts and software routines to synchronize to this random timing.

26. The most common serial transmission line is the one which meets the Electronic Industries Association (EIA) standard RS-232. An RS-232 signal is above +3 V for a logic 1 and below −3 V for a logic 0.

27. The RS-232 standard also specifies a number of handshaking lines to control the data flow. These lines are in parallel to the transmit and receive data lines in an RS-232 cable.

28. An RS-232 signal can drive a 50-ft cable at 9600 baud.

29. An RS-422 data communications signal uses a differential signal. The transmit and receive differential signals are sent over two twisted pairs in the data communications cable.

30. The twisted pair and differential signaling give RS-422 good noise immunity and the ability to drive cables as long as 4000 ft at 38.4K baud.

31. The 20-mA current loop originated with teleprinters, but is still used today for some industrial applications and to interface microprocessor-based equipment with older devices.

32. The 20-mA loop signaling makes and breaks the current loop. The system has only a single wire, so all control information must be sent as data.

33. A modem is used to signal over long distances and through transmission systems which are ac-coupled. Modem is a contraction of the names modulator and demodulator, which are the two main circuits in a modem.

34. The modulator generates a different tone for logic 1s and logic 0s. The demodulator detects these different tones and converts them back into the logic signals.

35. The Bell 103 standard uses two tone sets, one for originate and one for answer. The tones are frequency modulated to carry the data. Usually, this modem is used at either 110 or 300 baud.

36. A modem is full-duplex when the two tone sets work at different frequencies so that they both can be sent over the same transmission line at the same time.

37. The Bell 212 modem can work at 1200 or 2400 baud using QPSK (quadrature phase shift keying). This allows the user to transmit data at 120 or 240 characters/s.

38. Because a modem has no control lines which are parallel to the data line, all control signaling from modem to modem uses control characters mixed into the data stream. This increases the data communications overhead.

39. When microprocessor-based devices communicate with people, they often use displays and keyboards with alphanumeric characters.

40. The simplest keyboards and displays only work with numeric data. A 10-key keyboard and a seven-segment display perform this function.

41. If a device needs to communicate with full alphanumeric data, the keyboard becomes much more complex and the display often forms the characters using a 5 × 7 or 5 × 9 dot matrix. The terminal is a common example of this kind of I/O device.

42. Special keyboards and displays are often used with microprocessor-based systems which perform a special-, instead of a general-purpose, job.

43. Keyboards are usually constructed in a row-column matrix.

44. Real-world signals are usually analog signals. Therefore, we need analog-to-digital converters to input analog data into the microprocessor-based system and digital-to-analog converters to output analog data to the real world.

45. Transducers convert physical parameters into electrical signals and electrical signals into physical parameters. Microprocessors are often hooked to transducers.

46. A D/A converter uses a switched binary resistive ladder to generate switched binary currents. These currents are connected to the summing junction of an operational amplifier to produce a binary stepped voltage.

47. The binary weighted resistor network is either a number of binary-related resistors or an R-2R network.

48. Digital-to-analog converters are often made as integrated circuits to give fast settling times and precisely matched parts.

49. A very common form of A/D converter uses a D/A converter, voltage comparators, and logic. A measurement of the unknown voltage is made by detecting when the output of the D/A converter is equal to the unknown voltage.

50. The continuous balance A/D converter tracks the unknown voltage so that the binary word driving the D/A converter always represents the unknown voltage.

51. The successive approximation A/D converter finds the unknown voltage by comparing the D/A converter output to the unknown voltage for each bit starting with the MSB. Every bit which is less than or equal to the unknown voltage is kept.

52. The flash converter is a very fast A/D converter. It uses $n-1$ voltage comparators to compare the unknown voltage to an $n-1$-step voltage divider for a 2^n-bit measurement.

53. A sample-and-hold circuit is used to take a ''snapshot'' of the unknown voltage. This makes sure that the unknown voltage remains steady during the A/D conversion process, because many A/D converters give incorrect results if the unknown voltage changes during the measurement process.

54. Sound can be used as a special computer I/O device.

55. A simple tone generator can be used to send information by turning the tone on and off in different ways.

56. If you can change the tone frequency and control the tone's envelope, you can produce complex sounds like music or human speech.

57. Some speech synthesizers work by generating sounds called phonemes. A number of phonemes can be put together to make a word.

58. Speech recognition is a very complex form of computer I/O because each person pronounces a word differently from the next person and the sound pattern may change from day to day.

59. The joystick, track ball, and mouse are used to convert hand motions into computer input. These devices convert motion into x axis and y axis motions which are converted into proportional electrical signals.

60. Typically, the joystick, track ball, or mouse are used to give relative motion information to a cursor or character on a display.

61. A slotted disk is used to interrupt one or two light beams which hit photodetectors. The output of the photodetectors is used to tell if the shaft connected to the slotted disk is turning, how fast it is turning, and in what direction it is turning.

CHAPTER REVIEW QUESTIONS

Answer the following questions.

13-1. ASCII is the acronym for American standard code for information interchange. ASCII is a
   a. Code which computers can use to communicate with each other, giving predefined meanings to each character
   b. 7-bit code for alphanumeric, control, punctuation, mathematical operators, and a few other symbols
   c. Set of 128 characters often used for data communications
   d. All of the above

13-2. EBCDIC is the acronym for extended binary-coded decimal information code. EBCDIC is a
   a. Code which computers can use to communicate with each other, giving predefined meanings to each character
   b. 6-bit code for alphanumeric, control, punctuation, mathematical operators and a few other symbols
   c. Set of 256 characters often used for data communications
   d. All of the above

13-3. One way that a microcomputer can communicate with another device is to use the ACK and NAK control characters to acknowledge or not acknowledge each block of data characters transmitted. This system is called
  a. A powerful error-checking system
  b. A communications protocol
  c. An ASCII transmission
  d. An EBCDIC transmission

13-4. If you are using synchronous data communications, and you receive a single data word, you know the exact time when the next data word is to come. If you are using asynchronous data communications, you know that the
  a. Next character is an ASCII XON (Control Q)
  b. Next character is an ASCII character
  c. Time to the next character is random
  d. All of the above

13-5. Most _____ data communications are based on asynchronous transmissions.
  a. Mainframe-to-microcomputer
  b. Mainframe-to-terminal
  c. Voice
  d. Microcomputer-related

13-6. If you look carefully at the circuits which make up almost any kind of I/O port, you will find
  a. A parallel I/O port
  b. A UART or USART
  c. 4 bits of memory address decoding
  d. An 8-bit data latch to hold incoming data

13-7. A parallel I/O port is likely to have multiple I/O addresses because
  a. It may be used for different purposes
  b. Separate addresses are needed for the data and status registers
  c. More than 8 bits of data may be needed
  d. All of the above

13-8. A parallel port status register is used to let
  a. The microprocessor have an alternative place to store its status register information during an interrupt
  b. The programmer monitor and control the parallel I/O port
  c. Fast data transfers happen without interrupts
  d. TTL outputs drive long transmission lines

13-9. The _____ are used to control the flow of data into and out of the parallel I/O port external connections.
  a. Address decoders   *c.* High-order address bits
  b. Output drivers   *d.* Handshaking lines

13-10. A simple TTL output from a parallel I/O port can be expected to drive
  a. A long-distance telephone line over the switched telephone network
  b. An IEEE-488 bus
  c. 1- to 2-m cable
  d. A GPIB

13-11. You would use _____ parallel I/O bus to program and retrieve data from a group of electronic instruments.
  a. A Centronics   *c.* A SCSI
  b. A simple TTL   *d.* The IEEE-488

13-12. _____ interface is often used by personal computers to send data to a printer.
  a. A Centronics   *c.* A SCSI
  b. A simple TTL   *d.* The IEEE-488

13-13. Communications between a microcomputer and its addressable mass storage devices often use _____ interface.
  - *a.* A Centronics
  - *b.* A simple TTL
  - *c.* A SCSI
  - *d.* The IEEE-488

13-14. Serial data communications are often used because
  - *a.* It is not practical to build parallel drivers to work long distances
  - *b.* They reduce the need for auto equalization circuits
  - *c.* It is the simplest way to communicate over long distances, including the switched telephone network
  - *d.* All of the above

13-15. A UART (universal asynchronous receiver-transmitter) is used to
  - *a.* Convert parallel data into serial data
  - *b.* Convert serial data into parallel data
  - *c.* Add start and stop bits to the serial transmission
  - *d.* All of the above

13-16. In serial data communications, we use the term *baud rate* to
  - *a.* Tell the long-term average number of bits per second we are sending
  - *b.* Tell the signaling rate of an individual character
  - *c.* Indicate that the 5-level (Baudot) code is being used
  - *d.* Indicate transmissions over 600 baud are coming

13-17. The normal ASCII characters are 7 bits long. Normally, we use 8-bit data transmissions. The extra bit is used
  - *a.* To indicate overflow
  - *b.* For added data when 7 bits are not enough
  - *c.* For a parity (error-checking) bit
  - *d.* All of the above

13-18. A UART can be used to generate and check for
  - *a.* Overflow errors
  - *b.* Odd or even parity
  - *c.* The wrong baud rate
  - *d.* All of the above

13-19. The clock signal for a UART is usually 16 or _____ times the baud rate.
  - *a.* 10
  - *b.* 128
  - *c.* 64
  - *d.* 100

13-20. The microprocessor uses a software routine or a combination of interrupts and a software routine to synchronize with
  - *a.* Asynchronous data communications
  - *b.* The memory address bus
  - *c.* ASCII transmissions
  - *d.* Modems using QPSK modulation

13-21. The EIA standard for a serial transmission which swings at least from +3 V to −3 V on separate transmit and receive lines and may include a number of different handshaking lines is
  - *a.* RS-422
  - *b.* IEEE-488
  - *c.* RS-232
  - *d.* 20-mA current loop

13-22. If you are required to install a serial transmission line which is to operate at 19,200 baud at a distance of 0.5 mile, you would use
  - *a.* RS-422
  - *b.* IEEE-488
  - *c.* RS-232
  - *d.* 20-mA current loop

13-23. You are working on an older data acquisition system in a factory. The main I/O device for the system is an old teleprinter which is connected to the microprocessor-based controller with a two-wire cable using an odd connector. This leads you to believe that the data communications lines are
  - *a.* RS-422
  - *b.* IEEE-488
  - *c.* RS-232
  - *d.* 20-mA current loop

CHAPTER 14

An Introduction to Programming

CHAPTER OBJECTIVES

This chapter will help you to:

1. *Understand* the basic steps in programming.
2. *Solve* a word problem using programming steps.
3. *Name* and *explain* the essential steps of programming development.
4. *Define* and *use* algorithms, data, and programming constructs.
5. *Recognize* the importance of documentation in the programming process.

A program is used to tell the microprocessor-based system what it is to do. The process of developing the set of instructions which tell the microprocessor what to do is called *programming*. This process can be broken down into four steps: developing the specification, creating the design, coding the program, and testing and debugging the design. Not only is the process of developing a program very structured, but the program itself is very structured. It is this structure which allows programmers to create very large and complex programs with a high degree of success.

14-1 WHAT IS PROGRAMMING?

Often programming is treated as if it were "black magic." Nothing could be further from the truth. Programming is, very simply, the process of telling the processor exactly how to solve a problem. To do this, the programmer must "speak" to the processor in a language which the processor understands.

As you have learned from earlier chapters, the processor faithfully performs each instruction that the programmer gives it. However, the processor will do nothing that the programmer does not instruct it to do. The process of programming, therefore, must be very exact. The programmer must tell the processor absolutely everything that it must do and exactly how to do everything, step by step.

There is one part of the programming process that you must understand thoroughly before you begin to learn how to program. *Programming a processor cannot make the processor solve a problem if you, as the pro-* *grammer, do not know how to solve the problem.* Of course, this does not mean that you must know the answer to the problem in advance. What you must know and what you must tell the processor is how the problem can be solved. It is the programmer's job to tell the processor how to solve the problem.

For example, if you can express a problem as the equation

$$X = A + B$$

and you can supply input for all the equation's variables (A and B), the processor can compute the answer. However, the processor cannot tell you that you need to use the equation $X = A + B$ to solve the problem.

To program a processor, you must organize the problem so that the processor can solve the problem. In some cases, organizing the problem may be as simple as converting the problem stated in words into an algebraic expression. In other cases, you start with the desired result and what you know about the inputs which cause this result. From this information you can develop a set of logical operations which make the result you want.

The first step in the programming process is finding out what job is to be done. This is called *specifying the program*. If the programmer does not understand what is to be done, the programming process cannot begin. Frequently, lack of a complete specification is one of the major problems with a program design.

Organizing the problem so a computer can solve it is the major task in programming. This is called *designing the program*. During this process, the programmer must decide exactly how to approach the problem. In the design process, the exact step-by-step process that will be followed is developed and written down. There will usually be a number of ''right'' ways to solve the problem.

Once the program is specified and designed, it can be implemented. Implementation starts with the process of *coding* the program. To code a program is to state it in computer language. The computer language lets you put the design into specific instructions that the processor can follow. Each computer (or microcomputer or microprocessor) language has a set of instructions that you must choose from. Some languages have very English-like instructions and some have very machinelike instructions. The instructions to solve a certain problem are called *source code*. A special computer program translates source code statements in a given language into the binary instructions which can be executed by the processor. These binary instructions are called *object code*.

Once the program or a part of the program is coded, the next step is *debugging* the code. Debugging is the process of testing the code to see if it does the job you wanted it to and, if it does not work properly, fixing the errors.

Debugging is done by loading the code into the target processor and attempting to execute the program. Programmers often break programs into small parts. Each part is then debugged individually. After each part is debugged, more parts are added until the entire program is tested and debugged.

Let us look at an example of programming. To start with, we use a story problem. A *story problem* is a problem expressed in words. For example:

The customer requires a table of data points which will be plotted on an *x-y* graph. The

relationship between the points is such that the *y* axis data are the square of the *x* axis data. The table of data points should be assembled for integer values of *x* lying between -3 and $+3$ inclusively.

Fortunately, we can easily reduce this long problem to a simple equation. The equation and the limits of the data become the specification. The equation is

$$y = x^2 \quad \text{for} \quad x \text{ (integer)} = -3 \text{ to } +3$$

The design of our program is

1. Read the first element of data (-3)
2. Set *y* equal to x^2 (*x* times *x*)
3. Print the computed value of *y* and the current value of *x*
4. Test: IF Data = 3
   THEN End the program
   ELSE Repeat the program using the next data element
5. The data are to be $-3, -2, -1, 0, 1, 2, 3$

To implement this program, we will code the program in the computer language BASIC. The initially coded implementation is shown in Fig. 14-1(*a*) on page 410.

To debug this program, using BASIC, we run (execute) the program. The first time we run the program, we are told there is an error in line 40. The word THEN is spelled TH3N, and the computer cannot interpret it. The exact command is needed, so we change the program as shown in Fig. 14-1(*b*). With this error corrected, the program runs successfully, and the programming process is complete.

As you see, this is a very simple program. Figure 14-1(*c*) shows the printed data table from this program. Figure 14-1(*d*) shows a handmade plot of the data. The curve is the familiar parabola.

Reviewing this example, you can see that the story problem was first converted into an equation with data limits. This is the specification. The design outlines an approach to solve this problem on a computer. The design is then coded in instructions for the processor to follow. In this case, we used the high-level language BASIC.

The first time the program was executed, errors were found. These errors were debugged, and the program run again. The second time the program was run, it was error-free. This finished the implementation process. Final ex-

From page 408:

Programming

On this page:

Specifying the program

Designing the program

Coding

Source code

Object code

Debugging

Story problem

```
10      READ X
20      LET Y = X * X
30      PRINT Y,X
40      IF X = 3 TH3N 50 ELSE 10
50      END
60      DATA −3, −2, −1,0,1,2,3
```

(a)

```
10      READ X
20      LET Y = X * X
30      PRINT Y,X
40      IF X = 3 THEN 50 ELSE 10
50      END
60      DATA −3, −2, −1,0,1,2,3
```

(b)

X	Y
−3	9
−2	4
−1	1
0	0
+1	1
+2	4
+3	9

(c)

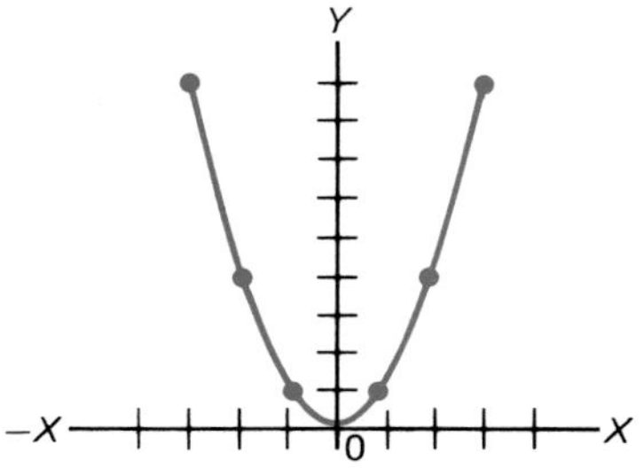

(d)

Fig. 14-1   A simple program to develop data for a graph. This program is written in the computer language BASIC. (a) The program with an error. (b) The program with the error corrected. (c) The data output printed when the program is executed (RUN). (d) A plot of the x and y data done by hand.

ecution of the program resulted in a data table. The data were then plotted by hand.

Let's look at another example. This one does not solve an equation but uses logic. The story problem is

Clear 4096 memory locations by setting each bit to logic 0. The range to clear starts at memory location 8000 hex.

The specification is

Clear memory locations 8000H to 9000H. Clear is defined as 0000H.

To solve this problem, we use the microprocessor's data-handling ability. We store data words consisting of all logic 0s in successive memory locations. We will take words with all logic 0s from the microprocessor's accumulator. The design is this idea and the following sequence of operations:

1. Set the contents of a register to all logic 0s (register B).
2. Point the memory address register to the first memory location (8000H).
3. Store the contents of the register in the memory location pointed to by the memory address register. This clears that memory location because it now has all logic 0s stored in it.
4. Point the memory address register to the next memory location.
5. Test to see if the memory address register is pointing to location 9000H.
6. If "yes," halt. If "no," return to Step 3. Repeat until done.

You can see that this program will continue executing until logic 0s are placed in all 4096 memory locations. Once the memory address register points to location 9000H, the program halts.

To implement this program, you could choose to use *assembly language* for the particular target microprocessor. Figure 14-2(*a*) shows the source code for this program using the Z80 8-bit microprocessor's assembly language.

Once the coding is complete, the program is assembled, creating machine-readable object code from the source code. The object code is then loaded into the target processor and we attempt to execute the program. Figure 14-2(*b*), on page 412, shows the assembled listing for this program. Note that the assembled listing shows the program is to be loaded into memory locations 7000H to 700DH. The listing gives hex memory location and the hex value for each instruction in addition to the source code.

The process of testing the program continues until we are sure that it is error-free. When the program is shown to be error-free, the programming process is complete.

In this last example, we converted a problem into a series of logical instructions that move data. In the earlier example, we calculated the data points for the parabola by using mathematical instructions. In both examples you can see that the task is to convert a written problem description into instructions the processor can follow. To implement these programs, we code them in a computer language.

The words, punctuation, and organization of the statements in a computer language follow exact rules. This is an important point to understand. We all use language to communicate with each other. However, our language is not as exact as a computer language. Communicating with other human beings, especially when we are speaking, gives us a great

```
            ORG     7000H       ;Load the program into
                                ;memory starting at memory
                                ;location 7000 Hex
START       MVI B,00H           ;Put 00 Hex into Register B
            MVI A,90H           ;Put 90 Hex into Register A
                                ;Later we will compare the
                                ;HI byte of the HL Register
                                ;pair with this value
            LXI H,8000H         ;Load the HL (Memory Pointer)
                                ;with the starting address for
                                ;the "clear" operation (8000 Hex)
CONT        MOV M,B             ;Move the contents of
                                ;Register B to the memory
                                ;location pointed to by the
                                ;HL register pair. This
                                ;"Clears" this location
            INX H               ;Increment the HL Register
                                ;Pair so that it points to
                                ;the next memory location
            CMP H               ;Compare the HI byte of the
                                ;HL Register Pair with the
                                ;accumulator. If the HI
                                ;byte equals 90 Hex, 4096
                                ;locations have been cleared.
                                ;Note: this sets the status
                                ;register's Zero bit to 1.
            JNZ CONT            ;If the Zero bit is not 1,
                                ;jump back to CONT. If the
                                ;Zero bit is 1, execute the
                                ;next instruction
            HLT                 ;End the program
            END                 ;End the assembly
```

(a)

Fig. 14-2 (a) Source code for a Z80 assembly language program to clear a series of memory locations.

deal of freedom to change the structure of the language and the words we use. For example, you can mispronounce a word, and there is a good possibility that the person you are speaking to will understand you.

As a programmer, you cannot do this. Each word in a computer language has a specific meaning. You must use the word exactly if the computer is to understand the word. When used correctly, the word causes its translator to generate a specific set of instructions for the target processor. Likewise, the punctuation, the organization of the words, and the objects the words act on must be put together following the exact rules of the particular computer language.

Self-Test

Answer the following questions.

1. Software refers to a microprocessor's
   a. Read-only memories
   b. Magnetic tapes
   c. Program routines
   d. Paper tapes

2. The basic purpose of programming is to convert a problem into
   a. A series of instructions in the computer's own language
   b. An equation that expresses the problem in mathematical terms
   c. A hardware equivalent
   d. A curve that graphs the data versus the result

3. The first step in the programming process is
   a. Coding
   b. Specifying
   c. Designing
   d. Debugging

4. The second step in the programming process is
   a. Coding   c. Designing
   b. Specifying   d. Debugging

5. The third and fourth steps in the programming process are interactive. They are _____ and _____ . (Choose two of the four answers.)
   a. Coding   c. Designing
   b. Specifying   d. Debugging

```
7000                              ORG    7000H    ;Load the program into
                                                  ;memory starting at memory
                                                  ;location 7000 Hex
7000 0600      START     MVI B,00H                ;Put 00 Hex into Register B
7002 3E90                MVI A,90H                 ;Put 90 Hex into Register A
                                                  ;Later we will compare the
                                                  ;HI byte of the HL Register
                                                  ;pair with this value
7004 210080              LXI H,8000H               ;Load the HL (Memory Pointer)
                                                  ;with the starting address for
                                                  ;the "clear" operation (8000 Hex)
7007 70        CONT      MOV M,B                   ;Move the contents of
                                                  ;Register B to the memory
                                                  ;location pointed to by the
                                                  ;HL register pair. This
                                                  ;"Clears" this location
7008 23                  INX H                     ;Increment the HL Register
                                                  ;Pair so that it points to
                                                  ;the next memory location
7009 BC                  CMP H                     ;Compare the HI byte of the
                                                  ;HL Register Pair with the
                                                  ;accumulator. If the HI
                                                  ;byte equals 90 Hex, 4096
                                                  ;locations have been cleared.
                                                  ;Note: this sets the status
                                                  ;register's Zero bit to 1.
700A C20770              JNZ CONT                  ;If the Zero bit is not 1,
                                                  ;jump back to CONT. If the
                                                  ;Zero bit is 1, execute the
                                                  ;next instruction
700D 76                  HLT                       ;End the program
700E                     END                       ;End the assembly
```

(b)

Fig. 14-2 *(b)* **The assembled output which shows the memory locations and data in hexadecimal form generated by the program. Note that the assembled output includes the original source code and the comments.**

6. Briefly explain what happens in each of the following steps in the programming process.
   a. Coding   c. Designing
   b. Specifying   d. Debugging
7. Once the problem is converted into a series of tasks that can be done by a processor, you then must
   a. Generate software
   b. Code the program into the computer's language
   c. Be sure you have a correct story problem
   d. Run the program.
8. Explain why a processor cannot solve a problem if you do not know how to solve the problem.
9. A program can be written to use _____ to solve a problem.
   a. Logic
   b. Mathematical formulas
   c. Logic and mathematical formulas
   d. External interrupts
10. Briefly explain why the words, punctuation and organization of the statements in a computer language must follow exact rules.

14-2 THE PROGRAMMING PROCESS

In the previous section, we learned what a program does, what it can do, and what it cannot do. We looked at two examples which showed the basic steps needed to write a program. These three basic steps are:

Writing a specification
Creating a design
Implementing the design in code and testing and correcting the implementation

The size of a computer program can vary from a few lines which you quickly create to solve an immediate problem to programs which take many thousands of lines of source code and many years of software engineering to complete. In every case, you must follow each of these steps. If you are part of a team creating a large program, the time spent on each step is very formal and very clear. For example, people may be assigned to write the specification. This may take weeks. What they are doing during this time is quite clear.

On the other hand, you may need to write a short five-line program to prepare a report on your experiment in the physics laboratory. In this example, the time spent in writing the specification is very short and difficult to separate from some of the other things you have to do. However, if you look carefully at the process you used to create and write the program, you will find that you followed each of the basic steps in the programming process.

In the following sections, we will look at each of these steps. We will learn what is to be accomplished in each of these steps, and understand why it is important that one step be finished before we move on to the next step. Once you understand the process, you then can spend as much time as you need performing each step in the process as you create a new program.

14-3 THE PROGRAM'S SPECIFICATION

The purpose of a *specification* is to make sure there is a clear understanding of what the program is to do. As we learned in the two examples, most programming requirements are first given as story problems. Typically, a story problem is not a complete specification. Frequently, story problems are not complete nor are they well organized. A really well organized and complete story problem can be a specification.

A written specification is usually only needed when the program is large or when you are writing a program for someone else. If you write a specification, it should be reviewed and approved by the person asking you to write the program. This makes sure that:

The person asking for the program tells you what it is to do
You understood what they wanted well enough to write the requirements
The person asking for the program knows you know what the program is to do

Often the story problem does not answer all the questions. You may be able to supply the answers and include them in the specification, or you may need to go back for the answers to the person wanting the program.

For example, you have been asked to write part of a program for a microprocessor-based shortwave receiver. You are told that the receiver tunes from 1.500 to 30.000 MHz in 1-kHz steps, that there is a beep each time a

key on the keyboard is depressed, and that the keyboard and accompanying display look like the ones shown in the sketch in Fig. 14-3. The up-arrow key and down-arrow key "tune" the receiver to a frequency higher or lower than the current setting.

Your part of the software program is to do three things. First, to read frequency entries made by the user at the receiver's keyboard. Second, to generate a 16-bit binary number which controls the receiver's frequency synthesizer. The receiver's frequency synthesizer tunes the receiver to the frequency the user enters. Third, to generate the liquid crystal display (LCD) driver information from the keyboard entries.

You are assigned the task of writing the specification for this part of the overall design. As you review the story problem, you can see that, although most of the desired results are covered, there are some unstated results and some unanswered questions.

For example, the story problem includes a front-panel sketch. The front-panel sketch shows numeric keys 0 to 9, a decimal point, the up-arrow and down-arrow keys, an Enter key, and a Clear key. From this information, you presume that once the user keys in a frequency, the Enter key tells the receiver to tune to this frequency. You also presume that if the user makes a mistake during frequency entry, the Clear key is used to start over. Further, you presume that all frequency entries must use the decimal point key.

On the other hand, there are some requirements for which you cannot assume the an-

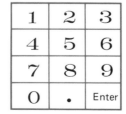

Fig. 14-3 A sketch of the keyboard and display for a microprocessor-based shortwave receiver. This sketch is part of the original story problem from the person asking for the programming effort. A great deal of information about the software can be inferred from this sketch.

swers. For example, nowhere in the story problem is there a table which shows you the 16-bit synthesizer control number which must be generated for each frequency. Also, there is no similar table to tell you the information which must be sent to the LCD driver. You will get these tables from the hardware engineers designing those parts of the receiver. This is the kind of information that is known by engineers, not by the person specifying the receiver.

After some further thinking, you find that you are not told what happens if the customer enters an out-of-band frequency. You know that the receiver tunes from 1.500 to 30.000 MHz. For example, what do you do if the customer enters 45.050 MHz? There are different ways an error of this type could be handled. The choice of which one to use is a marketing question.

In this case, you go back to the person who gave you the original story problem and ask what display is wanted if an out-of-band frequency is entered. You suggest that one of the following displays is easy to present as soon as the out-of-band frequency is entered.

1. Show the word ERROR either blinking or steady
2. Display the wrong frequency as blinking numbers
3. Return the display to the previous frequency
4. Sound the beeper three times to alert the user

After discussion, it is decided that the display will show the word ERROR blinking on and off in half-second intervals for 3 s, that the beeper will sound each time the word ERROR blinks on, and that after the 3-s interval, the display will return to the frequency that the receiver was tuned to before the out-of-band frequency was entered.

As you can see from the above example, a specification for what initially appeared to be a rather simple task is quite important. The story problem left some important questions unanswered. It is the programmer's job to carefully review the story program to be sure that all the different situations are covered.

From the above example you can see that the story problem has specifications presented in many different ways. Some specifications are detailed in the original narrative, and some are in the form of drawings. For example, you

learned that the display is made up of seven-segment characters by looking at the drawing. Some specifications are implied. For example, the use of the Enter, Clear, and decimal point keys are not documented anywhere. You just understood how they are to be used by looking at the drawing.

Other specifications, such as the table of synthesizer control codes, must come from other technical areas within the development process. Some specifications, such as the error handling, must come from the individual specifying the product, but may not be thought of until the programmer asks the questions.

It may seem that some of the questions raised here could be answered just as easily during the design or implementation processes. However, that may be too late in the development process to reach a satisfactory answer.

For example, the programmer gave the originator several options for responding to out-of-band frequency entries. If this discussion had not occurred until code was being written, the option of displaying a blinking ERROR message might have been eliminated because of the work already done. The result might have been simply a failure to accept the out-of-band frequency entry, leaving the user puzzled because the entry was not accepted and there was no explanation.

You should make every effort to create a complete specification before the design process starts. However, it is almost impossible to be totally complete except on relatively simple projects. This is why the specification must be a working document. A working document is regularly updated. If during the design or implementation, you find that the specification is incomplete or cannot be implemented as called for, the specification should be changed to show what is really going to happen.

Self-Test

Answer the following questions.

11. Briefly state the purpose of creating a specification for a program.
12. When should you use a written specification?
13. Why should the specification go back and forth between the person who wants the program written and the person writing the program?
14. You are given a story problem which explains a job to be done by a microproces-

In this section we have looked at four universal programming constructs. It is said that all programs can be created using these four constructs. Actually, all programs could be created using the first two constructs; however, the addition of the two extra constructs makes programming a great deal simpler. Once again, the use of these standard constructs makes describing the design of a program, maintaining the program, and debugging the program much simpler. The use of these constructs also leads to very cohesive program modules, reinforcing the requirement for creating cohesive program modules in a highly structured program.

Self-Test

Answer the following questions.

36. Modern programming practice uses fundamental programming constructs to build programs. This is another example of
    a. The use of data structures
    b. The use of algorithms
    c. Modular programming techniques
    d. Good programming for microprocessor-based clocks
37. The programming construct which places one program step after the next with no decision or branching is called the _____ construct.
    a. CASE
    b. IF-THEN-ELSE
    c. DO-UNTIL/DO-WHILE
    d. Sequential
38. The programming construct which includes a test and allows the program flow to branch in one of two directions is called the _____ construct.
    a. CASE
    b. IF-THEN-ELSE
    c. DO-UNTIL/DO-WHILE
    d. Sequential
39. The programming construct which includes a test and allows the program flow to branch to one of several different procedures is called the _____ construct.
    a. CASE
    b. IF-THEN-ELSE
    c. DO-UNTIL/DO-WHILE
    d. Sequential
40. The programming construct which runs a process until a test condition is met is called the _____ construct.
    a. CASE
    b. IF-THEN-ELSE
    c. DO-UNTIL/DO-WHILE
    d. Sequential
41. If you could only have two programming constructs to use to create a program, you would choose the _____ and _____ constructs.
    a. CASE
    b. IF-THEN-ELSE
    c. DO-UNTIL/DO-WHILE
    d. Sequential

14-8 DOCUMENTATION

Documentation is a very important part of the software development process. Simply, the documentation tells us what the software is to do, how we have decided to do it, and it explains some of the details of the code. Without documentation, the development of a larger program may tend to wander. This is because the developer does not remember everything done before. At some time in the future, maintenance will be very difficult because there is no record of the work done.

In this section, we will look at the documentation used for the various sections of the software development process. We will find there are different ways to document software, and that each has its advantages and disadvantages.

One important point to remember is that *documentation must be performed as the development process takes place*. It is very difficult to go back and "document" it later. Often this may seem the easiest thing to do. Documentation seems like a lot of work, especially when there is pressure to get the software development job done.

There are two major problems with documenting the work later. First, documentation done after the work is finished is usually incomplete. This is because the designer forgets details which can be included if the documentation is done as the development process takes place. Second, documentation that is done later is often never done. It is too easy to say the documentation will happen tomorrow. This excuse is probably the cause of more undocumented software than any other reason.

What documentation is used for the specification? Typically, the documentation of a specification is free-form. That is, the form of the documentation is very much up to the designer. A typical specification has

Narrative descriptions of the software functions

Tables of data which the software is to use

Samples of input data

Samples of output data and the form of output data

Sketches of equipment front panels

Sketches of keyboards

Data input screens and data output screens

Lists of formulas used by the process

Timing specifications which indicate how quickly the software must perform a function once the inputs happen

Lists of responses which occur in certain situations

Descriptions of each command which can be given to the software and the responses expected from it

Frequently commands are given using a semiformal notation such as that shown in Fig. 14-13.

A *specification* is nothing but documentation. The only rule which must be followed is that the documentation must be thorough. If the documentation is not thorough, the specification is weak. A weak specification leads to a weak design, which leads to weak software.

The specification is frequently passed back and forth between the person asking for the work and the software designer. Each time it comes from the person asking for the work, more detail is added to the request. Each time the designer returns a copy of the specification to the person requesting the development work, more detail is added to the specification. This allows the person asking for the work to see that the developer understands what must be done.

If there is a misunderstanding of the job to be done, the misunderstanding is documented in the specification. When the person asking for the work reviews the specification, the misunderstanding is found. This means the misunderstanding can be corrected early in the project rather than after the software is developed.

The documentation of a software design is usually more structured. There are, however, various documentation methods in common use. If the design documentation is thorough, the major difference between them is how comfortable the designer feels in using the particular design documentation. In some special areas, one particular type of documentation may be better than another.

Pseudocode is popular for design documentation. Pseudocode is sometimes called structured English. It uses the four basic constructs we studied earlier. In fact, the descriptions of sequential construct, the IF-THEN-ELSE construct, the CASE construct, and the DO-UNTIL construct all used pseudocode. Pseudocode also uses indentation to show the hierarchy of the operations. The design description in Fig. 14-14 uses pseudocode. This is the top-level design for the microprocessor-based telephone logger with long-distance lockout.

One of pseudocode's major advantages is that you do not need any special equipment to document software. As you will learn, most other software documentation techniques use drawings. Neat drawings are difficult to create without drafting equipment or a special computer-aided design (CAD) software. Pseudocode can be created using a text editor and printer. If pseudocode is being used to document a software program, it is much more likely that the documentation will be done and complete because it is easy to do.

As we also saw in Sec. 14-4, structure charts are used to describe the relationship of the dif-

MODE COMn: baud[, [parity] [, [databits] [stopbits] [,P]]]]

where n = 1 or 2, names the COM (Serial I/O) port

(*a*)

baud = baud rate to/from I/O port

parity = sets odd, even or no parity

databits = word length (5, 6, 7 or 8)

stopbits = 1 or 2 for number of stop bits

P = causes continuous retry on timeout error

[and] = optional parameter

(*b*)

Fig. 14-13 (*a*) **An MS-DOS command line given in formal notation.** (*b*) **An instruction showing how to read the command line given in this notation.**

Procedure: Telephone Time Use and Number
            Called Logger
Read_Account_Number
CASE of
  Account_Number = 12
    Open_John's_Record
  Account_Number = 13
    Open_Fred's_Record
  Account_Number = 14
    Open_Mary's_Record
  Account_Number = 15
    Open_Jane's_Record
  Account_Number = Other
    Beep_Error_Tone
    Close
Read_Telephone_Number
  IF First digit dialed is a 1
    THEN IF Next three digits are 800
      THEN Read_Next_Seven_Digits
      ELSE Close
    ELSE Read_Next_Seven_Digits
Dial_Phone_Number
  IF Telephone is answered
    THEN Store_Number_In_Buffer
         Store_Time_In_Buffer
         Start_Call_Timer
    ELSE Put_Number_In_Redial
DO-UNTIL Disconnect Signal Received
  Idle
Close
  Hang_up
  Compute_Call_Cost
  Store_Record_Of_Call
  Print_Record_Of_Call

Fig. 14-14 **A software design expressed in pseudocode or structured English. Note that pseudocode design in this example uses the basic sequential, IF-THEN-ELSE, CASE, and DO-UNTIL constructs.**

ferent modules. The structure chart, which looks very much like a corporate organization chart, allows the designer to show which modules make up the different functions. The structure chart shows the hierarchical relationship of all the different modules.

Pseudocode and structure charts do not do a good job of showing the flow of control or data. For some programs, especially those used in industrial control systems, the flow of control and data is so important that it needs to be documented thoroughly. In this case, the program's designer often uses other methods to document a design.

The use of the data flow diagram (DFD) lets the designer show the flow of data through the system. An example DFD is shown in Fig. 14-15 on page 426. The DFD allows the program designer to diagram how the data flow from the input through the different processes to the output. It shows where data is stored and the results of processing, so you can see how the data changes as it passes through the program. The DFD focuses on data; however, the processes are usually described using pseudocode.

The DFD allows the designer to check the design to make sure it obeys the rules of conservation of data. The *conservation of data rule* states that a software system does not create or use up data. It only takes data, changes its form, combines it with other data, and gives processed data as an output.

If the software program is generating data by itself, without any input, it is performing an unintended function. If the software system is taking in data and never doing anything with it, it is either not doing its job or it is asking for unnecessary input. The data flow diagram is an excellent way to test for the conservation of data as well as for diagramming the flow of data through the system.

Flowcharting is one of the oldest methods of documenting a software design. Today, many programmers use pseudocode, structure charts, and data flow diagrams rather than flowcharts. However, there still are many people who use flowcharts, and they are often the best documentation form when it is very important to document the flow of control.

Flowcharting uses four basic symbols to diagram what is to happen (see Fig. 14-16 on page 426). Each symbol has one input and one or more outputs. Inputs and outputs are shown as lines with arrowheads, which tell the direction of control flow.

The rectangular symbol represents a sequential *process*. The rectangle tells you that the program is to do something to the data. A short description of the process to be done is usually written inside the symbol.

The diamond is the *decision symbol*. The decision symbol shows places in the program where choices are made. The choice determines the direction the program flows from that point. The choice is made based on the result of a test. A description of the test to be done is written inside the diamond. Possible results of the test are written on the lines flowing out of the decision symbol.

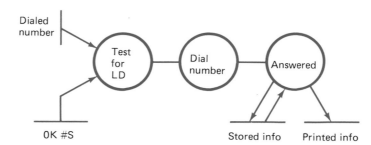

Fig. 14-15 A data flow diagram (DFD). This design documentation allows the programmer to visualize the processes developing the data as the data flow through the program.

The oval symbol is used as a *terminator*. In most cases, the terminator contains either the word *start* or the word *stop*. The terminator makes it easy to indicate where the program begins and where it ends.

Often the flowchart will not fit on a single page. This means that you must use the connector symbol. The *connector symbol* lets you draw large flowcharts. It also helps make flowcharts clear when a simple drawing would have many lines crossing one another. A number or a letter in the connector circle shows the symbols that are connected together. That is, all circles with a 1, for example, represent a common point in the program. The connecting symbol simply makes the flowchart much easier to read.

The flowchart shown in Fig. 14-17 is an excellent way to show the decision process and the alternative routes. For example, the diamond clearly shows a decision point. This decision is the same as the pseudocode IF-THEN-ELSE statement. For many people, the flowchart is easier to understand than pseudocode.

Once the design is complete, the implementation process begins. This needs a new kind of documentation. The implementation process, as you know, starts with the creation of source code. That is, the exact language of the computer program implementing the design is written down.

A few computer languages are almost self-documenting. That is, if a software developer who understands the computer programming language reads the source code, it is clear what job is being done. On the other hand, there are many computer languages whose source code is not very clear. For example, most assembly language programs require additional documentation before the job being done by the program can be understood.

To document source code, the computer programmer adds *comments*. Comments can be added in many different ways. In some cases, pseudocode from the design document can be added to the source code as the documentation. In other cases, additional comments must be supplied so that the particular function being performed by the software is clear.

Figure 14-18 on page 428 is a marked-up example of an earlier Z80 assembly language source code listing and its accompanying documentation. You should note two things. First, all the documentation is marked (the first character is a semicolon) so that it is clear to the assembly language program that it is documentation and not source code. Second, for every line of source code, there are two lines of documentation. This ensures that, at a later date, it will be clear to another programmer not familiar with the details of this particular implementation just what was being done and why it was being done.

During the course of implementation, there is another bit of documentation required. This

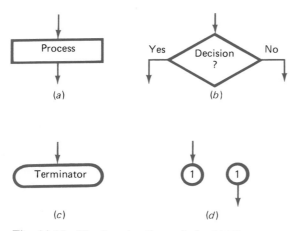

Fig. 14-16 The flowchart's symbols. (*a*) The process rectangle. (*b*) The decision diamond. (*c*) The terminator oval. (*d*) The circle connector. Note that the process rectangle and the decision diamond are really ways to graphically express the sequential and IF-THEN-ELSE constructs.

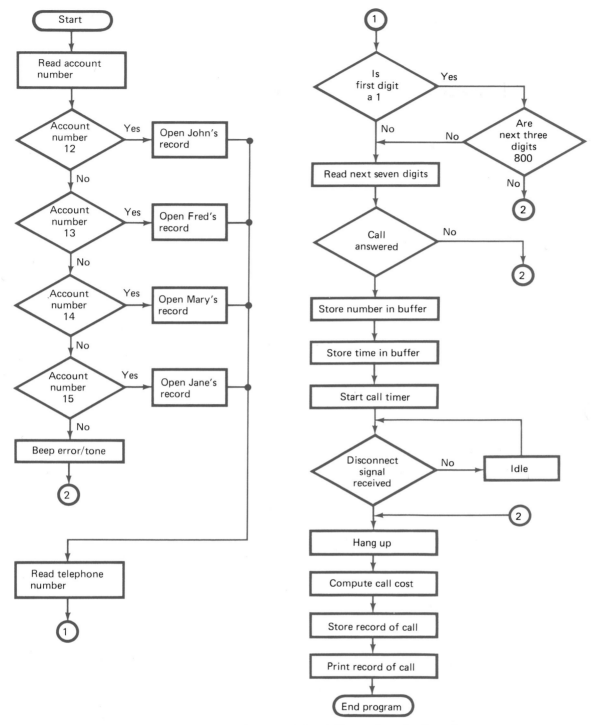

Fig. 14-17 A flowchart. The flowchart's advantage is its ability to show the flow of control in the program. Its disadvantage is that the drawings are not easily produced without drawing skills, templates, or some form of computer-aided drawing system.

documentation is the notes taken during testing. For example, during the testing and debugging process, a particular error is found. The documentation for this program should include a record of this error, who discovered it, what conditions caused it to happen, when it was fixed and how it was fixed.

On large programming efforts, a standard form is used to note the errors. After the software is retested, this set of notes becomes a

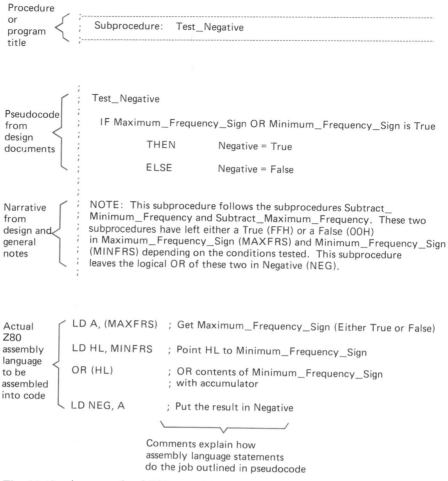

Procedure or program title
```
; ---------------------------------------------------------------------------
; Subprocedure:   Test_Negative
; ---------------------------------------------------------------------------
```

Pseudocode from design documents
```
; Test_Negative
;
;       IF Maximum_Frequency_Sign OR Minimum_Frequency_Sign is True
;
;               THEN            Negative = True
;
;               ELSE            Negative = False
;
```

Narrative from design and general notes
```
; NOTE:  This subprocedure follows the subprocedures Subtract_
; Minimum_Frequency and Subtract_Maximum_Frequency.  These two
; subprocedures have left either a True (FFH) or a False (00H)
; in Maximum_Frequency_Sign (MAXFRS) and Minimum_Frequency_Sign
; (MINFRS) depending on the conditions tested.  This subprocedure
; leaves the logical OR of these two in Negative (NEG).
;
```

Actual Z80 assembly language to be assembled into code
```
LD A, (MAXFRS)      ; Get Maximum_Frequency_Sign (Either True or False)

LD HL, MINFRS       ; Point HL to Minimum_Frequency_Sign

OR (HL)             ; OR contents of Minimum_Frequency_Sign
                    ; with accumulator

LD NEG, A           ; Put the result in Negative
```

Comments explain how assembly language statements do the job outlined in pseudocode

Fig. 14-18 An example of Z80 assembly language source code and the documentation which often goes with the source code. The margin notes show you the different parts of a well-documented assembly language source code listing.

record showing the work done to the original implementation to make a working software package. If, for example, at a later date someone finds a new problem, these notes can help locate a change which may have caused the new problem to appear.

As you can see, documentation is an important and large part of the overall process of developing a software package. Although it is easy to "document it later," or not to document it at all, that is very dangerous. There is very little software, even short test programs, developed and never used again or modified in the future. Therefore, there is never a good reason not to document a program.

One of software's advantages is the ease with which it can be reused and modified. This is not true of hardware. As you know, hardware can only be modified a few times and then it becomes useless. Typically a well-written, well-designed, well-documented software package

can be modified time and time again for new and different uses. In this case, it is essential that it be thoroughly documented to help those who are attempting to modify it.

Self-Test

Answer the following questions.

42. Software documentation tells us
    a. What the software is to do
    b. How we have decided to do the job
    c. Some of the details of the code which implements the job
    d. All of the above

43. Documentation is used
    a. Only as an aid to maintaining the software
    b. During the course of development to keep the project on target
    c. Only on large programs with six or more programmers for three or more months

d. Both as an aid for future maintenance and to help keep the project pointed toward its original goals

44. If the documentation for a software programming effort is to be good, it must be done while the various parts of the programming effort are being done. If the documentation task is not started until the project is over, it is quite possible that
   a. Much or all of the documentation will never be done because there will "never be any time."
   b. The documentation will be weak because the programming staff will forget many details which should have been included
   c. The programming effort will wander because of a lack of direction
   d. All of the above

45. Documentation of a software specification can
   a. Usually be a free-form document
   b. Use pseudocode, structure charts, data flow diagrams, and/or flowcharts
   c. Include part of the design documentation along with notes which explain how certain parts of the code are implemented
   d. Include test results from running the ATP

46. A software specification does not typically include
   a. A narrative description of the software functions
   b. Samples of input and output data formats
   c. Working code developed from the design
   d. Lists of formulas used by the processes called for

47. The documentation of a software design is typically more formal than the specification document. Usually it can
   a. Be a free-form document
   b. Use pseudocode, structure charts, data flow diagrams, and/or flowcharts
   c. Include part of the design documentation along with notes which explain how certain parts of the code are implemented
   d. Include test results from running the ATP

48. Pseudocode (structured English) is a very good way to document the logic of an algorithm. We use _____ statements when documenting a design with pseudocode.
   a. IF-THEN-ELSE
   b. CASE
   c. DO-UNTIL and DO-WHILE
   d. All of the above

49. Flowcharts are often used as design documentation where it is important to show
   a. Sequential flow
   b. The flow of control in a program
   c. The process (algorithm's) logic
   d. The relationship of one module to another

50. Documentation added to the source code shows the design and explains how the code implements the design. The documentation is added to the source code in the form of
   a. Pseudocode
   b. Comments
   c. Flowcharts
   d. Data flow diagrams

SUMMARY

1. Programming is the process of telling the processor (microprocessor) exactly what to do to solve a particular problem.

2. To program a processor, the programmer must speak to the processor in the exact words that the processor understands.

3. The programmer must tell the processor how to solve the problem; the processor cannot create a solution by itself.

4. The first step in the programming process is finding out what job is to be done.

5. When you design the program, you organize the problem so that the computer can solve the problem.

6. When the program is designed, the implementation process starts with coding the design. Coding the design means that the steps of the design are expressed in terms of a computer language.

7. Implementation continues with the debugging process. During the debugging process, the code is executed to see if it has

any errors; if it does, they are fixed and retested.

8. Often, a programming project starts with a story problem. A story problem is a problem expressed in words.

9. Some computer programs solve mathematical problems which can be expressed as algebraic equations. Other computer programs solve logical problems which involve the movement and manipulation of data.

10. The words, punctuation, and organization of the statements in a computer language follow exact rules which you must follow every time or the computer will not understand the instructions. You cannot use a computer language as freely as you use English.

11. The programming process can be divided into three steps
    a. Writing a specification
    b. Creating a design
    c. Implementing the design in code and testing and correcting the implementation

12. The programming process always follows these steps no matter what size the program. For very small (short) programs, it may be difficult to see each of the different steps, but they are there.

13. The purpose of writing a specification is to make sure you have a clear understanding of what the program is to do.

14. A written specification is usually reviewed and approved by the person who wants the program written. This makes sure that the programmer knows what is wanted and that the person asking for the work knows that the programmer knows what needs to be done.

15. The first description of the job to be done is often in the form of a story problem. When the programmer converts the story problem into a specification, there is often missing information. It is the programmer's job to find this information by asking the originator, by asking other experts, and by intuition. It is the originator's job to review the completed specification for accuracy and completeness.

16. It is important that as many questions as possible are answered in the specification and not left to be answered later during the design or implementation stages of the project. It is usually much more difficult to include new information during these later

stages of the program's development than it is to include them in the specification.

17. Because it is impossible to completely specify everything about a complicated program, the specification must be a living document. This means that the specification must be added to and changed as the program is developed so that the specification always shows the latest information about what the program is to do.

18. In the design process, the programmer moves from thinking what the program is to do to how the program is to do the job outlined in the specification.

19. The easiest way to develop a design is to start with the overall requirement. Once the overall requirement is written down, it can be broken into the next level of tasks. This process of breaking the design down into lower- and lower-level tasks is called a piecewise refinement or a top-down design.

20. A bottom-up design starts by designing individual modules and then making them into a complete system. This approach is dangerous because the modules may not fit together after they have been designed.

21. The design of a program should be very modular. Each module should be self-contained, have a uniform entry and exit point, and be short enough to be easily understood. A modular design ensures an organized approach to the software.

22. A well-designed module can be replaced with another module which performs the same function, but has a completely different internal design, without changing how the software operates.

23. The implementation process starts when the programmer begins to convert the design into source code statements written in the computer language chosen to do the job.

24. Once the program is coded in the computer language, the source code is processed by a program which converts the source code into computer-executable object code. The object code is then loaded into the target processor and the debugging process begins.

25. Each module may be tested independently and then combined with the rest of the software package. This is called unit modular testing.

26. When we test modules independently or test the software system without all of its modules, we need to make the missing parts seem as though they were there. To

do this, we create programming stubs. These stubs replace the missing parts so that the part being tested can do its job as if the whole system were there.

27. When the implementation process is finished, a thorough demonstration or test of the software package may be run to show the person requesting the job that the software development effort is complete. This is called an acceptance test procedure (ATP).

28. A program is made up of data and algorithms.

29. The processes used to manipulate the data are called the algorithms. They test data, calculate new data from other data, and move data.

30. Good programming practice keeps the algorithms and the data separate for ease of design and later for maintainability.

31. The program's data structures are where the data is assigned to different variables and where we define how the data is put together.

32. If the data structures are all in one place in the program, and later there is a change needed, it is easy to find the data which should be changed. If the data-to-variable assignments are made randomly throughout the program, it is easy to miss one of the assignments when making a change.

33. A well-trained programmer knows many of the standard ways of organizing data and algorithms for processing data. The organization and algorithm best suited for the job is then chosen as a part of the design process.

34. A standard programming construct is a program (software) building block.

35. A program can be written using just a few standard constructs. The commonly used constructs are

> The sequential construct
> The IF-THEN-ELSE construct
> The CASE construct
> The DO-UNTIL construct

36. The sequential construct consists of procedures which follow one another. The sequential construct does not permit any decision processing.

37. The IF-THEN-ELSE construct allows the programmer to create a test followed by a two-branch decision path. After executing an IF-THEN-ELSE procedure, the program flow follows either the procedure following THEN or the procedure following ELSE.

38. The CASE construct allows the programmer to test for one of several different conditions and to cause the program to execute one of several different procedures, depending on the condition found by the test.

39. The DO-UNTIL construct allows the programmer to execute a procedure until a certain test condition is met. The DO-WHILE construct is a variation which allows the programmer to execute a procedure while a certain test condition is true.

40. The CASE and DO-UNTIL constructs are special versions of the IF-THEN-ELSE construct, but they are very useful.

41. The software documentation tells us what the software is to do and how we have decided to do it; and it explains some of the details of the code which was written to implement the design.

42. It is important to create the program documentation while the development process is taking place so all of the details are remembered and so the documentation really takes place. The documentation serves as a guide to the people programming and later helps inform people who are asked to maintain the software.

43. The documentation of the specification is free form, but the specification should include
   a. Narrative descriptions of the software functions
   b. Tables of data which the software is to use
   c. Samples of input data
   d. Samples of output data and the form of output data
   e. Sketches of equipment displays
   f. Sketches of keyboards
   g. Data input screens and data output screens
   h. Lists of formulas used by the process
   i. Timing specifications which indicate how quickly the software must perform a function once the inputs happen
   j. Lists of stimulations and responses which occur
   k. Descriptions of each command which can be given to the software and the responses expected from it

44. There are different ways to document the design of a software project. Each way has

its own advantages and disadvantages; however, the one used is usually the one most familiar to the programmer.

45. Pseudocode, sometimes called structured English, is a popular documentation technique that uses the four basic programming constructs. One of pseudocode's major advantages is that it can be written at a standard computer terminal and printed using a standard printer.

46. Structure charts, whick look like corporate organization charts, are used to show how the different program modules are related to each other.

47. The data flow diagram (DFD) lets the programmer document the flow of data through the software system. It is a good way to document how the data changes as it flows through the system.

48. Flowcharts are one of the oldest forms of documentation. They require special drafting equipment to draw them. Flowcharts are very good at showing the flow of control in a program.

49. During the course of writing the source code, the programmer adds notes to tell what is being done. These notes are specially marked so that the programs which convert the source code to object code will ignore them. They are called programming comments.

50. Part of the program documentation is a complete set of notes describing all the problems found during testing and the fixes which were made to correct the problems.

CHAPTER REVIEW QUESTIONS

Answer the following questions.

14-1. The process of programming can best be described as
    a. Writing code in your favorite computer language
    b. Debugging the program
    c. Telling the processor exactly what to do to solve a problem
    d. The process of specifying and designing the work so that the coding effort can take place

14-2. A processor can only solve a problem when
    a. You have told it how the problem is to be solved
    b. The solution to the problem uses an algebraic formula
    c. The solution to the problem is a logical series of operations
    d. There are at least two variables involved

14-3. The first step in the programming process is
    a. Finding out exactly what must be done
    b. Coding the design
    c. Debugging the code
    d. Organizing the problem for solution by a processor

14-4. The second step in the programming process is
    a. Finding out exactly what must be done
    b. Coding the design
    c. Debugging the code
    d. Organizing the problem for solution by a processor

14-5. The third step in the programming process is
    a. Finding out exactly what must be done
    b. Coding the design
    c. Debugging the code
    d. Organizing the problem for solution by a processor

14-6. The fourth step in the programming process is
    a. Finding out exactly what must be done
    b. Coding the design
    c. Debugging the code
    d. Organizing the problem for solution by a processor

14-7. Briefly explain the difference between a story problem and a program specification.

14-8. Computer programs can solve mathematical problems or logical problems or combinations of both. Briefly explain the difference between these two types of problems.

14-9. Why do we have to use a computer language in a much more precise manner than we do our spoken language?

14-10. Briefly explain the purpose of writing a specification for a program.

14-11. Why should the person who asks for the program review the specification?

14-12. If the programmer who is preparing the program specification finds there is information missing from the original story problem, where can the programmer look for answers?

14-13. Why is it important that the specification answer as many questions as possible rather than leaving them to come up in the design or coding process?

14-14. What do we mean when we say that the specification should be a living document?

14-15. During the design process, the programmer moves from thinking about what the program is to do to thinking about
   a. How to code the program in the given language
   b. What computer language should be used
   c. How the bugs can be fixed with little work
   d. How to make a processor do the job

14-16. The best way to develop a program design is to
   a. Start with the individual modules and build up a system
   b. Start with the overall system and break it into increasing levels of detail
   c. Create one big module if at all possible
   d. Use a mixture of these three processes as needed

14-17. The technique described in choice *a* of question 14-16 is called _____ , and the technique described in choice *b* is called _____ . (Choose two of the four answers.)
   a. A bottom-up design   *c.* A highly modular design
   b. A top-down design   *d.* A horizontal design

14-18. In a very modular design, each software module
   a. Has a uniform entry and exit point
   b. Is short enough to be easily understood
   c. Is self-contained
   d. All of the above

14-19. One of the characteristics of a well-designed module is that it
   a. Is not longer than two pages
   b. Does not use any assembly language
   c. Can be replaced by another module which does the same job
   d. Can perform at least two functions without change

14-20. The implementation process begins with _____ , but also includes _____ . (Choose two of the four answers.)
   a. Specification
   b. Design
   c. Coding
   d. Debugging

14-21. A special computer program is used to convert source code into
   a. A specification
   b. Object code
   c. A test routine
   d. A straight-line program

14-22. Often the modules of a large program are written, tested individually, and then added to the rest of the program for system testing. To do this, we must have a way to run the program with many of the lower-level modules missing. One way is to use
  a. An assembly language   *c.* Programming stubs
  b. A debugger   *d.* A top-down design

14-23. The purpose of an acceptance test procedure is to
  a. Have a formal way to demonstrate that the job is done
  b. Give the end user a place to sign
  c. Have a way to check the program design before coding
  d. All of the above

14-24. A program is made up of _____ and _____ . (Choose two of the four answers.)
  a. Formulas   *c.* Logical tasks
  b. Algorithms   *d.* Data

14-25. An algorithm is
  a. A formula
  b. A logical sequence of operations on the data
  c. The procedure used to solve a problem
  d. The way that the data in a program is stored

14-26. The reason that all of a program's data assignments are grouped in one place is
  a. That it makes the design easier and more reliable
  b. That it makes program maintenance easier
  c. Because scattered data assignments can be missed during changes
  d. All of the above

14-27. A standard programming construct is
  a. A software building block
  b. A program module
  c. The only way to develop a top-down design
  d. The only way to develop a bottom-up design

14-28. A series of procedures in a program which follow one another without any way to have decision processing is called the _____ construct.
  a. IF-THEN-ELSE   *c.* Sequential
  b. CASE   *d.* DO-UNTIL/DO-WHILE

14-29. The _____ construct allows the programmer to process a procedure until a test condition is met and then continue with the next procedure in sequence.
  a. IF-THEN-ELSE   *c.* Sequential
  b. CASE   *d.* DO-UNTIL/DO-WHILE

14-30. The construct which allows the program to branch to one of two conditions based on the outcome of a test is called the _____ construct.
  a. IF-THEN-ELSE   *c.* Sequential
  b. CASE   *d.* DO-UNTIL/DO-WHILE

14-31. You would probably use the _____ construct to branch to the correct procedure for 1 of 10 different inputs.
  a. IF-THEN-ELSE   *c.* Sequential
  b. CASE   *d.* DO-UNTIL/DO-WHILE

14-32. Briefly explain the purpose of software documentation.

14-33. Why is it important to document the software thoroughly at each step of the development process rather than having the documentation process become a separate activity which is done when the program is working completely?

14-34. What do we mean when we say that the documentation of the specification is usually free-form?

14-35. List four different things which should be included in a software specification.

14-36. What do we mean by pseudocode, and what is it used for?

14-37. What does a structure chart tell us about the design of a program?

14-38. The data flow diagram and the flowchart are both graphical ways to show the design of a program. The data flow diagram is very good for showing how the data are processed, and the flow chart is very good for showing the flow of _____ .

14-39. The notes that a programmer adds to the source code are called _____ . Briefly explain why they are used.

CRITICAL THINKING QUESTIONS

14-1. We say that a programming effort consists of four steps: specification, design, coding, and test and debug. How would these four steps be different if you were using a high-level language like BASIC rather than an assembly language?

14-2. Before structured programming processes were developed, it was not uncommon to find a program with numbers scattered throughout it. When changes were to be made, each of these numbers had to be found and changed. If one was missed, the new program did not work all the time. This is an example of failing to recognize that all programs are composed of certain basic elements. What are these elements? How does the failure of the program demonstrate that fact?

14-3. In the section on constructs, we looked at a CASE construct for separating the days of the week into different procedures. How would you create this same effect using the IF-THEN-ELSE construct?

14-4. Although there are a number of different ways to document a program, there is one common rule: the documentation must be easy to change. Why is this true?

14-5. What do you think caused programming to be thought of as an art or magic? What do you think happened to make this perception go away?

Answers to Self-Tests

1. *c*
2. *a*
3. *b*
4. *c*   5. *a d*
6. **a.** During the coding process, the design of the program is converted into a series of exact instructions to the processor by writing them in a language that the processor can understand.
   b. The specification process makes sure that the programmer understands fully everything which the program must do.
   c. The design process converts the requirements in the specification into a logical series of processes that the computer can follow
   d. The debugging process is used to find and correct errors in the code.
7. *b*
8. The processor can only follow the procedure you tell it to follow. If you do not know the procedure needed to solve a problem, the processor cannot be told how to solve the problem.
9. *c*
10. The program which converts the words, punctuation, and organization of the statements into binary words meaningful to the processor only knows one exact meaning for them. If you do not use them exactly as the program understands them, you will be telling the processor to do the wrong thing.
11. A specification is created so that the programmer and the person requesting the work both know exactly what is to be done.
12. A written specification is used when the program is large enough to do more than one simple job.
13. The specification goes back and forth between the programmer and the requester to be sure that all the different requirements are fully documented and that the programmer understands them and the requester

knows that the specification is complete and accurate.

14. If the initial story problem explaining a programming requirement is missing some information, the programmer can find this missing information from the person making the request for the work, other technical people working on different parts of the project, and from inferences taken from the original material.

15. Questions should be answered as early as possible because the answers may call for changes to the work. The longer you wait for the answers, the more work is done and the larger the amount of rework which may have to be done to make the needed changes.

16. It is very difficult to answer all of the questions at the beginning, and therefore the specification must be changed as the program is developed to make sure it is always a correct specification.

17. A specification tells you what the program is to do. The design tells you how the job is to be done using a processor.

18. Breaking the design down into smaller and smaller pieces is called a piecewise refinement or a top-down design. The major benefit of this way of designing software is that you know that the complete system is growing as a working design and that each

module works as it is added to the system.

19. If the individual modules in a program are designed first and then fitted together later, there is a chance that the system will not work when all the modules are fitted together. If this happens, one or more of the modules must be reworked until they all work together as a system. This approach is called a bottom-up design.

20. If a module is removed from a software package and is replaced with another module which has a different design but performs the same function, the software should still continue to perform all of its functions.

21. A module which is no longer than a page is simple enough so that anyone studying the module later can easily understand what the module is to do.

22. If the module uses pieces of other modules and those modules are changed, the original module may no longer work.

23. A single point of entry ensures that every time the module is used the same initialization processes are used.

24. Coding and debugging

25. Writing source code means that we are converting the ideas in a design into statements in a computer language which must be converted into

machine-executable steps by another program.

26. The source code must be converted to object code by a special computer program.

27. If each new module is added to the system one module at a time and the system is tested as each new module is added, problems which are caused by a design or coding error must be in the module just added.

28. Programming stubs are short programs which are added to modules so that the entire program appears to be complete even though some modules are not finished.

29. An acceptance test procedure is a formal document which describes a series of tests used to demonstrate that a project is complete.

30. *c*
31. *b*
32. *d*
33. *c*
34. *d*
35. *a*
36. *c*
37. *d*
38. *b*
39. *a*
40. *c*
41. *d, b*
42. *d*
43. *d*
44. *d*
45. *a*
46. *c*
47. *b*
48. *d*
49. *b*
50. *b*

CHAPTER 15

Operating Systems and System Software

■

CHAPTER OBJECTIVES

This chapter will help you to:

1. *Explain* the term system software.
2. *Differentiate* between an operating system and an executive.
3. *Name* and *explain* the major parts of an operating system.
4. *Recognize* and *compare* the most common microprocessor operating systems.
5. *List* and *describe* basic programming tools.
6. *Name* and *compare* common low- and high-level programming languages.

─────────

When a microprocessor-based system is designed to do many different jobs depending on the programs which are loaded into the system, system software is often used to make the job easier. A number of different types of packages make up systems software. Operating systems let the applications programmer have a common way of talking to the microprocessor-based system's I/O devices. Programming languages allow the programmer to write instructions which speak to the processor in complex terms, and debuggers let the programmer test the new software on the system. All of these different system software elements are part of the tool box which is available for software development.

■

15-1 WHAT IS SYSTEM SOFTWARE?

A microprocessor-based system uses two different kinds of software. First, there is the software specially written to do the job at hand. This is called the *application software* or the *application program*. For example, this chapter was written with a word processor running on a personal computer. The word processor is an application program. It performs a specific-, not a general-purpose, function.

System software is a collection of programs or subprograms used over and over again by many different application programs. Many different kinds of programs make up system software. However, they all share the common characteristic of frequent usage by many different types of programs. You will find that most system software provides general-purpose support for application programs.

For example, when I saved this paragraph, a program called a *disk driver* managed the transfer of information from the microprocessor memory to disk. The disk driver is just one piece of the microcomputer system software. Many different application programs need to use a disk driver. Because you should not have to write a new disk driver program each time a new application is written, disk drivers are provided as a part of the system software.

Why do we want to use general-purpose system software? This is a very reasonable question because general-purpose software cannot be as efficient as software specifically written to do a job, and application programs are often written for efficiency. For example, why should the word processor use a general-purpose disk driver with universal capability rather than a specific driver written to handle the particular transfer from memory to the disk drive?

Programming efficiency is one of the major reasons for using general-purpose software. If the person writing the application software must write what can be done by standard programs, a great deal of time is wasted. This means the programmer becomes less productive. Since there are only a limited number of people who can write software and since there is a great deal of software to be written, it is important the people writing software be as productive as possible. Using the functions found in system software helps the programmers become productive.

A second reason is that using standard system software lets the programmer use software written by someone who knows a great deal about the job done by that software. Often, writing parts of the system software requires very special knowledge about the hardware, such as disk drives, displays, communication ports, and so forth. If every programmer must understand all of the different parts of a computer system, it is hard for the software developer to become an expert in his or her own application area.

Third, the use of system software forces standardization. For example, if the word processor performed its own disk access, it could write its own style of files. If system software is used, the word processor uses a standard file structure. This makes the word processor a more universal software program, because its files can be used for many purposes.

What different kinds of programs are included in system software? There are three major categories: operating systems, programming languages, and utilities. There is no exact list of functions performed in each of these three categories. The list of functions depends on the particular system being discussed. For example, an operating system can be no more than a very simple set of programs to help application programs use the hardware in a dedicated microprocessor-based product. On the other hand, the MS-DOS operating system is a sophisticated system. It has many different programs and subprograms to accommodate the wide range of hardware which may be added to a personal computer.

One simple form of system software is an *executive*. An executive is used on dedicated systems to manage a few special programs or just a single program. On simple systems, the executive does nothing more than schedule the application programs and manage data transfers between application programs, I/O ports, or mass storage devices. If the microprocessor-based system only uses a single program, the sequence of module activity may be controlled by an executive or *scheduler* module. Often these very simple executives or schedulers are written by the person developing the application package as a module in the package.

For more complex systems, a commercial executive may be purchased. The programmer only writes the application programs and may customize the executive to fit the particular hardware environment. If the system needs become sufficiently complex or general-purpose, the executive grows to the point that it is called an *operating system*. Because operating systems are used on most general-purpose microprocessor-based systems, we will study them in the next section.

Self-Test

Answer the following questions.

1. The software which runs most microprocessor-based products can be broken into two categories: system software and application software. Which of the following is not system software?
   a. The display management program which converts ASCII data into row-column information for a 640×200 dot-matrix display
   b. The file manager which opens files, keeps track of the file while it is open, and closes the file
   c. The HVAC (heating, ventilating, and air-conditioning) program which controls the environment in a plant
   d. The program which copies a file on one device to another device
2. One of the major reasons a particular software program may be included as part of a system software offering is that it is
   a. Used frequently and should not be rewritten time and time again
   b. Very technical in nature and requires someone with special knowledge to properly write it
   c. Part of what standardizes the application program so it is usable with a wide variety of other application programs
   d. All of the above

3. Rewriting standard programs has a negative impact on
   a. Programmer productivity
   b. The program efficiency (code size for the job done)
   c. Both a and b
   d. Neither a nor b
4. We are interested in improving programmer productivity because
   a. Writing software consumes expensive programming time, and lack of productivity increases the cost of developing software
   b. Improved productivity means we can develop more software in a given amount of time
   c. We have more software to write than there is time to write it, so improved productivity is needed so we can get as much software written as possible
   d. All of the above
5. System software includes
   a. An operating system and its associated application program
   b. The development tools such as languages and the source code for the application program being developed
   c. A word processor and its disk drive handler
   d. An operating system and the software development tools

15-2 OPERATING SYSTEMS

An operating system works with a computer system. It gives application programs easy and uniform communications with the microcomputer hardware and with other application programs. Operating systems provide a standard high-level way for application programs to perform these communications. This means individuals writing application programs do not have to understand device addresses, status bits, command words, and other such hardware-specific details. It also means they do not need to know the internal details of other application programs they are communicating with.

Operating systems can support either ROM-based or disk-based microcomputers. However, you will most often find executives used with ROM-based systems and operating systems used with microcomputers with disk-based mass storage. Thus most operating systems are called *disk operating systems* (DOS).

Most operating systems are made up of two different kinds of programs. These are *resident* programs and *transient* (nonresident) programs. The resident program is loaded into memory when the operating system is booted. It uses a fixed amount of memory and is in memory any time the microcomputer is used. The resident program provides functions which must respond quickly or are used frequently.

A memory map for a microcomputer with an operating system loaded is shown in Fig. 15-1. For example, the easiest way to move data between a disk and an I/O port is to copy the data file from the disk to the I/O port. The operating system provides the copy function as part of the resident program, because this is a very commonly used function.

The transient part of the operating system provides functions not used as often. Each transient function is a separate program which is loaded when it is needed. For example, before a floppy disk can be used it must be formatted. Formatting is done by using the operating system disk format utility. Because format is only used once in a while, it is a transient function.

The resident part of the operating system is the first program loaded into memory when the computer is booted. *ROM-based boot software* is used to load the operating system. One of the first functions the operating system per-

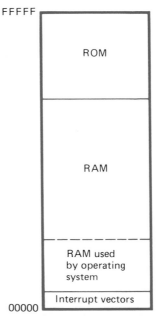

Fig. 15-1 **A memory map for a microprocessor-based system showing the memory used by the operating system.**

Disk operating systems (DOS)

Resident programs

Transient programs

ROM-based boot software

forms is to initialize the microcomputer hardware. This makes sure each device works in a known way. For example, the serial ports might be set to operate with 7 data bits, one odd-parity bit, and at 9600 baud.

After initialization, control is given to a section of the operating system called the *command line interpreter*. The command line interpreter monitors the keyboard for instructions from the person using the computer. The command line interpreter first error checks and then executes all commands.

A command is executed by running part of the resident software, by loading and executing one of the operating system transient programs or by loading and executing an application program. Figure 15-2 shows the sequence of operations for the MS-DOS operating system after you boot a personal computer.

Once the command execution is complete, system control returns to the command line interpreter. The operating system keeps a system stack, which tells it where to return once the current command or task is complete.

As you know, the microprocessor can only do one thing at a time. How does the operating system maintain system control after it has, for example, turned the microprocessor over to an application program? The operating system maintains control because it ran first. This means it has the first information on the stack. When the application program is finished, the information left on the stack puts part of the operating system (usually the command line monitor) back into operation.

While the application program is running, it may use parts of the operating system. How does it do this when the application program has control? The application program makes *calls* for system services. When the application program calls for a system service, it uses one of the operating system memory resident subprograms or transient programs. Remember, the operating system resident code is still loaded in memory.

When this subprogram is complete, program execution returns to the point in the application program which called for the system service. Part of the documentation included with the operating system or program development software tells an application programmer how to call the various system services. Figure 15-3 is a memory map of a microprocessor-based system loaded with an operating system and an application program. The application program makes calls to the operating system subprograms as shown by the arrows in the diagram.

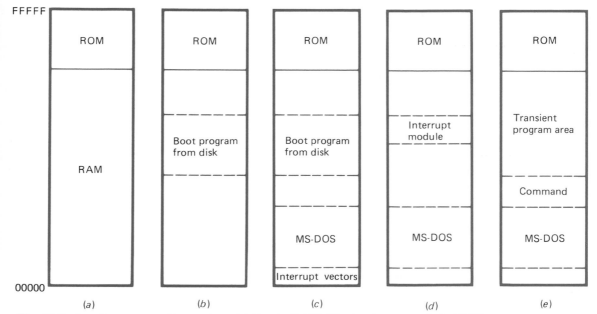

Fig. 15-2 **Loading an operating system.** (*a*) **Empty RAM. The boot program is in ROM.** (*b*) **The boot ROM loads the operating system boot program.** (*c*) **The boot program loads MS-DOS into lower memory above the interrupt processing vectors.** (*d*) **MS-DOS loads the initialization program which initializes the system hardware.** (*e*) **The command processor is loaded and takes over.**

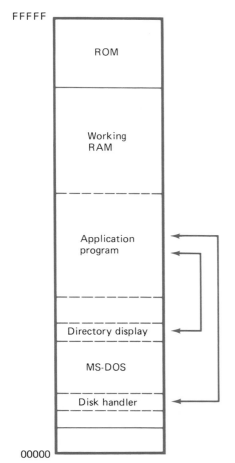

FFFFF

ROM

Working
RAM

Application
program

Directory display

MS-DOS

Disk handler

00000

Fig. 15-3   The application program makes calls on the MS-DOS system services.

The application program does not need to use system services if a better job can be done directly. For example, this is done when the application programmer decides the system services accessing the serial I/O ports or the display are too slow. If this is done, it must be done with care. One operating system job is to keep track of information about changes to system hardware, and this process bypasses the operating system.

For example, an application program accesses a serial I/O port and changes the parity from odd to even. However, the device initialization information in the operating system is not changed. This means the I/O port is left in a different condition than the operating system indicates. This leads to problems when another application program or a different part of the application program attempts to use the serial I/O port by accessing it through the system services. In this case, the operating system shows an incorrect device status.

There are two different kinds of operating systems. These are called *single-tasking, single-user* operating systems and *multitasking* and/or *multiuser* operating systems. As the names tell you, the difference between these operating systems is the number of simultaneous tasks (application programs) and/or users the operating system supports. Most personal computers and microprocessor-based products use the simpler, single-user, single-task systems.

Once again, you will remember the microprocessor can only do one thing at a time. That is, it can only execute a single instruction at a time. Therefore, it can only execute one program at a time. This makes us must ask "how can a microprocessor operate in a multitasking or multiuser mode if it can only really be executing one program at a time?" The answer is the microprocessor only *appears* to operate in a multitasking or multiuser mode. Its real mode of operation is to quickly switch from one task to another. It does this so fast that it appears multiple tasks are running at one time.

What is the difference between a multitasking and a multiuser operating system? The multitasking operating system lets a single user run more than one application program at one time. A multiuser operating system lets more than one user work with the system at a time.

Let us start by looking at an example of a multitasking operating system. In this example, a microcomputer is used in an industrial situation which runs two application tasks: a word processor and a data acquisition program. A diagram of this dual-application system is shown in Fig. 15-4 on page 442.

The data acquisition program starts first. It does not need to communicate with the operator except when it is initialized. During normal operation, it polls the analog-to-digital converters connected to the microcomputer I/O port once every 2 min. At the microcomputer data is converted into engineering units (temperature in degrees and speed in revolutions per minute) and is stored in a file. Each file entry is given a time tag which tells when the data was acquired. The word processor is started second. This is because it needs constant operator attention. In fact, it is not doing anything at all when the operator is not typing.

With the application programs loaded, a special part of a multitasking operating system called the *task scheduler* takes over. The task scheduler controls when each application program executes. In operation, the task

**Single-tasking
operating system**

**Single-user
operating system**

**Multitasking
operating system**

**Multiuser
operating system**

Task scheduler

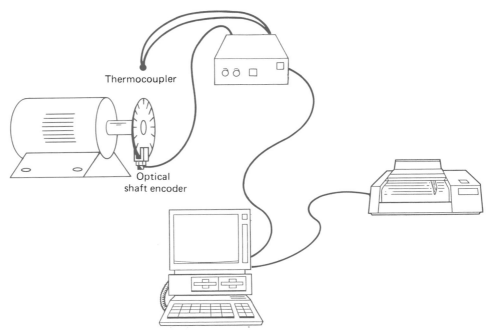

Fig. 15-4   A microcomputer in multitasking service. The analog I/O module collects temperature and rpm data while the microcomputer also runs word processing.

scheduler runs each program for a short while. After the first program runs for a while, the task scheduler stops program execution. It saves the current processor status and starts the second application program. The task scheduler might, for example, run each application program for a few milliseconds before the next application program is scheduled. This is a time-shared, multitasking operating system. Each application program gets an equal share of the processor time.

Other multitasking systems use *significant events* to cause the change from one task to another. A significant event might, for example, be the occurrence of an I/O interrupt.

These programs do not necessarily get an equal share of the processor time. Each application program completes a specific task while it is running, and high-priority tasks may get most of the processor time.

The task scheduler causes both application programs to appear to run at the same time. As you can see, each program really has only part of the microprocessor time. If there are two application programs, each program has less than half of the CPU time. This is because there must also be time for the task scheduler to run. The price for multitasking operations is efficiency. Figure 15-5 shows a *time line* for a multitasking system. As you can see, part of

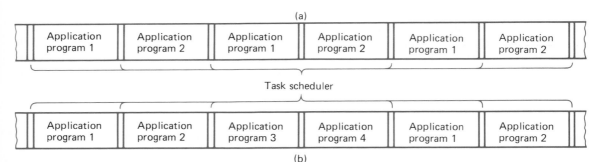

Fig. 15-5   A multitasking system time line. (*a*) A system with two application programs. (*b*) A system with four application programs. Note the task scheduler takes processor time from the application programs.

the processor time is spent managing the multitasking effort instead of processing the application program.

If two application programs run at the same time, and both programs must interact with the person using the program, a multiuser operating system is needed. What is the difference between a multitasking and a multiuser operating system? The multiuser operating system has more than one user interface. This means that the system supports two or more sets of keyboards and displays. You do this by connecting two or more terminals to the microcomputer. Each terminal is connected by a separate serial I/O port.

Figure 15-6 shows two different microcomputer systems. The one shown in Fig. 15-6(a) supports a multitasking operating system. It uses the PC memory-mapped display. The different application programs each have a part

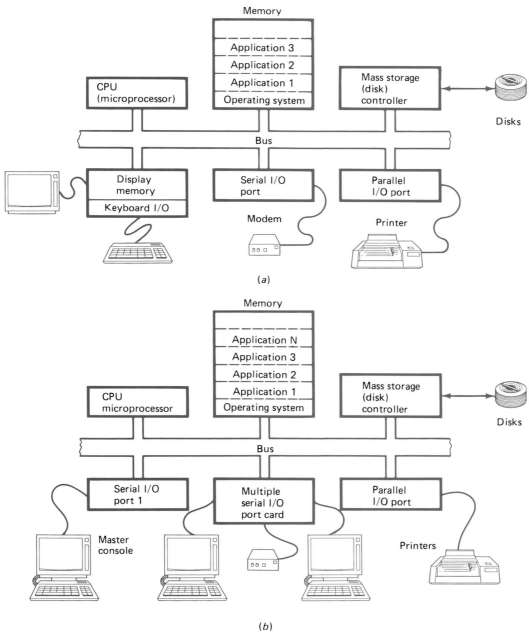

(a)

(b)

Fig. 15-6 (a) A single-user microprocessor-based system used in the multitasking mode. Note multiple application programs in memory. (b) A multiple-user microprocessor-based system. Note the multiport serial card which connects the multiple VDTs to the system.

of the PC RAM. If the programs are so big they cannot fit in RAM at one time, the multitasking operating system swaps the programs into and out of memory when it is their turn to run. The microcomputer serial port is connected to a modem, and the parallel port is connected to a printer to show typical connections for these I/O ports.

Figure 15-6(b) shows a microcomputer configured for multiuser operation. A multiport serial card is added to this system which gives the microcomputer multiple serial I/O ports. Each of the ports is connected to a terminal. No memory-mapped display is used by the system. The system operator manages the multiuser operating system and runs any application programs from serial port 1. If a modem is needed, it must use one of the serial ports as well.

Self-Test

Answer the following questions.

6. An operating system provides a
   a. Common way for application programs to access I/O devices in microcomputers
   b. Common way for application programs to communicate with each other
   c. Collection of utilities which the user or the application programmer may use to work with the microcomputer system
   d. All of the above
7. Usually, a disk operating system has two different kinds of support programs. These are called _____ and _____ programs. (Choose two of the four answers.)
   a. Transient   c. BIOS
   b. CP/M   d. Resident
8. If you are using a microcomputer system which uses an operating system, the operating system is some of the first software loaded into memory from disk. This is because
   a. The disks do not know how to load software into memory until the operating system is loaded
   b. You use the operating system to initialize your system hardware and to load and control the execution of your application programs
   c. An operating system takes dedicated memory space and has sections of memory reserved for its transient programs
   d. Not true; the application program is loaded first and then the operating sys-

tem is loaded after the application program is executing and needs help
9. The operating system uses the command line interpreter to
   a. Check for errors in operator-issued commands
   b. Interpret system commands and to initiate those operations
   c. Let the operator start executing an application program
   d. All of the above
10. Once the application program is done executing, control is typically returned to
   a. The next application program
   b. The command line interpreter
   c. The boot ROM
   d. There is no return of control; the user must reboot the system to initiate execution of another application program
11. When an application program is running, it may make use of the system services provided by the operating system by
   a. Making a system call
   b. Dropping back to the command line interpreter and requesting the user to initiate the system function
   c. Requesting the user to load the desired system utility from disk
   d. The application programs do not have access to system functions; these are reserved for the system user
12. An application program can access any of the peripheral devices directly rather than going through the operating system. If this is done, there is a potential for problems if
   a. The device is being used by another application program which is also running
   b. It is not possible to access the peripheral devices in any other way than going through the system services
   c. The application program does not leave the peripheral control registers in the same condition as they were set up by the operating system
   d. The system services are too slow for the application program to get its job done
13. A multitasking operating system allows
   a. The hardware to support two CPUs so two tasks can take place at the same time
   b. Two tasks (application programs) to apparently run at the same time by time-sharing the CPU between them
   c. Multiple tasks (application programs) to apparently run at the same time by time-sharing the CPU between them
   d. Two or more users to run application

programs on the CPU at the same time, each with his or her own terminal

14. The main difference between a multi-tasking and a multiuser operating system is that the multiuser operating system
    a. Is a single-tasking system with multiple users
    b. Just another name for a multitasking system
    c. Cannot be run on a personal computer
    d. Is a multitasking operating system which supports multiple users at multiple terminals

15. A multitasking or multiuser operating system uses a special program which allocates CPU time to each of the different application programs or users currently active on the system. This program is called the
    a. Device driver
    b. Task scheduler
    c. BIOS
    d. Command line interpreter

15-3 TWO MICROCOMPUTER DISK OPERATING SYSTEMS

There are two popular operating systems used with microcomputers. They are MS-DOS and Unix. In this section we will take a brief look at each of these operating systems.

MS-DOS (Microsoft disk operating system) is a 16-bit operating system developed by Microsoft Corporation for use with the 8088 family microcomputers which have the IBM PC architecture. It is the single-tasking, single-user operating system supplied with most 16-bit computers today.

MS-DOS has grown from the original MS-DOS V1.0 to a sophisticated operating system with many features.

Figure 15-7 shows a memory map of an 8088 system with MS-DOS. MS-DOS works with the PC's *BIOS* (*Basic Input/Output System*) functions which are contained in ROM. The BIOS software manages communications between the higher-level operating system software and the PC's specific I/O hardware. The BIOS is customized for the particular hardware architecture used.

Figure 15-8 on page 446 shows a typical hardware architecture for an MS-DOS system. Each block in the diagram has specific hardware requirements. For example, the serial I/O port must use one of a few specific asynchronous communications controllers, and the video controller must be one of a few specific video controllers. Some changes in this hardware can be allowed by rewriting the MS-DOS BIOS. But hardware changes frequently create a microcomputer which is not exactly compatible with the IBM PC. This means that certain software (application programs) will not run on the modified hardware architecture even with a modified MS-DOS BIOS.

Fig. 15-9, on page 447 and 448, is a summary of some MS-DOS resident and transient commands.

One major feature which MS-DOS introduced to personal computers is a hierarchical directory structure for mass storage devices

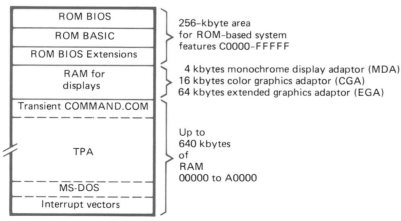

Fig. 15-7 An MS-DOS system memory map. Note the TPA lies between the transient COMMAND.COM and the balance of MS-DOS. The size of the TPA depends on the amount of memory (RAM) in the system.

MS-DOS inverted tree

Root directory

Subdirectories

Path

OS/2

Unix

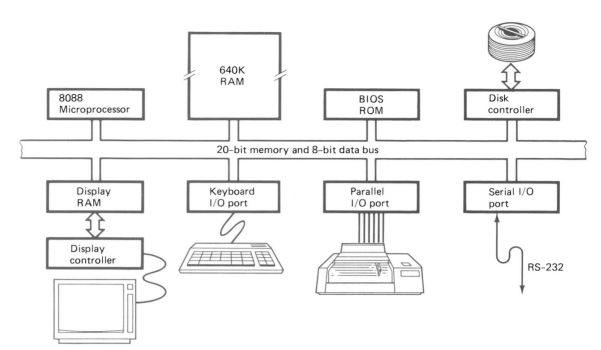

Fig. 15-8 **A block diagram of a microprocessor-based system which supports MS-DOS operation.**

This is needed because you can only have a limited number of files in the directory of a mass storage device formatted for MS-DOS files. Although this is not much of a limitation with a floppy disk (224 files in a directory), it becomes a significant limitation when 80-, 120-, or 240-Mbyte hard-disk systems are used. It is very easy for these large mass storage devices to have many hundreds of files.

With hierarchical directories, you can have directories within directories (see Fig. 15-10 on page 448). This structure is called the *MS-DOS inverted tree*. When an MS-DOS disk is formatted, a single file directory is created. This is called the *root directory*. The additional directories shown in Fig. 15-10 are called *subdirectories* and are actually MS-DOS files. The flow from the root directory to the directory in use is called the *path*.

MS-DOS also supports an advanced batch file command language. The batch file language allows you to make a file which is a series of commands to be executed. For example, a batch file may be used to assemble a series of files, link them together, and load the final object file. Figure 15-11 on page 449 shows a simple MS-DOS batch file. This one is used to initiate the programming language translator BASIC and to run a program called END when the operator is finished with BASIC.

OS/2 is a version of MS-DOS. This product adds multitasking features to MS-DOS. With multitasking capability, the user does not need to exit one application program to run another. Note that this is not a multitasking, multiuser operating system. It is controlled only by a single user but that user can have more than one program running at one time.

Unix is a multitasking, multiuser operating system. It was developed by Bell Telephone Laboratories during the late 1960s. Originally, Unix was developed to operate on 16-bit minicomputers. A great deal of the original development took place on the Digital Equipment Corporation (DEC) PDP-11 series of minicomputers. These minicomputers were the forerunners of microcomputers in laboratory and industrial situations.

The later versions of Unix were all written in a high-level structured programming language called C. Because Unix is written in C, it is fairly easily transported to any CPU with a C compiler. Because of this capability, Unix has been transported to a wide range of different CPUs. One family of CPUs Unix was transported to is the Intel x86 series of microprocessors. Unix has also been transported to the 68000 family of microcomputers.

Unix is an excellent programming environment. Unix supports a wide range of program-

Command	Command meaning
	Resident general-use commands
BREAK	Turns on or off MS-DOS check for CTRL-break or CTRL-C
CLS	CLear Screen—clears the screen (display RAM)
COPY	Copies files from one place to another
CTTY	Change TTY—Changes device used as command console (TTY)
DATE	Display or change the date
DEL	DELete—deletes named file from directory listing
DIR	DIRectory—lists requested directory entries
ERASE	Same as DELete command
EXIT	Exits from existing application program to command processor
PROMPT	Used to change MS-DOS command prompt
REN	REName—used to change a file name
TIME	Display or change the time
TYPE	Displays the contents of an ASCII file on console device
VER	VERsion—displays current MS-DOS.SYS and IO.SYS version
VERIFY	Verifies data is correctly written to disk
VOL	VOLume—displays disk volume label
	Resident hierarchical directory commands
CHDIR	CHange DIRectory—display or change current directory
MKDIR	MaKe DIRectory—make a hierarchical subdirectory
PATH	Specifies directories to be searched for executable files
RMDIR	ReMove DIRectory—delete a hierarchical subdirectory
	Resident special-purpose (batch) commands
ECHO	Turns batch file echoes on or off
FOR	Batch command for loop statement
GOTO	Batch command GOTO branch statement
IF	Batch command IF conditional statement
PAUSE	Suspends execution of batch file until any key is pressed
REM	Marks a batch file comment line
SET	Sets one string equal to another
SHIFT	Lets the system use more than 10 batch parameters
	Transient commands
APPLY	Used to execute a command more than once
ASGNPART	ASsiGN PARTition—assigns drive letter to hard disk partition
ASSIGN	Temporarily reassign logical drive names
ATTRIB	ATTRIBute—display or change read-only file status
BACKUP	Make a backup file or files
BOOTF	BOOT Floppy—Boot from floppy
CHKDSK	CHecK DiSK—Show status of disk contents
COMMAND	Lets you see COMMAND.COM features from a program
COMP	COMPare—compares two files and shows differences
CONFIGUR	CONFIGURes MS-DOS to match existing hardware environment
DISKCOMP	DISK COMPare—compares two disks (to verify copy)
DISKCOPY	Makes an exact copy of a disk
EXE2BIN	EXEcutable 2 BINary—converts executable to binary files
FC	File Compare—compares two files and makes difference file
FIND	Finds a string of text
FORMAT	Formats disk to receive MS-DOS files
GRAFTABL	GRAphics TABLe—loads graphics characters for ASCII 128–255
JOIN	Joins disk drive to a pathname
KEYBxxxx	Loads foreign keyboard tables

Fig. 15-9 **The Version 3.0 MS-DOS commands. The resident commands are divided into those for general use, those for the creation, use, and deletion of hierarchical file directories, and those for the batch command capability.** *Note:* **More commands appear in later versions.**

Command	Command meaning
	Transient commands (*cont'd*)
LABEL	Create, change, or delete disk volume label
MODE	Configures MS-DOS to match existing hardware environment
MORE	Displays output one screen at a time
NODEBUG	Turns resident debugger off
PRINT	Sends ASCII file to printer port
PSCxxxxx	Outputs alphanumeric and graphic characters to printer
RDCPM	ReaD CP/M—reads files written by CP/M
RECOVER	Used to salvage files from disks with bad sectors
RESTORE	Brings back files archived with BACKUP
SEARCH	Finds files in hierarchical directory/file structure
SELECT	Selects a foreign keyboard and date format
SHARE	Used to share files on a network
SORT	Used to sort data numerically or alphabetically
SUBST	SUBSTitute a virtual drive name for a path name
SYS	SYStem—transfers system files to new disk
TREE	Displays subdirectory paths on disk

Fig. 15-9   *Continued.*

ming tools and other features which are of great assistance to people developing software. Its ability to operate as a multiuser system and to provide multitasking capabilities is a great help. This is especially true when more than one programmer is needed to develop the software.

For example, a Unix-based microcomputer can have three terminals, each being used by an individual programmer (see Fig. 15-12). Each programmer can be using an editor, a compiler, a linker, and/or a debugger at the same time. The system only has one copy of each of these software development tools. However, as programmers call for these tools, a copy appears in their individual work spaces.

Each Unix programmer may have more than one tool working at one time. For example, a programmer may start compiling the source code for one program and, while the compilation is taking place, start editing another program. If three programmers do this, the system runs six application programs simultaneously. The multitasking, multiuser operating system keeps track of each program to ensure a balanced overall operation.

Other operating systems are currently being used on microprocessor-based systems. However, MS-DOS and Unix are the most popular. You will also find that once you become familiar with one operating system, you will be able to move easily to another operating system because they all have similar character-

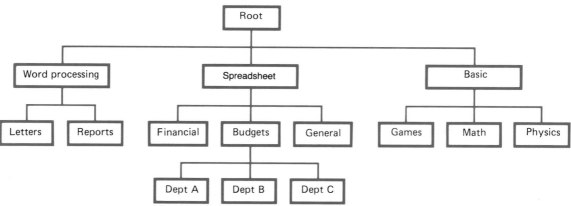

Fig. 15-10   A typical MS-DOS hierarchical file directory. Each level on the chart is one more subdirectory. To display the directory of department B spreadsheet files, you use the command C>DIR C:\SPREADSH\BUDGETS\DEPTB

```
CLS              } Clear the screen
ECHO OFF         } Do not echo batch file commands
CD\BASIC         } Change directory to one named BASIC
BASICA           } Run the application program BASICA
CD\              } When BASICA is done, return to root directory
END              } Run program (batch file) called END.BAT
```

Fig. 15-11   A simple batch file. The comments to the right are not normally included.

istics and they all provide essentially the same features.

Self-Test

Answer the following questions.

16. DOS
    a. Stands for disk operating system
    b. Is an operating system designed to work with the Intel x86 microprocessors
    c. Is a single-tasking, single-user operating system
    d. All of the above

17. MS-DOS uses the BIOS as a series of subprograms
    a. To interface the operating system and application programs to the memory-mapped display
    b. Which are loaded from disk to control the I/O devices
    c. To interface the operating system and application programs to the I/O devices
    d. All of the above

18. MS-DOS runs on a microcomputer which uses an Intel x86 CPU. However, each different hardware system implementation requires a new BIOS implementation because
    a. The I/O devices are interfaced through different microprocessors on different systems
    b. The BIOS translates system calls into actions based on the specific hardware architecture for the individual system I/O
    c. This is the only part of the software where the individual manufacturer can leave a personal mark
    d. Wrong; there should be no difference between the BIOS for one I/O implementation and the BIOS in another I/O implementation

19. Unix is a _____ operating system for microcomputers and minicomputers.
    a. Multitasking
    b. Multiuser
    c. Single-tasking, single-user
    d. Both *a* and *b*

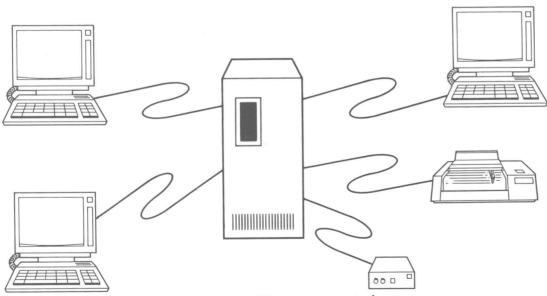

Fig. 15-12   A three-user microcomputer system. This system supports three terminals, a printer, and a modem from a multiport serial card.

15-4 PROGRAMMING TOOLS

Like any other form of development, software development must be supported with tools. Also, software development is made more productive by the addition of special tools which speed up the programmer's work. In this section, we will look at some of the tools which you may expect to find if you pursue a career which involves software maintenance or development.

The most basic software development tool is a programming language. A programming language converts the program design into machine language instructions the computer can execute. We tell the programming language what is to be done with statements written in the words and form of the language. This is called the *source code*. A special application program converts the source code statements into an appropriate set of executable binary instructions for the target CPU. This set of executable binary statements is called the *object code*. Figure 15-13 is a flow diagram showing how these tools work together to produce an executable program.

In this section, we will look at the software development tools which help the programmer develop the source code and debug the object code. In the next section we will look the different programs (languages) used to translate source code into object code.

As we learned earlier, the process of developing a program follows several steps:

Writing a specification
Creating a design
Implementing the design in code and testing and correcting the implementation
Acceptance testing

What are the different tools used to help the programmer in each of these steps? As you know, the specification is usually a narrative document supported with charts, tables, formulas, drawings, and so on. As you might guess, this document is generated by using the same tools for writing any scientific, technical or engineering report. The major tool used in this area and in other stages of the development process is an editor or word processor.

An *editor* is usually one of the programs included with the system software for a micro- or minicomputer. It is used to create and modify machine-readable text. *Machine-readable* means that the material is in a binary format,

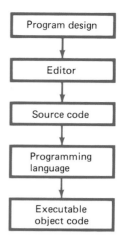

Fig. 15-13 **The flow from design to the generation of excecutable object code.**

usually ASCII characters. It is not absolutely necessary to have the specification written in machine-readable text. However, it is often a great help because it leads to specifications which are easily modified and therefore easily maintained.

There are two different kinds of programs used to create machine-readable text with a microcomputer: editors and word processors. An editor is the simpler of the two. Editors let the user generate lines of text, use tabs to put the text in columns, find lines of text by searching for *strings* (a sequence of characters) or by line number, and search for and replace one string with another. *Word processors* have additional features which let them manage paragraphs and pages, justify type, vary line spacing, and match their output to fit the needs of a wide range of letter quality printers. Either a word processor or an editor can be used to prepare and maintain a specification.

Design documents can also be prepared with either an editor or a word processor if the design is described in pseudocode or narrative form. For example, a number of designers use a *program design language (PDL)*. To use a PDL, designers describe the design in pseudocode by using an editor. The design source code is then processed by the PDL processor, which formats the information. Figure 15-14 shows the PDL processor for a demonstration program. This software tool allows the programmer to organize, modify, and communicate the design much more easily than could be done by simply using an editor.

Designs described with flowcharts, structure charts, or other graphic documentation may

PROCEDURE_ALPHA

```
This is the Text Segment for PROCEDURE_ALPHA. Note
that this example also contains an Extended Logic
Segment with the same name, which demonstrates that
PROCEDURE_BETA can properly reference it.
```

PROCEDURE_ALPHA

REF PAGE	
	IF First Condition
14	FIRST_PROCEDURE
	IF Second Condition
14	SECOND_PROCEDURE
	IF Third Condition
15	THIRD_PROCEDURE
	IF Fourth Condition
15	FOURTH_PROCEDURE
	IF Fifth Condition
16	FIFTH_PROCEDURE
	ELSEIF Fifth First Alternate Condition
16	FIFTH_FIRST_ALTERNATE_PROCEDURE
	ELSE Fifth Second Alternate Condition
16	FIFTH_SECOND_ALTERNATE_PROCEDURE
	ENDIF
	ELSE Fourth Alternate Condition
15	FOURTH_ALTERNATE_PROCEDURE
	ENDIF
	ELSE Third Alternate Condition
15	THIRD_ALTERNATE_PROCEDURE
	ENDIF
	ELSE Second Alternate Condition
14	SECOND_ALTERNATE_PROCEDURE
	ENDIF
	ELSEIF First First Alternate Condition
14	FIRST_FIRST_ALTERNATE_PROCEDURE
	ELSE First Second Alternate Condition
14	FIRST_SECOND_ALTERNATE_PROCEDURE
	ENDIF

Fig. 15-14   Output from a PDL (program design language) processor.
(Courtesy of Advanced Computer Concepts, Inc., Sarasota, FL.)

use a special microcomputer-based design documentation tool. Special computer-based graphic systems are used to draw these charts. Many designs are documented with handdrawn charts. The disadvantage of handdrawn charts is that it is difficult to maintain them, especially for large systems which have many paper diagrams.

Once a design is complete, source code is prepared by using an editor or word processor. Editors were made to write assembly language source code. Today, some programmers still use the editors for this work. However, many programmers use the more complex editing features provided by word processors.

Once the source code is written, a special program translates the source code into object format. Depending on the language used, this translation program is called an *assembler, compiler,* or *interpreter.* We will learn more

about these tools in the section on computer languages.

The *debugger* is an important software tool. It is used to aid the process of fixing problems designed or coded into the program. Such problems can come from many different places. However, there are two major sources of error. *Semantic errors* are errors in the way the design describes the job to be done. For example, a simple mistake in the design calls for a DO-UNTIL when a DO-WHILE is needed. Worse yet, the program logic just does not solve the problem. *Syntax errors* are mistakes in the coding. The coding error can be caused by a wrong implementation (DO-UNTIL written instead of DO-WHILE called for) or by a typographical error.

You should note that both of the errors in the above examples will not be caught by the language translator. This is because the DO-

WHILE and DO-UNTIL statements can be translated by the programming language. The programming language may catch a typographical error in the source code. This is because the typographical error does not produce any statement the programming language recognizes.

The debugger helps the programmer catch many different kinds of programming errors. The debugger is a program (software tool) supplied with the system software. Debuggers are loaded into memory with the program to be tested and provide two main testing features.

Breakpoints are a major debugger feature. Other debugger features let the programmer examine or fill a range of memory locations, examine or change (set) the contents of one or all of the microprocessor registers, start program execution at a particular address, single-step the program, trace the program, or disassemble the program. In the following paragraphs, we will look at some of these features offered by debuggers and how these features help the programmer find problems in the program.

A breakpoint is used to stop program execution and return control to the programmer. When a breakpoint stops program execution, it gives the programmer information about the system-program status. A typical breakpoint display is shown in Fig. 15-15.

Breakpoints can be set in a number of different ways, depending on the debugger flexibility. For example, a breakpoint can be set to stop program execution each time a particular instruction is executed, or a breakpoint can stop execution after a particular instruction executes a certain number of times. This allows the programmer to examine the contents of the CPU registers, memory, I/O port status, data registers, and so forth, as they are left after the particular instruction executes. Breakpoints can also be set to show when the program reads or writes a particular memory or I/O location.

The *debugger fill command* loads all memory locations in a given address range with data given by the programmer. For example, a programmer may use the fill command to put all logic 1s in memory locations between 4000H and 4100H. To do this, a command such as

$$\rightarrow \text{FILL 4000:0,100,FF}$$

is used. This is used before a program executes. The examine command is used after the program executes. It gives the display shown in Fig. 15-16. This shows the programmer a picture which tells how many memory locations the stack used.

The *go command* starts program execution with a particular value loaded in the microprocessor program counter. For example, the command

$$\text{GO 4200H}$$

instructs the microprocessor to start executing the program beginning with the instruction at memory location 4200H. Once the go command is used, program execution continues until a breakpoint is reached, the program stops because it is naturally finished, or the program crashes.

Often the last step (crashing) is what happens. The debugger may or may not provide enough information after the program crashes to let you find the problem. However, it is common that a program which crashes destroys data in memory and in the registers. This data could tell you what caused the crash.

To avoid destroying the data, you use the debugger to execute the program one instruction at a time. This is called *single-stepping* the program. Some debuggers refer to this feature as *trace*. After each instruction executes, you can look at the contents of critical registers and memory locations to see if the data you ex-

```
→
AX = 1F13 BX = 0003 CX = 0607 DX = 1102 SI = 0036  DI = 0000 BP = 0000 SP = 0382
CS = F000 DS = 0040 SS = F000 ES = 0400 IP = CO53 FL = NC PE NA ZR PL EI UP NV
F000:C053      EBDD           JMP           C032
→
```

Fig. 15-15 The output from an 8088 debugger which has stopped at a breakpoint. When this debugger stops, it displays the current content of all the 8088 registers (AX, BX, CX, DX, SI, DI, BP, SP, CS, DS, SS, EX, and IP) in hexadecimal and the flags (FL). It also displays the content of the memory location which caused the break. The memory location is F000:C053; that is, the code segment register (CS) is set to F000 and the current program counter value (IP) is C053. The content of this location is EBDD, which is a JMP C032 (jump to location C032 within the existing page).

→ d1800:0, ff

```
1800:0000   00 00 00 00 00 00 00 00-00 00 00 00 00 00 00 00     . . . . . . . . . . . . . . . .
1800:0010   00 00 00 00 00 00 00 00-00 00 00 00 00 00 00 00     . . . . . . . . . . . . . . . .
1800:0020   00 00 00 00 00 00 00 00-00 00 00 00 00 00 00 00     . . . . . . . . . . . . . . . .
1800:0030   00 00 00 00 00 00 00 00-00 00 00 00 00 00 00 00     . . . . . . . . . . . . . . . .
1800:0040   00 00 00 00 00 00 00 00-00 00 00 00 00 00 00 00     . . . . . . . . . . . . . . . .
1800:0050   00 00 00 00 00 00 00 00-00 00 00 00 00 00 00 00     . . . . . . . . . . . . . . . .
1800:0060   00 00 00 00 00 00 00 00-00 00 00 00 00 00 00 00     . . . . . . . . . . . . . . . .
1800:0070   00 00 00 00 00 00 00 00-00 00 00 00 00 00 00 00     . . . . . . . . . . . . . . . .
1800:0080   00 00 00 00 00 00 00 00-00 00 00 00 00 00 00 00     . . . . . . . . . . . . . . . .
1800:0090   00 00 00 00 00 00 00 00-00 00 00 00 00 00 00 00     . . . . . . . . . . . . . . . .
1800:00A0   00 00 00 00 00 00 00 00-00 00 00 00 00 00 00 00     . . . . . . . . . . . . . . . .
1800:00B0   00 00 00 00 00 00 00 00-00 00 00 00 00 00 00 00     . . . . . . . . . . . . . . . .
1800:00C0   00 00 00 00 00 02 F4 E3-90 E4 F5 A2 44 DE A2 12     . . . . . . t c . d _ " D ^ " .
1800:00D0   3E D3 B6 C7 77 23 45 B2-A5 66 02 F4 E3 90 E4 5F     > S  G w # E 2 % f . t c . d _
1800:00E0   A2 44 DE A2 12 3E D3 B6-C7 77 23 45 B2 A5 66 02     " D ^ " . > S 6 G w # E 2 % f .
1800:00F0   F4 E3 90 E4 5F A2 44 DE-A2 12 3E D3 B6 C7 77 23     t c . d _ " D ^ " . > S 6 G w #
```

→

Fig. 15-16   Using the fill and examine debugger commands to determine how much memory is used by the stack. First, the fill command puts logic 0s in all memory locations from 1800:0000 through 1800:00FF (FILL 1800:0,FF,0). The program was run. This figure shows the display which results where the examine command (d1800:0,ff) is used to see what memory locations have been used by the stack. Each pair of characters shows the hexadecimal contents of a memory location. Sixteen memory locations are shown per line, and their ASCII values are shown at the right. In this example, the stack started at memory location 1800:00FF and extended down to memory location 1800:00C5.

pected to be there is really there. Between the single-stepped instructions, the programmer uses other debugger features to examine or modify the microprocessor system.

Each debugger feature lets the programmer step through the program in a controlled manner. This is very necessary when a program bug causes an uncontrolled crash. Frequently, crashes overwrite memory locations and registers with random data which covers up what actually happened. This means the programmer has no way to see what caused the crash. Breakpoints and single-step along with the examine and fill features help the programmer to see where the program execution went wrong and to find the mistake.

There are simple and complex debuggers. Simple debuggers can be used with any object code, although a source listing always helps. Complex debuggers also use the program source code. These more sophisticated debuggers let the programmer debug the program using the source code rather than absolute addresses, for example, which are required with the simpler debuggers.

At times the program only crashes when it is executed at normal speed. In other words, it will not crash when it is single-stepped, but it does crash when it executes at normal speed.

In this case a special tool is used to find the problem. This tool shows the sequence of instructions executed before the crash. This tool is called a *logic analyzer*.

The logic analyzer is a separate electronic instrument used to solve both hardware and software problems in microprocessor-based systems. Chapter 17 looks at the logic analyzer in more detail. When the logic analyzer is used as a software debugging tool, it is connected to record bus operations in its special memory. These bus operations include all parts of the microprocessor fetch and execute cycle. The logic analyzer memory shows a history of the bus transactions which led to a crash. Because it is a separate hardware system, its memory is not erased or written over by the crash. Figure 15-17 on page 454 shows a typical disassembler display. Both logic analyzers and debuggers offer disassembly functions. Each fetch operation is trapped (captured) and the disassembler shows the sequence of instructions which occurred. The logic analyzer can also be used to monitor operations on a microprocessor-based system which does not have a terminal. To use a debugger, you must be able to connect a terminal to the system.

The number and sophistication of the software development tools provided as part of the

M:	Move memory block	M <range>,<dest>
O:	Output to port	O <port>,<value>
R:	Examine registers	R [<register>]
S:	Search memory	S <range>,{<byte>¦"<string>"}...
T:	Trace program	T [<count>]
U:	Unassemble program	U [<range>]
V:	Set video/scroll	V [M<mode>][S<scroll>][100][150]
TEST:	Extended diagnostics	TEST
Where	<range>is:	<addr>{,<addr>¦L<length>}

→ UF000:8F01,20

F000:8F01	2E	CS:	
F000:8F02	C7063A0000B8	MOV	WORD PTR[003A],B800
F000:8F08	A01000	MOV	AL,[0010]
F000:8F0B	2430	AND	AL,30
F000:8F0D	3C30	CMP	AL,30
F000:8F0F	7507	JNZ	8F18
F000:8F11	2E	CS:	
F000:8F12	C7063A0000B0	MOV	WORD PTR [003A],B000
F000:8F18	A04A00	MOV	AL,[004A]
F000:8F1B	2E	CS:	
F000:8F1C	3A064200	CMP	AL,[0042]
F000:8F20	754D	JNZ	8F6F

→

‿‿‿‿‿   ‿‿‿‿‿   ‿‿‿‿‿   ‿‿‿‿‿
Address   Hex contents   Instruction   Operand

Fig. 15-17 Using the disassembler display to examine a series of instructions.
Adapted from Zenith Data Systems.

system software vary widely. In addition to being part of the system software provided with a microcomputer, a wide range of software tools are also available from independent sources. These tools are not absolutely necessary to develop or maintain software; however, they do make the job of developing software much easier. That is, they allow the programmer to work more productively.

Self-Test

Answer the following questions.

20. Software development tools are supplied
    a. As part of the system software for most microcomputers
    b. To allow the programmer to develop software for the target microprocessor-based system
    c. To let the programmer work in the most productive way possible
    d. All of the above
21. The most fundamental software tool is
    a. A program design language
    b. An editor
    c. A programming language
    d. A debugger

22. To generate the narrative part of a specification or a design document or to write source code, you often use
    a. A program design language
    b. An editor
    c. A programming language
    d. A debugger
23. A translation program called _____ converts the source code statements into object code executable by a CPU.
    a. A program design language
    b. An editor
    c. A programming language
    d. A debugger
24. _____ is one software development tool which allows the programmer to organize the detailed approach to solving a problem using a CPU.
    a. A program design language
    b. An editor
    c. A programming language
    d. A debugger
25. _____ uses breakpointing, single-stepping, filling, and examining to help the programmer find errors in the program.
    a. A program design language
    b. An editor

c. A programming language

d. A debugger

26. It is very difficult to find problems in a program which crashes due to an unknown error because

   a. You can't fix what you don't know about

   b. Crashing often covers up the problem

   c. An improperly coded loop instruction often causes the problem

   d. All of the above

27. Because the debugger must be loaded into memory along with the program being debugged and because the debugger must interact with the programmer, debuggers

   a. Are seldom used on general-purpose systems which have terminal I/O

   b. Must be used with general-purpose systems which have terminal I/O

   c. Work best on systems with 100 percent ROM

   d. Only work on mainframe computers

15-5 PROGRAMMING LANGUAGES

As you have learned, the design for a computer program is implemented by writing statements in a programming language. These program statements are called *source code*. We have also learned that a special program translates the source code into *object code*. This special program must understand the programming language statements. It must understand the words used, how the statements are put together, and what binary words must be used for each instruction.

There are many different programming languages and translators for these languages. In this section we will look at low-level and high-level languages. We will also look at the assemblers, compilers, and interpreters which translate source code into object code.

One of the oldest computer languages is assembly language. *Assembly languages* are low-level languages. They are called low-level languages because there is one mnemonic for each instruction in the processor instruction set. That is, a low-level language generates one machine instruction for each statement in the source code. A single high-level language statement often generates many object code instructions. To do a good job of assembly language programming, the programmer must know the

details of the processor architecture and each instruction.

The assembly language source code, with one mnemonic for each CPU instruction, is written in fixed columns so that the instruction mnemonics and the data they are to act on are clearly separated. The translation program which converts this source code into equivalent binary instructions is called an *assembler*.

As you know, each CPU instruction has a unique mnemonic used to write programs in that assembly language. A table in the assembler matches the mnemonics with their binary values. If the table which matches mnemonics to binary instructions is changed, the assembler generates object code for a different microprocessor. This lets developers of software tools create similar assemblers which generate object code for different microprocessors.

Assemblers also perform some functions other than simple translation of mnemonics to binary instructions. One of the most important functions is keeping track of instruction addresses. As the assembler processes a list of instructions in mnemonic form, it puts the first instruction at the address it is told to use as the beginning (origin). From then on, it keeps track of all other instruction addresses relative to this starting point.

The assembler accepts labels (names) rather than the memory location numbers as addresses. These names are called *symbolic addresses*. For example, the program shown in Fig. 15-18 on page 456 starts at memory location 7000H. The assembler is told that the first instruction is to be placed at this address (ORG = 7000H). It is also told that the symbolic name or label for this address is START. Later, the program jumps to CONT. The assembler knows that CONT is at memory location 7007. When the source code is assembled, the assembler produces a list of symbolic addresses showing the absolute memory locations assigned to each symbolic address.

Assemblers may also provide other programming aids. For example, Fig. 15-18 shows the op code ORG. ORG is not a mnemonic for any instruction in the microprocessor. These instructions are called *pseudo-ops* or *assembler directives*. The pseudo-op ORG is recognized by the assembler and causes it to perform the special function of setting the beginning address for the program being assembled. Likewise the pseudo-op END terminates the assembly process. Depending on the assembler used, there may be many other pseudo-ops in addition to the instruction mnemonics.

Source code

Object code

Assembly language

Assembler

Symbolic addresses

Pseudo-ops

Assembler directives

Macro

Macroassemblers

Absolute object code

Absolute address

Relocatable object code

Linking process

Relocatable assemblers

Linker

Cross assemblers

```
7000                        ORG    7000H      ;Load the program into
                                              ;memory starting at memory
                                              ;location 7000 hex
7000 0600     START         MVI B,00H         ;Put 00 hex into register B
7002 3E90                   MVI A,90H         ;Put 90 hex into register A
                                              ;Later we will compare the
                                              ;HI byte of the HL register
                                              ;pair with this value
7004 210080                 LXI H,8000H       ;Load the HL (memory pointer)
                                              ;with the starting address for
                                              ;the clear operation (8000 hex)
7007 70       CONT          MOV M,B           ;Move the contents of
                                              ;register B to the memory
                                              ;location pointed to by the
                                              ;HL register pair. This
                                              ;"Clears" this location
7008 23                     INX H             ;Increment the HL register
                                              ;pair so that it points to
                                              ;the next memory location
7009 BC                     CMP H             ;Compare the HI byte of the
                                              ;HL register pair with the
                                              ;accumulator. If the HI
                                              ;byte equals 90 hex, 4096
                                              ;locations have been cleared.
                                              ;Note: this sets the status
                                              ;register's zero bit to 1.
700A C20770                 JNZ CONT          ;If the zero bit is not 1,
                                              ;jump back to CONT. If the
                                              ;zero bit is 1, execute the
                                              ;next instruction
700D 76                     HLT               ;End the program
700E                        END               ;End the assembly
```

Fig. 15-18 An assembly language program which demonstrates labels.

Some assemblers allow the programmer to write a section of code and then include that section of code each time a special pseudo-op called a *macro* is used. Assemblers which permit this form of coding are called *macroassemblers*.

If the assembler output is a body of object code which can be loaded and executed, the output is called *absolute object code*. This is because each instruction has an absolute address. An *absolute address* is an address which has an assigned memory location. Assemblers do not have to produce absolute object code. Some can produce *relocatable object code*. Before relocatable object code can be used, it must be processed or *linked*. The linking process converts relocatable object code into absolute (and therefore executable) object code. The linked relocatable object code starts at an absolute memory address given by the programmer at the time of linking.

Assemblers which allow you to produce a group of modules which must be linked before they are executable are called *relocatable assemblers*. Relocatable assemblers may produce either relocatable code or absolute code, depending on pseudo-ops included in the source statements. When a relocatable assembler is used, many modules can be produced and tested independently. Later, the individual modules are linked together to produce the final body of object code without modifying each module.

A linking editor or *linker* is used to combine a group of relocatable modules. The relocatable modules can include modules from a standard library of modules, as well as modules produced by assemblers and compilers. This allows the programmer to use a high-level language where the high-level language is efficient and assembly language where assembly language is most efficient. The linking process is shown in Fig. 15-19.

Assemblers which run on one CPU but which generate object code for another CPU are called *cross assemblers*. They do the same job; they just run in a different program devel-

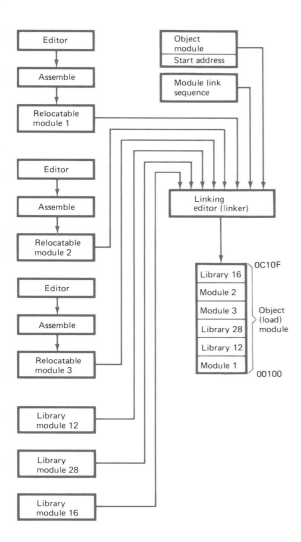

Fig. 15-19 Using the linking editor (linker) to produce a body of executable object code (an object or load module) which is made up of object modules from a number of different sources.

different from each of the other high-level languages. Some are fairly close to assembly-level programming. Most, however, use statements which are very far removed from the one-to-one relationship between assembly language statements and the processor instructions.

We use high-level languages to let programmers state problems with English-like or mathematical expressions. The programming job is always simpler when the programmer works in easily understood expressions. The programming expressions are easily understood when they are very much like the words the programmer uses to tell someone else about the problem being solved. When a programmer must solve a complicated problem, the problem statement is always clearer in a high-level language than in assembly language. When the purpose of the program is to control external hardware by moving bits and bytes to and from the I/O ports, assembly language expressions may be used to clearly express the actual problem to be solved.

High-level languages let the programmer work much more efficiently because the programmer does not have to convert the algorithm design directly into CPU instructions. In fact, a single high-level language statement may generate many equivalent assembly-level language instructions.

High-level language source code is converted into object code by one of two kinds of translators—a compiler or an interpreter. The *compiler,* like the assembler, is used one time to translate the source code into object code. It translates all of the high-level statements into a body of object code which, when executed on the target microprocessor, performs the needed function. A compiler is a much more complicated program than an assembler. This is because the compiler must translate complex statements written in English-like or mathematical terms to a series of machine instructions which perform the function. There is no one-for-one translation as is performed by an assembler. Like the assembler, there are both compilers and *cross compilers*.

The *interpreter* is different from the compiler because it translates source code statements one line at a time. This is done each time the program is executed. No body of executable object code is created. The advantage of an interpreter is that each line of source code is error-checked as it is executed. This means that problems can be detected and corrected on a line-by-line basis rather than on a module-by-module basis.

opment environment. For example, a Z80 assembler which runs on a PC (x86 processor) is a Z80 cross assembler.

Programs written in assembly language run very efficiently. This is because the programmer can make sure there are no wasted instructions. It is often said that an assembly programmer can write "tight" code. However, programming in assembly language is not efficient because the programmer must code each action the processor is to perform. High-level languages, on the other hand, do not run as efficiently, but they are much easier to program.

There are many different levels of high-level languages. Each high-level language is slightly

Compiler

Cross compiler

Interpreter

Because the source code is interpreted line by line, the interpreter must be resident while the program is executing. Programs written for interpreters usually run slower and take more memory space than do programs which are compiled. They are easier for the nonprofessional programmer to write and debug.

As you would expect, each microprocessor must have its own assembler. An assembler or cross assembler is usually the first programming language developed for a new microprocessor. In fact, many popular microprocessors have several native assemblers and cross assemblers. A *native assembler*, or cross assembler, is one writer for that particular microprocessor.

Likewise, there are usually different compilers which generate object code for a particular microprocessor. Some microprocessors have both native and cross compilers. However, many microprocessors do not have native compilers because a compiler is a complex general-purpose program, and many microprocessor architectures were not designed for this purpose. For example, there are cross compilers written for the Intel 8051 single-chip microcomputer. However, its control architecture does not easily support native compilers.

There are many different compilers with different levels of processing sophistication that process different languages. Each programming language must have its own compiler, and different compilers may be available for the same programming language. The differences between these compilers are programming features and differences in the sophistication of the programming language. Again, any microprocessor may or may not support different programming languages, depending on the compilers which have been written for that microprocessor.

Some commonly used high-level languages for microprocessors include: BASIC, FORTRAN, Pascal, PLM, C, and COBOL. BASIC (Beginners All-purpose Symbolic Instruction Code) is a simple language for general purpose computing often offered as an interpreter. FORTRAN (FORmula TRANslator) is an old programming language primarily used for solving scientific problems. Pascal, PLM, and C are structured languages which make extensive use of the programming constructs we learned in Chap. 14. They are principally used to solve logic programs. COBOL (COmmon Business Oriented Language) is a language developed specifically for business applications.

Self-Test

Answer the following questions.

28. The purpose of a programming language is to
    a. Let the programmer use BASIC
    b. Let the programmer use mnemonics and symbolic addresses to tell the microprocessor what to do
    c. Convert high-level statements into executable object code
    d. Give the programmer a way to instruct the processor to perform the functions described in the program design without directly using binary instructions

29. The translator which converts source code written in the programming language into executable object code is called
    a. A compiler   c. An interpreter
    b. An assembler   d. All of the above

30. A programming language which uses mnemonics for each processor instruction is called
    a. A high-level language
    b. An assembly language
    c. An interpreter
    d. FORTRAN

31. If a mnemonic used by an assembler is not used as a symbol for a CPU instruction, it is a special instruction called
    a. A pseudo-op   c. Symbolic address
    b. Label   d. Any of the above

32. The assembler keeps track of addresses so that the programmer does not have to know the absolute memory location for each instruction. Additionally, the programmer can give some addresses a name so they can be called by their name rather than by the memory location. This name is called _____ or _____ . (Choose two of the four answers.)
    a. A pseudo-op
    b. A label
    c. A symbolic address
    d. An assembler directive

33. Some assemblers let you write sections of assembly language code and include them by referring to their name rather than including all of the instructions in the source listing. An assembler with this capability is called
    a. An absolute assembler
    b. A macroassembler
    c. A linker
    d. A relocatable assembler

34. Some assemblers let you write and assemble code in small modules. The final object code is made up of a series of these modules produced by _____ and which are combined into one executable body of object code by _____ . (Choose two of the four answers.)
    a. An absolute assembler
    b. A macroassembler
    c. A linker
    d. A relocatable assembler
35. The major difference between a high-level language and an assembly language is that
    a. The object code generated by an assembly language processor (assembler) is specific for a particular microprocessor; it is not for a high-level language processor (compiler or interpreter)
    b. An editor can be used to generate source code for either the assembler or the high-level language processor
    c. An assembler uses one statement to generate one instruction, whereas a high-level language processor may generate many instructions for each statement
    d. The high-level language processor produces much more efficient object code than the assembly language processor
36. BASIC programs are implemented by _____ the sources.
    a. Assembling   c. Compiling
    b. Interpreting   d. Either b or c
37. FORTRAN programs are implemented by _____ the sources.
    a. Assembling   c. Compiling
    b. Interpreting   d. Either b or c
38. The process of converting the statement MOV A,C into machine-executable object code is called _____ the source code.
    a. Assembling   c. Compiling
    b. Interpreting   d. Either b or c

SUMMARY

1. There are two kinds of software. Applications software is written to do the job at hand. System software is general-purpose software provided to make the application programming job easier.
2. The purposes of system software are to avoid rewriting often-used programs, provide software which requires special knowledge to write, and force the use of standard methods of communicating and storing data.
3. System software is usually made up of operating systems, programming languages, and utilities.
4. A simple program or subprogram used to control the operation of other programs and connect their data with the hardware of the microprocessor-based system is called an executive.
5. An operating system is a general-purpose executive which gives the application programs easy communications with each other and with the system hardware.
6. A disk-based operating system is called DOS.
7. A disk-based operating system uses resident and transient programs. The resident programs always stay in system memory, and the transient programs are loaded when needed.

8. One of the first jobs the operating system performs after it is loaded is to initialize the microprocessor-based system and its peripheral hardware.
9. The command line interpreter intercepts commands from the keyboard, interprets these commands, and then requests the correct action from different parts of the operating system.
10. The operating system is loaded first so that it is always the highest-priority software on the processor stack. If you start an application program executing from the operating system command line interpreter, control returns to the command line interpreter once the application program finishes.
11. An executing application program can make use of system functions by making calls on the operating system features.
12. Sometimes a programmer does not use the system services to access the system hardware. When this is done, there is a possibility of confusing the operating system because the hardware setup is changed without the operating system knowing about the change.
13. Most PCs and microprocessor-based systems use a single-user, single-tasking operating system.

14. A multitasking operating system appears to let the system programmer execute two programs at the same time. Because the microprocessor can only execute one instruction at a time, the multitasking is really done by time-sharing the microprocessor.

15. Because a multitasking operating system must have a task scheduler in addition to the application programs, the application programs get even less processor time.

16. A time-shared multitasking operating system switches between application programs based on time. A significant-event multitasking operating system switches between application programs when a programmer-defined event happens.

17. A multiuser operating system is a multitasking operating system with the software and hardware support needed to allow two or more programmers to interface with application programs at the same time. Each programmer interfaces with the application programs with a terminal.

18. MS-DOS (Microsoft disk operating system) is a single-tasking, single-user operating system designed for IBM-compatible PCs. It is one of the popular 16-bit operating systems in use today.

19. MS-DOS depends on a special software package called the BIOS (basic input/output system) which processes all of the communications with system hardware. The BIOS can be rewritten to let the operating system work with a slightly different hardware configuration.

20. Unix is a multitasking multiuser operating system which is a very good programming environment.

21. The most basic software development tool is a programming language. A programming language converts source code into executable object code.

22. Specifications may be prepared with handwritten documents or with machine-readable documents prepared with an editor or word processor.

23. Design documents using a program design language or pseudocode are prepared with an editor or a word processor. Designs documented with drawings use either handdrawn documents or documents produced with special software packages developed to document software designs.

24. A debugger is used to test software under enough control so that the programmer can find any errors.

25. There are two basic kinds of errors. Semantic errors are design errors and syntactic errors are coding errors.

26. Debugger functions include fill, display or examine, single-step, and breakpoints.

27. Breakpointing lets the programmer run the program until a particular memory or I/O location is read or written. When this happens, the program execution stops and the programmer is shown the current contents of a few memory locations and the microprocessor registers.

28. Often a program crash puts bad data into memory and the registers, making it very difficult for the programmer to discover what caused the program crash.

29. A logic analyzer is a separate electronic instrument which is used to record the flow of fetch and execute data on the microprocessor bus. It can show the sequence of instructions which lead up to a crash.

30. A low-level programming language produces one microprocessor instruction for one source code statement.

31. A high-level programming language may produce many microprocessor instructions for one source code statement.

32. Low-level languages are quite efficient and are frequently used when the microprocessor function is to control external hardware.

33. High-level languages do not produce as efficient object code, but are much easier to work with because the statements are English-like or mathematical expressions.

34. An assembler is a low-level language which translates mnemonics into machine instructions, keeps track of symbolic addresses, and includes a few special mnemonics which perform special tasks.

35. A macroassembler allows the programmer to predefine a segment of code and to include it in the source listings by referring to its name.

36. An absolute assembler produces object code which starts at a given absolute (numerically defined) address.

37. The output of a relocatable assembler must be processed by a linking editor before it becomes absolute (executable).

38. A linker can be used to combine many relocatable modules into a single executable module.

39. Assemblers which run on one CPU but produce object code for another CPU are called cross assemblers.

40. High-level source code is converted into object code by a compiler or an interpreter.

41. A compiler is used to translate the high-level source code into an executable object module. After the conversion is complete, the compiler is no longer needed.

42. An interpreter converts the source code into executable object modules one line at a time. The interpreter must be used each time the program is executed.

CHAPTER REVIEW QUESTIONS

Answer the following questions.

15-1. Software which is written to perform a specific task is called
  a. MS-DOS
  b. BIOS
  c. An application program
  d. All of the above

15-2. The basic function of system software is to
  a. Perform a specific job
  b. Provide a hardware-independent I/O function
  c. Improve programmer productivity
  d. Ensure that a standard file format is used

15-3. System software is made up of
  a. Operating systems
  b. Utilities
  c. Programming languages
  d. All of the above

15-4. An executive is a software package which performs
  a. Task scheduling and I/O communications
  b. Editing of executive letters and memos
  c. File copying and disk formatting
  d. All of the above and more

15-5. Typically executives are ROM-based, and operating systems are
  a. Very general purpose
  b. Disk-based (DOS)
  c. Single- or multitasking
  d. All of the above

15-6. A DOS package uses both resident and _____ programs.
  a. Application
  b. Permanent
  c. Transient
  d. All of the above

15-7. One of the first jobs that the operating system performs after booting is to
  a. Load and run the application program
  b. Clear memory
  c. Communicate with the console device
  d. Initialize the system peripheral hardware

15-8. The purpose of the command line interpreter is to
  a. Initialize the system peripheral hardware
  b. Accept, interpret, and pass on commands from the user
  c. Provide system services to application programs which need to use functions available in the operating system
  d. All of the above

15-9. The _____ is always the first program loaded in a microcomputer, so it always has the highest priority on the system stack.
  a. Operating system
  b. Command line interpreter
  c. Application program
  d. Boot program

15-10. An executing application program can make use of system services by
  a. Exiting to the command line interpreter and requesting the system operator to enter the command
  b. Making a system service request using the techniques specified for the operating system
  c. Rebooting the system from disk
  d. The application program cannot make system calls

15-11. If the system services are bypassed
   a. The application program may be able to perform the function quicker by direct access
   b. Later problems may develop if the application program modifies the hardware setup during this direct bypass
   c. The operating system does not know about any setup changes to the hardware
   d. All of the above

15-12. MS-DOS is an _____ operating system.
   a. Single-tasking   *c.* 8-bit
   b. Multitasking   *d.* Multiuser

15-13. A multitasking operating system
   a. Allows two application programs to run at the same time
   b. Allows the system to support multiple-user interfaces by using multiple terminals
   c. Does not require a time allocation for the task scheduler
   d. Allows two or more programs to appear to run at the same time

15-14. A time-shared multitasking or multiuser operating system uses _____ as its significant event to trigger a switch between tasks.
   a. Time   *c.* The boot ROM
   b. An I/O interrupt   *d.* The resident software

15-15. _____ is a single-tasking, single-user operating system designed for use with the Intel x86 microprocessors in a PC architecture.
   a. OS/2   *c.* Unix
   b. MS-DOS   *d.* Windows

15-16. _____ is a multitasking, multiuser operating system designed by Bell Laboratories for use with 16-bit minicomputers.
   a. MS-DOS   *c.* OS/2
   b. Unix   *d.* Windows

15-17. MS-DOS uses special software called the _____ to configure the operating system to match the exact hardware environment in which the operating system is installed.
   a. TPA   *c.* DOS
   b. BDOS   *d.* BIOS

15-18. The most basic software development tool is the
   a. Debugger   *c.* Linker
   b. Editor/word processor   *d.* Programming language

15-19. The _____ is used to find problems with the object code.
   a. Debugger   *c.* Linker
   b. Editor/word processor   *d.* Programming language

15-20. The _____ is used to create a machine-readable specification, design, or source code.
   a. Debugger   *c.* Linker
   b. Editor/word processor   *d.* Programming language

15-21. To create an executable object module from a series of relocatable object modules which have been developed over time, use the
   a. Debugger   *c.* Linker
   b. Editor/word processor   *d.* Programming language

15-22. A low-level programming language
   a. Has one line of source code for each processor instruction generated
   b. Can generate very efficient object code (fewest introductions to do the job)
   c. Is the most difficult source code to understand
   d. All of the above

15-23. A high-level programming language
   a. Is probably the best for generating object code which is used to control hardware connected to a microprocessor

b. Generates the most efficient object code (fewest instructions to do the job)
c. Presents the source code statements in an English-like or mathematical form
d. All of the above

15-24. _____ can produce object modules which can either be loaded and executed or ones which must be processed by a linker.
a. A debugger
b. An absolute assembler
c. A relocatable assembler
d. All of the above

15-25. If you want to assemble or compile source code on a CPU other than the one which will execute the object code, you need _____ assembler or compiler.
a. A relocatable
b. A cross
c. An absolute
d. A macro

15-26. A compiler is more like _____ than _____ . (Choose two of the four answers.)
a. An interpreter
b. A cross-compiler
c. An assembler
d. A cross assembler

CRITICAL THINKING QUESTIONS

15-1. Many early industrial microprocessor-based products did not use system software. Today, many industrial microprocessor-based products use system software and are based around PC architectures. What role do you think PC system software helped in making this change to PC-based systems?

15-2. Most PCs use one of the Intel x86 microprocessors for their CPU. However, each PC manufacturer has a BIOS which is customized to its PC. Why is there some difference in the BIOS if they all use an x86 microprocessor?

15-3. Why do you think that MS-DOS, the most popular PC operating system, is a single-user single-task operating system rather than a multiuser, multitasking operating system?

15-4. UNIX is considered to be a highly portable operating system; that is, it has been written to operate on a wide range of processors. As you learned, UNIX is written in the programming language C, which makes it much easier to modify and to transport to other processors. If UNIX is written in C, what other piece of systems software must be available for the new processor, before UNIX can be transported to that processor?

15-5. From time to time, a programmer will write most of an application in a high-level language such as C, but certain parts will be written in assembler. Why do you think the programmer will use two different types of languages to develop the application?

Answers to Self-Tests

1. *c*	11. *a*	21. *c*	31. *a*
2. *d*	12. *c*	22. *b*	32. *b, c*
3. *c*	13. *c*	23. *c*	33. *b*
4. *d*	14. *d*	24. *a*	34. *d, c*
5. *d*	15. *b*	25. *d*	35. *c*
6. *d*	16. *a*	26. *b*	36. *d*
7. *a, d*	17. *c*	27. *b*	37. *c*
8. *b*	18. *b*	28. *d*	38. *a*
9. *d*	19. *d*	29. *d*	
10. *b*	20. *d*	30. *b*	

CHAPTER 16

Servicing Microprocessor-Based Products

■

CHAPTER OBJECTIVES

This chapter will help you to:

1. *Name* and *explain* the four steps in servicing a defective product.
2. *Find* the area of failure in a microprocessor-based product.
3. *Predict* the problem using basic troubleshooting techniques.
4. *Perform* basic repairs.
5. *Use* test equipment to troubleshoot microprocessor-based products.

───────

As the number of microprocessor-based products grow, the number of microprocessor-based products that fail will also grow. Because the microprocessor-based product is often expensive and depended on, it must be placed back in service quickly. The process of servicing microprocessor-based products is very much like servicing any other product. You have to find out what is really wrong, isolate the problem, and correct the problem. If a product is microprocessor-based, some special techniques will be needed to isolate the problem.

■

16-1 REVIEWING SERVICE PROCEDURES

Any time we service a defective product the job has four steps.

1. Understanding the need for service
2. Finding the area of failure
3. Identifying short- and long-term repair options
4. Implementing a repair

The following paragraphs look at each of these topics in greater detail. We must understand how each step applies to the service of electronic products, including microprocessor-based products.

With any product it is most important you understand the need for service and repair. Always ask yourself, "what does the customer believe is wrong?" Then be sure you know the answer to "what will make the customer feel the problem is corrected?" These two important questions are even more important when you are servicing microprocessor-based products.

Why do you need to know the answers to these questions? First, there can be numerous problems with a product. This is very true with a complex product like the PC. It is not uncommon to find that some problems were always there. They are not really problems—they do not need to be fixed. It is important you know what problems the customer wants fixed. If there are other problems, you can tell the customer about them once you understand the complete situation.

Second, you must know what the customer expects you to do. For example, you are called to fix a PC. When you arrive at the office, you are told the screen has become too dim. Investigating, you find the PC uses an old monitor. Over time it is quite possible the CRT lost some of its original brightness. This can be corrected by installing a new CRT. Will this fix the problem?

With further investigation, you find the office was just rearranged. The monitor was moved in the process. Now, light from a window falls directly on the CRT. New CRT or old, the monitor will not work very well in this light. Although the CRT may need replacement, proper servicing in this situation is to shield the CRT from the outside light or to reposition the monitor so it can be used. In this example, the customer really called because the monitor did not work in this situation, not because the monitor failed, although that was the original complaint.

Once you understand the problem to be corrected and what correction will satisfy the customer, you can begin to diagnose the technical problem, if one exists. Remember, the personal computer is a universal product. This means it can do many different things with no more than a new software package. It also means the product is difficult to understand and can be easily misused. Or someone may be asking the PC to do something it cannot do. Both situations cause service calls.

Understanding what the customer knows is important. If you are called to service a TV or hi-fi system, you can reasonably expect the average consumer knows what a hi-fi does. On the other hand, many people in consumer, commercial, and industrial environments do not know what to expect of their computer. They may well ask it to solve problems at the push of a button.

Additionally, many people find microprocessor-based products complex and difficult to understand. For example, the VCRs in use today have complex programming capabilities, and the sophisticated copiers found in offices have many features—both are difficult to operate. All of these issues are important to understand before you start to service a product. It is very important that you service the real problem.

It is also quite possible you will not know the exact function the product is to perform. The wide variety of software and applications for PCs and other industrial and commercial microprocessor-based products is almost limitless. It is almost impossible to know in advance what all of them do and how they are to perform. To do a good job of servicing these products, it is much more important that you thoroughly understand the components which make up these products than it was in the past, because previous electronic products had more limited functions.

This is also a good time to consult the manufacturer's service manual. Such a manual may well have a suggested troubleshooting procedure as well as a number of helpful hints suggesting areas to be attacked first and how to attack them. If there is no manual, or if the manuals at hand do not have any suggested service procedures in them, you must develop the approach yourself. It is worth your time to think out an approach.

A logical approach has many benefits. First, you know where you are going and what you are to do next. Second, you have a way to track what you have done. This avoids repeating earlier work. Third, it helps you to avoid frustration which can limit your ability to think. Following an organized plan avoids the panicky "what do I do next" feeling.

Self-Test

Answer the following questions.

1. The first step in approaching any service job is
   a. Isolating the fault
   b. Identifying the options
   c. Finding out what is really wrong
   d. Making sure the customer can pay the bill
2. Briefly explain why it is important you understand what the customer *thinks* the problem is before you begin work on the product.
3. Briefly explain how it is possible that a product can have a failure which is not the problem to be fixed and which never needs to be fixed.
4. You are called to service an industrial specific-purpose product. It is based on the personal computer. Even though you have never seen this product, you can expect this product to have
   a. An Intel x86-family-base CPU
   b. A memory-mapped video system
   c. Auxiliary cards which fit a standard bus structure
   d. All of the above

5. It is much easier for the customer to have inappropriate expectations for a microprocessor-based product because the products
   a. Tend to be universal in nature and therefore can do many different things
   b. Tend to have a great deal of functionality and this may confuse the customer
   c. Are newer and therefore the customer has not had as much time to become familiar with the product
   d. All of the above are valid reasons for different microprocessor-based products

16-2 FINDING THE PROBLEM

Finding the area of failure in a microprocessor-based product is much like finding the area of failure in any other electronic product. It is important you have a *systematic approach*. The basis for a systematic service is often the *divide-and-conquer approach*. First you isolate those parts which work from those which do not work. Then you concentrate on those parts which are not working.

The best starting place is to observe the symptoms. What is the product telling you about itself? What can the people who work with the product tell you about it? Observing the symptoms begins with knowing what happened. Did the product just die? Was it working properly or was it showing signs of failure? Did it ever work? Is the power turned on and off every day, each time it is used, or hardly ever? Has it ever been serviced before? If so, who did it and what information did that individual leave? How often is the product used—every day, every week, or every month?

Once you determine the original status and when it failed, you should then try to find what event caused the failure. Was the failure caused by an outside event? For example, was there a lightning storm? Did someone pour water on the unit while watering a nearby plant? Was there smoke or a noise when it failed? Has the unit been moved recently? Did it fail while in use or on power-up?

The next question you must answer is, "how does the failure show itself?" What symptoms does the product show? What doesn't the product do that it should do? Does it have any life at all or is it completely dead? If it is only partially dead, does the part that is alive work properly or is it also having problems? Does the product always exhibit the problem or does the problem come and go (an intermittent problem)?

It is also helpful to know about things that go with the product. Where are the accessories, if there are any? What technical manuals exist?

You must ask these questions until you are sure you completely understand the problem you are to fix. If you start to fix a problem you do not really understand, you may fix the wrong problem or spend a lot of time finding an answer which could have been easily obtained with just a few preliminary questions.

Now you know what the problem is, it is time to analyze the possible causes. It is very important to start with the obvious. Is the completely dead product plugged in? Do you have power to the outlets for the product? Have other products also failed at the same time? Has someone else attempted to service the unit? This is very important to understand, because they may have made adjustments to the product which must be corrected before the product can ever operate properly again.

Are all of the peripherals and I/O devices connected? For example, some products "hang" when a printer or a modem is not connected to the I/O port. In these cases the system software attempts to initialize the peripheral, finds it is not connected, and halts. In other cases, the peripheral is connected when the system initializes, but it becomes disconnected. A program expecting to find the peripheral connected hangs when the peripheral cannot be found. Good software should not do this without informing the operator. However, not everyone is using well-written software. You will find software and system problems are often a cause for service calls.

Limiting the possibilities reduces the work you need to do to find the problem. Again, do not overlook the very obvious. It is embarrassing to spend 20 minutes searching for a problem only to find the unit is unplugged.

When you limit the possibilities, eliminate the obvious first. Figure 16-1 is a checklist of obvious possibilities you should always review before beginning any detailed product analysis. This is all part of organizing your attack on the defective unit. Remember, an organized approach will make the difficult troubleshooting jobs much easier in the long run.

If the product passes the obvious tests, it is time to move to more detailed *troubleshooting techniques*. Usually this means the product

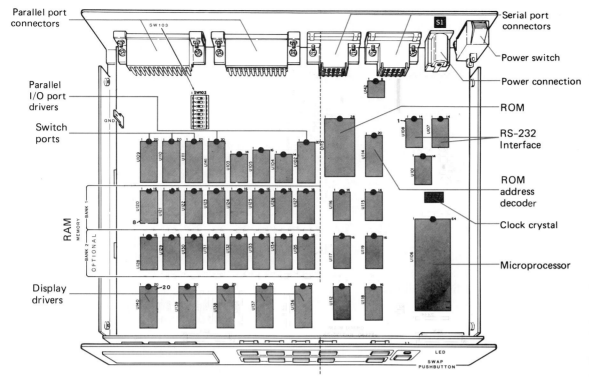

Fig. 16-5 Identifying parts in a microprocessor-based product. This product uses a modified Z-80 CPU which includes a serial port, ROM, two Centronics parallel ports available to the user via two rear panel connectors, an internal parallel port to read the 8-bit DIP switch (SW103), and up to 512 kbytes of RAM. *(Courtesy of Heath Company.)*

If you have the proper test equipment and a schematic or the microprocessor pinout, you can check for clock, address, data, memory control line, I/O control line, and even interrupt activity. Your first step is simply to see if there is any activity.

If you can find the microprocessor reset input, force the RESET line so you reset the processor. When you release the RESET line, monitor a low-order address line for activity. How you check for activity and what information you can gain depends on your test equipment. The section on test equipment tells what you can expect from different pieces of test equipment.

If there is no sign of activity from the microprocessor, there are a few other steps available to you.

1. Is the processor getting power? With the voltmeter negative clip connected directly to the power supply return line, check for V_{CC} on the microprocessor power pin. Also check to make sure that the V_{ss} (ground) pin is at ground.
2. Does the microprocessor activity change significantly when you remove socketed ROM, RAM, or latches? A significant change may tell you one of these parts has a major failure.
3. If the microprocessor is socketed and you have a replacement available, try a substitution. If you cannot replace the IC, replace the CPU board if possible.
4. Start checking signals on each pin to make sure there are no incorrect signals connected to the microprocessor.

Self-Test

Answer the following questions.

17. Making sure the CPU is running must be one of your highest priorities because
    a. Nothing else runs until the microprocessor runs
    b. You cannot address memory if the CPU does not run
    c. You cannot run I/O diagnostics without the microprocessor
    d. A failed microprocessor causes the power supply to crowbar and lose output voltage

18. You can perform a preliminary "health check" on a microprocessor by checking for
    a. A clock signal
    b. Data on the data lines
    c. Data on the address lines
    d. All of the above
19. You are servicing a PC which doesn't work. When you try to boot the computer the screen displays

    + + DISK ERROR: Drive not ready! + +

    You now know
    a. There is a problem with the disk drive
    b. The microprocessor is operating
    c. The memory-mapped video system is operating
    d. All of the above
20. The reason you force the RESET line when trying to find out if the microprocessor is working is
    a. This resets any memory or I/O which may be connected
    b. When RESET is released, a working microprocessor addresses the RESET vector memory locations and you can look for that activity.
    c. Because a microprocessor will hang if it cannot find a valid instruction at the RESET vector
    d. All of the above

16-6 TROUBLESHOOTING MEMORY

When you have determined the CPU is running, your next area to attack is main memory. If possible, the best check for RAM failure is a RAM exercise program run by the CPU. This is called a *memory diagnostic*. This program may be ROM- or disk-based. Most diagnostics write a variety of patterns into memory and then check to see if the memory can reproduce all of the patterns.

RAM diagnostic programs are the best approach to memory testing because they catch one of the most difficult memory failures to check—pattern sensitivity. *Pattern sensitivity* is a RAM failure which only shows up when a certain pattern of bits is written to certain RAM locations. The complexity of RAM diagnostic programs depends on the number and complexity of the pattern sensitivity tests the program produces.

If you do not have access to a memory diagnostic, there are some alternatives. Simple substitution is always a reasonable approach. There are three ways to substitute.

1. Replace suspect RAM ICs with known good parts.
2. Swap RAM ICs to see if the problem moves or changes the way it shows up. This approach is always more feasible with memory since there are usually several identical ICs in any product.
3. If the product uses memory cards, swap or substitute memory cards.

ROM is not as easy to check as RAM because each product uses a ROM with a different bit pattern. If you have access to an EPROM programmer and the EPROM programmer will program the ROM or EPROM in your product, you can use the EPROM programmer to verify the EPROM. *Note:* the majority of products you will work with today will use EPROMs. Without an EPROM programmer, you must substitute to check the ROM or you can check the ROM using a processor-based diagnostic routine if the processor is running.

Swapping ROMs does not provide much information. Even though the ROMs are the same part (for example, both 27128s) they have different programs. *Be sure ROMs are replaced in their proper sockets.* If the ROMs are installed in the wrong sockets, the product will never work. The only troubleshooting information you get by swapping ROMs is to see if removing one from the bus lets the processor begin to run. If this is the case, you may have a major ROM failure.

Self-Test

Answer the following questions.

21. RAM diagnostic software checks memory by
    a. Exercising all of the memory data lines
    b. Enabling and disabling the CE (CHIP ENABLE) line
    c. Exercising all of the memory address lines
    d. Writing bit patterns to the memory and then making sure they can be read
22. You are working on a microprocessor-based product which uses eight 1 × 256-kbit dynamic RAMs. When you write any data to memory location 30FFF, the result comes back as a number between 80 and FF. To verify a memory failure, you swap the high- and low-order memory ICs. You

would expect the results to show that the high-order memory IC has a(n)
a. Permanent logic 1 bit at address 30FFF
b. Permanent logic 0 bit at address 30FFF
c. Operating bit at address 30FFF
d. None of the above
23. You perform the reset test on a microprocessor and nothing happens. Later, you remove one of the two 27256 EPROMs, try the reset test, and the microprocessor starts actively addressing memory, the product runs for a while, and then crashes. You now suspect
a. You goofed the initial reset test
b. There is a software error in the ROM-based program
c. An EPROM failure keeps the microprocessor from working
d. Somehow the two EPROMs became mixed up

16-7 TROUBLESHOOTING MASS STORAGE AND I/O

If you believe the RAM and ROM are working properly, make sure the mass storage systems are functioning. For most products this means floppy-disk drives, Winchester-disk drives, and occasionally a tape backup system.

Again, isolation is your greatest tool. A typical PC may have one or two floppy-disk drives in addition to an 80-, 120-, or 240-Mbyte Winchester drive. The first step is to disconnect all the drives from the controllers and remove the Winchester controller from the product. Now you can see if one of the two floppy-disk drives works. If one does, you have at least one good drive and a good floppy-disk controller. If you cannot get the floppy-disk systems to work, try to boot the product from the Winchester drive with the floppy-disk system removed.

Most products with floppy- and Winchester-disk drives are supplied with drive diagnostic software. Some advanced PCs have ROM-based diagnostics. These ROM-based diagnostics let you check the floppy-disk drives to see if they perform basic read-write functions. If you run the *ROM-based diagnostics* and still feel there are drive problems, run the more extensive floppy-disk-based diagnostic packages. These packages will help you pinpoint problems to the controller or the drive. They often

show controller problems to the nearest IC. Drives and controllers are often best serviced by replacing the drive or card. *Remember, a Winchester-disk drive may have a great deal of valuable information stored on it.*

With the mass storage isolated and operating, eliminate problems with the serial and parallel I/O ports. Again, the approach you take depends on the product. If you are working with a PC, the diagnostic software usually has an I/O port section. However, this software may not exercise the entire interface, or it may require a *loopback connector*. The loopback connector routes the port output signals back to its input. The diagnostic software then writes a signal to the output and looks for the same signal at its input.

Without diagnostic software, there are different approaches you can take. First, try writing a few simple characters to or reading a few simple characters from the device normally connected to the I/O port. If, for example, this is a printer, try to print a single character. If you can print a single character, send the printer a stream of the same characters. Does the printer generate and the I/O port receive a signal telling the host to shut off the flow of data? This may be an XOFF or DTR signal on an RS-232 system, or it may be the BUSY line on a parallel port. If you have an oscilloscope, logic probe, or any form of communications monitor, it is much easier to see which communications port lines are being exercised.

Setup is one of the major problems often encountered with devices using serial communications with an external peripheral. Before two devices can communicate, the baud rate, word length, parity, and handshaking must agree. It is common to find a peripheral with a limited range of communication parameters. For example, a printer serial port might be permanently fixed at 1200 baud. In this case, the host communication port must be set up to match the peripheral. Although this seems obvious, it is frequently the source of a service call.

If you are called to service a product that was hit by lightning, the communications ports are always suspect. Often, the cables between the computer and its printers, modems, and terminals run through the walls and ceilings. There they contact electrical conduit, steel beams, water and gas pipes, and so forth. During a lightning strike, these can couple high-voltage spikes into the communications lines. When these voltage spikes arrive at the computer and the peripheral, they often cause damage. With luck, you may find the damage is limited to destroying the communications interface ICs. Often these are socketed and can easily be replaced. The 75188/1488 and 75189/1489 are common ICs used for RS-232 interface.

Other forms of I/O require some ingenuity on your part. First, you must understand what the interface does. Is it an analog port with 0- to 10-V signals from 16 different lines? This is quite a different problem from servicing a PC game port with two different analog signal inputs (x and y) and the potentiometer drive signals. For most special I/O ports, you need basic electronic service equipment such as an oscilloscope and a DMM.

Many built-in I/O devices also use microprocessors. For example, it is very common for a PC keyboard to use a single-chip microprocessor to provide the scanning and encoding. There are two approaches to servicing these simpler microprocessor-based products. You can try simple substitution, or you can use instruments. With substitution, you can work with a minimum of information. All you need to know is what part should be substituted. Usually a properly operating identical unit is the best source for these parts. With instruments you need a schematic and part identification.

You can substitute parts in one of two ways. First, you can take the suspect part from a defective unit and put it in a working unit. Second, you can take a known good part and put it in the defective product. In both cases, you run some risk of damaging a good part. However, the risk is usually slight if you only substitute after you have checked to make sure the major risks are gone. For example, do not substitute known good ICs in a defective product until you have determined the power supply voltage is within the correct limits.

Module substitution is always a good approach. For example, if you are not getting response from a PC keyboard, substitute a keyboard from a working PC which uses the same keyboard. This is a quick check to see if you have a keyboard problem or if the problem is somewhere else. Remember, PC keyboards should not be unplugged after the processor initializes the PC. You must turn the PC off and reboot the product each time you change the keyboard. Keyboards are not the only modules which need to be initialized with the product. It is good practice to turn off the power to the unit when changing any modules.

Self-Test

Answer the following questions.

24. You are working with a PC-based medical instrument. When you boot the system, the drive activity indicator on both floppy-disk drives light, and shortly thereafter the system fails to boot, showing a disk error message. Your isolation troubleshooting technique tells you to
   a. Order a new floppy-disk drive controller
   b. Replace both floppy-disk drives, because one of them is more than likely defective
   c. Disconnect both floppy-disk drives from the controller and then reconnect each individually, trying to boot a single floppy system to see which floppy is bad
   d. Disconnect both floppy-disk drives, boot the system from ROM, load the floppy-based diagnostic programs, and check out the floppy controller and then the floppy drives

25. The main PC in an office complex has four floppy-disk drives driven by two floppy controllers. Diagnosis shows the main controller (drives 0 and 1) has failed. The system boots from this controller, which also has the video controller circuits. However, the second floppy controller (drives 2 and 3) is OK. It is a stand-alone controller. The short-term repair for this product is
   a. To order a new controller/video card, reconfigure the working controller to boot the system, and temporarily disable the main controller. This leaves the customer with two floppy disks while the new card is coming.
   b. To pull the controller/video card and return to the shop where you can attempt to service the controller. This leaves the customer with two floppy disks.
   c. Replace the defective controller with a second stand-alone controller; take the first one to the shop for repair. This leaves the customer with four floppy disks.
   d. All three of the above solutions give short- and long-term service solutions

26. You are called to repair a printer which "just started printing garbage characters." When you arrive you find the problem started right after the PC was used for a

computer-to-computer data transfer which "did not involve the printer at all." On inspection you find the printer connects to serial port COM1:, and it is permanently configured for 1200 baud, 1 start bit, 1 stop bit, 7-bit words, and no parity. Suspicious, you check the setup file and are not surprised to find COM1: is now configured for

a. 9600 baud, 1 start bit, 1 stop bit, 8-bit words, and no parity

b. 1200 baud, 1 start bit, 1 stop bit, 7-bit words, and even parity

c. 1200 baud, 1 start bit, 2 stop bits, 7-bit words, and no parity

d. 1200 baud, 1 start bit, 2 stop bits, 7-bit words, and even parity

27. You are told a PC has a keyboard input circuit failure. When you are also told the keyboard input failure was diagnosed by unplugging the original keyboard and plugging in a known good keyboard which did not fix the problem, you immediately ask if

a. The substitute keyboard used the same connector

b. The PC printed a keyboard error message

c. Any other keyboard from the other PCs in the office had been tried (maybe there were two defective keyboards)

d. The PC was powered-off before the substitute keyboard was plugged in and tried

16-8 OTHER TROUBLESHOOTING HINTS

Most microprocessor-based products are susceptible to *transient electrical noise*. Transient electrical noise can come from voltage spikes on power lines, on a telephone line, over RS-232 lines or other communications cables, or directly from static discharge to the product. These transients cause a wide variety of problems.

The simple and relatively harmless ones only modify a few bits in memory. This may cause a program to generate incorrect data or to crash. If the transient reaches the processor, it may change the program counter (or some other register) values. This usually results in an immediate crash. In some cases it may cause serious problems such as making a disk drive write incorrect data to a disk.

The more serious transients actually cause damage. We have discussed some of the damage which can occur when lightning strikes are coupled into the product via the communications ports. Additionally, these transients can be coupled into the product over the power lines. One serious problem they can cause is power supply failure. If the power supply places unregulated high voltage on its +5 V output, many ICs can be damaged. A poorly filtered power supply may pass the transient on to the load circuits. Again, these transients can either cause circuit damage or just temporarily modify the contents of a memory location or register.

Many power supplies have *crowbar* protection. Crowbar circuits cause power supply shutdown if the output voltage exceeds some desired limit. This protects the load circuits from damage due to excessive supply voltage. If the product regularly receives transient voltage spikes and these cause the power supply to crowbar, it may be necessary to install filtering on the ac lines to block transients before they reach the power supply. Similar problems due to transients on the communication lines or from static discharge may indicate the need for filters or the use of antistatic mats to prevent regular transient-caused shutdowns.

It is important that you understand a microprocessor-based product is much more susceptible to transients than is other electronic equipment. Remember, the microprocessor depends on exact register and memory values. Even a single bit error in a program counter can make the difference between correct actions and totally wrong actions. If the contents of a memory location or register change outside of program control, it will cause a problem. For example, each time the program counter increments, it must point to the next instruction. It cannot skip a few instructions as a transient might make it do. On the other hand, an audio amplifier may make a popping noise if a transient couples into one of its stages. But after the pop, there is no further effect on the amplifier's performance.

When you are servicing a microprocessor-based product, it is important that you also pay attention to the routine maintenance issues. For example, most of the larger products use a fan for cooling. Fans have two problems. First, rotating parts have a finite life. When they fail, the equipment overheats. Second, fans pull dirt and dust into the unit. This means the equipment needs regular cleaning.

Digital multimeter (DMM)

Volt-ohm-milliammeter (VOM)

Cable and card connectors can be damaged by dirt, dust, and corrosive substances in the air. A corroded connector does not conduct electricity. Corroded connectors can cause intermittent problems which often show up when the unit is moved. The movement causes the corroded connection to fail and the unit stops operating. Connectors can be cleaned by scraping or by using a light abrasive such as a pencil eraser. Often the simple operation of unplugging and replugging the cards and cables restores an inoperative unit to working condition. If this happens, you must thoroughly clean the connectors or you will soon return for a second similar service call.

When you are servicing, you must always be ready to find the unexpected. Products are used in odd ways and in odd places. This can result in conditions which sooner or later cause the product to fail. For example, you might find a PC in an office situation also used as a plant stand, a place to pile books and reports, or a footrest. These uses may lead to water or dirt in the PC, blocked ventilation so the unit overheats, or excessive shock and vibration. In these situations you must diplomatically explain to the users why the unit must not be treated this way and suggest a way to properly use the equipment.

Self-Test

Answer the following questions.

28. You are called to service a PC which "overheats after about 7 hours of operation." They are confident of the 7 hours because people noted the time the PC dies is just after the 3:00 P.M. welding shift comes on at the new building being constructed next door. You decide the first task is to
    a. See if the PC ventilation ports are blocked
    b. Check for transients from the welding operation
    c. Make sure the PC is really turned on at 8:00 A.M.
    d. See if this is a PC or disk drive problem
29. You are called to service a microprocessor-based product which often "shocks" the operator and then quits working. The trouble seems to happen most frequently in the winter. You might recommend
    a. Replacing the power supply module
    b. Disconnecting the printer
    c. Using an antistatic mat in the work area to keep static charges from building up on the operator
    d. All of the above
30. You are servicing a PC-based controller used in a steel mill. The place smells like someone spilled a bottle of vinegar near the computer. The reported problem is that the unit becomes intermittent when heavy equipment vibrates the test stand the controller is mounted in. You suspect you will have to
    a. Eliminate the arcing from the nearby electric furnace
    b. Suppress the ignition noise on their trucks
    c. Clean the cable and card connectors to remove corrosion built-up from acid fumes
    d. Correct the overheating caused by dust buildup
31. A manager in your office claims one of the PCs "eats" floppy disks. You run the disk drive diagnostics and find that there are no problems with the controller or the floppy-disk drive. Also you notice the manager stores the floppy disks in a desk drawer with small tools. You recommend
    a. Replacing the floppy-disk drive just in case
    b. Keeping the floppy disks away from the tools which may be slightly magnetized and erasing the disks
    c. Using an antistatic mat
    d. Replacing both the floppy-disk drive and its controller

16-9 USING TEST EQUIPMENT

Although there is a wide variety of electronic test equipment available today, there are only a few pieces which are really valuable for microprocessor-based product service. If the microprocessor-based product is used to control or measure analog quantities, you may find a need for many more conventional pieces of electronic test equipment.

For general microprocessor-based product troubleshooting, the following basic equipment is usually be more than enough.

1. A general-purpose DMM or VOM
2. A 35- to 100-MHz dual-trace oscilloscope
3. A logic probe
4. An RS-232 breakout box

Although the logic analyzer, the EPROM programmer, and the microprocessor in-circuit emulator can be used for sophisticated service and troubleshooting efforts, they are rarely used for this purpose. For this reason, these instruments are covered in the chapter on development systems.

The DMM or VOM is used to ensure proper supply voltages are present, ground connections are made, and cables and other conductors have continuity. A DMM and VOM of the type normally used for service work are shown in Fig. 16-6. Typical voltages found in microprocessor-based products are +5 Vdc for the logic and ±12 Vdc for communication circuit interfaces and disk drive motors. Other special voltages may be required by products that are used for industrial control. A DMM is a good way to make sure that the ac line voltage is within the limits for the product power supply.

Either a DMM or VOM works well. You do not need exceptional accuracy or resolution, so the analog instrument does well if you have one. On the other hand, you do not make trend measurements with these circuits, so the analog instrument has no advantage.

Occasionally you may use a meter to see if a logic circuit is working. Theoretically, a 50 percent duty cycle square wave produces a reading of 2.5 V. However, the speeds and pulse widths commonly found in microprocessor circuits are often well outside the meter's capability. At best this indication is only a rough sign the circuit is working.

An oscilloscope is the best general-purpose tool you can have. Figure 16-7 on page 478 shows a typical 50-MHz *dual-trace oscilloscope* commonly used to troubleshoot logic circuits. With the oscilloscope, you can see waveforms to make sure they are of the proper shape, timing, and amplitude. For example, Fig. 16-8 on page 478 shows the points to measure on a Z80 microprocessor to find out if the microprocessor is running. In this example, you check to see the clock is running, the $\overline{\text{M1}}$ timing output line is active, and the address and data lines are active.

Using the oscilloscope to analyze address and data lines only provides limited information. You can tell the lines have proper amplitude and timing. What you cannot see is what data is being transmitted over the particular address line. To know that information, you need to look at all eight data lines together. That is the job of the *logic analyzer*. On the other hand, it is rare you need to look at the data. When you are troubleshooting a microprocessor-based product, what you need to know is whether or not the parts are working. If they are working, it is likely the correct data is there.

An oscilloscope can be used to analyze circuits which have a burst of data. To do this, you use the triggered-sweep mode. For example, you may ask the question, "is the microprocessor receiving an interrupt from the pe-

(a)

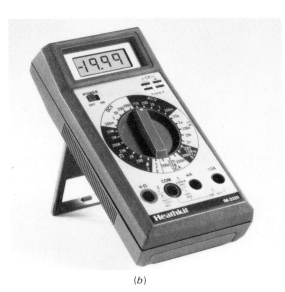

(b)

Fig. 16-6 Two common multimeters used for service work. Figure 16-6(*a*) is an analog multimeter which typically supplies 1.5 to 3 percent accuracy. Figure 16-6(*b*) is a digital multimeter (DMM) with typical accuracies of 0.2 to 0.5 percent. (*DMM photograph courtesy of Heath Company.*)

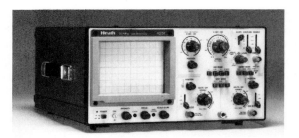

Fig. 16-7   A 50-MHz dual-trace delayed-sweep oscilloscope. This is commonly used for service work which requires comparing waveforms with each other or with known wave shapes.   (*Photograph courtesy of Heath Company.*)

ripheral device?'' Because the interrupts only come once in a while, you cannot see them by using the oscilloscope in the conventional modes. However, by setting the oscilloscope to trigger when a pulse is present, you can leave the triggered oscilloscope monitoring the interrupt line. When the interrupt happens, the oscilloscope triggers and you can see that it triggered by observing the control panel.

The *logic probe* can be used as a simple oscilloscope. Logic probes such as the one shown in Fig. 16-9 are small enough to fit easily in a toolbox with the DMM. The oscilloscope will not; it is large and heavy and not always easy to carry to the service location. The logic probe tells you if a circuit is at a logic 1, a logic 0, or pulsing. Typically, the pulses are coming so fast and are so narrow

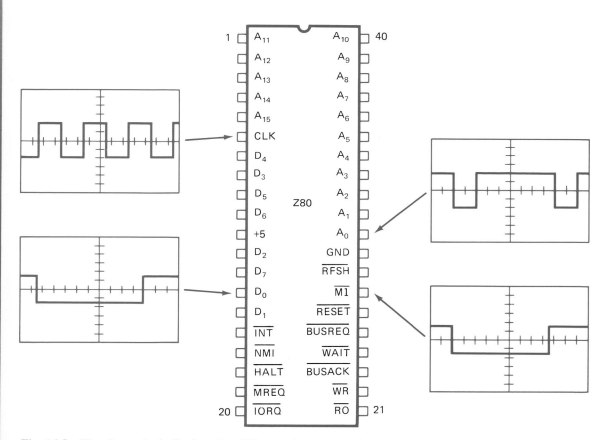

Fig. 16-8   Waveforms typically found at different pins on the Z80 microprocessor. Although many more of the pins will also show waveshapes, waveshapes on these pins usually show that all of the pins are working.

the logic probe cannot tell you their width or number. Once again, knowing the width of the pulses or how many there are is not really necessary. You need only know the pulses exist.

Suppose you are servicing an intermittent fault in the unit shown in Fig. 16-10 on page 480. You only have your DMM and logic probe. First, make sure that the power supply is working. You find +5 V on pin 40 of the microprocessor, pin 20 of the LS373 latch, and pin 28 of the memory IC. It also shows 0 V on pin 20 of the microprocessor, pins 10 and 11 of the latch, and pin 14 of the memory IC. The ICs are getting power. But is the microprocessor running? You start by looking at the CLOCK pin (pin 18—XTAL 2). The logic probe shows a signal here. A second check at ALE (pin 30) shows a signal here as well. You are now fairly sure the microprocessor is running. What about the other parts?

Is the processor accessing memory? You use the logic probe to check the $\overline{\text{CHIP ENABLE}}$ (pin 20) on the memory IC. The steady indicator on the logic probe tells you it is not getting any pulses. You check back at the microprocessor (the memory $\overline{\text{CE}}$ pin is tied to P2.5). The logic probe indicator winks, indicating there are pulses on P2.5. Where is the fault? Looking closer at the memory IC, you now see pin 20 is bent under the IC. Depending on the temperature, the pin can expand just enough to touch the socket contact occasionally. This explains the intermittent operation.

The front panel of a typical breakout box is shown in Fig. 16-11 on page 481. An *RS-232 breakout box* provides four different functions. First, it lets you intercept a 25-wire RS-232 cable and open or close any of the conductors. Second, it lets you cross-patch a number of those connectors. For example, if you must connect pin 11 on a printer serial I/O port to pin 5 on the computer serial port, you can use the RS-232 breakout box to do this.

Third, the RS-232 breakout box has indicators on the critical data lines. By looking at the breakout box indicators, you can see if there is receive and/or transmit data. You can also check the status of the TD, RTS, DTR, CTS, DSR, DCD, TD, RC, SQ, RI, and BUSY lines. Fourth, most RS-232 breakout boxes are supplied with both male and female DB-25 connectors on both sides of the box. This allows you to use the box to match cables which otherwise might require a gender adapter.

The RS-232 breakout box becomes a very handy tool when you are determining just what RS-232 cable is needed to connect a peripheral to the computer. Using the RS-232 breakout box, you can quickly verify the proper connections, which control lines are needed, and which are not. Once verified, you can build the cable, knowing the difficult job of installing the connectors does not need to be redone. It is also a very handy tool to verify data flow between two serial ports.

When you are working with today's microprocessor-based products, especially PCs, it is very valuable to have a collection of cables on hand. These cables allow you to make 9-pin to 25-pin RS-232 conversions, perform a loop-back function, move data from one device to another, substitute a suspected cable, etc.

There are many other electronic instruments which come in handy once in a while. For example, a *function generator* is useful when checking out a product with built-in A/D converters. Likewise a *pulse generator* can come in handy when you need to exercise some logic to see if it is performing to specification. However, these are not common requirements, and often you will find a simple substitution of an IC or module gives the same answer in less time.

It is always important to remember that your job, when servicing a product, is to place the product in operation as soon as possible at the lowest cost to the customer. This means you should always look for the most efficient service techniques possible.

Fig. 16-9 **A simple logic probe. The probe can tell you if a circuit is active or stuck at a logic 0 or logic 1.** *(Photograph courtesy of Heath Company.)*

RS-232 breakout box

Function generator

Pulse generator

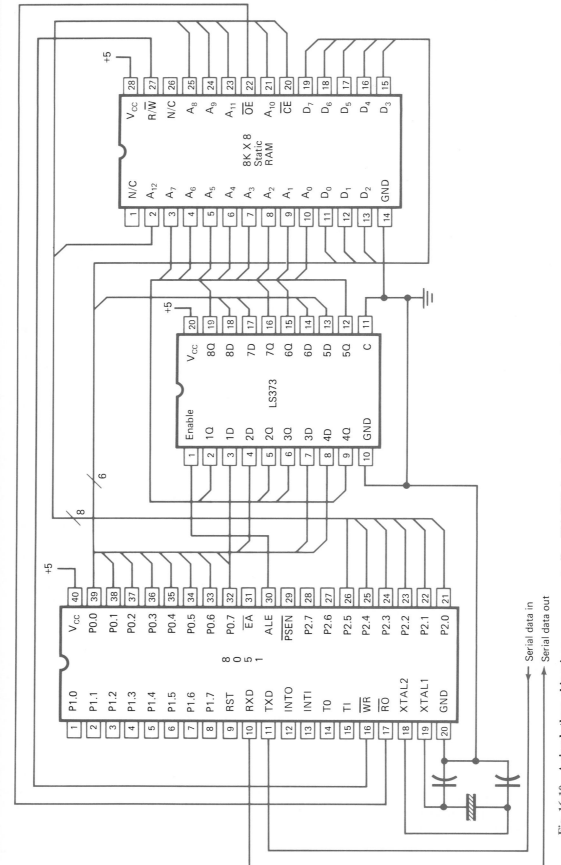

Fig. 16-10 A simple three-chip microprocessor controller. This 8051-based product uses an 8K × 8 static RAM IC to provide additional read-write storage. The LS373 octal latch stores the lower-order address bits as the 8051 uses a multiplexed address data bus.

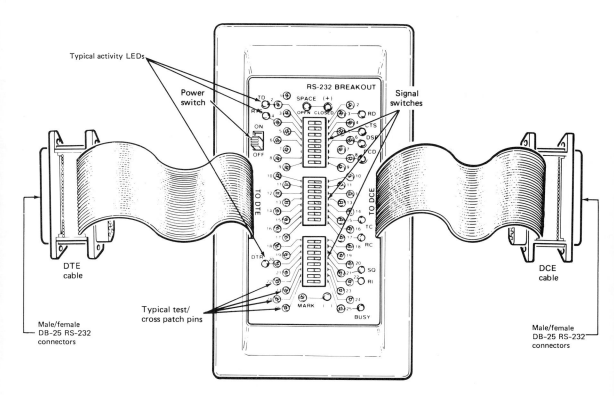

Fig. 16-11 The front panel of an RS-232 breakout box. The breakout box is provided with the jumper cables shown so that the user may open the direct connections with the switches and patch special connections using these jumper cables. *(Courtesy of Heath Company.)*

Self-Test

Answer the following questions.

32. The specifications for an industrial monitoring system show the power supply should generate +5 V ± 3 percent. Which of the following readings may not be within specification?
   a. A 3.5-digit DMM specified at ± 0.05 percent of reading ± 1 digit reads 5.1 V
   b. A VOM on the 10-V range (± 2 percent of full scale) reads 5.0 V
   c. The DMM above reads 4.9 V
   d. They are all well within specification

33. You transmit the [character to serial ports COM1: and COM2: on a PC which is reported to have problems when communicating with other devices. Figure 16-12(*a*), COM1:, and (*b*), COM2:, are two oscilloscope presentations of the two RS-232 signals. For each display, the centerline is 0 V and the vertical sensitivity is set to 5 V/div. Based on these readings, you
   a. Assume there are no problems with the serial ports, and you should look elsewhere for the problem

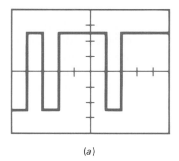

(*a*)

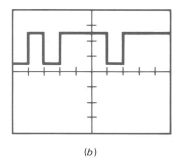

(*b*)

Fig. 16-12 See question 33.

b. Note COM2: is not swinging below 0 V. You can assume COM2: is probably defective.

c. Note COM1: is swinging both above and below 0 V and therefore assume COM1: is probably defective

d. Cannot see any problems at these points and therefore move to the peripheral as the next possible cause

34. Which of the following is not a valid oscilloscope measurement?

a. The verification of a CLOCK signal at the microprocessor clock input

b. A check of the I/O WRITE line to make sure the I/O devices are being selected for output

c. A check of the +5-V line for ripple

d. A check of the data lines to make sure the correct instruction is being placed on the data bus

35. The logic probe cannot be used to

a. Determine if the microprocessor is getting a CLOCK pulse

b. Check the data lines to see if there is bus activity

c. Check the +5-V power rail for intolerance voltage

d. Make sure a ROM is getting a $\overline{CE}$ signal

36. You are called to service a "dead modem." You place your RS-232 breakout box between the modem and the computer. When you send data from the computer to the modem, there are no echo characters shown on the computer screen. However, the RS-232 breakout box TD and RD lights show data being transmitted to and from the modem. You suspect the problem is

a. Just what the customer told you, a "dead modem"

b. A completely dead serial port

c. A serial port with a dead transmit port

d. A serial port with a dead receive port

37. You need to build an RS-232 cable to go between a PC and a printer. The data is sketchy, but it looks like the cable should be configured as shown in Fig. 16-13. The best instrument to confirm this configuration before building the cable is your

a. Oscilloscope   *c.* Breakout box

b. Logic probe   *d.* DMM

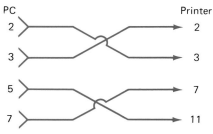

Fig. 16-13   **See question 37.**

SUMMARY

1. Basic service technique involves
   a. Understanding the need
   b. Finding the failure
   c. Identifying short- and long-term repair options
   d. Implementing a repair

2. To do a good job you must know what the customer thinks is wrong.

3. You must also know what the customer expects you to do.

4. The customer will not always fully understand a microprocessor-based product and therefore may have unrealistic expectations.

5. You cannot know in advance what every microprocessor-based product does, so you must be ready to learn.

6. Always try to find and review the manufacturer's operation and service manuals on an unfamiliar product.

7. A complex product requires an organized approach to service.

8. Part of a systematic approach is to isolate the problem area and separate the areas that still work from those which do not.

9. Before you start to attack the product, find out what went wrong. What should the product do that it does not?

10. Ask "What caused the problem?"

11. Look for the obvious. There are more simple problems than there are difficult ones.

12. Be sure to observe basic safety practices when servicing. Unplug the product and discharge large capacitors.

13. Find or draw a block diagram for products which you do not know well. Use the block diagram to help isolate the area of failure.

14. A failed power supply will cause the complete product to be inoperative.

15. Many of the larger microprocessor-based products (especially PCs) use switching power supply modules. They have many line-connected parts and operate from 20 to 100 kHz.

16. Often a switching power supply will shut down if its load demands more current than it can supply. This can happen if the microprocessor circuits develop a short circuit. This may kill all of the power supply outputs.

17. You can look for overcurrent shutdown by disconnecting the power supply from the load to see if the outputs return to their normal voltage. Some switching power supplies require some load, or they will not operate.

18. Switching power supplies are difficult to service because of their line-connected components and closed-loop operation. Also, many are not well-documented.

19. If the microprocessor in a product is not operating, nothing will operate. Getting the microprocessor operating is your first priority.

20. If possible you should disconnect all the memory and I/O devices when trying to make the microprocessor operational. All you need is the microprocessor, ROM, and some RAM on single-board systems.

21. If the video portion of a PC is working, you know the microprocessor is working.

22. Look for diagnostic indicators that might tell you the CPU is running. Also check to see if displays or keyboards are multiplexed by the CPU. If they are and they work, there is a high probability the CPU is running.

23. Forcing the RESET line on a microprocessor will cause it to fetch an instruction from the RESET vector. You should be able to see this activity on the address and data buses. This activity tells you that the microprocessor is working and that the problem is elsewhere.

24. If the microprocessor does not appear to be working,
    a. Make sure it is getting power.
    b. Unplug socketed latches, RAM, ROM, and so on, to see if any one of them releases activity.
    c. Substitute a known working microprocessor.
    d. Start a signal check.

25. The best possible RAM check is a CPU-driven diagnostic. It can check for complex pattern sensitivity problems.

26. Without a processor-based diagnostic, you can perform IC substitutions or swaps.

Use the swaps to cause changes in the way the problem exhibits itself.

27. A ROM problem can be detected by using any device which will read the ROM, if you know what the code should be. Substitution is a good check for ROM problems, but remember each ROM in a microprocessor-based product is different from all of the others.

28. Substitution and isolation are the best service techniques for mass storage devices. Often the failure of a single drive will cause other drives on the same controller to stop operating.

29. You may need a loopback connector to check out a serial port by using the diagnostic software supplied with the product.

30. Some I/O port problems can be located by reducing the communications to a single character or transaction. This often shows up handshaking problems.

31. Improper setup is also one of the major causes of I/O problems. Before a serial port can work properly, its baud rate, word length, and parity must match the serial port it is to communicate with.

32. Lightning damage is often coupled into a product via the communications lines. Frequently, the lightning will destroy the communications interface ICs. The damage may or may not be limited to these ICs.

33. You may need to use more conventional instruments to check out analog I/O ports and other special-purpose ports.

34. Component and module substitution is a good isolation and immediate repair technique. There is some possibility of damage to the good parts, so care must be taken to ensure they are not being placed in dangerous situations.

35. Transient voltage spikes can cause a microprocessor to process incorrect data or instructions. It can also change register values, including the program counter, which may cause an immediate crash.

36. If transients are a problem, you may need to provide filtering on the power lines, telephone lines, and communications lines as well as preventing static discharge from operators.

37. Cables and connectors are susceptible to dirt, dust, and corrosive vapors. These contaminants may cause intermittent connections which are difficult to find and which may show up when the product is moved.

38. A good service technician is prepared to find the unexpected. Products are used in odd ways and are often not technically understood.

39. The DMM or VOM is used to provide basic voltage checks, ensure modules and components have ground and V_{CC}, and to perform continuity testing. High accuracy and trend indication are not required.

40. An oscilloscope is your best general-purpose instrument. You can use it for waveform analysis, logic analysis, voltage checks, pulse detection, and many other checks.

41. The oscilloscope is difficult to use for parallel data analysis, but you can analyze data on serial lines.

42. The logic probe lets you detect the presence of logic 1s, logic 0s, and pulses. It is too crude to provide timing information except on the slowest circuits.

43. The RS-232 breakout box lets you
  a. Intercept the cable and make or break any conductor
  b. Cross-patch different conductors
  c. View data activity or logic status on critical lines
  d. Make cross-gender connections

44. The breakout box is very useful for diagnosing communication circuit problems and verifying the proper construction of RS-232 interconnection cables.

CHAPTER REVIEW QUESTIONS

Answer the following questions.

16-1. List the four basic steps you must use when servicing a product.

16-2. Briefly explain why it is important to know what the customer thinks is wrong with the product and why this is more important than usual with a microprocessor-based product.

16-3. Briefly explain why there could be a difference between what the customer expects from your service call and what you might expect.

16-4. Briefly explain why microprocessor-based products may require you to learn a new product function more frequently than for a conventional electronic product.

16-5. List at least five questions you should ask the customer before you actually start working on the product you have been called to service.

16-6. List at least five obvious areas you should check if you have been called to service a PC.

16-7. In order to help isolate the problem area in a microprocessor-based product, you should have access to
  a. Its diagnostic software
  b. A block diagram of the unit
  c. A schematic for each card and the backplane
  d. Disassembly instructions

16-8. One major area of a microprocessor-based product which, if it has failed, will keep all others from working is the
  a. Serial I/O ports
  b. Parallel I/O ports
  c. Mass storage devices
  d. Power supply

16-9. Which of the following is not a characteristic of switching power supplies?
  a. All of the outputs may shut down if one of the power supply outputs is shorted
  b. They are typically used in very low power products
  c. They are not easy to service because of line-connected components and closed-loop circuits
  d. Typically they switch at 20 to 100 kHz

16-10. Once you know you have power, the next most critical area to have operating is the
  a. Microprocessor
  b. Memory
  c. Serial I/O
  d. Disk drives

16-11. In order to check out the microprocessor, it is best to
  a. Run CPU diagnostics
  b. Have a full memory map
  c. Reduce the system to the minimum components possible
  d. Remove all ROM

16-12. Which of the following is not a good indication the microprocessor is working?
  a. Finding +5 V on the V_{CC} pin
  b. Seeing an error message printed on the memory-mapped screen
  c. Data-line activity
  d. Address-line activity

16-13. Briefly describe the process you might use to check a microprocessor by forcing its RESET line. Assume you have a logic probe.

16-14. Briefly describe the special checks that are made by a memory diagnostic and why these are difficult to make without processor-based testing.

16-15. The process of swapping memory ICs may help you find which IC has a problem, but it will not fix the problem. Why?

16-16. Briefly explain why swapping like ICs does not work to find problems with EPROMs. What can you learn by connecting and disconnecting them?

16-17. Which of the following is not likely to be a problem with a mass storage device?
  a. The media   c. The controller
  b. The drive   d. They all can be problems

16-18. If you can get one mass storage device running through isolation, this will
  a. Unload the power supply so everything else runs
  b. Let you have a part to order
  c. Let you load mass storage diagnostics to check out the other devices and their controllers
  d. Make sure the serial I/O devices can be worked on next

16-19. A loopback connector is used to check I/O ports so that
  a. You can check for any problems between the receiving and transmitting communication controllers
  b. You can completely check the I/O ports, including the line drivers, connectors, and any cables
  c. The parallel signals are kept separate from the serial signals
  d. All of the above

16-20. If a serial port will work when you send one character at a time, but it does not work when the CPU tries to send a stream of characters to the peripheral, you suspect there is a _____ problem.
  a. Line driver   c. Baud-rate
  b. Handshaking   d. Parity

16-21. If you have diagnosed an I/O port setup problem which was causing complete garbage characters to appear on the output device, the problem was probably with the
  a. Line driver   c. Baud rate
  b. Handshaking   d. Parity

16-22. It is relatively common for lightning to damage a serial port's
  a. Line driver   c. Baud rate
  b. Handshaking   d. Parity

16-23. An effective approach to component and module substitution is
  a. To substitute known good components or modules from your spares supply in the defective product
  b. To replace components and modules in a known working product with components and modules from a defective product
  c. To replace components and modules from a known working product with those in a defective product
  d. All are effective approaches, but choice a has the least risk

16-24. List three sources of transients which can cause a microprocessor-based product to crash or produce incorrect data. What would you do to prevent these transients from causing problems in the future?

16-25. Briefly explain why a corroded cable may not cause a problem until just after the unit has been serviced for a different problem.

16-26. Why does the person servicing microprocessor-based products need to be more prepared to deal with the unexpected than the person servicing more conventional consumer electronic products?

16-27. List three measurements you might make with a DMM or VOM when initially checking out a PC at a customer site.

16-28. You find an EPROM in a microprocessor-based controller which is hot to the touch. Additionally, you find 1.2 V on its V_{CC} pin. Briefly explain what you think the problem might be with this product.

16-29. Briefly explain why a service person might say, "if I could only have one instrument, it would be my oscilloscope."

16-30. Using an oscilloscope, what are two measurements you can make to know the microprocessor clock is operating?

16-31. Briefly compare the uses for the logic probe versus the oscilloscope.

16-32. Briefly describe the RS-232 breakout box and three major functions you can perform with it.

CRITICAL THINKING QUESTIONS

16-1. Why is a systematic approach toward troubleshooting preferred to a "shotgun" or random approach? Why do we say that an experienced person who looks at five to six key suspect components scattered randomly about the unit is using a systematic rather than a random approach?

16-2. A complex device such as a PC is often easier to troubleshoot by swapping known good modules with suspected bad modules. Why is this technique best for PCs and not so good for a hi-fi amplifier?

16-3. You are troubleshooting two different pieces of microprocessor-based equipment. One has most of the ICs in sockets, and the other has most of the ICs soldered directly to the printed circuit board. This observation tells you that two different troubleshooting techniques are called for. Why?

16-4. Your troubleshooting of one of three identical microprocessor-based products leads you to believe that one of the EPROMs has failed. How can you verify this diagnosis? What piece of equipment do you need if you are to make a repair on site (once you have proved that the EPROM is indeed defective)?

16-5. Make up a list of the things you might do as part of a regular 6-month "routine maintenance" program to keep the PCs in a school lab running properly.

Answers to Self-Tests

1. c
2. You must understand what the customer thinks the problem is to be sure you are addressing the correct problem.
3. There may be some functions on a microprocessor-based product which are never used. If one of these is defective, there may be no need to repair the problem.
4. d
5. d
6. a
7. c
8. c
9. b
10. d
11. b
12. d
13. a
14. c
15. d
16. d
17. a
18. d
19. d
20. b
21. d
22. a
23. c
24. c
25. a
26. a
27. d
28. b
29. c
30. c
31. b
32. b
33. b
34. d
35. c
36. d
37. c

CHAPTER 17

Developing Microprocessor-Based Products

■

CHAPTER OBJECTIVES

This chapter will help you to:

1. *Understand* the steps of a complete design development process for a microprocessor-based product.
2. *Understand* how a design specification is prepared and what it tells its audience.
3. *Describe* the basic engineering design process.
4. *Learn* to implement and test a design.
5. *Understand* the need for compliance with government regulatory body rules.
6. *Recognize* the common test equipment used in the design process.

───────────

The development of a microprocessor-based product requires the development of the programs (software) and the hardware (electronic circuits) which allow the software to perform the intended task. The design effort for an electronic product follows the same basic steps used in the development of software. A specification is prepared, a design is created and then implemented, and a final product is demonstrated. For large products a number of specialized engineering people are assigned to the product. For smaller products, one person may complete the entire product. Part of the effort requires the use of electronic test equipment such as the logic analyzer and microprocessor development system (MDS) to complete the design effort.

■

17-1 AN INTRODUCTION TO THE DESIGN PROCESS

In Chap. 14, we learned the steps used to develop software. That is, we learned the steps used to write a program. This chapter introduces the process used to develop a complete microprocessor-based product. The outline for the design process does not change. We first saw this outline in Chap. 14. There we saw the design steps are:

- ■ Writing a specification
- ■ Creating a design
- ■ Implementing the design in code and testing and correcting the implementation
- ■ Demonstrating a final product

There is no difference in the names used for these steps for a complete product design or just the software design. The difference is what must be done in these steps and the tools used to make the job easier and to make sure you can perform the job properly.

The development of a microprocessor-based product, just like the design of software, is a major effort. The purpose of this chapter is to show you the steps used so that you can help the people assigned to completing the development task.

Who are these people and what jobs do they do? The number of people needed to develop a microprocessor-based product depends on the size and complexity of the project. Some products are small enough so a single individual does the whole development task. Other, larger projects have one or more engineers assigned to the different tasks.

What are the different tasks which must be done to develop a microprocessor-based product? As an example, let's look at the tasks and team required to develop the microprocessor-based shortwave radio from Chap. 14. The major development areas are electronic, software, and mechanical.

The electronic development task is broken into microprocessor-based control system design and RF (radio frequency) circuit design. The mechanical design effort creates the package. The term *package*, in this case, does not refer to the box used to ship the product but to the mechanical housing (case) which protects the electronic parts and gives the radio a pleasing appearance.

The development of a shipping box for the radio is also a separate task. This too must be completed before the product can be sold, and requires a design effort to make sure the radio is protected from damage during shipping.

Each development area is often assigned to an engineer trained to develop the part of the product which calls for this particular technology. The radio in this example may have four engineers assigned to the effort. They are:

An RF design engineer who develops the analog circuits, including the radio circuits, the audio amplifier, the power supply, and the synthesizer

A digital design engineer who develops the microprocessor-based control circuits

A software design engineer who develops the software for the microprocessor which allows the user to control the radio operation

A mechanical or packaging design engineer who develops the radio housing

Additionally, one of the engineers is normally assigned the role of project manager or project engineer. The *project engineer* is the individual responsible for the overall project.

The project engineer decides what parts of the project the other engineers are responsible for based on their technical expertise and the best sectioning of the project.

Once the project is sectioned, the project engineer works with the other engineers to develop a schedule. The schedule shows management when the project will be complete. It also shows the other engineering team members when the parts they are working on must be complete and when other parts will be complete.

The project manager is responsible for regular project status reports to management and marketing. The project engineer is marketing's principal engineering contact and shows them the design meets the functions and specifications marketing asked for.

Typically, each engineer has completed a four-year course of study at a university leading to a bachelor's degree in his or her field. For example, RF and digital design engineers often complete a BSEE (bachelor of science in electrical engineering) and software engineers complete a BSCS (bachelor of science in computer science) or a BSEE with a specialty in software design. Likewise, mechanical engineers often hold a BSME (bachelor of science in mechanical engineering).

Because engineering is a complex field and advanced studies beyond a four-year course of study are often necessary to be thoroughly trained in some engineering areas, it is not uncommon to find engineers who have completed a total of six years of study. The additional two years beyond their BS degree led to the MS (master of science) degree.

The engineers are often supported by engineering technicians and product developers. These individuals assist the engineers with the completion of their assigned tasks. Typically, engineering technicians and product developers complete a course of study similar to the one that includes this course. Often the product developer is an engineering technician who has completed additional studies leading to an AS (associate of science) degree in engineering.

Engineering technicians are often assigned such tasks as:

Creating printed circuit boards from schematic diagrams

Building models of the product

Testing the model to confirm that it works according to the specifications

Developing documentation such as parts lists (bills of material) and sketches of component specification drawings
Assembling prototypes, parts, documentation, and reports which must be supplied to manufacturing, technical publications, quality assurance, customer service, field sales, marketing, and so on, to "transfer" the completed product from engineering into production

The rest of this chapter concentrates on the activities assigned to the engineers developing our microprocessor-based shortwave radio. In particular, we will concentrate on the activities assigned to the *digital design engineer*, since we have spent time in Chap. 14 studying the tasks required to complete the software development. The completion of RF and mechanical designs is beyond the scope of this text.

In this example, J. Black is the digital design engineer and, due to experience and interest, is assigned the role of project engineer.

J. Black has created the engineering development schedule shown in Fig. 17-1. This

schedule is created at the beginning of the project and is continuously updated as the project progresses. Usually such a schedule is updated once every week or once every two weeks.

The engineering staff developing a product such as the shortwave radio does not work alone. There are many different organizations and support personnel to help them. For example, the mechanical engineer usually receives a sketch such as that in Fig. 17-2 on page 490. This comes from the industrial design department. The industrial designers conceive an appearance which meets marketing desires for the intended marketplace. The mechanical engineer must then develop drawings for the different plastic and metal parts used in production to house a product which looks like the concept drawings from industrial design.

Often individual component drawings and assembly drawings are finalized in a drafting department. Here individuals trained in the creation of drawings called blueprints convert engineering sketches into formal drawings. These drawings are used by purchasing to precisely tell a supplier the details about a part that

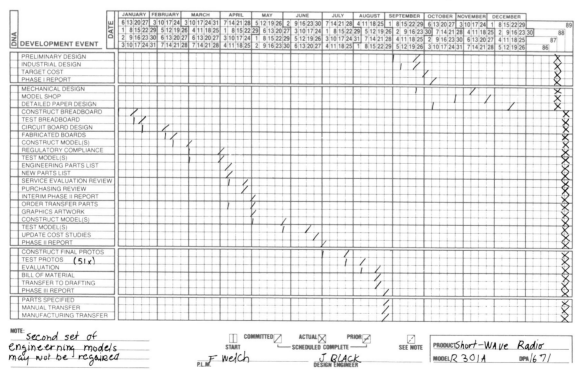

Fig. 17-1   An engineering schedule for the microprocessor-controlled shortwave radio. This is the preliminary schedule put together by the project engineer. As the project continues, this schedule is updated to show actual progress and new estimates for completion dates. As you read this chapter, you can check this chart to see how the scheduling for the various development events has been done.

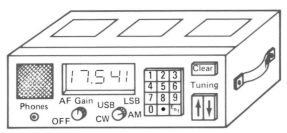

Fig. 17-2 An artist's sketch showing what the microprocessor-controlled shortwave radio should look like. The mechanical engineer uses this "specification" to develop the detailed mechanical drawings specifying the mechanical parts which will house the electronics.

is to be purchased, by quality assurance to check the purchased part to be sure it meets all of the specifications assigned by engineering, and by manufacturing to tell how the parts go together to create the final product.

The technical publications department uses the materials from engineering to develop an operator's manual. The end user reads this manual to determine how the product operates, whom to contact if repairs are required, what accessories are needed, and so on.

Self-Test

Answer the following questions.

1. The number of engineering or development people assigned to a particular project
   a. Depends on the size and complexity of the project
   b. Depends on the skills required and the skills available from different people in the organization
   c. May be as few as one or as many as there are different technological areas to be developed
   d. All of the above

2. Often the design of the electronics in a project is broken into the digital and analog sections. Usually, the microprocessor hardware development is assigned to a _____ engineer.
   a. RF   *c.* Digital
   b. Mechanical *d.* Software

3. You would expect the project leader for a moderate complexity development program to hold a _____ degree.
   a. BSEE   *c.* MSEE
   b. BSCS   *d.* Any of the above

4. As the engineering technician assigned to the development team for the shortwave radio, you could expect to be assigned the task of
   a. Developing a schedule which shows the time when each of the engineers on the project will complete the design of his or her portion
   b. Running experiments defined by the digital design engineer which show how much time is required for the shortwave radio to change frequency after a new frequency is entered
   c. Selecting the programming language to be used for the control software
   d. You could be assigned any of the above tasks

5. You have been assigned the task of delivering a prototype of the shortwave radio to the quality assurance evaluation engineering department. They will evaluate the prototype to determine if it meets marketing's requirements and if it complies with the company guidelines for product quality. After the transfer the quality assurance engineer tells you some details of keyboard entry sequences are missing or incomplete. To provide these, you contact
   a. The RF design engineer
   b. The mechanical engineer
   c. Marketing
   d. The software engineer

6. You are contacted again by the quality assurance engineer, who asks for the range of line voltage the shortwave radio is designed to work over. On this project you contact _____ for this information.
   a. The RF design engineer
   b. The mechanical engineer
   c. Marketing
   d. The software engineer

7. The purchasing department contacts you requesting a change in the part specification drawing for a molded foot on the shortwave radio case. A vendor has requested an angle be changed by 0.5°. You mark the drawing so that drafting knows how to change it and take the marked-up drawing to _____ for approval before giving the change to drafting.
   a. The RF design engineer
   b. The mechanical engineer
   c. Marketing
   d. The software engineer

17-2 PREPARING THE SPECIFICATION

The first step in developing a product is to create a *specification*. Just as we learned when developing a software specification in Chap. 14, the specification from marketing may require expansion before it can be used by engineering to develop the product. The specification shown in Fig. 17-3 is typical of the specification which engineering might receive to develop the shortwave radio.

Each engineering group must expand the specification to provide the details needed for its area. For example, Fig. 17-4(*a*) shows the shortwave radio operating power specifications. In Fig. 17-4(*b*), the RF design engineer, who is responsible for the power supply design, expanded these specifications so that the design requirements can be determined.

```
Power:      120 Vac @ 60 Hz
            240 Vac @ 50 Hz
                    (a)

Power:      105–135 Vac  }
            210–270 Vac  }  user switchable

            48–62 Hz
            15 W maximum
              (target)
            Detachable line cord
Regulatory
compliance: UL, VDE, CSA
                    (b)
```

Fig. 17-4 (*a*) **Marketing specifications for the shortwave radio power specifications.** (*b*) **The power specifications as expanded by engineering. Note engineering added a power target and a line cord detachability specification. Both of these specifications impact the mechanical design. Engineering has also added the regulatory compliance requirements (UL, VDE, CSA).**

Proposed specifications
R-301A
Advanced shortwave receiver

Frequency range:	1.5 MHz – 30 MHz
Tuning steps:	1 kHz
Frequency stability:	less than 50 Hz per hour drift after warmup
Modes:	AM
	USB (upper sideband)
	LSB (lower sideband)
	CW
Frequency display:	See attached drawing
Frequency entry:	Keyboard (see attached drawing)
Sensitivity:	0.35 µV for 10 dB S/N (SSB/CW)
	3.5 µV for 10 dB S/N (AM)
Selectivity:	SSB/CW AM
	2.7 kHz @ −6 dB 6 kHz @ −6 dB
	5.4 kHz @ −60 dB 12 kHz @ −60 dB
Audio power:	2 W @ 10% maximum total harmonic distortion
Antenna:	50 Ω with diode protection to prevent static damage, UHF connector
Dimensions:	4″ high
	10″ wide
	6″ deep
Weight:	12.5 lb
Power:	120 Vac @ 60 Hz
	240 Vac @ 50 Hz
Accessories:	Antenna kit
	Earphones
	External speaker

Fig. 17-3 **The marketing specification for the microprocessor-controlled shortwave radio. Note this specification only identifies the major marketable features and specifications about the product.**

You can see the specifications are expanded in two ways. First, the engineer added a tolerance specification, since marketing only indicated the radio was to operate from 120 Vac, the most common voltage throughout the United States. However, the engineer knows line voltage throughout the United States can vary ± 10 percent. So, the specification is expanded to reflect this need. The engineer used a range specification rather than a percentage. Alternatively, the specification might read 120 Vac ± 15 Vac or 120 Vac ± 12 percent.

Marketing intends to sell this shortwave radio in Europe as well as the United States so they also show the operating voltage of 240 Vac @ 50 Hz. Again engineering expanded this specification to voltage range. The power-line frequency specification is expanded to include the U.S. line frequency of 60 Hz and the European line frequency of 50 Hz and some reasonable tolerance. With the specification rewritten this way, engineering is also telling marketing they intend to design a single dual-voltage power supply, not two single-voltage power supplies.

Engineering also added a target power consumption specification of 25 W. Note this is marked as a target specification. The actual specification will be determined once the design is complete and a number of models are actually tested.

As with the software, there are many unspecified issues covering the digital design. In fact, a check of the specification shows the marketing specification does not include any requirements about the digital design, including the fact the radio is controlled by a microprocessor.

How is the specification for the digital design developed? The digital design specification for this shortwave radio must be developed by the engineering staff. Not all digital design specifications are developed by the engineering staff. If the digital design is important to the product features or specifications, the digital specifications are given by marketing. However, in this case, the reason for a microprocessor is that it is the best and lowest-cost way for engineering to provide the product which marketing is asking for.

What questions must be answered by the engineering digital specification? A few of the many which must be answered are:

Does this design need a microprocessor?
How much ROM and RAM are needed?

Are any special I/O ports needed?
Should a single-chip microprocessor be used, or does the product require a larger microprocessor system?
What integrated-circuit technology is required (NMOS or CMOS?)
Which microprocessor architecture best fits the design requirement?
Which microprocessor do we have development capability for?

As you might expect, the *digital design engineer* does not make these decisions alone. For example, we learned in Chap. 14 that the ROM and RAM requirements are mostly software engineering decisions. How to implement a special serial port, for example, may seem to be a hardware design decision. However, the "should we do it in software or should we perform the function with hardware" is a decision the software engineer and the digital design engineer must make together. Likewise the choice of using a single-chip microprocessor or a microprocessor system is usually a joint decision made by the engineers responsible for the software and digital designs.

Usually the digital engineer selects the IC technology used. However, this decision is not made without discussing the overall power demands with the engineers responsible for the other portions of the circuit design. Often major design considerations are made based on a power budget. Figure 17-5 shows a typical power budget table which may be assembled as part of developing the design specifications. Using the information found in this table, the engineers decide which IC technology will let them create a product which meets the power and cost targets set by marketing.

Often it is difficult to see the difference between the end of the specification phase of a development project and the beginnings of the design effort. Like the development of software, the development of hardware is an interactive process. This means engineering develops as much of the specification as possible and then begins the design process. When unanswered specification issues are found, the design engineers return to the process of developing a specification.

The specification process can be quite formal or very informal. On very simple projects the specification may be totally unwritten. On very complex projects, the specification process may take many engineering-years to com-

Shortwave radio preliminary power budget

	Nominal current (mA)	Maximum current (mA)	Operating voltage (Vdc)
RF circuits			
Front end	5	7	12
IF	20	25	12
Synthesizer	45	70	5
Synthesizer	10	12	12
Detector	5	7	12
BFO	10	12	12
Audio	300*	600**	12
Digital circuits			
Micro	8	11	5
LCD driver	1	2	5
LCD	5	7	12
Total current @ 5 V	54	83	
Total current @ 12 V	355	670	
Total current all	409	753	
Power @ 13.8 V	5.6 W	10.4 W	

* With 1.0 W audio output
**With full 2.0 W audio output

Fig. 17-5   A preliminary power budget for the shortwave radio. This first pass at the power requirements shows there is a possibility of creating a radio with as little as 10 W power requirements. Engineering will update this as the design develops, but may suggest the feasibility of battery operation with marketing.

plete and result in a formal document which is carefully reviewed before it is agreed to by engineering and marketing. Often the specification and preliminary design are done together. When the specification and preliminary design are complete, the project engineer can check off the steps leading to the Phase I reports on the engineering schedule.

Self-Test

Answer the following questions.

8. The reason engineering often must develop an expanded version of the specification supplied by marketing is
   a. Marketers tend to develop incomplete specifications, and engineers tend to develop precise specifications
   b. The selection of hardware specifications must be recorded so that all of the development team know of the decisions

   c. Although the microprocessor selected, for example, is very important to what the product can do, marketing does not have enough knowledge to make that decision
   d. Most companies have a written policy directing this activity to the engineering department

9. One reason for providing an expanded specification is to convert a statement of purpose into a measurable performance. Expanding the power specification from simply 120 Vac to a range of 105 Vac to 135 Vac is an example of such an expansion because
   a. The voltage over the United States can vary by as much as 10 percent and the product design must allow for this
   b. Marketing's statement of 120 Vac only indicates the product is to work in the United States. Expanding the specification to 105 Vac to 135 Vac meets company standards for line-operated voltage range
   c. The specification 120 Vac is impossible to meet because it implies an exact voltage which you will never get. The range of 105 Vac to 135 Vac gives limits which you can test.
   d. All of the above are examples of the reason stated

10. Although the original shortwave radio specification from marketing did not include a power consumption specification, engineering added one. They probably added this specification because
    a. Different power levels will affect the IC technology used and can affect the cost which is a target specification
    b. Failure to specify a power consumption level may let another engineer make a bad decision about the power consumption in circuits included in the design
    c. The mechanical engineer needs to know how much power is warming up the inside of the box so that proper ventilation can be included in the package design
    d. All of the above are valid reasons for adding the specification

11. Typically the digital design engineer must create the digital circuit specifications by working with
    a. The RF design engineer
    b. The mechanical engineer
    c. Marketing
    d. The software engineer

12. For the following list of products, mark each to show it probably requires an informal (I), simple (S), or formal (F) engineering specification.
   a. A 27-in. digital television
   b. A digital alarm clock intended to retail for $9.99
   c. A 486-based computer-aided engineering color graphic workstation
   d. A single-chip microprocessor-based box which converts parallel data to serial data
   e. A two-station intercom
   f. A three-digit monitor which displays the line voltage and gives a warning if the voltage goes above or below user-programmed values

17-3 DEVELOPING A DESIGN

As with software, the product is designed after the specification is complete and before implementation starts. In this section we look at some of the activities which must take place in the various engineering areas, with special emphasis on the digital design.

As noted in the section on specifications, it is often difficult to develop some of the engineering specification without developing some of the product design. For example, setting the power specification is often done while selecting the IC technology used to implement the product.

What does the digital design process look like? With a specification, the digital design engineer knows what functions the digital circuits must perform and what specifications the digital circuits are to perform to. The purpose of the design process is to develop the circuits to perform these functions.

This *design process*, like the software design process, begins with an overview and expands the requirements into greater and greater detail. Often the design is described by using a series of blocks showing how the various parts are interconnected. This is called a *block diagram*. When the block diagram is expanded so that each block represents an individual component, it is called a *schematic diagram*.

Figure 17-6 shows a preliminary block diagram for the shortwave radio. In the diagram the controller is a single block. All controller connections to other functions in the shortwave radio are shown. The controlled sections are the frequency synthesizer, which sets the radio's frequency, the keyboard used to enter the desired receive frequency, and the display, which shows the receiver's frequency and displays any errors which occur.

Once the basic block shortwave radio diagram is complete and all engineers agree the blocks are correct, the digital design engineer begins to refine the design. The next step is to break the controller into smaller parts. This is shown in Fig. 17-7. This expansion shows the details of connections between the controller

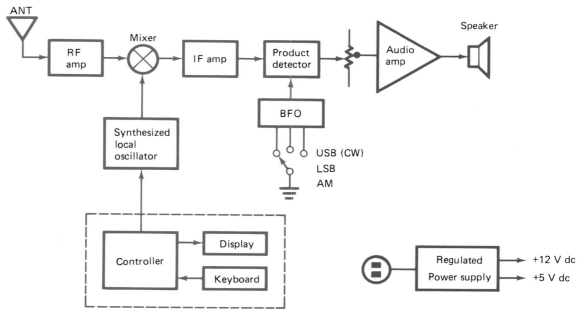

Fig. 17-6   **The preliminary block diagram design description for the shortwave radio. The digital circuits are all within the dotted lines.**

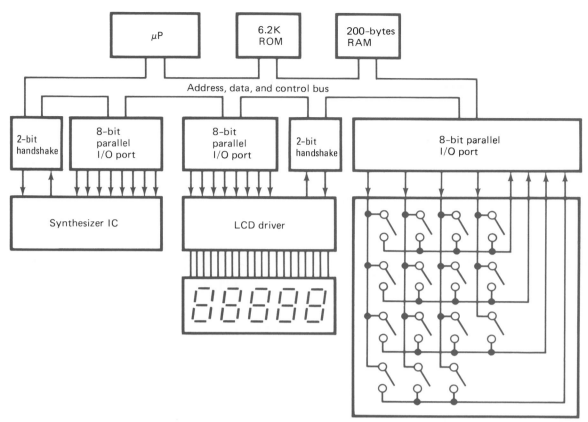

Fig. 17-7 An expanded block diagram for the shortwave radio controller. Note the digital engineer has decided the LCD should be managed by an LCD driver IC. This IC converts data from the microprocessor into properly scanned segments and planes on the LCD. The synthesizer control IC requires some setup words followed by data words to set the desired frequency.

and the controlled sections. Additionally, this diagram shows preliminary ROM and RAM requirements from the software engineer.

After some discussion, the engineers agree the controller design can be simplified as shown in Fig. 17-8 on page 496. In this diagram, the 8-bit parallel I/O port which controls the synthesizer and the 8-bit parallel I/O port which controls the LCD (liquid crystal display) driver are combined. Discussions between the engineer responsible for the synthesizer design and the digital design engineer conclude the eye will not recognize a slight hesitation in the display while the synthesizer IC receives its 10 bytes of new frequency information.

The final digital design is shown in Fig. 17-9 on page 497. In this figure the microprocessor, ROM, RAM, and 20 bits of I/O are combined into one IC—a single-chip microprocessor, the 80C52. The 80C52 provides 32 I/O lines, 8 kbytes of ROM, and 256 bytes of RAM. Before the digital design engineer could make

this choice, the software engineer had to estimate the ROM and RAM required to perform this function. A CMOS version of the 8052, the 80C52, was chosen to limit the amount of power drawn by the microprocessor.

Each engineer with design responsibility must make similar choices about their areas in the shortwave radio. For example, the mechanical engineer must decide on the internal construction. Talking with the RF design engineer, the mechanical engineer decides the digital controller circuits and the RF circuits should be separated by the maximum possible distance. Dividing the shortwave radio design into two circuit boards as shown in Fig. 17-10 on page 497 creates this separation and lets the display and keyboard components be mounted in their natural vertical plane. The mechanical engineer must find out how many square inches of *printed circuit board (PCB)* are required for each of the two areas to make sure the radio fits into the desired cabinet size.

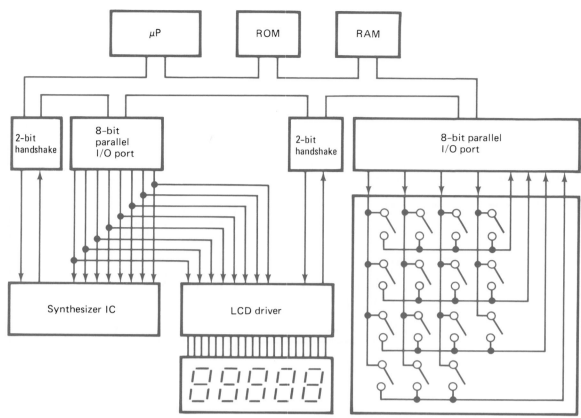

Fig. 17-8   A simplified controller design. After analyzing the time needed to set up the synthesizer IC and to manage the LCD driver, the digital engineer decides these two can be driven by the same 8-bit data port. However, each needs separate control lines. A single 8-bit port is dedicated to scanning the keyboard. Four lines output a scan sequence at the top of the matrix, and four additional lines monitor the four rows for contact closure.

Other issues entering into the design decisions made by the mechanical engineer include:

The power used (ventilation requirements).

The weight of larger components such as the power transformer. This tells the mechanical engineer how to mount these components so that they do not break loose when the product is shipped to the customer.

The number and required placement of external connections. For example, the short-wave radio requires an antenna connection, a line cord connection, and a place to connect an external speaker. Although both the line cord and the external speaker connections can be placed where it is convenient, the antenna connection must be placed near the radio input circuits and away from noise generators such as the microprocessor.

The kind of plastic which must be used for the case. This must have the look desired by industrial design, have the strength required to support the components, and meet the fire resistance requirements of agencies such

as UL (Underwriters Laboratories), CSA (Canadian Standards Association), and VDE (Verband Deutscher Elecktrotechniker).

When each engineer completes a section of the design, a mathematical model can be created. This mathematical model provides *worst-case analysis*. This analysis lets the engineer evaluate what happens under extreme conditions such as when the unit is at its hottest, coldest, highest line voltage, lowest line voltage, and so forth. The engineer uses such tools as computer-aided design systems to help create the model and perform the worst-case analysis.

The completed design effort produces numerous documents. These are:

A bill of materials, which is a complete list of all parts used to build the product. A final bill of materials includes each washer, screw, PCB, transistor, and resistor, as well as the microprocessor.

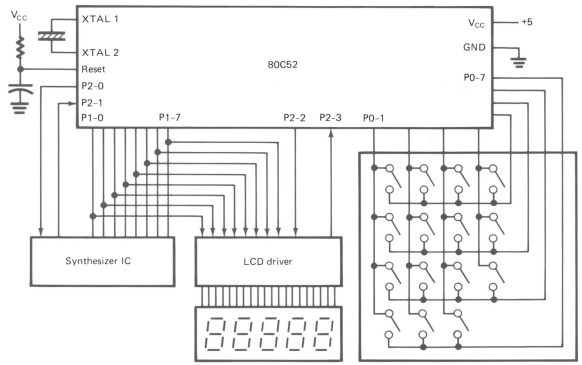

Fig. 17-9 A block diagram of the final digital design showing the selected 80C52 single-chip microprocessor and how it is connected to the other parts in the system. At this time temporary I/O port assignments are made; however, these may change when the PCB layout begins. At that time, the I/O assignments may be redone to give the best layout.

A complete set of component specifications. These drawings describe each part listed in the bill of materials. The drawings provide enough information for purchasing to buy the part and for quality assurance to verify the parts received are correct.

An electrical interconnection diagram (usually called a schematic diagram) showing how each part is connected in the system.

A method to test the product to ensure it complies with its functional requirements and specifications.

When the product development is being performed in a formal environment, each phase usually ends with a design review. At the design review, each engineer presents the completed design to a group of other engineers,

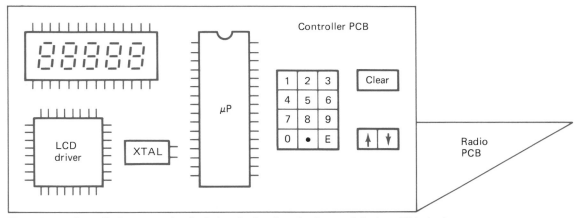

Fig. 17-10 A preliminary mechanical sketch showing the internal design. This is the mechanical engineer's block diagram. It shows the controller PCB and the radio PCB are separate and oriented at 90°. A separate sketch shows how the connections are to be made between these two PCBs.

Implementation
phase

Breadboard

Engineering model

Engineering
prototype

engineering managers, and marketing managers. Questions are encouraged to bring out any possible flaws in the design before the implementation phase is started. When the design phase is complete, the steps between the Phase I and Phase II reports on the engineering schedule are complete.

Self-Test

Answer the following questions.

13. The main purpose for the design phase of any development process is to
    a. Describe a way the functional requirements and specifications in the specification can be implemented
    b. Select components which connected together make a product meeting the required functions and specifications
    c. Provide a model which can be reviewed analytically to ensure theoretical compliance with the specifications under worst-case conditions
    d. All of the above
14. The process of developing a digital design begins with
    a. Creating a schematic diagram
    b. A high-level description of the digital system showing its I/O connections, usually documented as a few-part block diagram
    c. A circuit simulation using a CAD workstation
    d. Any of the above may be used, depending on the design being developed
15. The block diagram design document for the shortwave radio shows the _____ is the interface between the radio circuits and the digital control circuits.
    a. USB (CW), LSB, AM control switch
    b. Synthesizer control lines
    c. Common power supply lines
    d. Audio volume control
16. One of the best design techniques for software development is a piecewise or top-down design. This process is also applicable to the design of
    a. Analog circuits
    b. Digital (including microprocessor controller) circuits
    c. Mechanical designs
    d. All of the above
17. When designing the shortwave radio, the digital design engineer worked closest with

the _____ to decide the 80C52 was the microprocessor for this design.
    a. Mechanical engineer (heat considerations)
    b. RF design engineer (noise considerations)
    c. Software engineer (ROM and RAM space)
    d. All of the above
18. One possibility the mechanical engineer considered was putting the entire shortwave radio on a single printed circuit board. This was probably rejected because
    a. It would not meet the styling requirements generated by marketing and industrial design
    b. There is the possibility of increased noise coupling with both RF and digital circuits on a single board
    c. You cannot build a PCB large enough to put both the digital controller and RF circuits on
    d. All of the above

17-4 IMPLEMENTING AND TESTING THE DESIGN

When the design is complete, including documentation, the engineering team can turn to the job of implementing the design. The *implementation phase* converts the design into a set of final models (prototypes). Each time a model is constructed, it is carefully tested to the specifications, and any differences initiate design corrections before the next model is built. Final models accompanied by finished documentation are turned over to the manufacturing, procurement, and support services to put the product into production.

The first implementation effort is to create working models. There are various ways to do this, and different names for the various stages of models. Some commonly used terms are:

Breadboard or engineering breadboard: The breadboard usually does not look a great deal like the product. That is, it is an electrical model, but not a physical model.
Engineering model: The engineering model is usually an electrical and physical model of the product. Generally, many nonstandard parts are handmade, thus giving the model an unfinished appearance.
Engineering prototype: The engineering prototypes are usually constructed from tooled parts and therefore should closely approxi-

mate the final product. Typically prototypes are not salable since they are made from sample parts which still may not meet all of the requirements shown on the drawing.

Pilot run units: The pilot run units are generally produced in manufacturing by using production equipment, tooled parts, and final test procedures. Pilot run units are used to check out the manufacturing processes to make sure the product can be built. Often pilot run units are sold or used for sales demonstration units.

If the product is simple, or if there are special circuits which need to be checked out before the total product is assembled, breadboarding is often used. A simple breadboard is shown in Fig. 17-11.

There are many different ways to create a breadboard, and different people think of different things when they hear the term *breadboard*. Basically, the term refers to an assembly of electronic parts to perform a certain function or subfunction connected together without use of a final PCB. The objective of breadboarding is to make sure the circuit performs the desired functions before undertaking the expense and effort of designing a final PCB.

Wire-wrap is a special form of breadboarding often used for digital circuits and, in some cases, for low- or moderate-volume production. A wire-wrap model is shown in Fig. 17-12 on page 500. There are several advantages of prototyping with wire-wrap.

The wire-wrap model is fairly easy to change. Modifications to the circuit can be made without redoing the model.

A wire-wrap model can be made as small, although not as thin, as the PCB version.

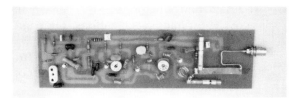

Fig. 17-11 Modeling an electronic circuit using breadboard techniques. As you can see, the physical implementation of a breadboard does not represent the final design. It is simply the quickest, lowest-cost way to build an electronic model of the product or part of the product. This breadboard is implemented on a crude PCB which is cut and drilled until the parts fit and the breadboard works.

The wire-wrap technique is very reliable when properly constructed. This means the model may be used in more rugged environments than other circuit modeling techniques allow.

There are many standard wire-wrap tools and prototyping aids which give a lot of flexibility using standard parts.

Wire-wrapping is relatively low-cost. The price of sockets is slightly more than direct mount on PCB, but the overall cost is usually close to the PCB cost.

The major disadvantage of wire wrap is that it does not do a good job of prototyping some analog circuits. The capacitive coupling between wires in the wire-wrap system is not as well-controlled as the coupling between traces on a PCB. Additionally, wire-wrapping takes a lot of labor. This means it is expensive to make on a repeated basis.

Often engineers use a combination of wire-wrap and other breadboarding techniques. For example, the breadboard for the shortwave radio could use two different techniques. Wire-wrapping is the best way to model the digital controller circuits. However, the radio portion of the product is not easily wire-wrapped. The most common breadboarding technique for this kind of circuit connects individual components together above an unetched (blank) PCB. The copper sheet makes a ground shield and is a place to solder parts, terminal strips, sockets, and other components.

An alternative modeling technique, often used today, is to go directly from the schematic to a PCB. There are several reasons for using this technique.

The PCB creates a model, especially for analog circuits, much closer in performance to the desired product performance than other breadboarding techniques.

Many quick-turnaround prototype PCB shops make 2 to 10 prototype PCBs in the course of a few days.

Most PCB design is now being done on *computer-aided design and drafting (CADD) systems*. The use of CADD cuts down on the impact of changes discovered during the initial prototype testing.

The use of a PCB for early models means these models are physically the proper size. This allows the mechanical engineer to become involved in the prototyping much earlier than could be done if the initial models did not meet the target size and shape due to prototyping techniques.

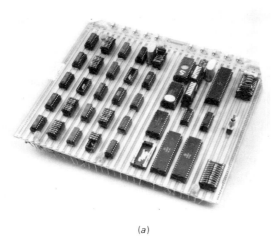
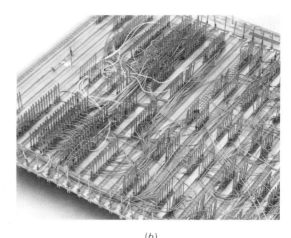

(a)   (b)

Fig. 17-12   Modeling using wire-wrap. Wire-wrap lets you interconnect very densely packed ICs and similar components. It is best applied to digital circuits. The wire-wrap model in the photograph is a dual 6802 system with 32K of RAM, 32K of ROM, two parallel ports, and two serial ports. (*a*) Top side. (*b*) Wire-wrap side.

Before a breadboard or PCB model is built, you must select the parts. There is a lot of work from the time an engineer identifies a general type of part to be used in a circuit and when the specific part is chosen. Often an engineering technician spends significant time reviewing many different parts from many different vendors to select the best one. Selection is based on the electrical requirements, the physical characteristics, price, and availability, among other criteria.

During the engineering effort, many models are built. It is not uncommon for 25 or more models to be constructed before manufacturing begins its pilot run. Each time a model is built, it must be fully tested. A careful record is kept showing the performance to the functions and specifications. Often a review committee sets the actions needed after the testing of a particular model. When a fully compliant model is complete and tested, engineering transfers its information to other departments and prepares the Phase III report, as shown on the engineering schedule.

Self-Test

Answer the following questions.

19. A principal task in the implementation phase is the
    a. Testing of various models to make sure they comply with the specifications

b. Finalizing the drawing package so that there is sufficient documentation for purchasing to buy the parts and for manufacturing to assemble and test the product
c. The construction of different levels of models to show that the design makes a product meeting the desired specifications
d. All of the above

20. The major difference between an engineering breadboard and engineering model is
    a. The engineering breadboard physically represents the product, but it may not be fully functional
    b. Breadboards are often constructed in a "haywire" fashion to show the circuits perform the intended function, but often do not show the intended physical characteristics
    c. Engineering breadboards rarely meet all of the product specifications, but engineering models always do
    d. Engineering models are only built for testing; engineering breadboards cannot be tested

21. The main difference between an engineering model and an engineering prototype is that
    a. The engineering prototype is often incomplete, but the engineering model is a final implementation
    b. The engineering model is often hand-built, whereas the engineering proto-

type is often built with early samples of tooled parts

c. The engineering prototype uses PCB construction, but the engineering model does not

d. An engineering model is a salable unit; the engineering prototype is not

22. Breadboarding is often used to
a. Provide manufacturing with a sample unit
b. Show purchasing what parts to buy for production
c. Check out certain subsections of the design
d. Verify the product can work in wire-wrap form

23. Wire-wrapping is often used when
a. You need a limited production of a very high density digital circuit
b. A conventional breadboard would be too fragile to withstand the rigors of testing
c. There are no critical analog signals which would vary from one unit to another due to variations in stray capacitance on a wire-wrap unit
d. Any of the above is a valid reason for wire-wrapping a circuit

24. You would expect a product design for _____ to be most likely to go directly from schematic diagram to a nearly final PCB model.
a. A fully synthesized 144-MHz radio receiver
b. A television set with digital controls
c. An add-in serial port card for a PC
d. A digital multimeter

25. One of the advantages of using a CADD system is that it has a "library" showing the PCB layout patterns for all of the parts previously used for circuit board layout. This means each part in the design does not have to be researched for a new PCB design. You would expect that this fact has contributed to
a. A tendency to design with the parts already in the library
b. Significantly reduced PCB layout time when a design uses mostly common parts such as the DIP packages, which house most digital ICs
c. The tendency to convert a digital design directly into a PCB layout rather than going through earlier modeling techniques
d. All of the above are reasonable expectations

26. The construction of each set of engineering models is followed by
a. An update to the engineering documentation
b. Additional testing to see if the problems found in the previous model were corrected
c. Additional testing to see if the new model complies with the product functions and specifications
d. All of the above

17-5 REGULATORY COMPLIANCE TESTING

One of the major challenges facing a digital engineering team today is making sure the product meets *Federal Communications Commission (FCC)* requirements. The FCC is the United States government agency regulating the use of radio transmissions and other interstate communications.

Because digital circuits generate many RF signals, FCC rules limit the signal level these products can radiate. Without controls, products could radiate so much energy it might block reception of licensed radio transmissions by a nearby radio receiver.

The FCC rules and regulations, part 15, subpart J for Class A and Class B computing devices, cover microprocessor-based products. These rules specify the maximum signal strength at different frequencies that may be radiated by a product or conducted out of the product through such connections as the line cord and interface cables. The rules also tell how the measurements are to be made.

Before any product covered by these rules can be sold to the general public, it must demonstrate *compliance* with the FCC regulations part 15. Figure 17-13 on page 502 shows a typical label which must be attached to any device covered by part 15 subpart J for Class B computing devices. The final demonstration must be performed by a testing laboratory certified to make FCC part 15 measurements. However, the project engineering team can do a great deal of testing and design to make sure the unit passes when submitted; resubmissions are expensive and time-consuming.

In addition to meeting the FCC part 15 rules for computing devices, there are other regulations a product may have to meet. For example, FCC rules cover devices connected to telephone lines (part 68). There are also rules

> # HEATH / ZENITH
>
> FCC I.D.
>
> CERTIFICATION OF COMPLIANCE WITH FEDERAL COMMUNICATIONS
> COMMISSION RULES AND REGULATIONS, PART 15, SUBPART J.
>
> This equipment is certified to comply with the limits for a Class B computing
> device pursuant to Subpart J of Part 15 of FCC Rules. See instructions if
> interference to radio or TV reception is suspected.
>
> Heath Company, Benton Harbor, Michigan 49022

**Fig. 17-13 A sample of the certification label which must be attached to
a product falling under the FCC rules and regulations, part 15, subpart
J for Class B computing devices. Before such a label can be attached to
the product, technical data must be provided to the FCC showing compliance with the rules. The FCC then issues the certification number
which is to be included on the label.** *(Courtesy of Heath Company.)*

covering radio receivers (part 15), remote-
control transmitters such as garage door open-
ers (part 15), and licensable devices such as
radio transmitters. All of these require additional testing and some level of certification
showing compliance with the FCC rules and
regulations.

Many companies require compliance with
safety standards established by UL in the
United States, CSA in Canada, and TUV and/
or VDE in Europe. Each organization has minimum specifications for electrical components
connected to the power line, the flammability
of case materials, and case construction techniques which keep users from electrical shock.
In some cases they also set comfort requirements for people using certain products (for
example, terminals).

Compliance with each of the various standards is a major effort in the engineering development process. An understanding of the
requirements placed on the design by these
standards makes the job easier, but not simple. All through the design project, each member of the engineering team must constantly
keep these requirements in mind as various decisions are made.

For example, suppose a particular digital circuit design may easily be expanded to provide
an external input at very low cost. This extra
input gives the product another feature. Such
a suggestion is welcomed by marketing because it allows them to offer a product with
more features than competitive models. However, a hidden cost of adding this extra input
is making sure the option does not cause the

product to lose its FCC part 15 compliance.
The input option may cost very little, but if it
significantly adds to the time required to obtain FCC part 15 certification, it may not be
worth the effort.

Compliance with the various regulatory
agency requirements and with the desired product functions and specifications is only verified
with extensive testing. This testing is a major
part of the engineering effort. If the testing is
to be of value, it must be thorough and well-
documented.

In many companies, product compliance is
further verified by the *quality assurance department*. Their function is to ensure the company produces a product which meets
management quality standards. Because quality must be designed into a product (it cannot
simply be inspected in), the quality assurance
department must review design models to be
sure the required quality is in the design. Note
that regulatory compliance testing is a separate
activity on the engineering schedule, as is evaluation (testing by the quality assurance evaluation engineering department).

Self-Test

Answer the following questions.

27. The reason microprocessor-based products
    used by the general consumer must com-
    ply with part 15, subpart J for Class B com-
    puting devices of the FCC rules is that
    a. This allows government control of
       transmitters

b. The FCC can report any problems with foreign imports to the customs officer in charge

c. An improperly designed product with digital circuits can radiate sufficient energy to interfere with the legitimate use of radio receivers or TVs

d. The telephone network is a public utility, and if an improperly designed device is connected to it there is a possibility it could keep many people from using their telephones

28. Although final testing to determine compliance with part 15 must be performed by a certified laboratory, any work done to minimize radiation before the compliance testing is performed

a. Is a wasted effort because any change from the FCC testing will render obsolete any previous effort

b. Saves the potential expense of retesting

c. Is not required because most products meet part 15 with little or no design effort

d. All of the above

29. You would expect an externally connected intelligent modem which operates from 120 Vac to meet requirements set forth by the

a. FCC part 15, subpart J

b. FCC part 68

c. Underwriters Laboratories

d. All of the above

30. Often the company quality assurance department performs additional product testing

a. To check up on the engineering department, which may try to get away with developing a product that does not meet marketing's requirements

b. Because they are responsible for ensuring that the company ships quality products, and you must design quality in if you want quality to ship

c. To make sure that marketing has not specified a product which cannot be built

d. All of the above

17-6 DESIGN TOOLS FOR MICROPROCESSOR DEVELOPMENT

When you develop products built with digital circuits, including microprocessors, you require electronic test equipment to analyze and troubleshoot the models. Conventional test equipment is required for many measurements. Some common test equipment found on the engineering bench includes

Digital and analog multimeters
Dual-trace 50- to 250-MHz oscilloscopes
Digital frequency meters or counters-timers
Logic probes
Power supplies
Function generators
Pulse generators
EPROM programmers and erasers
Personal computers

In addition, there are two instruments used frequently in the development of digital electronics and microprocessor-based systems. They are the logic analyzer and the microprocessor development system.

The *logic analyzer* looks like an oscilloscope. However, its operation gives many new features. First, the logic analyzer displays signals from many inputs. Logic analyzers for microprocessor-based product development have 16, 32, 48, or more inputs. Second, logic analyzer inputs respond to logic levels, not to the analog signal value. That is, the logic analyzer displays the logic 1 or logic 0 state of the input signal, not its real amplitude. Third, the logic analyzer stores this digital information in its own memory. This gives you time to analyze events that only happen once.

Fourth, a logic analyzer displays data in two ways: with timing displays or with state displays. The user selects the best display for the measurement being made.

An example of a *timing display* is shown in Fig. 17-14. A timing display looks like an oscilloscope display. In the timing mode, data at the input is sampled at regular intervals established by a "clock." For the timing display, data is sampled at five or more times the greatest data rate being measured. The timing display sampling rate determines the display time resolution. For example, if the sampling period is 20 ns, you can only measure and capture events longer than 20 ns. Often this is referred to as a 50-MHz sampling rate, because $f = 1/T = 1/(20 \times 10^{-9}) = 50$ MHz.

The timing display shows timing relationships in a system. Looking at the timing display, you can easily see which event happened first. You can also use the timing display to measure timing relationships in the system. For example, the display in Fig. 17-15 shows 101 happened on the last three traces 3 clock cycles after the trigger signal (01000000). If the

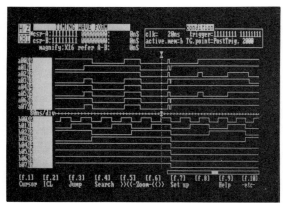

Fig. 17-14   A logic analyzer timing display. Much like a multiple-trace oscilloscope, the timing display shows timing relationships between the different points of a logic signal or between different points of different logic signals in a multisignal system. (*Courtesy of Tektronix, Inc.*)

clock signal is a 1-MHz signal (1 μs), then the 101 data pattern happened 3 μs after the trigger.

Displaying multiple traces lets you see timing relationships of many signals at the same time. For example, a logic analyzer displaying 24 traces shows the logic levels of all 16 memory address lines and all 8 data lines on an 8-bit microprocessor. With this information, for example, you can see the data returned when a particular memory location is addressed by the microprocessor.

The sweep on an oscilloscope is triggered by a signal on one of its input channels or on an external trigger input. How do you trigger the logic analyzer timing display when there is no one channel which is the most important?

There are two ways to trigger the logic analyzer.

Triggering from an external trigger input. For example, the external trigger line might be connected to the microprocessor MEMORY-READ or MEMORY-WRITE line. This lets the user examine bus addresses and data relative to this event.

Triggering from a match between a previously stored or user input word and a matching word on selected inputs. This lets the user examine bus activity relative to addressing a certain memory location or relative to the time a certain instruction is fetched. An example is shown in Fig. 17-15.

Logic analyzer triggering is more sophisticated than oscilloscope triggering, but there are many common features. For example, delayed sweep

is a common function for laboratory-grade oscilloscopes. This allows you to start the sweep some time after the trigger happens. This lets you see events following the trigger which would be lost without delayed sweep. Delayed sweep is shown in Fig. 17-16.

Likewise, the logic analyzer offers a "delayed sweep." Because the logic analyzer "sweep" is really a clock signal, the delay is usually given in clock cycles. For example, the logic analyzer clock might come from a microprocessor clock. If, for example, you wish to see the bus transactions which follow the addressing of a certain memory location by 200 clock cycles, you use the post-trigger function (see Fig. 17-17 on page 506).

Because the logic analyzer is looking at digital information stored in memory, you can perform an even more sophisticated triggering function called *pretriggering*. For example, suppose you want to know what events caused a particular memory location to be addressed. First, you set the logic analyzer to trigger when the selected address appears on the memory address bus. The logic analyzer continually stores the measured data in its 4096-word memory. Second, when the trigger occurs, you stop writing to memory. The 4096 *previous* bus transactions are stored in memory waiting for you to look at them (see Fig. 17-18 on page 506). You cannot perform the pretrigger function with a conventional analog oscilloscope because all of the information which happened before the trigger occurred is lost.

Using the memory storage feature, the logic analyzer offers pretriggering, post-triggering, and mixed triggering. These trigger forms are shown in Fig. 17-19 on page 507.

The *state display* (Fig. 17-20 on page 507) shows the binary status of the logic analyzer inputs. It is updated every time an input changes state. This captures each "state" change in the measured system. Normally, you can present such information in binary, octal, hexadecimal, or ASCII form.

State displays are very useful for watching data flow on the microprocessor bus, especially if the data is coming at a variable rate. If there are no changes for a while, the display waits until there is a change. This is different from the timing display, which shows the passage of time. The state display is extremely useful, because frequently you do not need to know when (time) something happened but just that it did happen, and what its relationship to other signals was at the time.

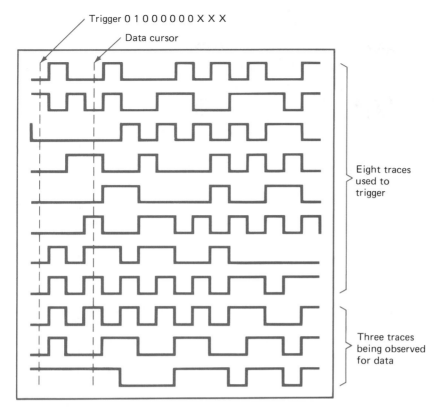

Fig. 17-15   Looking for the occurrence of a particular set of data following the trigger. In this case the system is set to trigger on 01000000XXX (X = don't care). The bottom three traces are from a data source, and we are looking for the first time that the data is 101 following the trigger. The second cursor is positioned over the data.

Many logic analyzers also offer a diagnostic tool called a *disassembler*. The disassembler is targeted at a specific microprocessor. The disassembler converts data in a state display into a listing of instruction mnemonics and their associated memory addresses. Obviously, this generates an undocumented listing, and you must start at the beginning of an instruction sequence for the disassembler to work properly. However, this can be a great aid to

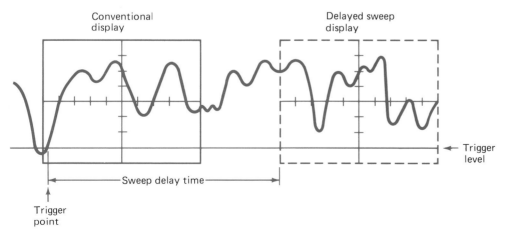

Fig. 17-16   Delayed sweep on a conventional oscilloscope. In this example, the oscilloscope is triggered by the positive slope of the signal returning from an negative excursion. Because the portion of the signal you are interested in comes too late to be displayed if the sweep started at the trigger, you use delayed sweep to see the portion of interest.

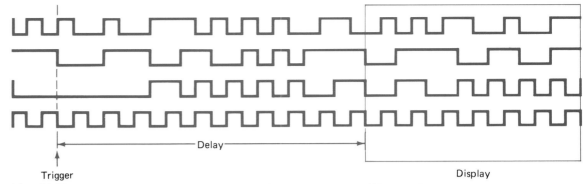

Fig. 17-17   Post-triggering. Triggering is taken from a change to 1000. The time base or clock is set so that there are 10 samples between any known changes in logic state. Because the area of interest is delayed (shown in the area marked display), a delay of 200 clock counts is used before the display starts.

the microprocessor software and hardware developers who are working on a system which is closely controlling hardware.

A second major development tool is the *microprocessor development system (MDS)*. This instrument answers the question, "how do I develop and test software which controls hardware for a microprocessor-based product when the target system is not large enough to run software development tools?"

Of course, one approach is to develop the software on a separate computer system, load the object code into an EPROM programmer, program an EPROM, plug the EPROM into the target system, and, using logic analyzers, oscilloscopes, and other test instruments, try to debug the code and the hardware. This approach is very frustrating and slow.

The MDS plugs into the target system using the same socket the microprocessor uses. The MDS circuits at the other end of this connection let you *emulate* the microprocessor. The MDS emulation microprocessor is connected to extra memory and special logic to capture bus data while the microprocessor is executing the test software. This lets the operator see what is happening inside the microprocessor, I/O devices, memory locations, on the buses, and elsewhere.

A modern MDS is controlled by a microcomputer system such as a PC or a dedicated system. Some MDS implementations use microcomputers built with the same microprocessor they are emulating. MDS commands are entered at a keyboard, and the status is displayed on a CRT.

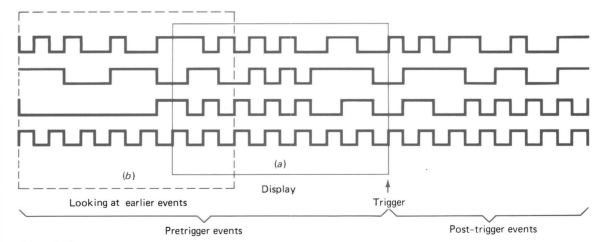

Fig. 17-18   Using the logic analyzer pretrigger function. (*a*) The logic analyzer is set to trigger on the first occurrence of a 1001. When the logic analyzer is triggered, no more information is written into the memory. Therefore, the 4096 previous data words are captured in memory and can be viewed by "sliding" the display window as needed. (*b*) Sliding the window to look at earlier events.

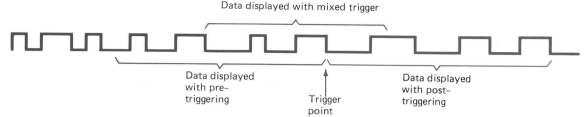

Data displayed with mixed trigger

Data displayed with pre-triggering

Trigger point

Data displayed with post-triggering

Fig. 17-19 The three different kinds of triggering on a logic analyzer. Pretriggering, mixed, and post-triggering. Once the data is captured in memory, the user can look up to the point that triggered the data capture (pretriggering), or the user can look after the point that triggered data capture (post-triggering), as well as both sides of the trigger point (mixed).

The MDS includes extensive software development capability. If the MDS is PC-based, the PC hosts an editor, cross compilers and cross assemblers, linkers, debuggers, and other software development tools. An MDS built around the target microprocessor uses compilers and assemblers designed to run on the target microprocessor.

The MDS loads your newly developed software into target memory or MDS memory. It can execute this code from MDS memory, target system ROM or RAM, or a combination of locations. Some of the possible divisions are shown in Fig. 17-21 on page 508.

Common MDS analysis features include

Single-step mode which allows you to execute the object code one instruction at a time and observe the effect on various registers, memory locations, I/O ports, and the microprocessor after each instruction executes.

Breakpoint the object program. This mode lets you execute instructions at a normal rate un-til a specified condition happens. When the specified condition happens (a particular memory location is addressed, for example), program execution halts and the operator can observe system conditions.

Trace the operations which happen after a particular trigger occurs. This mode lets the MDS act like a logic analyzer. Depending on the MDS system you are using, the operations may be captured by a hardware accessory which acts as fast as a logic analyzer, or the operations may be captured by special software which provides the same information but slows the overall operation by the time required to capture the data.

Wherever possible, the MDS is there to let you debug your hardware and software in real time. That is, its job is to give an operation which is as close to actual performance as is possible. True real-time operation lets the microprocessor operate at normal execution speed. However, many MDS debugging and measurement

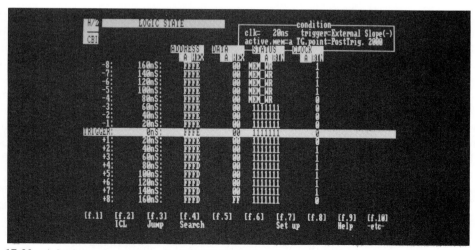

Fig. 17-20 A logic analyzer state display. This display shows the logic state (logic 1 or logic 0) of each data word in memory. A new data word is written into memory each time there is a trigger signal on one of the tested lines. (*Courtesy of Heath Company.*)

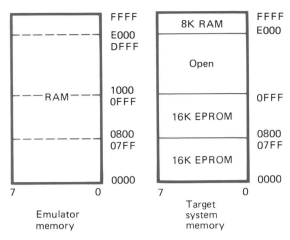

Fig. 17-21 Alternative memory mapping available with an MDS. In this example, the 8-bit microprocessor-based system uses two 27128 EPROMs and one 8-kbyte static RAM. Using the MDS, object code which will finally reside in one of the two EPROMs can be loaded into the equivalent memory space in the emulator and debugged without programming EPROMs. If, for example, the code is temporarily too large to fit in the 32 kbytes allocated to EPROM, the code can execute from the larger emulator memory. Once the code is debugged, an effort can be made to reduce its size to fit the EPROM space. Note that the emulator can execute code between 1000H and DFFFH even if the microprocessor-based system does not decode those memory addresses.

operations slow down the operation—how much depends on the requested action and the MDS design.

For example, one way to breakpoint software is to insert a conditional jump instruction in the object code. When this is done, the timing of a loop is lengthened by the time required to execute the conditional jump instruction each time the loop executes. If this loop only has a few instructions, the addition of one more instruction can lengthen the operation significantly. If the loop has many instructions, the addition of one more instruction will not lengthen the loop significantly.

Real-time testing helps find problems that happen when the microprocessor is running at the maximum speed. Often, problems do not show up if the microprocessor does not execute the code in the shortest time possible.

For example, suppose your microprocessor system is to input data through its parallel I/O port. Each time you slowly step your microprocessor system through its instructions, they all work well. However, when you run the system at full speed, the system goes into an endless loop.

An MDS helps you solve this problem. In this case, you set the MDS to trace data transfers on your microprocessor memory address bus and data bus for a few instructions before and after the I/O instruction that causes the trouble.

Using the MDS, you can start tracing (storing) each bus transaction after some trigger point you have entered into the system. The trigger point is usually when a certain instruction is executed or when a certain memory location is addressed. Later you can review all the data captured in the MDS trace memory.

In this example, you find that, at high speeds, the I/O port status register is always reset to 0 by a timing error. Because an all-0 status register prevents the microprocessor from inputting any data, the software goes into an endless loop. If the system is run at less than real-time speeds, the extra delay allows the I/O port status register to become set and the software acts properly.

Microprocessor development systems have many features other than the trace mode. For example, after each instruction is executed, the MDS can display the status of each register in the microprocessor plus a few selected memory locations. The MDS cannot do this at full speed, but the MDS still performs this operation on a limited amount of code at fairly high speeds.

Other MDS features include sophisticated triggering to help find a fault in the tested system. An MDS strong point is that it permits quick correction of software errors. When a software error is found, it is corrected by switching into the software development mode, editing the correction into the source code and documentation, and recompiling or reassembling a new object module. You can begin testing this revised object code almost immediately, knowing the "fix" is a completely integrated part of the source code and it has been properly documented.

The alternative is to "patch" the object module and hope you remember to correct the source code at a later date. Even then, the new object code must be retested to make sure it performs the same as the patched object code.

For special applications, there are many other special-purpose electronic instruments which may be required to complete the development process. However, the ones reviewed in this chapter should be sufficient for most microprocessor-based all-digital product development.

31. _____ is only used to troubleshoot design errors in microprocessor-based products.
    a. A logic analyzer
    b. A 50- to 250-MHz dual-trace oscilloscope with delayed sweep
    c. An MDS
    d. A 150-MHz counter-timer
32. One major difference between the logic analyzer and the oscilloscope is
    a. A logic analyzer displays many more channels of information
    b. An oscilloscope displays signal amplitude versus time, whereas a logic analyzer displays logic level versus time
    c. A logic analyzer stores the input data in memory and then shows the requested portion of its memory; an oscilloscope has no memory
    d. These are all major differences
33. A logic analyzer has two main display modes: timing and state. The information displayed in the timing mode
    a. Shows the logic levels for all inputs at the time any of the inputs changes its logic level
    b. Shows the events which happen after the trigger occurs
    c. Looks very much like a multiple-trace display on an oscilloscope, but is really a view of the data stored in the logic analyzer memory
    d. Can be presented in binary, octal, decimal, hexadecimal, or ASCII form
34. The shortest event shown on a logic analyzer display in the timing mode depends on the
    a. Kind of logic being used
    b. Logic analyzer disassembler
    c. Sampling rate
    d. Any of the above
35. If you need to know how long address line A_9 is at logic 0 following the leading edge of the $\overline{M1}$ pulse, you would not use a
    a. Dual-trace oscilloscope with delayed sweep
    b. Logic analyzer set to present data in a state display
    c. Logic analyzer set to the timing display mode
    d. Universal counter-timer in the time interval mode

36. The logic analyzer timing mode can be triggered by a signal at an external input or by
    a. Issuing a I/O read pulse on the system bus
    b. Addressing a certain memory location
    c. Fetching a particular instruction from ROM
    d. Any of the above can be used to trigger the logic analyzer by setting certain conditions and connecting the logic analyzer inputs to the correct signals on the microprocessor buses
37. You might set certain bits in the logic analyzer trigger word to "don't care" so that
    a. The logic analyzer ignores any data you do not care about
    b. The logic analyzer triggers on a match between the set bits and any combination of logic levels for the bits marked don't care
    c. The logic analyzer will trigger on any level signal instead of a fixed level
    d. Don't care is used on oscilloscope triggering but not logic analyzer triggering
38. You set the logic analyzer to pretriggering when you want to see
    a. The data which follows the trigger word
    b. The data which immediately precedes the trigger word
    c. The data which follows the trigger word by a predetermined number of clock cycles
    d. Each of the above is displayed with pretriggering, depending on the slope switch setting
39. You are debugging a microprocessor-based product your company has under development. As you execute a section of the program, you find memory location FA24 contains the proper data (4C) most of the time it is read, but occasionally when it is read incorrect data is returned. You know it must have been changed some time before, but you still don't think the memory location is being written to. To check what is happening, you set your logic analyzer to
    a. Trigger on a memory-write pulse
    b. Trigger on a memory-read pulse
    c. Post-trigger when the microprocessor writes to memory location FA24
    d. Pretrigger when the microprocessor reads from memory location FA24
40. You are working on another design project which seems to be having problems with an I/O port. The product inputs very slow,

almost randomly generated data through its parallel I/O port until an ASCII EOF or EOT character is found in the data. Either EOF or EOT end the data input routine. The system is not working, and a review of the software indicates the routine is checking for the EOF and EOT characters. Someone suggests the external device sending the data may not be generating the EOF or EOT characters. To check this theory, you

  a. Use a logic analyzer in the timing mode connected to the data bus; the logic analyzer is set to trigger any time either an EOT or EOF data word is placed on the bus
  b. Use a logic analyzer in the state mode and connected to the I/O port input connector; it is set to display ASCII characters, and you set it to trigger when EOT or EOF is found.
  c. Use your MDS to examine the file in memory to see if somehow the EOF or EOT character got through the software trap and was stored in memory
  d. Any of the above proves the theory if the EOF or EOT character is found by the test

41. An MDS is different from a logic analyzer because the
  a. Logic analyzer is a general-purpose tool, and the MDS is built to emulate a single microprocessor.
  b. MDS has the ability to breakpoint the object code and disassemble code, and the logic analyzer can only disassemble code
  c. Logic analyzer does not have built-in software development tools

  d. All of the above are differences between the MDS and the logic analyzer

42. You are debugging some new code for a microprocessor-based controller. There is a short conditional loop in the code which just does not seem to be branching properly. To see exactly what is happening, you use your MDS in the
  a. Single-step mode so that you can execute the loop while watching the results of executing each instruction and see the contents of each register to note what conditions the instructions are acting on
  b. Examine memory mode so that you can see the code in memory to determine if it matches the output of your compiler
  c. Trace mode so that you can look at the sequence of instructions executed to determine if the program is accidentally branching off to another loop
  d. None of the above will catch this problem; you should be using a logic analyzer in the state mode.

43. The advantage of using an MDS with a software development system is that
  a. You know that the code you develop is compiled for your microprocessor
  b. Compilers and assemblers resident on the MDS produce more efficient code than nonresident compilers
  c. You can quickly edit corrections into the source code, regenerate object code, and test the new object code without going through the process of transferring code to the MDS from an external system
  d. Each of the above is a valid reason, but choice b carries the greatest weight

SUMMARY

1. The steps to develop a complete microprocessor-based product are the same as those to develop the software for that product:
  a. Write a specification
  b. Create a design
  c. Implement the design and test and correct the implementation
  d. Demonstrate the final product
2. Only small microprocessor-based products are entirely designed by one individual. Larger products require a team effort.

3. Typical skills required to develop a microprocessor-based product include electronic (digital and analog), software, and mechanical.
4. If there are several engineers assigned to a development project, then one of them is usually assigned the job of project manager.
5. The project manager develops the project schedule, assigns work effort to the other engineers, and makes sure the overall ef-

fort produces a product which meets the functions and specifications defined by marketing.

6. A typical engineer assigned to a development project has completed a four-year course of study leading to a bachelor's degree or a six-year course of study leading to a master's degree.

7. Engineering technicians and product developers assist the engineer with the development project. They often are responsible for
   a. Circuit board layout
   b. Constructing models
   c. Testing
   d. Documentation (drawings and bills of material)
   e. Interdepartmental transfers

8. Other support organizations assist engineering with the development of a project. Some of these organizations are
   a. Purchasing for component selection
   b. Industrial design for aesthetic design
   c. Quality assurance for compliance testing
   d. Drafting for final drawing completion

9. Often the marketing specification must be expanded by engineering to provide a detailed description of the target product and to make sure compliance to requirements can be measured.

10. Often a section of the product will be completely unspecified by marketing. This may be because there is no marketing need to specify this topic, but once engineering decides on a method of implementation the topic must be specified.

11. Topics not specified by marketing must be specified by engineering if there are several individuals working on the project to ensure the complete specification is known by the whole engineering team.

12. Many specification and design topics must be answered by two or more engineers responsible for different areas of the project. They must agree on a common solution to the topic.

13. Often it is difficult to tell the difference between the early parts of the specification section of a project and the early parts of the design effort.

14. The larger the project, the more common it is to have a major written specification. Very small projects may only have a simple written specification or just an oral specification.

15. The design process begins with an overview design which is expanded into greater and greater detail.

16. The final expansion of an electronic design is expressed as a schematic diagram, a list of parts needed to build the design, and drawings specifying the parts used in the design.

17. Each part of the design can be modeled and tested for worst-case conditions. This is how engineering makes sure the product will work with all of the variations in parts and over the environmental extremes.

18. Often each design phase is concluded with a design review where the engineers present their design, explain how it works, and answer questions from a panel of other engineers, marketers, and management personnel.

19. The implementation phase converts the design information into a series of models which are tested to show the design implements the functions and specifications of the original requirement.

20. The different kinds of models of the product are
    a. The engineering breadboard
    b. The engineering model
    c. The engineering prototype
    d. Pilot run units

21. Usually the breadboard is only an electrical implementation of the product or critical parts of the product and does not model the physical characteristics of the product.

22. Wire-wrap is a special form of breadboarding which is often used with digital circuits. It is rugged and compact, and may be used for very low volume production.

23. Often a product breadboard uses a combination of modeling techniques. The techniques used are those which can generate the most accurate electrical model as soon as possible with the least investment of nonproductive effort.

24. The use of CADD (computer-aided design and drafting) systems lets many products go directly from a schematic to a PCB (printed circuit board) model.

25. Before a PCB can be designed, each of the parts used must be selected for its electrical and mechanical characteristics.

26. As each model is built, it is thoroughly tested to see if it complies with the specification and to see if errors found in ear-

lier models have been corrected in the latest implementation.

27. A major portion of the testing required today is to determine compliance with the FCC rules and regulations.

28. Because digital circuits use high-frequency oscillators and fast rise-time pulses, they radiate a great deal of energy in the RF spectrum. The FCC rules and regulations in part 15 specify maximum levels for radiated and conducted signals.

29. Before a microprocessor-based product may be sold to the general public, it must be tested to show that it complies with the part 15 requirements and have a label showing that it completed this testing and certification.

30. The FCC rules require tests to show compliance for devices connected to the telephone lines, which have radio receivers over a certain frequency, and which are transmitters by design.

31. Many companies also require testing to show that the product complies with the safety standards issued by the test laboratories UL, CSA, and VDE.

32. The development of microprocessor-based products requires the use of many conventional electronic test instruments.

33. The logic analyzer is a special-purpose instrument developed to test digital circuits. It looks like an oscilloscope, but it is different because it
   a. Displays signals from 16, 32, or 48 inputs
   b. Displays logic levels, not signal amplitudes
   c. Stores the data words in a digital memory
   d. Can show data in either the timing or state displays

34. A timing display looks like an oscilloscope display with many inputs.

35. The timing display uses a clock which sets the minimum resolution of an event which will be recorded in the logic analyzer memory.

36. Timing displays are used to show timing relationships between the data on different inputs and at different times on the same input.

37. A logic analyzer can be triggered by a signal at an external trigger input or by matching a bit pattern in a trigger register with a bit pattern at the data inputs.

38. The logic analyzer can be pretriggered or post-triggered because the data is stored in memory. This means the memory can be filled after the trigger happens (post-trigger) or be filled until the trigger happens (pretrigger).

39. State displays show the condition (state) of each data source each time one changes state.

40. Usually state displays can be presented in binary, octal, decimal, hexadecimal, or ASCII format.

41. If a logic analyzer has a disassembler option, it can be used to convert data in a state display into a series of instruction mnemonics for a particular microprocessor.

42. The MDS (microprocessor development system) lets you replace the microprocessor with a powerful tool which emulates the microprocessor and allows the operator to see the internal register and memory conditions.

43. The MDS allows you to execute object code for the target microprocessor from emulation memory, target system RAM, or target system ROM.

44. Common MDS features include
   a. A single-step mode
   b. Breakpointing
   c. Tracing
   d. Software development

45. The better the MDS, the closer it comes to letting you execute the object code in real time.

46. An MDS with fully integrated software development capability lets you correct software errors very quickly and make sure they are a documented part of the corrected source code.

CHAPTER REVIEW QUESTIONS

Answer the following questions.

17-1. Name the basic steps which must be carried out to develop any hardware or software product or any product which uses a combination of hardware and software.

17-2. List the three major skill areas which are required to develop a micropro-cessor-based product. What is a major subdivision of one of the skill areas?

17-3. Briefly describe some of the functions and responsibilities assigned to the project manager.

17-4. The typical engineer assigned to the digital circuit development of a microprocessor-based product will hold
  a. An associate degree in engineering
  b. A bachelor's degree in electrical engineering
  c. A master's degree in electrical engineering
  d. A digital circuit engineer may hold any of the above degrees, but a bachelor's is most common

17-5. The engineer and technician on a project form a very tight team which is responsible for designing and implementing their portion of the project. Match the following duties with the engineer (E) or the technician (T).
  a. Development of the signal-processing equations
  b. Construction of the input amplifier breadboard
  c. Initial screen-room testing to determine compliance with FCC part 15
  d. Selection of the microprocessor
  e. Preparation of the final bill of material
  f. Presenting the digital circuit design to the design review committee
  g. Layout of the printed circuit board
  h. Selection of the IC technology to be used

17-6. Briefly describe the functions performed by the following support groups to aid engineering development of a product.
  a. Purchasing
  b. Drafting
  c. Industrial design
  d. Quality assurance
  e. Technical publications

17-7. Briefly explain why a marketing specification such as ''operates from 12 Vdc'' must be expanded by engineering to ''operates from 11 to 15 Vdc.''

17-8. Why would marketing leave out an important part of the specification, such as indicating that the product is microprocessor-based?

17-9. Briefly explain why many specification and design issues must be resolved by two engineers with different technical responsibilities. Give an example of such a situation for a microprocessor-based design.

17-10. What is the major reason a large project development effort may require a formal written specification with a review?

17-11. A hardware design effort goes from a design concept documented by one large block showing its real-world interfaces, to a multiple-block diagram, to a schematic diagram. Briefly explain how this design effort is similar to a structured software design effort.

17-12. What is the purpose of worst-case testing?

17-13. You could say that the major features of the implementation phase are the construction of_____ and the _____ of those _____ .

17-14. Briefly explain the major difference between an engineering breadboard and an engineering model.

17-15. Briefly explain the major difference between an engineering model and a pilot run unit.

17-16. Your company has contracted to build six special digital processors. These use a large quantity of ICs, must be packed into the smallest box possi-ble, and are subject to some customer modifications once the six are built and installed. What is the best implementation technique. Why?

17-17. Briefly explain why the general availability of low-cost PC-based CADD systems has caused a significant increase in the number of first models implemented with PCBs.

17-18. What are the major characteristics of digital circuits which cause them to be subject to FCC rules and regulations?

17-19. How must you show the general consumer that a digital product complies with FCC part 15, subpart J?

17-20. What is the basic function of testing performed to show compliance with UL, CSA, or VDE requirements?

17-21. What are the major characteristics of a logic analyzer which make it different from an oscilloscope even if it looks somewhat like an oscilloscope?

17-22. What logic analyzer mode would you use if you want to find out how many nanoseconds are required between the time memory is addressed and the time data is available on the bus? Why?

17-23. What hexadecimal trigger word would you use to detect any memory accesses in the top 25 percent of a Z-80's memory map?

17-24. Briefly explain what it means when you say you are using a logic analyzer with pretriggering, and compare this to the similar trigger mode in a conventional oscilloscope.

17-25. Why would you use the logic analyzer's state mode to make sure that a microprocessor was addressing sequentially higher and higher memory locations?

17-26. Briefly explain why the disassembler on a logic analyzer cannot document the disassembled code to the extent the original programmer can.

17-27. We say the MDS emulates or does the same thing as the microprocessor it replaces. Why do we use the MDS if it only emulates the microprocessor? If it does more, what more does it do?

17-28. Briefly state one reason why you might wish to execute object code from emulation memory instead of target machine memory.

17-29. List four key functions performed by an MDS.

17-30. Briefly explain what is meant by real-time operation of an MDS.

17-31. Briefly explain the value you get from having the MDS easily able to perform software development as well as emulation.

CRITICAL THINKING QUESTIONS

17-1. Why do you suppose the general outline for developing new electronic circuits is the same as the general outline for developing new software?

17-2. Why is it so important to start a design effort with a complete specification? Why do you suppose incomplete specifications occur more frequently than they should?

17-3. If the electrical block diagram for a microprocessor-based product is equivalent to the software's flowchart, what is the schematic diagram equivalent to?

17-4. Often the initial prototypes of a complex microprocessor-based product are checked out a section at a time and then the checked-out sections are put together to build toward a complete prototype. How does this process compare with the development of a software package?

17-5. Would you expect the microprocessor development system (MDS) to be used more frequently during the design process for a microprocessor-based control system or a microprocessor-based personal computer? Why?

Answers to Self-Tests

1. *d*	11. *d*	16. *d*	26. *d*	36. *d*
2. *c*	12. **a.** F	17. *c*	27. *c*	37. *b*
3. *d*	**b.** S	18. *a*	28. *b*	38. *b*
4. *b*	**c.** F	19. *d*	29. *d*	39. *d*
5. *d*	**d.** S	20. *b*	30. *b*	40. *b*
6. *a*	**e.** I	21. *b*	31. *c*	41. *d*
7. *b*	**f.** S	22. *c*	32. *d*	42. *a*
8. *b*	13. *d*	23. *d*	33. *c*	43. *c*
9. *c*	14. *b*	24. *c*	34. *c*	
10. *d*	15. *b*	25. *d*	35. *b*	

CHAPTER 18

New Developments in Microprocessor Technology

■

CHAPTER OBJECTIVES

This chapter will help you to:

1. *Explain* the differences between flash memory and conventional memory and *give* examples of when flash memory is used.
2. *List* the basic characteristics of a PCMCIA card and *explain* where and how the card is used.
3. *Give* a brief overview of RISC processors, *explain* their advantages, and *discuss* the PowerPC RISC processor.
4. *Recognize and list* the major features of the different wireless communications systems used in conjunction with microprocessor-based systems.

This chapter examines some of the emerging technologies as they affect microprocessors and microprocessor-based systems. Many of these new technologies were first used on or are predominantly found on PCs. However, the PC is not the only application for these technologies, and this chapter should not be viewed as a "PC accessories" chapter. The PC simply happens to be a large user of microprocessor-based or related technology and therefore often is the initial platform or a common platform for these technologies.

■

18-1 FLASH MEMORY

One of the significant limitations of modern semiconductor-based microprocessor systems has been the lack of nonvolatile read-write memory. Increasingly, microprocessors are found in situations where low power and/or battery operation is required. Additionally, today's microprocessor-based systems often require large amounts of main memory and mass storage. Common examples are palmtop computers and industrial data-gathering devices intended for remote-site operation. In these applications, the power consumption of a hard or floppy disk drive is prohibitive. The drives may also be physically too large for the application.

Erasable programmable read-only memory (EPROM) came early in the history of semiconductor memory systems. However, EPROM is read-only memory, not read-write memory; that is, with EPROM, there is a significant problem if you wish to change the contents. EPROM must be removed from the circuit, erased by means of a powerful ultraviolet light, and then reprogrammed.

Electrically erasable programmable read-only memory (EEPROM) was significant improvement over EPROM. Erasing EEPROM is done with a special electrical signal; it is a slow process. Write times are also slow because it is necessary to write multiple times to a memory cell to ensure that the data is permanently stored. Typical EEPROM de-

Table 18-1 Key Characteristics of Flash Memory ICs

Memory Organization	Access Time, ns	Number of Write Cycles	Sector Size	Pin Count	Programming Voltage
32 kbytes	70–200	10,000	Bulk	32	12 Vdc
64 kbytes	70–200	10,000	Bulk	32	12 Vdc
128 kbytes	45–120	100,000	16 kbytes	32	5 Vdc or 12 Vdc
512 kbytes	70–120	100,000	64 kbytes	32	5 Vdc or 12 Vdc
256K 16-bit words	70–120	100,000	64 kbytes or 32 kword	44	5 Vdc or 12 Vdc

From page 516:

Erasable programmable read-only memory (EPROM)

Electrically erasable programmable read-only memory (EEPROM)

On this page:

Flash memory

Programming the flash memory

vices are small (they hold a limited number of bits). They are used for nonvolatile storage of device setup information, special-purpose scratchpad memory, or program memory for very small systems.

Flash memories are an outgrowth of EE-PROM technology. *Flash memory* developed from the idea of creating a solid-state equivalent to the disk drive. Although flash memories do not have the storage capacity of hard disks, they offer many of the advantages of disk drive mass storage systems without the need for the mechanical aspects of a drive.

Fundamentally, flash memory is a special EEPROM. Before writing to a memory location in a flash memory, you must erase the memory. That is, each bit must be set to a logic 0. Then, as with EEPROM, each bit is written to a number of times to make sure that the semiconductor junction is permanently modified—if the new bit condition is to be logic 1. Nothing has to be done if the bit condition is to be logic 0. When the junction is permanently modified, a logic 1 is written to the memory location. Although the end result of this process is similar to writing to a normal read-write memory location, the process is usually called *programming the flash memory*.

Like EPROM, flash memory products cannot erase a single bit or word. In most cases, the entire memory must be erased, and then individual bits or words can be programmed. For some flash memory ICs, sections of the IC can be erased.

Flash memories are usually organized to work with a microprocessor. The most common are organized as bytewide devices. Table 18-1 shows the organization of typical flash memory ICs.

Table 18-1 also shows other characteristics of flash memories. First, flash memories have access times which do not slow down typical microprocessor systems. With access times in the 45- to 70-ns area, a flash memory is as fast as most DRAMs.

Second, flash memories, like EPROMS, do not have an unlimited number of write cycles. The older devices were limited to 10,000 write cycles. However, newer flash memories have 100,000 or more write cycles. Depending on the use, the number of write cycles may or may not be a limitation.

For example, consider a flash memory used for program and setup data storage in an industrial controller based on a 68XXX processor. This memory will probably be cleared and rewritten only a few times after it is installed. At the most, it will be changed only each time there is a major change on the production line it controls. If this happens four times a day, the normal 250-workday-year allows the controller to have a 100-year life for the flash memory—a highly unlikely situation! If the flash memory is used as the disk drive for a palmtop PC, a normal day's use might easily involve 50 to 100 disk sector writes. This could limit the flash memory life to 2 or 3 years.

The fourth column in Table 18-1 is sector size. As discussed earlier, flash memory must be erased before it can be programmed. Erasing a 2- or 4-Mbit flash memory and then rewriting the entire memory is wasteful and time-consuming. Some devices offer a segment erase. This allows the user to erase a 16- or 64-kbyte section of memory without affecting the other memory segments. In some cases, multiple segments can be erased without affecting the other memory segments. *Note:* Not all flash memory ICs have segment capability. Without segment capability, you must erase and reprogram the entire IC each time you want to change any information stored in the flash memory.

Sector size or flash memory system size has a great impact on the required system read-write main memory (RAM). For example, suppose that you are working with a 68040-

based industrial controller which uses a 4-Mbyte flash memory system for mass storage. The flash memory system is made up of eight 512-kbyte flash memory parts, and the main system memory is 2 Mbyte of dynamic RAM. If the flash memory system parts have only bulk erase, you must erase and reprogram 512 kbytes of flash memory at a time. This means that you must keep a copy of what is to be written to flash memory in 512 kbytes of dynamic RAM until the information is written to flash memory. For example, if you want to modify 1 byte in the 4 Mbytes of flash memory, you must first determine which 512-kbyte "sector" has the byte to be changed. Second, that 512 kbytes of information is copied to RAM. Third, the byte in RAM is modified. Fourth, the new 512 kbytes of RAM is copied into the flash memory section using the erase first and write second process.

If the flash memory has individually erasable sectors, the process in the above example becomes quicker. This is because you only need to determine which sector contains the byte to be modified, copy that sector to RAM, modify the byte in the sector, and write the sector back to flash memory.

The amount of time required to write to a flash memory IC varies depending on the erase and write needs and the sector size. A typical 128-kbyte device can be erased in 3 s and programmed in 3 s. Needless to say, these times drop significantly if sector erasure and programming are available. For a 128-kbyte device, erasure and programming times can be 300 ms for a single segment.

The fifth column in Table 18-1 shows device pin count. The flash memory pin count supports address lines, data lines, chip enable, output enable, power, and ground. If the chip has a byte or word function, a separate input makes this selection. Figure 18-1 shows a typical functional diagram and pinout diagram for a 4-Mbit flash memory IC. Flash memory packages include standard DIPs, PLCCs, and special flat packs designed for use in high-density surface-mount applications. Some devices are offered in normal and mirror-image pinouts. The mirror-image pinout is useful when the devices are in high-density layouts.

In certain packages, many flash memory devices have the same pinouts found in similarly sized EPROMs. This allows the manufacturer to replace EPROMS with flash memory

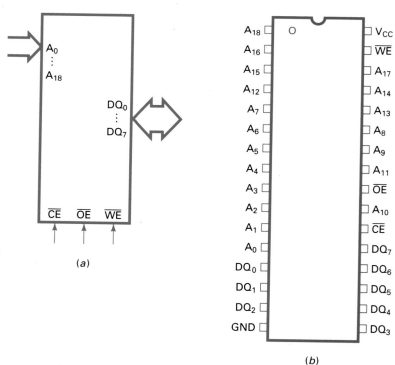

(a)

(b)

Fig. 18-1 Functional and pinout diagrams for a 4-Mbit flash memory IC. (a) The functional diagram. Note the relative simplicity of this device compared to an EPROM. (b) The pinout diagram for the 32-pin DIP. This device is also packaged in PLCC and flat packs.

devices without changing the circuit board configuration.

In Fig. 18-1, note that there are no special pins provided for placing the chip in the erase or program modes. Erasing or programming is accomplished by the sequencing of address, data, and chip enable information. Once the timing of address and chip enable information starts the programming function, data on the data lines tells the device which segments are to be erased or programmed.

Flash memory devices include internal logic to generate function decoding, programming pulse width timing, and the erase and program sequences. Figure 18-2 shows the internal structure of a flash memory device.

Often, multiple flash memory devices are combined to create a larger flash memory system. One common package for flash memory systems is the *Personal Computer Memory Card International Association* (PCMCIA) card (discussed in Sec. 18-2). Figure 18-3 on page 520 shows a diagram of a 4-Mbyte PCMCIA flash memory card which uses sixteen 256-kbyte flash memory devices. It has a separate 512-byte EEPROM memory which stores information about the card in accordance with the PCMCIA standard.

Self-Test

Answer the following questions.

1. The introduction of flash memory was driven by
   a. The need for higher-density EPROMS
   b. The lack of ultraviolet light sources in a PC
   c. The need for a low-power nonvolatile mass storage device
   d. All of the above, emphasizing *b*

2. Flash memory technology is an outgrowth of _____ technology.
   a. EPROM   c. Static RAM
   b. Dynamic RAM   d. EEPROM

3. Before you can program a flash memory, you must
   a. Totally erase all sectors of the flash memory device
   b. Erase the sector to be programmed (which may be the entire device)
   c. Make a backup copy in main memory in case the programming effort fails
   d. All of the above

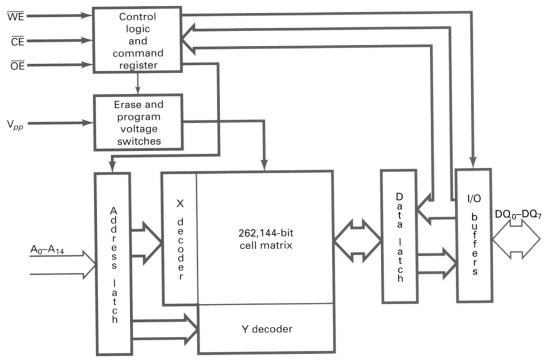

Fig. 18-2 A simplified block diagram showing the internal structure of a flash memory device. This 32-kbyte device uses a separate 12-Vdc programming voltage V_{pp} to erase and program the memory cells. New 5-Vdc-only devices generate the programming voltage on chip.

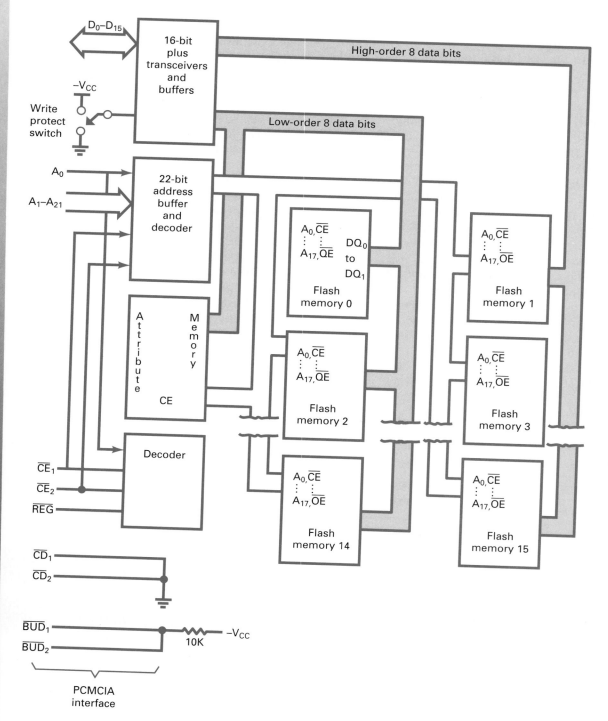

Fig. 18-3 A simplified diagram of a 4-Mbyte PCMCIA flash memory card. This card uses sixteen 256-kbyte flash memory devices. The attribute memory is a separate 512-byte EEPROM which stores information about the card in accordance with the PCMCIA standard (CIS). *Note:* A write-protect switch is provided so that the user can physically select a condition which keeps the card from being programmed and thus protects valuable data.

4. In terms of the number of useful programming cycles you can get from flash memory, it is most like
   a. A hard disk drive
   b. A floppy disk drive
   c. Dynamic RAM
   d. An EPROM
5. The use of sectors in a flash memory
   a. Is done to speed up the erase before reprogramming time
   b. Is very much like the use of sectors on a disk drive system
   c. Can involve either 16- or 64-kbyte sectors
   d. All of the above
6. You might guess that one reason flash memory devices have identical pinouts to the pinouts of EPROMs with a similar bit count is
   a. To convert a system so that it can have simple field modification of the memory contents by the device to which the flash memory is attached
   b. Simply a marketing tool used by IC manufacturers in order to get their users to switch from EPROMs to flash memory
   c. Because there are times when the fast programming speed of an EPROM is not needed and the user can substitute the slower flash memory devices
   d. All of the above
7. Flash memory devices do not require externally controlled programming pulse width because it is controlled by
   a. Logic inside the flash memory device
   b. The microprocessor software
   c. The natural microprocessor read-write timing cycles
   d. None of the above, because it does not need to be controlled

18-2 PCMCIA

Personal Computer Memory Card International Association (PCMCIA) is an emerging electrical and physical standard for peripheral devices which can be attached to PCs and other microprocessor-based devices. The PCMCIA card is approximately the same size as a standard credit card; the thickness depends on the PCMCIA type.

PCMCIA cards started as a way to provide a standard means to add memory to devices such as notebooks, palmtops, and other very small devices in which there was insufficient room to add a conventional memory card. Later, the PCMCIA card also became a way to add mass storage to devices in which a disk drive was too large. Today, the PCMCIA card has become a standard used for memory expansion and mass storage and to house many other useful attachments to a microprocesor-based system.

Currently, three different card thicknesses are specified by the PCMCIA standard (see Fig. 18-4 on page 522). The original standard was the Type I, which is 3.3 mm thick. Unfortunately, 3.3 mm is very thin and does not allow very much circuitry. The Type II card is 5 mm thick and is the most commonly used. The Type III card is 10.5 mm thick and can support a miniature disk drive.

In addition to card dimensions, the PCMCIA standard defines the card connector and the communications protocol which allow the microprocessor-based device to communicate with the PCMCIA card. The current standard allows multiple PCMCIA cards to be connected to a system. Frequently, a specification indicates a device that can accept one or two Type I or Type II PCMCIA cards or one Type III card.

Table 18-2 on page 523 shows the PCMCIA card pin assignments. A standard 68-pin connector, 34 on each side, is used. The connector is shown in Fig. 18-5 on page 524.

In addition to these physical specifications, a communications protocol specifies how a device will communicate with the PCMCIA card. Part of this communications protocol is the card identification structure (CIS). The CIS specifies information which the microprocessor-based device can address from the PCMCIA card. This information tells the microprocessor-based device about the PCMCIA card attached to the PCMCIA connector. The CIS is kept in a separate memory on the PCMCIA card called *attribute memory*. Attribute memory can be up to 8 kbytes; it supplies data to the microprocessor-based device in a standard format.

Because the PCMCIA card is used so frequently with PCs, MS-DOS drivers have been written specifically to communicate with a PCMCIA card. Additionally, custom ICs have been designed which connect with processors such as the Intel 386 and provide a direct interface with a PCMCIA card. Also, super I/O VLSI chips used to build PCs with a few chips typically have a PCMCIA interface.

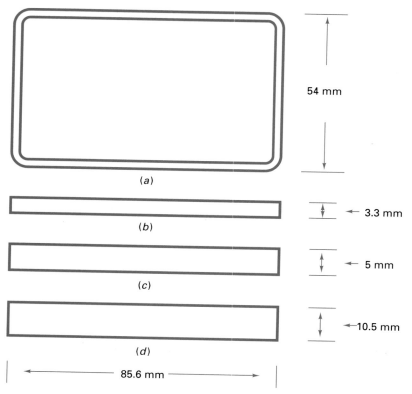

54 mm

(a)

3.3 mm

(b)

5 mm

(c)

10.5 mm

(d)

85.6 mm

Fig. 18-4 **The PCMCIA card physical standard.** (*a*) **The physical size. This can be extended by another 50 mm if needed for communications devices, etc.** (*b*) **The original standard was the Type I, which is 3.3 mm thick. The 3.3-mm card is so thin that it does not allow very much circuitry.** (*c*) **The Type II card is 5 mm thick. This is the most common in use today.** (*d*) **The Type III card is 10.5 mm thick and can support a miniature disk drive.**

A recent addition to the PCMCIA standard allows the cards to be changed when power is on (hot). This *hot swapping* ability allows a user to, for example, use the PCMCIA card as one would use a floppy disk to transfer data between a notebook computer and a desktop computer.

As discussed at the beginning of this section, the original PCMCIA cards were used to provide physically small memory expansions and mass storage for PCs, especially the portables. The initial products were ROM cards which could bring a certain set of programs or a specific database to the PC. The next expansion used static RAM with a battery backup. The difficulty with this card is the unknown point at which the battery will fail and the card will lose its data. DRAM versions of the PCMCIA have not become popular because of their need for refresh logic and their higher power consumption.

As discussed in the section on flash memory, the flash memory–based versions of the PCMCIA card are very popular. Although the

card addressing lines allow addressing of up to 64 Mbytes, most PCMCIA cards are in the few to low tens of megabytes.

The PCMCIA card is also commonly used as a communications interface. PCMCIA modem, fax-modem, and local area network (LAN) cards are very common. These cards allow a portable computer, for example, to communicate with other PCs, networks, or fax machines by simply plugging in a single card. Because these devices have to connect with an outside network (the telephone network or a LAN), they must support another connector as well as the PCMCIA connection to the PC. These connections are implemented in multiple ways, and there is no standard for them.

A few PCMCIA cards contain a radio transmitter-receiver. These cards allow the PC to "talk" to another transmitter-receiver and thus exchange data with the external device without having to be connected to that external device. These are often called "wireless" modems. They are discussed in more detail in Sec. 18-4.

Table 18-2   PCMCIA Card Pin Assignments

Function	Symbol	Pin		Function	Symbol	Pin
Card enable 1	$\overline{CE1}$	7		Address 8	A_8	12
Output enable	$\overline{OE}$	9		Address 9	A_9	11
Write enable	$\overline{WE}$/PGM	15		Card Reset	RESET	58
Ready/busy	RDY/$\overline{BSY}$	16		Data 0	D_0	30
Card detect	$\overline{CD1}$	36		Data 1	D_1	31
Card enable 2	$\overline{CE2}$	42		Data 10	D_{10}	66
Register select	$\overline{REG}$	61		Data 11	D_{11}	37
Battery voltage detect 2	$\overline{BVD2}$	62		Data 12	D_{12}	38
Battery voltage detect 1	$\overline{BVD1}$	63		Data 13	D_{13}	39
Card detect	$\overline{CD2}$	67		Data 14	D_{14}	40
Address 0	A_0	29		Data 15	D_{15}	41
Address 1	A_1	28		Data 2	D_2	32
Address 10	A_{10}	8		Data 3	D_3	2
Address 11	A_{11}	10		Data 4	D_4	3
Address 12	A_{12}	21		Data 5	D_5	4
Address 13	A_{13}	13		Data 6	D_6	5
Address 14	A_{14}	14		Data 7	D_7	6
Address 15	A_{15}	20		Data 8	D_8	64
Address 16	A_{16}	19		Data 9	D_9	65
Address 17	A_{17}	46		Extend bus cycle	WAIT	59
Address 18	A_{18}	47		Ground	GND	1
Address 19	A_{19}	48		Ground	GND	34
Address 2	A_2	27		Ground	GND	35
Address 20	A_{20}	49		Ground	GND	68
Address 21	A_{21}	50		Programming voltage 1	V_{pp1}	18
Address 22	A_{22}	53		Programming voltage 2	V_{pp1}	52
Address 23	A_{23}	54		Refresh	RFSH	43
Address 24	A_{24}	55		Reserved for future use	RFU	44
Address 25	A_{25}	56		Reserved for future use	RFU	45
Address 3	A_3	26		Reserved for future use	RFU	57
Address 4	A_4	25		Reserved for future use	RFU	60
Address 5	A_5	24		Supply voltage	V_{cc}	17
Address 6	A_6	23		Supply voltage	V_{cc}	51
Address 7	A_7	22		Write protect	WP	33

Self-Test

Answer the following questions.

8. The initial purpose of the PCMCIA card was to
   a. Provide a ROM card for portable PCs
   b. Allow a standard way to connect additional memory to a PC when there was not sufficient space for a standard memory card
   c. Provide a means to add a low-power nonvolatile mass storage device to a small PC
   d. All of the above

9. When the PCMCIA card is used to add flash memory to a PC, its purpose is to
   a. Provide a ROM card for portable PCs
   b. Allow a standard way to connect additional memory to a PC when there is not sufficient space for a standard memory card
   c. Provide a means to add a low-power nonvolatile mass storage device to a small PC
   d. All of the above

10. Part of the PCMCIA standard defines a 68-pin connector and the signals on each pin of the connector. You might guess that the signals connected to pins 18 and 52 (V_{pp1} and V_{pp2})
    a. Provide the required high voltage (+12 Vdc) needed by flash memory devices to erase and program the memory
    b. Are not required when the flash memory devices are single-voltage (+5 Vdc only) devices

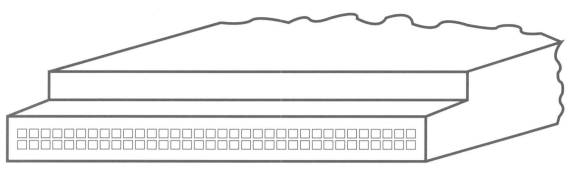

Fig. 18-5 **The standard 68-pin PCMCIA connector. There are 34 connections per side.**

c. Are not required when the device the PCMCIA card is connected to is reading data from the PCMCIA card rather than erasing segments or programming memory locations

d. All of the above

11. In many devices which have a PCMCIA slot, the PCMCIA card is specifically intended to go fully into the device in much the way a floppy disk goes fully into a floppy disk drive. You might suspect that this causes a problem with which of the following devices?

a. EPROM PCMCIA cards, because there is no room for the battery

b. Flash memory PCMCIA cards, because the chips are too thick to fit into either the 3.3- or 5-mm PCMCIA standard thicknesses

c. Modems and network adapters, which require a connection with the outside world

d. All of the above

12. The purpose of a custom-made PCMCIA processor interface chip is to

a. Reduce the number of devices (ICs) which are used to implement the microprocessor-based product that uses the PCMCIA slot

b. Convert the main microprocessor bus to the PCMCIA standard signals

c. Lower the cost of the product using the PCMCIA card

d. All of the above

13. The ability to "hot-swap" PCMCIA cards will let a user

a. Unplug one PCMCIA card and plug in another PCMCIA card without damaging either of the two cards or the device the cards are being plugged into

b. Use the PCMCIA memory card as one would use a floppy disk to transfer data between two machines

c. Use the PCMCIA slot for memory expansion at one time and as a modem slot the next time, thus expanding the capabilities of the device that has the PCMCIA slot

d. All of the above

18-3 RISC PROCESSORS AND THE POWERPC

The development of microprocessors is always toward faster and faster operation. Additionally, microprocessors are expected to handle increasingly complex software operations. This has led microprocessors to evolve from the 8- to the 16- to the 32-bit models. At the same time, the instruction sets and addressing modes have become much more complex and have expanded into many different specialized instructions and addressing modes.

The microprocessors we have studied so far in this text belong to a group which uses *complex instruction set computing (CISC)*. CISC processors have complex instruction sets, often implemented through the use of microcode, with many special instructions which are a part of the processor instruction set that has the sole purpose of handling specific situations when they happen.

In the mid-1970s computer scientists started looking for alternative ways to build computers to meet very fast computational requirements. One specific requirement was for a very fast computer to perform telephone switching operations. The speed requirements for this computer were well beyond anything which could be performed by a conventional computer. Scientists at IBM worked on this problem and, in the process, developed the underlying principles for a new type of computer architecture called *reduced instruction set computing (RISC)*. This new architecture was extended to microprocessor designs in the 1980s.

The basic concept behind RISC-based processors is to simplify the processor instruction set and to support it with very fast hardware. The result is a processor which runs much faster than the equivalent CISC models.

In 1991 IBM and Apple announced a joint effort to develop a RISC processor specifically aimed at the PC market. The objective was to provide a foundation for both IBM-compatible and Macintosh-compatible personal computers. The name for this new processor is the *PowerPC*.

For the PowerPC, or any RISC processor, to be a successful replacement for a current microprocessor, it must have a way of executing the existing software base for that processor. To do this requires an additional software layer which emulates (makes the processor look like) the processor it is replacing. This requirement is diagrammed in Fig. 18-6. Figure 18-6(*a*) shows the normal operating system layer which interfaces the applications program to the processor, and Fig. 18-6(*b*) shows the additional emulation layer. If the processor is to be an improvement over the processor it is replacing, it must run so fast that the loss of speed due to the emulation process is offset by the increased speed inherent in the RISC processor.

Ultimately, much of the software originally written for either the X86 processors or the 68XXX processors can be either partly rewritten or completely rewritten to take full advantage of the RISC processor. That is, it will be rewritten, reassembled, or recompiled as native-mode executable software. When this happens, the applications will begin to execute much faster in the RISC environment, because the need for the emulation layer will disappear.

How does a RISC processor gain significant speed improvement over a CISC processor? As noted at the beginning of this section, RISC processors take advantage of many new IC techniques which were not available when the current CISC processor designs were developed, so their basic architecture could not use them to their advantage.

Other ways of achieving significant speed improvements are to:

Make all instructions the same length. This means that

- Each instruction starts on a known memory boundary
- The logic is simpler and the processing time shorter than when variable-length instructions are decoded and executed
- Pipelining becomes much simpler and therefore code caches are much more effective
- Branching and branch prediction become simpler

Keep the instructions simple. This means that

- Each instruction has a known processing time

Reduced instruction set computing (RISC)

PowerPC

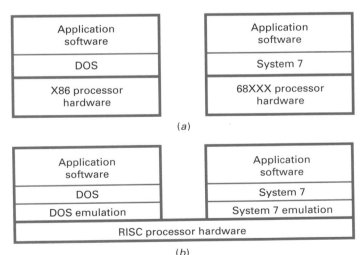

(*a*)

(*b*)

Fig. 18-6 The additional software layer which emulates (makes the processor look like) the processor it is replacing. (*a*) The normal operating system layer which creates the interface between the applications program and the processor. (*b*) The additional emulation layer. If the processor is to make an improvement over the processor it is replacing, it must run so fast that the loss of speed due to the emulation process is offset by the increased speed inherent in the RISC processor.

- If a complex action is needed, multiple instructions should be used
- Needed speed must be achieved from sophisticated hardware which allows parellel processing and pipelined instruction processing

Limit data modification instructions to register-to-register addressing. This means that

- Most instructions operate in a single processor cycle
- A separate set of load and store instructions is used to move data between the register set and memory
- Multiple execution units allow continuing interregister processing while, for example, the processor is waiting for memory access
- Addressing is simplified and therefore fast

Give the processor many general-purpose registers. Therefore,

- Support the registers with many execution units so that there is no limitation of processing speed while waiting for registers to become free
- Give instructions triple operands so that the result does not overwrite one of the operands, thus requiring a reload of this data

These basic principles are fundamental to any RISC processor design. As you can see, many of them rely on new IC technologies to give the RISC processor its speed. They also rely on the designer's ability to start with a clean slate well after the basic architecture for advanced microprocessors was defined. Note that a number of the techniques used in the RISC processors have also been applied to CISC processors. For example, multiple execution units (often called *superscalar architecture*) and pipelining are both incorporated in the latest versions of the X86 and 68XXX processors. However, extending the general-purpose register set to thirty-two 32-bit registers, as is found in the PowerPC, would require a major architectural change for both the X86 and the 68XXX. Likewise, using triple operand instructions would not be an advantage to these processors.

There are two versions of the PowerPC. There are also many other RISC processors. PowerPC RISC processors were specifically designed to be used in PC and Macintosh environments. Some RISC processors were designed as general-purpose RISC processors, and others were designed for specific environments, usually as the heart of an advanced workstation. The Intel i960 and the Motorola 88000 are two general-purpose RISC processors. Two examples of the latter are the Sun SPARC and the DEC Alpha. The basic characteristics of these different processors are shown in Table 18-3.

Table 18-3 Main Characteristics of Some Popular RISC Processors

Processor	Register Size	Number of Registers	Data Bus	Address Bus	Data Path	Code Cache	Data Cache	Speed
Advanced micro devices (AM29XXX)	32 bits	192	32 bits	32 bits	32 bits	512 bytes to 4 kbytes	0 bytes to 2 kbytes	16–33 MHz
DEC Alpha 21064-A275	64 bits	64	128 bits	64 bits	128 bits	16 kbytes	16 kbytes	275 MHz
Intel i960	32 bits	32	32 bits	32 bits	32 bits	1–4 kbytes	1 kbyte	16–40 MHz
Motorola PowerPC 601	32 bits	64	64 bits	32 bits	64 bits	32 kbyte combined		50–100 MHz
Motorola PowerPC 603	32 bits	64	32 or 64 bits	32 bits	32 or 64 bits	Two 8-kbyte combined caches		80 MHz
Sun SPARC	64 bits	168	64 bits	64 bits	64 bits	20 kbytes	16 kbytes	50–60 MHz

Self-Test

Answer the following questions.

14. The basic approach with RISC processors is to have
    a. A set of instructions which include specific instructions to perform special jobs which the processor might encounter from time to time
    b. A limited set of instructions combined with very fast hardware so that the speed shortfall from limited instructions is made up by processor speed
    c. Variable-length instructions
    d. All of the above
15. A PC based on the PowerPC RISC processor that is running a conventional copy of MS-DOS must also be running
    a. An emulator which makes the PowerPC look like a CISC Intel X86 processor
    b. One or more application programs
    c. Windows
    d. All of the above
16. Which of the following is *not* an attribute of RISC processors?
    a. The use of separate instructions to move data into or out of memory
    b. Many registers
    c. Variable-length instructions
    d. Simple instructions
17. One difference between data-modification instructions in a RISC processor and data-modification instructions in a CISC processor is that the instructions in a RISC processor
    a. Include register-to-register and memory-to-register operations
    b. Have three operands so that source data is not modified
    c. Do not include any arithmetic instructions
    d. All of the above
18. One reason why the DEC Alpha RISC processor is so powerful is its
    a. 64-bit register and data path organization
    b. Very high clock speeds
    c. Large on-chip data caches
    d. All of the above
19. The PowerPC RISC processor was specifically designed to be
    a. The main processor in a PC which runs IBM-compatible PC software
    b. The main processor in a PC which runs Apple Macintosh–compatible software

    c. Faster than a system running the Intel 486 processor
    d. All of the above

18-4 WIRELESS COMMUNICATIONS

Throughout this text, we have seen example after example showing that microprocessor-based systems are limited by their I/O. In many cases, connecting the I/O signals to the microprocessor-based system is one of the real challenges facing the system designer. For example, a number of microprocessor-based systems are able to capture data in the form of student responses to questions in a lecture hall. The objective is to give the lecturer immediate feedback. The difficulty is getting signals from the response keypads at each student's desk back to the data collection system. The signals can be sent over wire; however, this means that there are wires all over the floor (a safety hazard) or substantial construction to embed the wires in floors and walls. In another example, a microprocessor-based system called a *data logger* gathers data from a number of pumping sites at an industrial location. Again, sending the data to the data logger by wire involves difficult and unsafe temporary installation or an expensive permanent installation.

Both of these examples, and many others, are reasons why microprocessor-based system designers look for a better way to connect I/O (mainly input) signals to microprocessor-based systems without the use of wires.

Two technologies are used to eliminate wires. The most popular is the use of radio frequency (RF) transmissions. In special situations, infrared (IR) signals much like a TV remote control are used. In this section, we will look at some of the basic characteristics of these two wireless communications technologies. The basic characteristics will be considered because currently there are no standards in this area (one IEEE standard is in development for a certain class of RF signals). Each system is usually implemented according to the needs as seen by the system developer.

The common systems in use today are shown in Table 18-4 on page 528. The following paragraphs give a brief overview of the systems found in each of these major categories.

As you can see from Table 18-4, wireless digital communications systems which use RF transmissions are broken into three major

Table 18-4   Key Characteristics of RF and IR Wireless Communications Systems

Characteristics	Type of Wireless Signal			
	Very Low-Power RF	Moderate-Power RF	High-Power RF	Infrared
Distance	1–10 ft	100 ft to a few miles	A few miles to worldwide	10–50 ft
Data rate (bits per second)	100 bps to 10 kbps	10 kbps to 200 kbps	10 kbps to Mbps	100 bps to 10 kbps
License requirements	None (must comply with FCC regulations)	None (must comply with FCC regulations and operate in selected frequency ranges)	Must be licensed; design must comply with FCC regulations	None
Operating power source	Small battery, long life	Large rechargeable battery or power line	Large rechargeable battery or power line	Small battery, long life
Typical operating frequency	Unassigned; usually 100 kHz to 350 MHz	Assigned to industrial, scientific, and medical bands, 908–928 MHz, 2400–2450 MHz	Assigned (often to a specific frequency); depends on the service used, usually VHF or UHF radio	980-μm wavelength (near infrared)
Modulation	PCM or FM	Spread spectrum	Advanced FM	PCM
Open loop or closed loop	Open (one-way transmission)	Closed	Closed	Both

classes which are roughly dependent on the power level of the signal used. The *Federal Communications Commission (FCC)* categorizes RF signals by their power level as it is measured at standard distances from the transmitter and antenna.

Depending on the power level, the FCC may have no licensing requirements, or it may require the product design to conform to certain technical standards but not require a license, or it may require the product design to conform to specific technical standards and require that each individual station be licensed. Radio and TV stations, cellular telephone systems, two-way radio systems, amateur radio stations, microwave links, and satellite systems fall into the third class. The garage door opener control and the cordless telephone

fall into the intermediate class. A wireless (RF-based) mouse for a PC falls into the first category.

Products in the first category have such a low power level signal that they are unlikely to interfere with any other radio service. Designers are free to establish the operating frequency as best meets their need. Because the power is very low, the range is very short. The wireless mouse is an example of this category. Typically, these wireless I/O systems use base-band *pulse code modulation (PCM)*. This means that the logic 1s and logic 0s are represented by the transmitter being on (logic 1) and off (logic 0). If a number of different devices are to operate in the same area, usually they must be differentiated from one another by their operating frequency.

Operating frequencies are usually in the range of a few hundred kilohertz to a few megahertz.

The original product in the intermediate class was the garage door opener control. They started with tone-modulated AM signals and progressed to base-band PCM signaling and then to tone-modulated PCM signaling. A number of limited-use systems associated with microprocessor-based security systems and other such infrequent I/O (a requirement of the band to which the garage door opener controls are assigned) are in use in this nominal 310-MHz band. The garage door opener control is also a good example of one-way transmission. There is no feedback to tell the garage door opener control that the signal has been received and the door is being opened or closed. Needless to say, systems that have no feedback must be used only in noncritical applications and in very close proximity to the receiver.

A lot of development is going on in the *industrial, scientific, and medical (ISM) bands*. These are a number of UHF and SHF bands which the FCC has assigned for use by industrial equipment. The two popular bands in use today are 902 to 928 MHz and 2400 to 2450 MHz. There are other ISM bands, in the 5-GHz region and higher, but the difficulties of operating at these frequencies are significant. The ISM bands are often used for wireless modems or wireless LAN applications. You should note, however, that these bands are shared with other ISM applications. Part of the use stipulation is a recognition that there may be interference from the other uses. For example, the conventional microwave oven typically operates in the 2.4-GHz region. Many medical diathermy machines also use these two ISM bands. Both of these devices could interfere with a wireless modem operating in these bands.

In the late 1980s, the FCC permitted unlicensed operation in the ISM bands with power levels close to 1 W if *spread-spectrum modulation* was used. The intention was to develop spread-spectrum modulation (which had been primarily confined to military applications) and to give reasonable range for such applications as cordless telephones, wireless modems or LANs, and other consumer devices. Spread-spectrum modulation spreads the transmitted signal in what *appears* to be a random fashion over the entire band. Therefore, to another system listening to it, one spread-spectrum signal seems to be random noise rather than

interference. Spread spectrum substantially increases transmission reliability and provides a reasonable level of security because without the spreading code, the signal appears to be random noise.

Typical spread-spectrum systems are being used for remote data collection, as wireless modems that avoid wiring between two or more microprocessor-based systems, and to implement small wireless LAN systems in offices and industrial sites. As you can well imagine, the use of these systems (which typically cost a few hundred dollars per node) can save large sums of money which would be spent on either permanent or temporary wiring to make the same connections.

Fully licensed RF systems come in two categories: (1) those which use the public communications systems (cellular telephone system), and (2) specific license systems.

With the advent of cellular telephone systems (which operate at a frequency quite near the lowest-frequency ISM band), interest in using computers with these systems has developed. One approach is to simply connect the computer's modem to the cellular telephone system. In most applications this works reasonably well. However, when a cellular telephone system is in operation, the mobile station makes many frequency changes as it moves from cell to cell. Also, administrative signaling, in addition to the voice signals, takes place. Although neither of these factors may cause the voice user any problems, both may cause the interruption of a higher-speed modem signal.

Cellular telephone systems use two transmission technologies for the voice signal. The older systems use analog signals to carry voice signals. The newer systems convert voice signals to digital signals. Systems which use digital voice signaling and which are designed for direct digital input (typically an RS-232 input) are much more reliable than the analog signaling systems because the data is passed from the mobile unit to the telephone system as a stream of binary data. However, when analog voice data is used, a conventional modem must reject all the customary clicks, pops, and other noise which may be present in the analog voice channel.

There are also some specialized cellularlike systems designed strictly for handling digital signals. Because these systems are specifically designed to work with digital signals, they have the highest reliability. Although these specialized cellular-like systems work well for digital information, the user may be burdened by the requirement for two systems—one for the digital communications and one for voice communications.

Two-way radio systems such as those used by fire departments, police, taxis, and oil delivery services have been in use for a long time. Typically, each one is assigned a specific operating frequency or set of operating frequencies and operates under a specific license from the FCC. Systems are assigned frequencies within one of a number of VHF or UHF bands in the ranges of 30 to 50 MHz, 154 to 174 MHz, or 450 to 520 MHz. In addition, there are special frequency assignments for special situations.

With the advent of digital computers, fax machines, and other voiceless digital transmission devices, many frequency assignments cover digital transmissions. Typically, these systems are designed to operate within a particular city or other local geographic area. The range depends on the power used, the frequency assigned, and the antenna height of the base station. Most systems use FM with conventional modem signaling.

Infrared (IR) wireless systems are used in special situations. Relative to an RF-based system, IR is low-cost. The typical transmitter is an LED which operates in the near-infrared. The IR LED is simply turned on and off (baseband PCM) by the digital signals. An IR-sensitive phototransistor is used as the receiver. The IR TV remote control is the most common use of an IR wireless digital communications system. IR-based systems are also used to create a wireless mouse for small-area data-collection systems, and to couple a number of data-entry terminals in one area to a central computer.

A typical IR data system is shown in Fig. 18-7. The system in this figure is used to collect immediate responses from students in a classroom. Each response keypad has an IR transmitter. The student keys in the appropriate response (for example, yes, no, or 0 to 9). The microprocessor-based system (probably a PC) is connected to the main IR receiver. This receiver, however, also includes an IR transmitter. When a response is expected, the central system pools each keypad. The polling technique is used because only one IR transmitter can be used at a time. That is, if one

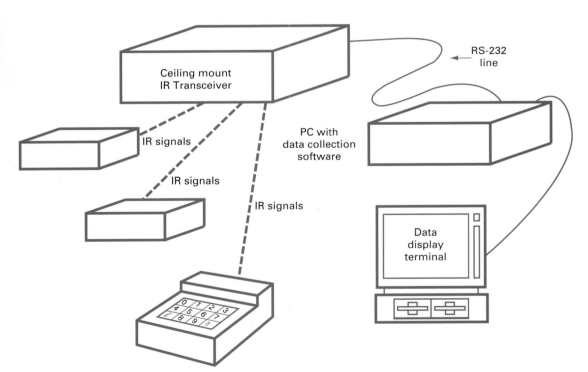

Fig. 18-7 **An IR data-collection system. This system collects immediate responses from students in a lecture hall. Each response keypad has an IR transmitter. The student keys in the appropriate response (for example, yes, no, or 0 to 9). The data-collection system is connected to the main IR receiver. This receiver includes an IR transmitter to poll each keypad for its response.**

keypad is transmitting and another keypad transmits, the receiver will not be able to tell them apart and the information from both will be lost. Although the polling technique is slow, if not much data is to be transmitted, transmission can be accomplished in a fairly short time.

Another disadvantage of the IR system is that it is subject to interference from bright light sources. For example, on a very sunny day, the IR signal from the sun may be enough to cause interference from the IR signals transmitted by the keypads or the central receiver. Additionally, in most cases, there must be line-of-sight transmission between the transmitter and the receiver.

Self-Test

Answer the following questions.

20. Wireless communications is becoming popular because
    a. PCs and other microprocessor-based products which need input from a wide variety of data sources are becoming increasingly common
    b. The cost of temporarily wiring a site for data collection may be very high
    c. The exact location for a data source may not be known
    d. All of the above
21. The three broad categories of wireless systems are based on
    a. Their operating frequency
    b. The type of permissible data and modulation
    c. The transmitter power and the potential interference they might cause to other people using the RF spectrum
    d. All of the above
22. The popularity of wireless communications as a part of a microprocessor-based system has caused a number of IC manufacturers to develop and market ICs specifically designed for implementing wireless digital communications. An IC which develops 100 mW at up to 250 kHz probably falls into the
    a. Unlicensed and unassigned frequency category
    b. Unlicensed but assigned frequency category
    c. Licensed and operating on an assigned frequency category
    d. IR transmitter category

23. RF-based wireless systems which fall into the unlicensed but assigned frequency category
    a. Do not have any feedback from the receiving system to the transmitting system
    b. May or may not have feedback from the receiving system to the transmitting system
    c. Always have feedback from the receiving system to the transmitting system
    d. Cannot have feedback because of the technical limitations of this category
24. One of the limitations of unlicensed but assigned frequency category wireless systems operating in the 310-MHz frequency range is that they can operate only intermittently like a garage door control transmitter. Another device which probably would fall into this category would be
    a. A data entry terminal in an office
    b. A door or window transmitter in a security system which tells the main system when the door or window is operated
    c. A detector that counts the number of bottles filled in a beverage-filling plant
    d. All of the above
25. The main difference between sending data over a cellular telephone system and using an individually licensed two-way radio system is that
    a. The cellular telephone system is designed to let the user communicate with a wide variety of other users who may be reached through the public telephone network
    b. The individually licensed two-way radio system is very good for point-to-point transmissions or mobile to fixed-location transmissions
    c. The individually licensed two-way radio system is usually restricted to a particular geographical area, whereas the cellular telephone system can be used virtually anywhere in the United States
    d. All of the above
26. The chief advantage of an IR-based wireless data transmission system is its
    a. Very low cost
    b. High-speed transmission
    c. Ability to connect to a wide range of systems because of the standard communications protocols
    d. All of the above

1. Flash memory developed with the specific idea of creating a solid-state equivalent to the disk drive.
2. Flash memories are an outgrowth of EE-PROM technology.
3. Before writing to flash memory, it must be erased (each bit is set to logic 0). Each bit is written to a number of times to make sure that the semiconductor junction is permanently modified.
4. Flash memory cannot erase a single bit or word. The entire memory must be erased. Individual bits or words can be programmed. For some flash memory ICs, sections of the IC can be erased. Usually these segments are 16 or 64 kbytes.
5. Flash memories are usually organized to work with a microprocessor. The most common organization is as bytewide devices.
6. Flash memories have access times in the 45- to 70-ns area.
7. Flash memories have a limited number of write cycles. Older devices were limited to 10,000 write cycles. Newer devices have 100,000 or more write cycles.
8. The time to write to a flash memory depends on the memory or sector size. A typical 128-kbyte device can be erased and programmed in 3 s. Erasing and programming a single segment can take 300 ms.
9. Initiating erasing or programming is accomplished by the sequencing of address, data, and chip enable information.
10. Multiple flash memory devices can be combined to create a larger flash memory system. A common package for flash memory systems is the PCMCIA card.
11. PCMCIA is an emerging electrical and physical standard for peripheral devices which can be attached to PCs and other microprocessor-based devices.
12. The PCMCIA card is approximately the same size as a standard credit card. The thickness depends on the PCMCIA type.
13. There are three different card thicknesses specified by the PCMCIA standard. Type I is 3.3 mm thick, Type II is 5 mm thick and is the most common, and Type III is 10.5 mm thick.
14. The PCMCIA card is connected to its host with a standard 68-pin connector.
15. A PCMCIA communications protocol specifies how a host device communicates with the PCMCIA card. The CIS provides information about the PCMCIA card attached to the PCMCIA connector. The CIS is kept in a separate memory on the PCMCIA card called the attribute memory.
16. MS-DOS drivers are available to communicate with a PCMCIA card.
17. Custom ICs give microprocessors such as the Intel 386 a direct interface with PCMCIA cards.
18. Hot swapping allows PCMCIA cards to be changed when power is on. This allows a user to transfer data between a notebook computer and a desktop computer.
19. The PCMCIA card modems, fax-modems, and LAN cards are very common. They allow a portable computer to communicate with other PCs, networks, or fax machines. These devices have a second connector in addition to the PCMCIA connection to the PC.
20. RISC is a new computer architecture extended to microprocessor designs in the 1980s. RISC architecture is different from the CISC architecture used with conventional microprocessors.
21. A key objective of RISC microprocessors is to achieve much greater computing speed.
22. The basic concept behind RISC-based processors is to simplify the processor instruction set and support it with very fast hardware. The result is a processor which runs much faster than the equivalent CISC models.
23. In 1991 IBM and Apple announced a joint effort to develop a RISC processor, called the PowerPC, for the PC market. This processor provides a foundation for both IBM-compatible PCs and Macintosh-compatible PCs.
24. For a RISC processor to be a replacement for an existing microprocessor, it must have a way to execute the microprocessor's software base. This requires software which emulates the processor it is replacing. To make an improvement, the RISC processor must run so fast that the loss of speed due to the emulation process is

offset by the increased speed inherent in the RISC processor.

25. Software originally written for the X86 or 68XXX processors can be partly or completely rewritten to take full advantage of the PowerPC RISC processor. Rewritten (native-mode) applications will execute much faster.

26. The RISC processor achieves a significant speed improvement over a CISC processor by taking advantage of many new IC techniques.

27. RISC speed improvements come from having:
    - All instructions of the same length
    - Only simple instructions
    - Data modification instructions limited to register-to-register addressing
    - A processor with many general-purpose registers.

28. A number of techniques used in RISC processors have been applied to CISC processors. Multiple execution units and pipelining are incorporated in the latest versions of the X86 and 68XXX processors.

29. Connecting I/O signals to a microprocessor-based system is one of the real challenges facing the system designer. Often, connecting I/O signals by wire can be costly and time-consuming.

30. Two technologies are used to eliminate wires. The most popular is RF transmission. In special situations, IR signals are used.

31. Wireless digital communications systems which use RF transmissions are broken into three major classes based on the power level of the signal.

32. FCC regulations classify these three levels as no licensing requirements, product design requirements but no license, or product design requirements and individual station licenses.

33. The range for very-low-power devices is very short. The wireless mouse is an example of this category. Typically, these systems use base-band PCM. Operating frequencies are usually in the range of a few hundred kilohertz to a few megahertz.

34. The original intermediate product was the garage door opener control. A number of limited-use systems associated with microprocessor-based security systems and other such infrequent I/O (a requirement of the band to which the garage door opener controls are assigned) are in use in the 310-MHz band. The garage door opener control and similar devices use one-way transmissions.

35. The two popular ISM bands are 902 to 928 MHz and 2400 to 2450 MHz. They are often used for wireless modems or wireless LAN applications. These bands are shared with other ISM applications, and there may be interference from these other uses.

36. The FCC permits unlicensed operation in the ISM bands with power levels close to 1 W if spread-spectrum modulation is used.

37. Spread-spectrum modulation spreads the transmitted signal in what *appears* to be a random fashion over the entire band. A spread-spectrum signal seems to another user to be random noise and substantially increases transmission reliability and provides a reasonable level of security.

38. Spread-spectrum systems are used for remote data collection, as wireless modems, and to implement small wireless LANs.

39. One approach to a wireless data connection is to connect a modem to the cellular telephone system, which makes a great number of frequency changes as it moves from cell to cell and has administrative signaling in addition to the voice signals. These may cause interruption of higher-speed modem signals.

40. Cellular telephone systems use analog or digital signaling for the voice signal. Systems which use digital voice signaling also may be designed for direct digital input. They have a much higher reliability than analog signaling systems. When analog voice data is used, a conventional modem must be used. It must reject the noise present in the analog voice channel.

41. There are specialized cellularlike systems designed for handling digital signals.

42. Two-way radio systems are assigned a specific operating frequency or set of operating frequencies under a specific FCC license. Frequency assignments are within one of a number of VHF or UHF frequency bands in the ranges of 30 to 50 MHz, 154 to 174 MHz, or 450 to 520 MHz.

43. Many two-way radio systems include digital transmissions. These systems are designed to operate within a particular city or other local geographic area. The range depends on the power used, the frequency

assigned, and the antenna height of the base station. Most systems use FM with conventional modem signaling.

44. IR wireless is used in special situations. It is very inexpensive. The transmitter is a near-infrared LED using base-band PCM. An IR-sensitive phototransistor is used as the receiver.

45. A polling technique is used when there are multiple IR transmitters because only one IR transmitter can be used at a time. Polling is slow, but if not much data is to be transmitted, transmission can be accomplished in a fairly short time.

46. IR systems are subject to interference from other bright light sources such as the sun. Most IR systems need line-of-sight transmission between the transmitter and the receiver.

CHAPTER REVIEW QUESTIONS

Choose the letter that best answers each question.

18-1. Flash memory fills the need for
 a. Nonvolatile semiconductor read-write memory
 b. Low-power mass storage
 c. Physically small mass storage
 d. All of the above

18-2. Flash memory is like an EEPROM because
 a. You must first erase the memory before you can write to (program) it
 b. It can be erased entirely or in sectors
 c. Programming time is very short
 d. All of the above

18-3. One of the major advantages of flash memories which are divided into sectors is that you can
 a. Program the memory 1 byte at a time
 b. Find organizations which are tailored to those needed for microprocessor-based systems
 c. Erase smaller sections of the memory and therefore have to reprogram only limited amounts of the device when modifying a previously written section of mass memory
 d. All of the above

18-4. To start the erase or program functions in a flash memory,
 a. Assert the program pin
 b. Sequence the signals on the address and chip enable pins
 c. Select the device and assert the program-erase pin
 d. First erase the flash memory before programming the device

18-5. _____ is a likely application for flash memory.
 a. A battery-operated microprocessor-based industrial control device used to collect data in the field over a long time
 b. A replacement for a 240-Mbyte hard disk drive in a desktop PC
 c. The program memory in a microprocessor-based electromechanical toy
 d. All of the above

18-6. The PCMCIA card is
 a. About the same size as an ordinary credit card
 b. About the same size as a 3½-inch floppy disk
 c. 10 mm thick for all modem units
 d. Never more than 3.3 mm thick

18-7. There are three types of PCMCIA cards: Type I, Type II, and Type III. The main difference between them is that
 a. Type I is used for memory expansion, whereas Type II is used for ROM expansion

b. Type II is used for communications devices

  c. They vary in thickness

  d. All of the above

18-8. The PCMCIA has a CIS protocol which

  a. Lets the PCMCIA card tell the host processor what kind of card it is

  b. Is kept in a separate memory on the PCMCIA card

  c. Is in a standard format

  d. All of the above

18-9. Because the PCMCIA card is becoming so popular, you will find that

  a. Custom ICs have been developed to connect a PCMCIA card to host microprocessors such as the 386 and 486

  b. DOS drivers have been written to allow PC operating systems to communicate with the card

  c. It is often used with desktop PCs so that they can share PCMCIA devices and data with palmtops and other such small devices

  d. All of the above

18-10. Although the PCMCIA started out as a memory-expansion device, it is now often used

  a. To expand the microprocessor's cache memory

  b. For compressors

  c. In communications devices such as modems, fax-modems, and LAN cards

  d. All of the above

18-11. RISC processors have come into being because

  a. Computer scientists are always looking for ways to make processors faster and faster

  b. The advent of new IC technologies like VLSI have made possible architectural concepts like multiscalar processing and pipelining

  c. It is now possible to make a simple instruction set run very fast and therefore outperform a complex instruction set

  d. All of the above

18-12. Like any processor, a RISC processor can run in its native mode (execute a program written in its instruction set), or it can emulate another instruction set. In the emulation mode

  a. DOS runs faster because of the emulator speed improvements

  b. The machine speed is slower than it would be if the program were written in native code

  c. The PowerPC cannot run DS or Windows

  d. All of the above

18-13. When both the operating system and the application program have been rewritten to run using a RISC processor's native-mode instructions, you would expect a RISC processor to

  a. Run at or slightly slower than the speed at which the programs had run on the processor they were originally designed for

  b. Run much faster than the speed at which the programs had run on the processor they were originally designed for

  c. Take about the same time for operating system utilities such as disk access

  d. All of the above

18-14. The effect of pipeline processing is to have instructions be processed in fewer and fewer clock cycles. Pipelined processing is enhanced in RISC processors because

  a. All instructions are of the same length

  b. All data-modification instructions are limited to register-to-register operations *c.* RISC processors use three operand data-modification instructions, thus leaving the original operand unchanged

    d. There are separate instructions for data modification and register and memory interchanges

18-15. One way a RISC processor gains speed is because the data-modification instructions are not slowed by the processor memory system. This is true because

    a. All instructions are of the same length

    b. All data-modification instructions are limited to register-to-register operations

    c. RISC processors use three operand data-modification instructions, thus leaving the original operands unchanged

    d. There are separate instructions for data modification and register and memory interchanges

18-16. One characteristic of RISC processor instructions that has not arisen with CISC instructions because of unsufficient numbers of registers is that

    a. All instructions are the same length

    b. All data-modification instructions are limited to register-to-register operations

    c. RISC processors use three operand data-modification instructions, thus leaving the original operands unchanged

    d. There are separate instructions for data modification and register and memory interchanges

CRITICAL THINKING QUESTIONS

18-1. In time, is flash memory likely to replace hard disk drives on all PCs? If this is to happen, what characteristics of flash memories must change or improve?

18-2. In the section on PCMCIA cards, you learned that the PCMCIA card started as a memory expansion device and then expanded to become a mass storage and communications I/O device. What other applications do you think the PCMCIA might be used for?

18-3. One of the reasons for developing the PowerPC was to find a way which might break the processing speed barriers which today's CISC processors face. What other industry limitation is imposed on both IBM-compatible PCs and Apple Macintoshes? How is this barrier broken by the introduction of the PowerPC?

18-4. Wireless communications as discussed in this chapter makes use of RF and IR. Do you think that there might be other communications mediums which, in special circumstances, might be used to implement wireless communications?

Answers to Self-Tests

1. *c*	8. *d*	15. *a*	22. *a*
2. *d*	9. *c*	16. *c*	23. *b*
3. *b*	10. *d*	17. *b*	24. *b*
4. *d*	11. *c*	18. *d*	25. *d*
5. *d*	12. *d*	19. *d*	26. *a*
6. *a*	13. *a*	20. *d*	
7. *a*	14. *b*	21. *c*	

Glossary

∎

Term	Definition
Accumulator	A register to hold data for manipulation and to receive the results of data manipulations.
Algorithm	The process to solve a problem. Usually expressed as a mathematical formula or a logical statement.
ALU	Arithmetic-logic unit. The logic which performs computation in a digital computer or microprocessor.
Application program	A program written to perform a specific task.
Architecture	The way the microprocessor's logic circuits are put together to build the calculating logic, memory circuits, and input/output structure.
Arithmetic and logic instructions	The group of microprocessor instructions used to arithmetically or logically combine or test data.
AS	Associate of Science. A two-year degree in engineering or science. This degree is often required for a career assisting engineers in product development work.
ASCII	American Standard Code for Information Interchange. An 8-bit binary word which encodes 96 printable characters, 32 control characters, and a parity bit for error checking.
Assembler	An application program which translates a source code file, made up of mnemonic representations of a microprocessor's instructions, into an executable binary file (object code) for a particular processor.
Asynchronous data	Data transmission where the time between data elements (words) is random (unknown). The time between the bits which make up the asynchronous data words is known (fixed).
A/D converter	An electrical circuit which converts an analog signal (voltage or current) into a proportional binary number (digital word).
Bank-switched memory	Memory which can be moved in and out of memory address space so the microprocessor system can address more memory than its addressing capability normally allows.
Base	The number of characters in a number system. The decimal system has a base of 10.
Baud rate	The data transmission speed for a serial transmission measured in bits per second. Baud rate is only constant for the transmission of a single word with asynchronous data.
BCD	Binary-coded decimal. A method to represent decimal numbers with 4-bit binary words. BCD uses the words from 0000_2 to 1001_2 to represent the decimal numbers 0 to 9.
Benchmark program	A short program used to compare different processors in terms of the time required to execute the benchmark.
Binary	A number system with two characters (usually represented as 0 and 1). The easiest number system to implement with digital circuits.
BIOS	Basic input/output system. The portion of the CP/M or MS-DOS operating systems which manages the input/output devices.
Bipolar memory	A fast semiconductor memory made with bipolar (NPN or PNP) transistors.
BISYNC	Binary synchronous. A synchronous data transmission protocol used to communicate with mainframe computers.
BIT	A binary digit. One piece (element) of information in a binary word. It may be either true (logic 1) or false (logic 0).
Block diagram	A design description tool where a series of interconnecting blocks show the flow of signal (processing) and control in an electronic circuit.
Block transfer instructions	The group of microprocessor instructions used to direct the movement of blocks of data from one set of locations to another set of locations.

Term	Definition
Boot program	A program which performs the microprocessor system initialization and loads the application program or operating system. Usually the boot program is stored in ROM.
Breakout box	A service tool used to find problems in serial communications lines. Normally a breakout box lets the user open or close each line in the cable, swap signals from one line to another, and see signal activity.
BSCS	Bachelor of Science in Computer Science. A 4-year degree in computer science engineering, usually with emphasis on software design and computer architecture and circuits.
BSEE	Bachelor of Science in Electrical Engineering. A 4-year degree in electrical/electronic engineering which may have an emphasis on analog or digital circuit design.
BSME	Bachelor of Science in Mechanical Engineering. A 4-year degree in mechanical engineering.
Bus	The common set of electrical connections which lets different parts of the microprocessor or a microcomputer communicate. A bus has address, data, and control signals.
Byte	A digital word of 8 bits.
Call instruction	Used to initiate execution of a subprogram (subroutine). A subroutine is terminated by a RETURN instruction.
CAM	Computer-aided manufacturing.
Case construct	A programming step which allows multiple answers to a question and gives multiple paths to follow based on the answer.
CD-ROM	A read-only mass storage device based on the compact disk.
Centronics interface	A parallel interface often used to connect a personal computer to a printer. Often the Centronics interface is a one-way data path.
Chip	An integrated circuit.
CHIP ENABLE	See CHIP SELECT.
CHIP SELECT	A memory or device (IC) input which enables the chip so it can accept address and data.
Clock	The signal generated by an oscillator used to provide the basic timing for a microprocessor and/or its associated circuits.
Compiler	An application program which translates a source code file, made up of statements written in a high-level language, into an executable binary file (object code) for a particular processor.
Condition code register	The status register.
Continuous balance A/D	A type of analog-to-digital converter which continuously balances a known signal generated by the digital system against an unknown signal.
Control logic	The logic circuits used to decode and carry out instructions. This logic may be microprogrammed to carry out its function.
Controller	A device which controls the mass storage drive and read/write circuits.
Coprocessing	A special form of parallel processing. Coprocessing normally refers to the work performed by a separate and special-purpose processor to solve certain problems. An arithmetic coprocessor is a common example.
Core memory	A nonvolatile memory system made with doughnut-shaped electromagnets.
Counter-timer	A digital circuit used to provide timing or counting functions in a microprocessor or one of its I/O devices.
CP/M	Control program/microprocessor. An 8-bit single-user single-tasking disk operating system used with many early personal computers.
CPU	Central processing unit. The logic which performs data handling and computation within a computer or microprocessor.
CRC	Cyclical redundancy check. An error-checking algorithm which treats data as the coefficients of a 16th-order polynomial. The remainder is transmitted with the data and compared to a recalculated remainder at the receiving end.
Cross assembler	An assembler which runs on one computer but generates object code for another computer.
Cross compiler	A compiler which runs on one computer but generates object code for another computer.

Term	Definition
CRT	Cathode-ray tube. The special-purpose vacuum tube used by televisions and video terminals to display images. The term CRT is used improperly to refer to a video terminal.
CSA	Canadian Standards Association. A Canadian agency which tests products for safety and electromagnetic radiation standards required in Canada.
Cylinder	The series of identical tracks on a number of platters (Winchester disk drives).
D/A converter	An electrical circuit which converts a digital (binary) word into a proportional current or voltage (analog signal).
Data bus transceiver	A bidirectional amplifier used to place data on or receive data from a bus. When not driving the bus, the bus transceiver outputs are at a high impedance state.
Data structure	The part of a program where data is assigned to variables and where the data organization is described.
Data transfer instructions	The group of microprocessor instructions used to direct the movement of data from one location to another.
Debugging	Testing hardware or software implementation to see if it performs to specification and correcting any errors found.
Dedicated processing	Using a microprocessor or computer to perform a single job rather than a wide range of general jobs.
Density	The number of bits of data written onto a unit length of mass media.
Design	The description of how a problem will be solved using specific devices or computer programs. The design references the specification which details how the design must act.
Direct addressing	A memory addressing mode which has the memory address as a word in the instruction.
Directory	The section of mass storage which holds information describing the files stored on the device.
Disassembler	A process which examines a series of digital words (object code) and interprets and displays the interpretation as instructions for a particular microprocessor. Both MDSs and logic analyzers may have disassemblers, and a disassembler may be an application program.
DMA	Direct memory access. A method of moving data directly between system memory and an external device. Under DMA, the external device takes over the address and data buses to manage the data transfer.
DOS	Disk operating system. An operating system which resides on a disk drive until booted and then uses the disk for mass storage operations.
Dot matrix	A rectangle of dots (usually either 5 × 7 or 7 × 9 dots) used to display alphanumeric characters.
Double-density recording	Writing 2 bits per clock cycle using MFM (modified frequency modulation).
Download	Transferring a program or data into a microprocessor memory, usually from another microprocessor system, using a serial communications device.
DRAM	Dynamic RAM. A random access read/write memory device which stores information as a charge on the input capacitance of a transistor. DRAMs must be refreshed to keep the information from fading.
Drive	An electromechanical device which moves mass storage media under the read/write heads.
Drive electronics	The electronic circuits which control media movement and head motion on a mass storage device.
EAROM	Electrically alterable read-only memory. A ROM which can be programmed or reprogrammed with special electrical signals.
EBCDIC	Extended Binary-Coded Decimal Interchange Code. An 8-bit code for alphanumeric characters and control codes which is mostly used on mainframe computers.
EDAC	Error detection and correction. A technique used to detect errors in data transmission that also provides a means to correct certain kinds of errors.
Editor	An application program which allows a programmer to enter source code in machine-readable form.

Term	Definition
EMS	Extended memory standard. A method used by personal computers to extend the amount of memory space available by mapping memory locations above 1 Mbyte into specially defined windows.
FCC	Federal Communications Commission. A United States government agency charged with the regulation of radio transmissions. Part of the FCC rules concern electromagnetic radiation from microprocessor-based products (Part 15, Sub-Part J).
Fetch/Execute Cycle	The process of retrieving an instruction from program storage memory (Fetch) and carrying out that instruction (Execute).
Flash converter	A high-speed A/D converter which makes the conversion in a single operation by making comparisons to each possible voltage level using a voltage divider and multiple comparators.
Flash memory	A nonvolatile memory device that is similar to EEPROM but must be erased in blocks.
Floating point number	A technique used to represent a wide range of numbers (very small to very large) by expressing the number in terms of a signed mantissa and signed exponent.
Floppy disk	Mass storage medium consisting of a donut-shaped piece of magnetic tape in a paper envelope.
Floppy disk drive	A mass storage drive designed to use floppy disks. The floppy disk drive is slow, so the head can physically contact the media without excessive wear. Typical floppy disk drives come in 8, 5.25, and 3.5 inch sizes.
Flow chart	A diagram which shows the flow of control. Usually used to document software designs.
FM	Frequency modulation. A method of transmitting intelligence by varying the frequency of the transmitted signal.
Format	A description of the way data is written on a medium.
Formatting	The process of writing blank data and/or clock information on blank media.
Framing error	A data transmission error which occurs when the start of the data transmission is either detected too early or too late.
FSK	Frequency shift keying. A method of transmitting intelligence by varying the transmitted signal between two frequencies.
Full duplex transmission	A data transmission system which permits data to be transmitted in both directions at the same time.
General-purpose registers	Any of the registers in a microprocessor accessible to the programmer and used for general temporary data storage.
GP-IB	General-purpose interface bus. See IEEE-488.
Half-duplex transmission	A data transmission system which only permits data to be transmitted in one direction at a time.
Hard copy	A paper printout created from an output device such as a printer.
Hard error	An error in writing or reading data which cannot be recovered by trying again.
Hard-sectored disk	A disk which uses a sector identification hole to identify the beginning of each new sector.
HDLC	High-level data-link control. An advanced synchronous protocol established by the International Standards Organization (ISO). Used to communicate with mainframes and as a LAN protocol.
Hex	Hexadecimal. A 16-character number system (0 to 9 and A, B, C, D, E, and F). Used as simple shorthand for the binary numbers 0000 to 1111.
High byte	The most significant byte in a multiple-byte word. (Also HI byte.)
HP-IB	Hewlett-Packard interface bus. See IEEE-488.
IC	Integrated circuit. A solid-state circuit implemented by creating and interconnecting a number of transistors on a single silicon wafer. Also called a chip.
ICE	In-circuit emulator. An instrument which replaces the microprocessor and lets the user display the microprocessor's internal operation as it executes a program under user control.
IEEE-488	A parallel data communications system normally used to communicate with and to control electronic instruments. Also called GP-IB and HP-IB.

Term	Definition
IF-THEN-ELSE	A programming construct which allows a decision and selects one of two paths to follow based on the result.
Immediate addressing	The addressing mode where the data to be acted on is part of the instruction.
Implementation	The process of converting a design into a working product—coding in software or prototyping in hardware.
Indexed addressing	A memory addressing mode where the memory address is calculated by adding the contents of an index register with the contents of a word which is part of the instruction.
...rect addressing	A memory addressing mode where the memory address is contained in a register or memory location addressed by the instruction.
...nerent addressing	The addressing mode where the address is part of (inherent in) the instruction.
...struction	A unique binary word which, when interpreted by a microprocessor as an instruction, causes the microprocessor to take a specific action.
...nstruction pointer	An alternate term for program counter.
...nstruction register	A special-purpose register to hold instructions fetched from memory. The instruction register holds the instruction for the instruction decoder.
Instruction set	The complete set of instructions a microprocessor uses.
Interface	A term which refers broadly to all of the microprocessor circuits and processes having to do with communicating with external devices.
Internal data bus	The data bus used to transfer data between the microprocessor's registers, ports, and ALU.
Interpreter	An application program which translates source code statements one line at a time into executable object code modules. The interpreter must reside in memory along with the source code.
Interrupt	A request to initiate execution of a special program caused by an external electrical signal.
Interrupt service routine	A special program called to service an interrupt request.
Interrupt vector	The memory location pointed to by the microprocessor in response to an interrupt.
I/O	Input/output. The circuits needed to move data into and out of the microprocessor.
I/O mapped ports	I/O ports which are addressed in separate address space from memory. I/O mapped systems have special instructions for managing I/O transfers.
I/O port	A port in a microprocessor system specifically used to transfer data into or out of the system.
Joystick	A device which converts hand motions into X- and Y- axis signals for the digital computer. The joystick lever is moved to generate the signals.
Jump instructions	The group of microprocessor instructions used to redirect the sequence of program execution to another point in the program.
LAN	Local area network. A data communications network used to interconnect a number of microcomputers and their peripherals with a high-speed data path.
LCD	Liquid crystal display. A display technology often used with microprocessor-based products because the display can be easily customized using a photographic process.
Level-activated interrupt	An interrupt triggered by either a high-level or low-level electrical signal at the interrupt input.
Low byte	The least significant byte in a multiple-byte word. (Also Lo byte.)
Logic analyzer	A design tool which lets the user examine and store data on 16 to 48 lines in a digital circuit for a period of time. The stored data is shown by either timing or state displays.
Logic probe	A simple service tool which shows logic 1s, logic 0s, or pulses in digital circuits.
LSB	Least significant bit. The bit position in a binary word with the least weight.
LSI	Large-scale integration. A term defining the size of an IC. Usually describing an IC with more than 100 equivalent gates to implement the circuit.
Main memory	Program and data memory which can be addressed by the microprocessor's instructions.

Term	Definition
Maskable interrupt	An interrupt input which can be masked (shut off) by the programmer.
Mass storage	A device used to store large quantities of programs and/or data accessible by the microprocessor.
MDS	Microprocessor development system. A system to assist with hardware and software development in microprocessor-based products. The MDS lets the user replace the microprocessor with an emulator to give control of program execution. See ICE.
Medium	The material used to store data in a mass storage system. Often a thin film of magnetic material on a nonmagnetic base.
Memory access time	The time a memory device takes to place data at its output or write data into a memory location after being addressed.
Memory address register	A special-purpose register used to hold the address of the current memory location interacting with the microprocessor.
Memory cycle time	The minimum time between consecutive reads from or writes to a memory device.
Memory diagnostic	An application program which checks memory systems by writing various patterns of logic 1s and 0s into memory locations and then confirming they are read back without error.
Memory-intensive architecture	A microprocessor architecture which performs most data transfer and computations with memory locations.
Memory map	A diagram used to show the addressable range of a memory system. The memory map is often used to show how the different areas of memory are utilized.
Memory-mapped I/O	Memory-mapped I/O devices are addressed as memory locations and can use the full set of memory instructions.
Microcomputer	A single-chip microprocessor or a microprocessor-based general-purpose computer.
Microcontroller	An IC which contains a microprocessor, RAM, ROM, and some I/O.
Microprocessor	An IC which implements the architecture of a digital computer's CPU.
Microprogram	A special program used to define the microprocessor control logic's response to the instructions it receives. Not accessible to a programmer.
MIPS	Million instructions per second. A unit of measure to show processing speed.
MMU	Memory management unit. The support hardware which a microprocessor uses to manage bank-switched memory.
Mnemonic	An abbreviation that reminds a person what it stands for. Mnemonics are often used to reference microprocessor instructions as their binary words are difficult to remember.
Modem	Modulator-demodulator. A device which turns data into changing tones (FSK or PSK) so data can be transmitted through the telephone network or other AC-coupled transmission systems.
MOS memory	A semiconductor memory made with field-effect transistors.
Mouse	A device which converts hand motions into X- and Y-axis signals for the digital computer. The mouse is moved over a surface to generate the signals.
MS	Master of Science. Two years of study in engineering or science beyond a BS.
MSB	Most significant bit. The bit position in a binary word with the greatest weight.
MS-DOS	Microsoft Disk Operating System. The single-tasking single-user disk operating system most frequently found on 16-bit 8088-based personal computers.
MSI	Medium-scale integration. A term defining the size of an IC. Usually describing an IC with 20 to 100 equivalent gates to implement the circuit.
Multiple precision	Using two or more binary words to represent a number greater than the maximum value of a single word.
Multiplexed addressing	The method of addressing a device by using the same inputs for both the row and column addresses at slightly different times.
Multitasking multiuser	An operating system which allows multiple users to appear to run multiple programs at the same time.
Nesting	The process which lets a subroutine be started while another subroutine is running. Nesting can be many levels deep.
Nibble	One half of a byte. A 4-bit binary word.

Term	Definition
Nonmaskable interrupt	An interrupt input which is always active. It cannot be masked (shut off) by the programmer.
Nonvolatile memory	A memory device which retains its data when it loses power.
Object code	The binary code generated by an assembler, compiler, or translator. Object code is executable by the target processor.
Op code	Operation code. The first part of an instruction word. This part of the instruction tells the microprocessor what to do. The op code tells the microprocessor to act on one or more operands.
Operand	The data or address to be acted on by the process called for by the op code.
Operating system	A collection of general-purpose programs (utilities) provided with a computer to provide standard interdevice and interfile communications, to sequence applications programs and to interpret and execute user commands.
Optical disk drive	A very high capacity mass storage device that stores information as pits in the surface of the disk.
OROM	Optical read only memory. An optical disk drive with prerecorded data.
Overrun	A data transmission error which occurs when one data word is written over a preceding unread data word.
Packed BCD	Putting two or four BCD words in an 8- or 16-bit word.
Parallel I/O port	An I/O port on a microprocessor system which communicates via a number of parallel data lines. The data transfer is usually controlled by a few handshaking lines.
Parallel processing	Any microprocessor architecture which permits two or more processors to work on the computing process at the same time.
Parity bit	An error-checking bit added to a data word. Parity bits can be added to make the number of logic 1 or logic 0 bits either odd or even.
Parking cylinder	A special cylinder which is a safe storage place for the heads on a Winchester drive.
Part 15	A section of the FCC rules describing how much electromagnetic radiation can be emitted by a product. Subpart J describes two classes of computing devices—Class A for commercial use and Class B for consumer use.
Patch	An object code fix made by changing the object code at the binary level. This is often done by replacing a section of code with a JUMP instruction to a patch area where the corrected code resides.
Pattern sensitivity	A common failure mode for semiconductor memory. When a memory has pattern sensitivity, it only fails when certain patterns are written into certain locations. Pattern sensitivity is found by using a memory diagnostic program.
PC	Personal computer. Any computer suitable for home or individual use. Today this usually means a device with an architecture similar to the IBM Personal Computer (based on the 8088, 80286, or the 80386 microprocessor).
PCMCIA	Personal Computer Memory Card International Association. A standard for PC add-on devices which started as a memory expansion and now includes a wide range of I/O devices.
Pin grid array	An IC package which has many pins sticking straight down from the bottom of the package. It is often used for high-pin-count devices such as advanced microprocessors.
Pipelining	An architecture which allows the processor to work on a job a piece at a time like an assembly line.
Platter	One of the disks in a hard (Winchester) disk drive.
PLCC	Plastic leaderless chip carrier. A package that snaps into a special socket and is often used for high-pin-count devices such as advanced microprocessors.
Polling	A method of servicing an I/O port. When an I/O port is serviced with a polling routine, the routine checks each I/O port to see if it needs service and then services those that require service.
Pop	Taking data off of the stack.
Ports	The circuits in a microprocessor system used to transfer data into or out of the system.
Prefetch	An operation which fetches instructions or data, ahead of the time they are needed, when there is idle time in the system.

Term	Definition
Program	A set of instructions which tells a microprocessor how to process the data to get a certain job done.
Program counter	A special-purpose register which points at the next instruction in the program being executed—also called the instruction pointer. The program counter keeps track of where the microprocessor is in the process of executing a program.
Programmed data transfer	A method of moving data into or out of a microprocessor system which uses one or more instructions to move a word of data into or out of the system via a register, usually an accumulator.
Programming model	A model of a microprocessor which only shows the parts a programmer can use.
Protocol	The rules describing the organization and flow of data (a data communication protocol).
Pseudocode	Also referred to as structured English. Pseudocode is used in documenting a design using English-like statements which follow the four basic programming constructs.
PSK	Phase shift keying. A method of transmitting intelligence by varying the phase of the transmitted signal.
Push	Putting data onto the stack.
QPSK	Quadrature phase shift keying. A method of transmitting intelligence by varying the phase of the transmitted signal in 90 degree steps with each step representing two bits of information.
Radix	The mathematical term for the base of a number system.
Radix point	A general term for the symbol used to separate the integer and fraction parts of a number in any base.
RAM	Random-access memory. A memory system where each element (bit, byte, or word) can be accessed as easily as any other. Often the term RAM is used to refer to solid-state read/write memory.
Refresh logic	The logic used to refresh dynamic RAMs so they keep their data.
Register	A part of the microprocessor used to temporarily store a digital word. Often the contents of a register are the source or destination of an ALU operation.
Register-intensive architecture	A microprocessor architecture with instructions to perform data transfer and computations with the set of registers inside the microprocessor.
Relative addressing	The addressing mode where data which is part of the instruction (the offset) is added (using two's complement arithmetic) to the current program counter value to give the desired memory location.
Reset	A special interrupt used to initialize the microprocessor. The reset vector holds the first program step.
Resident program	A program which is loaded into main memory and remains permanently in memory so it can be quickly accessed.
ROM	Read only memory. A memory device which holds program instructions or data. Once programmed a ROM cannot be easily changed. Usually a ROM is implemented as an IC.
Rotate and shift instructions	The group of microprocessor instructions used to change the bit positions of the object data to greater or lesser significant positions.
Row-column addressing	A memory device addressing technique which breaks a memory address into two words and loads the address into the device (IC) as two words—one word is the row address and the other is the column address.
RS-232	An electrical standard for transmitting serial data. This standard defines a logic 1 as greater than $+3$ Vdc and a logic 0 as less than -3 Vdc. (Also RS-232C.)
RS-422	An electrical standard for transmitting serial data. This standard defines a logic 1 as a positive voltage differential between two wires and logic 0 as a negative voltage differential.
Sample and hold	An electrical circuit which samples an unknown voltage at a given moment in time and holds that voltage until a new sample is taken.
Schematic diagram	A document which shows each electronic component in a circuit and shows how it connects to every other component in the circuit.

Term	Definition
SCSI	Small computer systems interface. An 8-bit parallel data bus which addresses up to eight external devices. SCSI is often used to connect disk drives to a microcomputer.
SDLC	Synchronous data-link control. A synchronous data transmission protocol often used to communicate with mainframes and as a LAN protocol.
Sector identification hole	A small hole in a disk which signals the start of a new rotation of the medium and therefore identifies the first sector of the current track.
Sectors	The division of tracks into short segments which contain a number of bytes of data. Single-density floppy disk tracks are broken into nine 512-byte sectors.
Sequential construct	A series of procedures—one following the other.
Serial I/O	A communications technique which converts binary words into a sequence of bits sent one after the other over a single wire.
Servicing an interrupt	The operation which manages the transfer of control information and data between an I/O port and the microprocessor.
Single-density recording	The lowest-density information on magnetic media. Data is written using FM (frequency modulation) with 1 bit per clock cycle.
Single-tasking single-user	An operating system which only allows one user and one program at a time.
Soft copy	A display which does not leave any permanent record.
Soft error	An error in writing or reading data which can be recovered by trying again.
Soft-sectored floppy disk	A floppy disk which uses a sector identification hole to identify the beginning of a revolution and timing to identify the rest of the sectors on that track (rotation).
Source code	The computer instructions (a program) expressed in the terms of a computer language. Source code is interpreted by an interpreter, assembled by an assembler, or compiled by a compiler to produce object code which can be executed by the processor.
Speech recognition	The process of converting human speech into recognizable patterns for the digital computer.
Specification	A detailed scientific statement of exactly what must be done to solve a problem.
SRAM	Static RAM. A random access read/write memory device which stores information as the state of a flip-flop. The information remains stored for as long as power remains on the flip-flop. Usually static RAM is implemented as a many-thousand-bit memory IC.
SSI	Small-scale integration. A term defining the size of an IC. Usually describes an IC with 1 to 20 equivalent gates to implement the circuit.
Stack	A special area in memory used to keep information on a LIFO (last-in, first-out) basis. The top of the stack is pointed to by the stack pointer.
Start bit	A data bit placed before a data word to indicate the start of a serial transmission.
Status register	A special-purpose register used to store the results of arithmetic and logical operations in a CPU or special conditions in a device. Each bit in the status register describes a different condition. It is also called a condition code register in some microprocessors.
Stop bit	A data bit placed at the end of a data word to indicate the end of a serial transmission.
Streaming tape	A special-purpose magnetic tape mass storage system designed to record a steady stream of data such as that found when backing up a large mass storage device.
Successive approximation A/D	An analog-to-digital converter which finds the balance between an unknown signal and a known signal by generating a signal which comes closer and closer to the unknown signal in binary steps.
Superscalar architecture	A processor architecture which uses multiple processors, each of which addresses a different problem in order to achieve faster operation.
Switching power supply	A power supply which develops regulated output voltages by filtering high-frequency-modulated pulses rather than using a linear regulator which converts unused power to heat. Typical switching power supplies operate from 20 to 100 kHz.

Term	Definition
Synchronous data	Data transmission where the time between data elements (words or bits) is known (fixed).
System software	A collection of programs provided with a general-purpose computer. They include the operating system, general-purpose utilities, and program development software.
Temporary registers	Special-purpose registers used to hold data for parts of the microprocessor's architecture which must have stable data to operate. The ALU uses temporary registers at its input.
Top-down design	The design process which starts with the overall requirement and breaks this requirement into smaller and smaller pieces until the last detail is described.
Track	A single line of data written onto or read from media.
Transient program	A program which is loaded into main memory when needed and written over when not being used.
Transition-activated interrupt	An interrupt triggered by the change of an electrical signal from either a high to a low level or a low to a high level at the interrupt input.
Tri-state output	A device output which can be logic 1, logic 0, or a high impedance. When at a high impedance, other devices can control the level of the output signal.
Two's complement	A method of representing positive and negative numbers which does not have two different representations of 0. The two's complement of a number is found by taking the complement (changing all logic 1s to logic 0s and all logic 0s to logic 1s) and adding 1.
UART	Universal asynchronous receiver transmitter. This device converts parallel data to serial data and serial data to parallel data. Usually implemented as a large-scale integrated circuit.
UL	Underwriters Laboratory. An independent organization which tests electrical and electronic devices for safety. UL certification is often required before a product can be sold in some areas or purchased by certain organizations.
Unix	A multiuser multitasking highly transportable operating system. Unix is found on both 80286/386- and 68020-based systems.
VDE	Verband Deutscher Elektrotechniker. A German agency which tests products to ensure they meet safety and electromagnetic radiation standards required for sale in Germany and other parts of Europe.
VDT	Video display terminal. An I/O device built to communicate with humans. The VDT displays alphanumeric characters on a screen (usually 24 or 25 lines by 80 columns) and inputs data from a 90 to 100 alphanumeric key keyboard.
VLSI	Very large scale integration. A term defining the size of an IC. Usually describes an IC with more than 1000 equivalent gates to implement the circuit.
Voice synthesizer	A device which produces sounds which mimic the human voice.
Volatile memory	A memory device which does not retain its data when it loses power. Semiconductor RAM is volatile memory unless it has battery backup.
Weight (column)	The decimal value of each unit of character in that column.
Winchester disk drive	A mass storage device which uses a fixed (nonremovable) hard medium (one or more platters) and a flying read/write head. Typically Winchester drives store 10 to 600 Mbytes of information.
WMRA	Write many—read always. An optical disk drive which can be written on and rewritten as required.
Word	A collection of bits (binary digits) used to represent some quantity. A word is usually characterized by its number of bits and usually has a fixed length for a system. Often words are expressed as an integer multiple of bytes.
WORM	Write once—read many. An optical disk drive which can be written on once and then read many times.
X.25	An advanced synchronous protocol established by the International Standards Organization (ISO). X.25 is often used by LANs.

Index

Absolute address, 456
Absolute instructions, 109
Absolute long address mode, 279
Absolute object code, 456
Absolute short address mode, 278–279
ACALL (absolute call), 212
Acceptance test procedure (ATP), 418
Accumulator, 73, 76–77, 115, 142, 196
  double-length, 77
  double-wide, 61
  floating-point, 42
  importance of, 76
  multiple, 77
Accumulator instructions, 161, 179
Accumulator-specific transfer instructions, 209, 244
ACIA (asynchronous communication interface adapter), 165–166
Add, 5
Addend, 29
Addition, binary, 28–30
Address, 11–12, 103
Address bus, 79, 292–293
Address latch enable (ALE), 200
Address-object data transfer instructions, 211, 244–246
Address range, 11–12
Address register, 269–270
Address register direct mode, 275–276
Address register indirect mode, 276
Address register indirect with displacement mode, 277
Address register indirect with index (base displacement mode), 277, 278
Address register indirect with postincrement mode, 276
Address register indirect with predecrement mode, 276–277
Addressable memory, 54–58
Addressing
  direct, 104, 114–117, 235–236, 238
  immediate, 114
  implied, 112–113, 114
  indexed, 118–119
  indirect, 104, 117–118, 236
  inherent, 113–114
  register indirect, 117–118
  relative, 119–120
Addressing modes, 112–121
  68XXX microprocessor, 273–279
  X86 advanced microprocessors, 235–242

Advanced microprocessors:
  Intel X86 family, 230–265
  introduction, 226–229
  Motorola 68XXX family, 267–297
AJMP instructions, 212
Algorithms, 419
Alternate register set, 142–143
ALU (arithmetic logic unit), 5, 72–73, 76, 89–90, 93, 236–237
American Standard Code for Information Interchange (ASCII), 11, 140, 366, 368
Analog-to-digital (A/D) interface, 390–394
AND, 5, 107, 212, 284
Answer tones, 384
Apple II, 140–141
Apple Macintosh, 53
Application software, 437, 438, 441–444
Architecture, 1, 49–66
  addressable memory, 54–58
  development and maintenance systems, 65–66
  8051, 192–194
  instructions, 63
  memory addressing, 64
  other characteristics, 59–61
  registers, 62–63
  support circuits, 64–65
  word length, 50–54
  (See also specific systems)
Arithmetic instructions, 107, 172–173, 246–248
  (See also Processor arithmetic)
Arithmetic logic unit (ALU), 5, 72–73, 76, 89–90, 93, 236–237
Arrowheads, 425
ASCII (American Standard Code for Information Exchange), 11, 140, 366, 368
Assembler, 66, 105, 417, 451, 452, 455, 456
Assembler directives, 455
Assembly language, 311, 410, 455–458
Asserted service routine, 133
Assigned registers, 144–147
Asynchronous bus control, 293–294
Asynchronous communication interface adapter (ACIA), 165–166
Asynchronous data transmission, 368, 377
Attribute memory, 521
Augend, 29
Autoequalization circuits, 385

Backup copy, 339
Bank-switched memory, 329

Base, 16
Base 10 system, 16–18
Base 2 system, 18
BASIC (Beginners All-purpose Symbolic Instruction Code), 43, 458
Basic disk operating system (BDOS), 445
Basic input/output system (BIOS), 57, 445
Basic instruction types, 106–111
Batch mode, 3
Battery, 5
Battery backup, 310
Baud rate, 202, 374
BCD instructions, 285
Bell 103 standard, 384
Bell 212 standard, 385
Bell Laboratories, 2
Benchmark programs, 13
BERR (bus error), 296
Bidirectional bus transceiver, 92, 148
Binary addition, 28–30
Binary-coded decimal (BCD), 9, 285
Binary division, 37–41
Binary multiplication, 35–37
Binary number system, 18–19
Binary point, 19
Binary subtraction, 30–31
Binary-to-decimal conversion, 19–20
Binary-to-hexadecimal conversion, 22–23
BIOS (basic input/output system), 57, 445
Bipolar memory, 311
Bipolar memory cell, 312
Bisync, 368
Bit, 4, 9, 19
Bit field instructions, 284–285
Bit instruction, 149
Bit manipulation instructions, 249, 284, 285
Bit set, 150, 175
Bit set instruction, 109
Block diagram, 71–72, 494
  Z80, 148
Block move instructions, 150
Block search instructions, 170
Block transfer instructions, 107, 170
Boot program, 56–57, 310
Boot-up routine, 134
Bottom-up design, 415
BOUND instruction, 253–254
BRA instruction, 286
Branch instructions, 79, 81, 180, 286
Branch prediction, 231
Branch to subroutine (BSR) instruction, 161
Breadboard, 498, 499
Breakout box, 476, 479
Breakpoints, 452

BSR (branch to subroutine)
 instruction, 161
Buffers, 318
Burroughs Corporation, 3
Burst data transfer, 293
Bus, 6
Bus control lines, 163
Byte, 9
 lower, 9
 upper, 9

C, 446, 458
Cache memory, 60
CAD, 53, 424
CADD, 499
Call instruction, 109–110, 150,
 177, 251
Calls, 440
CAM, 53
Card identification structure
 (CIS), 521
Carry bit, 81
Carry/borrow bit, 82
CASE construct, 422, 424
Cathode-ray tube (CRT), 6
Cellular telephone systems,
 529–530
Central processing unit (CPU), 4
 troubleshooting, 470–471
Centronics interface, 371
Chips, 9
CLA, 105
Clear accumulator instruction,
 87
Clear op code, 104
Clock, 12–13, 54–56, 91
Clock frequency, 13
Clock speed, 58
CMOS, 59, 312
CMP instruction, 284
COBOL, 458
Code caching, 234
Code segment (CS) register,
 237–238
Coding the program, 409
Command line interpreter, 440
Comments, 426
Commodore 64, 140–141
Common-mode interference, 379
Communications:
 interrupts, 131–135
 need for microprocessor I/O,
 127–130
 polling, 131–135
 (See also Data
 communications)
Compact-disk (CD) technology,
 338, 357
Compare function, 5, 107–108
Compiler, 451, 457
Complementary metal-oxide
 semiconductor (CMOS),
 59, 312
Complex instruction set
 computing (CISC),
 524–526

Compliance, 501–502
Computer-aided design (CAD),
 53, 424
Computer-aided design and
 drafting (CADD), 499
Computer-aided manufacturing
 (CAM), 53
Concentric rings, 344
Condition code instructions,
 288
Conditional branching, 81
Conditional call instruction, 110
Conditional instructions, 81,
 110, 286
Conditional jump, 109
Configuration, 355
Connect signal, 384
Connector, corroded, 476
Connector symbol, 426
Conservation of data rule, 425
Continuous balance A/D
 converter, 392
Control instructions, 150, 175
Control lines, 131
Control logic, 90–92
Control program, 56
Control transfer instructions,
 249–251
Control transfer operation, 212,
 214–215
Controller, 338, 345
 configuring, 355
 parallel input/output (PIO),
 153
 serial input/output (SIO), 156
Coprocessing, 60
Core memory, 310–311
Corroded connector, 476
Counter-timer circuit (CTC),
 153, 155–156
Counter-timer control register
 (TCON), 204
Counter-timer mode control
 register (TMOD), 204
CPU (central processing unit),
 4, 470–471
CPU bus control lines, 153
CPU card, 6
CPU control lines, 153, 162–163
Crashing, 452–453
CRC, 342
Cross assembler, 66, 456–457
Cross compilers, 457
Crowbar protection, 475
CRT (cathode-ray tube), 6
CSA (Canadian Standards
 Association), 496, 502
CTC, 153, 155–156
Custom display, 388
Cyclical redundancy check
 (CRC), 342
Cylinder, 353

Data available (DA), 376
Data bus, 148, 293
Data caching, 234

Data communications, 366–368
 analog-to-digital interfaces,
 390–391
 digital-to-analog interfaces,
 390–391
 input/output devices, 2,
 386–389, 395–398
 modems, 366, 383–385
 protocols, 368
 serial connection lines,
 379–382
 serial interface, 374–378
 special I/O devices, 395–398
 UART, 374–378
 (See also Communications)
Data flow diagram (DFD), 425
Data logger, 527
Data move instructions, 282, 283
Data movement instruction,
 287–288
Data pointer (DPTR), 198
Data processing, 5
Data processing unit, 4
Data protection, 228
Data register, 269
Data register direct mode, 275
Data segment registers, 238
Data structures, 419
Data transfer instructions,
 106–107, 209
Data words, 9
Debouncing, 387
Debugging, 409, 451–453
Decimal conversions, 24–25
Decimal number system, 16–18
 addition, 28–30
 floating-point arithmetic in,
 42–43, 287, 289
 subtraction, 30–31
Decimal point, 18
Decimal-to-binary conversion,
 20–22
Decision symbol, 425
Decrement, 5, 81–82, 84,
 276–277
Dedicated computers, 3
Demodulator, 383
Design language, 415–417
Design process, 487–490
Designing the program, 409,
 415–417
Destination, 76
DFD, 425
Diamond, 425
Differential signaling, 379
Digit, 19
Digital computers, growth of
 technology in, 1–3
Digital design engineer, 489
Digital electronics, binary
 number system in, 18
Digital magnetic-tape systems,
 356
Digital-to-analog (D/A)
 interface, 390–394
Direct addressing, 104, 114–117,
 235–236, 238